STANDARD LOA

Book

HANDBOOK
OF
DRUG ANALYSIS

APPLICATIONS IN FORENSIC AND CLINICAL LABORATORIES

Ray H. Liu

Daniel E. Gadzala

American Chemical Society
Washington, DC

Library of Congress Cataloging-in-Publication Data

Liu, Ray H.
    Handbook of drug analysis : applications in forensic and clinical
laboratories / Ray H. Liu and Daniel E. Gadzala.
        p.   cm.
    Includes bibliographical references and index.
    ISBN 0-8412-3448-5
    1. Drugs—Analysis. 2. Forensic pharmacology. I. Gadzala,
Daniel E., 1962–   II. Title.
RB56.L58   1997
615′.1901—dc21

96-38029
CIP

The paper used in this publication meets the minimum requirements of American National Standard for Information Sciences—Permanence of Paper for Printed Library Materials, ANSI Z39.48-1984.

PRINTED IN THE UNITED STATES OF AMERICA

Dedication

To: Ko-wang Mei
Robert F. Borkenstein
James W. Osterburg

Foreword

The problems associated with drug abuse have been reported in the national media for decades. The consequences of drug abuse and the resulting increases in criminal behavior affect us all. With a continuing increase in the political awareness of drug abuse, those responsible for taking action in our legislative system have made commendable efforts to address the problems with treatment and enforcement initiatives. Treatment programs are left to professionals specifically trained in the areas of sociology and medicine. Enforcement initiatives are the responsibility of professionals on the state, local, and federal levels whose mission is to stem the flow of illegal drugs.

Forensic science in general, and forensic science laboratories in particular, have played an integral part in the prosecution of those involved with the importation, distribution, and use of controlled substances. An essential factor in this enforcement effort is the requirement that the government establish the identity of a controlled substance with scientific certainty in order to proceed with the prosecution of a defendant. Providing documented proof is the responsibility of the forensic science laboratory.

In 1968 Congress passed the Omnibus Crime Control and Safe Streets Act (Public Law 90–351), which established the Law Enforcement Assistance Administration (LEAA). Under LEAA the federal government began a program of assisting forensic laboratories in upgrading technical capabilities with state-of-the-art scientific methods and instrumentation. State and local forensic science laboratories were given the financial support to hire and to train scientists in the latest forensic science techniques. Funding was also provided to ensure that scientific instrumentation available primarily for research facilities would also be available to forensic science laboratories for enforcement efforts. So began a 30-year evolutionary process of melding science and law enforcement.

From the early days of the analyst identifying controlled substances by using color tests, a microscope, and thin-layer chromatography, the profession has progressed to meet the requirements of timeliness and absolute specificity. Ultraviolet (UV) spectrophotometry and then infrared spectrophotometry (IR) became the norms in the middle to late 1970s. In the 1980s gas chromatography–mass spectrometry (GC–MS) provided additional capabilities for those laboratories overburdened by high caseloads. During this time there also arose a need for rapid screening of drug samples and quantitative capabilities. Capillary column gas chromatography (CCGC) and high-performance liquid chromatography (HPLC) were adopted for forensic purposes to address these requirements. Demands for increased specificity in controlled-substance examinations of analogs and homologs increased as the forensic sciences entered the 1990s. Among the developments were the evolution of nuclear magnetic resonance (NMR) spectrometry, which addressed molecular configuration determinations, and capillary electrophoresis (CE), which addressed isomer separations. Forensic science has come a long way since 1970, but the best years for improvement and refinement still lie ahead.

Some challenges faced by forensic drug laboratory managers over the past three decades were

1. The analyses of steroids, analogs and homologs of phenethylamines, and phencyclidine derivatives;

2. Clandestine laboratory investigations;

3. Sampling protocols;

4. Comparative analyses to determine commonality of source; and

5. Documented specificity in drug identification methods.

Another area that is gaining prominence for forensic science laboratories is the establishment of credible quality assurance programs (QAPs). A primary goal of these QAPs must be confidence in the results reported by forensic analysts. One of the first criteria for any program is a library of available texts that explain forensic instrumentation and analytical protocols. No one text will provide all the answers to all the questions involved in the assurance of a high-quality work product. However, a collection of credible texts taken together can and will continue to provide the source material and documentation so critical in maintaining a "high standard of excellence" in the forensic science laboratory.

These issues were discussed on the fringes 20 years ago. They are now becoming requirements for forensic science laboratories involved in controlled-substance examinations. One key improvement in our profession will be documented peer-reviewed drug analysis methodology.

Many fine publications address specific aspects of drug analysis. However, comprehensive in-depth treatises that focus on the forensic scientist's responsibilities in the laboratory, and technical descriptions of how the corresponding methodologies are utilized in the analysis of suspected controlled substances, have heretofore been few in number. Ray H. Liu and Daniel E. Gadzala have assembled a text bringing together many of the methods and concepts that are available to forensic scientists in the examination of controlled substances. The *Handbook of Drug Analysis: Applications in Forensic and Clinical Laboratories* provides a comprehensive examination of laboratory procedures and a compendium of technical data necessary for the analysis and identification of controlled substances. This publication addresses the classical (translated as "older") methods of analysis that now serve as screening tests. The discussion progresses up to the more sophisticated methods that involve highly technical instrumentation. Liu and Gadzala provide scientific discussions of scientific topics from relevant scientific perspectives that are appropriate for the 1990s and which will take forensic drug analysis into the next millennium.

The *Handbook of Drug Analysis: Applications in Forensic and Clinical Laboratories* should be a welcome addition to any forensic drug analyst's library. This publication will provide the practitioner and anyone interested in entering the practitioner's world of the laboratory with a plethora of information. There is a wealth of information in this text about what we do, how we do it, and why what we do works.

BENJAMIN A. PERILLO
Associate Deputy Assistant Administrator
Office of Science and Technology
Drug Enforcement Administration
U.S. Department of Justice

Preface

The analysis of drugs in various forms and media is one of the most common practices in today's clinical and forensic laboratories. This activity has been intensified by the recent escalation of drug abuse in our society and the implementation of workplace drug-testing policies. Drug analysis is, by its nature, an interdisciplinary applied science. Knowledge in several areas—instrumentation, the chemical properties and metabolism of drugs, and the principles of analytical procedures—is necessary for the successful practice of this science.

Most books that are related to drug analysis fall into the following categories:

1. Pure instrumentation texts for instructional use in chemistry departments;

2. Texts on analytical methods covering general and basic instrumentation, for use in training medical technologists and other readers with some chemistry background;

3. Training manuals developed by various practicing agencies, emphasizing step-by-step analytical procedures; or

4. Source books with extensive data compilations for reference purposes.

This book was prepared with the intention of building a bridge between instrumentation and drug-analysis practice, and to give those who perform step-by-step analytical procedures a broader understanding of the application of analytical principles so that new problem-solving avenues may be realized. The book may serve as a complement to training manuals for trainees, or as a reference for experienced analysts. It may also be used as a text on the subject for upper undergraduate or beginning graduate students.

All chapters were written with the assumption that readers already have a substantial background in chemistry and in the basic features of various analytical methods. Readers who desire to refresh or strengthen their background in these areas should consult general instrumentation texts or specialized treatises. The discussion of instrumentation in this book is brief, addressing essential and unique features; emphasis is placed on illustrating how certain characteristic features, techniques, and approaches of various types of instrumentation can be used to achieve specific analytical objectives.

With deep appreciation, the authors acknowledge the contributions and assistance of many colleagues and associates who have directly or indirectly, and knowingly or unknowingly, made invaluable contributions to the completion of this work. Ben Lin of the U.S. Department of Transportation critically read the earlier version of the first chapter; Sara Jane Sudmeier read the entire manuscript; S. K. Kim of the U.S. Environmental Protection Agency, A. Wu of Hartford Hospital (Hartford, CT), and B. A. Goldberger of the University of Florida (Gainesville, FL) provided much information on immunoassay; the U.S. Drug Enforcement Administration (Statistical Services Unit of the Intelligence Division) provided the secondary drug information summarized in Table 1.5; Franco Tagliaro of the University of Verona, Italy, made many valuable comments and suggestions on Chapter 10; and Hsiu-Lan Liu skillfully drew (with ChemIntosh) all chemical structures and patiently checked through every entry in the appendices.

R. H. Liu especially cherishes the assistance, encouragement, and opportunity provided by Robert F. Borkenstein, Gerald V. Smith, James W. Osturberg, Frederick P. Smith, Charles A. Lindquist, and A. S. Walia. He is also grateful to Tsung-Li Kuo (of the National Taiwan University College of Medicine) and the Taiwanese National Science Council for providing a visiting appointment (NSC 86-2811-M-002-001R- and NSC 86-2811-M-002-0026-) which facilitated the completion of the last phase of this book's publication. This book could not have been written without the patient support rendered by the members of R. H. Liu's family during his 30 years of study and applied research in the subject area.

RAY H. LIU
DANIEL E. GADZALA
Graduate Program in Forensic Science
School of Social and Behavioral Sciences
University of Alabama at Birmingham
Birmingham, AL 35294-2060

About the Authors

Ray H. Liu is a professor and the director of the Graduate Program in Forensic Science and the interdisciplinary Ph.D. Training Program in Forensic Science at the University of Alabama at Birmingham (Birmingham, AL). He completed a law degree at the Central Police University (Taipei, Taiwan) in 1965, studied forensic science under Robert F. Borkenstein and James W. Osterburg at Indiana University (Bloomington, IN) during 1969–1971, and received a Ph.D. in chemistry from Southern Illinois University (Carbondale, IL) in 1976.

Professor Liu taught forensic science at the University of Illinois (Chicago, IL) for about three years. He then served as a chemist for the U.S. Environmental Protection Agency (Chicago, IL) and as a mass spectrometrist in the U.S. Department of Agriculture's Eastern Regional Research Center (Philadelphia, PA) before joining the University of Alabama at Birmingham in 1984. He was an intern at the Center of Forensic Sciences (Toronto, Canada) and has been on visiting assignments for or received awards from the Taiwanese National Science Council, the U.S. Department of Health and Human Services' Addiction Research Center, the U.S. National Aeronautics and Space Administration, and the U.S. Air Force Summer Research Programs.

Professor Liu has conducted applicational studies and published in areas of NMR spectrometry, and electrochemical, spectrophotometric, immunological, and chromatographic methods, but has devoted most of his efforts to basic and applied mass spectrometry. His recent applied research related to the analytical aspects of abused drugs is on the differentiation of drug enantiomers, the correlation of immunoassay and gas chromatography–mass spectrometry test results, and the use of deuterated analogs as internal standards for quantitative determinations. Professor Liu has published about 70 papers in refereed journals, written six invited book chapters, and systematically compiled two specialized books based on his research articles: *Approaches to Drug Sample Differentiations* (Central Police University Press, Taipei, Taiwan; 1981) and *Elements and Practice in Forensic Drug Urinalysis* (Central Police University Press, Taipei, Taiwan; 1994). He has also co-edited, with Bruce A. Goldberger of the University of Florida, the *Handbook of Workplace Drug Testing* (American Association for Clinical Chemistry, Washington, D.C.; 1995).

Professor Liu's professional activities include the establishment of a commercial drug-testing laboratory qualified to report drug-test results under the regulations of the U.S. Departments of Defense and Health and Human Services. He has also served as a resource person for the U.S. Coast Guard's drug-testing programs. He is currently serving as a consultant to programs under the auspices of the U.S. Department of Health and Human Services, the U.S. Department of Justice's National Institute of Justice, and the Departments of Health and Justice of the Taiwanese government. He has participated in the organization of several local, national, and international workshops and symposia and currently serves as editor-in-chief of *Forensic Science Review*.

Daniel E. Gadzala is currently on surgical staff with the Sawyer Surgery Clinic at the Medical Center in Enterprise, Alabama. Before entering practice, Dr. Gadzala completed his B.Sc. at the University of Toronto, Canada, in 1985, and his M.S. in forensic science at the University of Alabama in 1987 under the mentorship of Dr. Ray H. Liu. Dr. Gadzala then entered medical school at the University of Alabama, completing his M.D. in 1991. He was subsequently appointed to a fellowship in general surgery at the Alton Ochsner Medical Foundation in New Orleans, completing his chief year in 1996.

Dr. Gadzala has presented to a diversity of disciplines including the American Academy of Forensic Science, the American College of Surgeons and the Society of Surgical Oncologists, continuing his activity and interests in forensic science and surgery. Awards include the American College of Legal Medicine Schwartz Award for excellence in medical-legal writing (1990) and Outstanding Graduate Fellow in Criminal Justice (University of Alabama, Birmingham, 1987).

Publications include "Mass spectra of derivatized ephedrine and norephedrine" in *Organic Mass Spectrometry*, and "Appropriate management of atypical ductal hyperplasia diagnosed by stereotactic core needle breast biopsy" in *Annals of Surgical Oncology*.

Contents

SECTION I NATURE OF DRUG ABUSE AND SAMPLE CHARACTERISTICS AND PRETREATMENT

Chapter 1. Nature of Abused Drugs and Sample Characteristics

Chapter 2. Sample Pretreatment

SECTION II   PRELIMINARY TEST AND CHROMATOGRAPHIC METHODS

Chapter 3. Color Tests

Chapter 6. Chromatographic Methods

Section III Identification Methods

Chapter 7. Molecular Spectrophotometry

SECTION IV   DEVELOPING TECHNOLOGIES AND ANALYTICAL ISSUES

Chapter 10.   Developing Analytical Technologies

Chapter 11.   Sample Differentiation

Chapter 12. Interpretation of Test Results

APPENDICES

Introduction

Easy access to licit and illicit drug preparations, the change in many population groups' perceptions toward drug use, and the countermeasures adopted by society toward combating drug abuse have rendered drug analysis one of the most important components of today's analytical sciences. Drug analyses are carried out under clinical or forensic settings, depending on sample origins and analytical objectives. Although sample-preparation approaches may vary, the technologies used to analyze both therapeutic and abused drugs are basically the same.

In addition to those drugs that have long been recognized as addictive and/or dangerous, abused drugs include ethanol and the recently controlled anabolic steroids. These drugs are also important topics, and comprehensive references on anabolic steroids [1–6] and ethanol [7–9] are available; this book does not address application features that are uniquely associated with the analysis of these two categories of controlled substances.

The contents of this book are in parallel with the most important topics associated with drug analysis and are divided into five Sections. The first Section (Chapters 1 and 2) provides an overview of the nature of drugs of abuse, drug-sample characteristics, and drug-sample pretreatment approaches. The second Section (Chapters 3–6) addresses analytical methods that do not necessarily depict the specific molecular features of the drug compounds under examination, and that therefore may not generate analytical findings specific enough for conclusive drug identification. More definite methods of analysis are included in the third Section (Chapter 7–9). The fourth Section (Chapters 10–12) is problem-oriented, addressing recent developments (Chapter 10), applications (Chapter 11), and test-result interpretation (Chapter 12) of many unique analytical methodologies and approaches that are pertinent to the solution of specific problems in the analysis of drugs of abuse. The fifth Section of the book (the Appendices) is intended for reference purposes; much useful information on commonly encountered drugs and their metabolites is included.

Sections 2 and 3 of this book are methodology-oriented. Immunoassay techniques can be fully automated and are therefore uniquely suited to the high-volume workplace urine drug-testing mandate. Immunoassay tests are also widely used in clinical and postmortem forensic laboratories because they can be rapidly and easily performed with adequate specificities for most intended applications. High-pressure liquid chromatography (HPLC) is extensively utilized in clinical laboratories for therapeutic drug monitoring and "stat" analyses because of its capability to detect many drugs, such as benzodiazepines, without derivatization and other extensive sample-preparation procedures. Gas chromatographic methods are popular in postmortem laboratories because of the volatility of most drugs of interest, the variety of detectors, and the greater ease of handling the gas carrier than the liquid mobile phase in HPLC. Among the established identification methodologies that are readily available and affordable today, mass spectrometry can be easily adapted as a detector for gas and, to a lesser extent, liquid chromatographic methods. The recent development of associated data systems has further improved the efficiency of gas chromatographic–mass spectrometric methodologies, and they are now an indispensable tool in routine drug analysis.

The information provided in Chapters 11 and 12 should be especially valuable to those who, in the practice of their profession, are associated with scientific drug investigation and workplace drug-testing programs. Chapter 11 looks at approaches that can be used for sample source differentiation, which is a unique analytical objective often pursued in crime-laboratory analysis. Chapter 12 provides essential information required for proper interpretation of test results produced by the now widely adopted workplace drug-testing practices.

References

1. Lin, G. C.; Erinoff, L., Eds. *Anabolic Steroid Abuse;* National Institute on Drug Abuse: Rockville, MD, 1990.
2. Park, J.; Park, S.; Lho, D.; Chung, B. In *Forensic Applications of Mass Spectrometry;* Yinon, J., Ed.; CRC Press: Boca Raton, FL, 1995; p 95.
3. Sample, R. H.; Baenziger, J. C. In *Analytical Aspects of Drug Testing;* Deutsch, D. G., Ed.; John Wiley & Sons: New York, 1989; p 247.
4. Chan, S. C.; Petruzelka, J. In *Analysis of Addictive and Misused Drugs;* Adamovics, J. A., Ed.; Marcel Dekker: New York, 1995; p 293.
5. Jansen, E. H. J. M.; van Ginkel, L. A.; van den Berg, R. H.; Stephany, R. W. *J. Chromatogr.* **1992,** *580,* 111.
6. Chan, S. C.; Nolan, S. L. *Forensic Sci. Rev.* **1993,** *5,* 53.
7. Dubowski, K. W. *The Technology of Breath-Alcohol Analysis;* National Institute on Alcohol Abuse and Alcoholism: Rockville, MD, 1992.
8. Jones, A. W. *J. Forensic Sci.* **1992,** *37,* 21.
9. Jones, A. W. *Forensic Sci. Rev.* **1996,** *8,* 13.

SECTION I

Nature of Drug Abuse and Sample Characteristics and Pretreatment

Chapter 1
Nature of Abused Drugs and Sample Characteristics

Drug Abuse and Drug Testing

Drug abuse is traditionally a social problem that involves only a limited number of subcultures, such as drug addicts and crime syndicates. Because of easy access to licit and illicit preparations and the changing perceptions of many population groups toward drug use, the drug problem has now become a "grave public affliction" and a "national crisis" [1]. The high percentage of positive drug tests observed in people involved in automobile fatalities [2], job applicants [3], and high-school students [4] reflects the urgency of the problem faced by modern society. The costs to society due to illegal drugs have been estimated to be $11 billion for federal drug expenditures in 1991, $5.2 billion for state and local drug-crime expenditures in 1988, and $2.3 billion for health care in 1985 [5].

The drug phenomenon has become more deadly by the utilization of knowledge possessed by "world-class medicinal chemists" [6]. The production of so called "designer" drugs [6, 7] and "look-alike" drugs [8] has not only reduced the effectiveness of drug laws, it has also resulted in serious injury and the death of users. For example, 1-methyl-4-phenyl-1,2,5,6-tetrahydropyridine, an analog of meperidine, was believed to have stricken users with symptoms resembling those of Parkinson's disease, and the extremely high potency of 3-methylfentanyl has resulted in death of heroin users by overdose. (*See* the section "Look-Alike and Designer Drugs" in this chapter for further details.)

Assessing the Extent of Drug Use

Obtaining reliable assessments of the extent of drug use is one of the priorities in combating the problem. Among the earliest attempts were surveys of students and young adults from high schools and colleges and surveys of the nation's households. A different category of data was obtained by documenting drug-related crises reported in the emergency rooms of the nation's metropolitan hospitals (Drug Abuse Warning Network or DAWN). Third, the U.S. Department of Justice established the Drug Use Forecasting program (DUF) by conducting drug testing of persons arrested for serious offenses.

Data from the 1993 National Household Survey on Drug Abuse, conducted by the National Institute on Drug Abuse, were available in the following age groups: 12–17, 18–24, 25–34, and ≥35. Among the available statistics, "past 30 days use" of marijuana ranged from 2% for the age group ≥35 to 11% for the age group 18–25; cocaine use ranged from 0.4% for the age groups ≥35 and 12–17 to 2% for the age group 18–25 [9]. DAWN data obtained in 1992 [10, 11] included approximately 120,000, 48,000, and 24,000 mentions of emergency room episodes related to cocaine, heroin–morphine, and marijuana–hashish, respectively. The corresponding numbers of death-related mentions were approximately 2,900, 2,900, and 360, respectively. Even more alarming, the 1992 DUF data [12] reported drug-test positive rates for male arrestees ranging from 48 to 78% in the 24 major U.S. cities included in the program. These data clearly indicate the severity of the drug problem in the nation's households and in the health care and criminal justice systems.

Evolution of the Nation's Antidrug Policy

The evolution of the federal government's antidrug agencies and policy reflects the worsening of the drug abuse situation faced by our society. During the earlier years, several drug-of-abuse related units were structured within various departments to handle the unique problems that they faced. For example, the Bureau of Narcotics and the Bureau of Drug Abuse Control were established within the Treasury Department and the Food and Drug Administration in 1930 and 1965, respectively.

As drug abuse became a serious social issue and a major law enforcement problem, the federal government merged the Bureau of Narcotics and the Bureau of Drug Abuse Control to form the Bureau of Narcotics and Dangerous Drugs (BNDD) under the Department of Justice in 1968. Following the formation of BNDD, the Controlled Substances Act (CSA) of 1970 was signed into law. This act systematically defined controlled substances and their control mechanisms. It also separated the enforcement responsibility (Department of Justice) from the scientific and medical aspects (Department of Health and Human Services) of controlled substances. In 1973, BNDD was merged with two other agencies in the same department to form the Drug Enforcement Administration (DEA) to streamline and unify drug-of-abuse related operations. The federal government's efforts to reduce drug supply is coordinated by the Office of National Drug Control Policy, created in 1988, whose director ("the drug czar") is appointed by the president.

Along with the evolution of federal agencies designed to combat drug abuse, various preventive and deterrent tactics were formulated and implemented. Earlier approaches emphasized conventional law-enforcement wisdom in punishing those involved in "trafficking" in controlled substances. These substances were rather precisely defined and classified into five Schedules in the CSA of 1970 [13].

As the drug-scheduling procedures became too slow to cope with the rapid and constant appearances of designer drugs, an expedited procedure was added into the CSA by the Dangerous Drug Diversion Control Act of 1984. This new procedure provides a rapid mechanism for placing an uncontrolled substance into Schedule I on a temporary basis for up to 1.5 years. The effectiveness of the act toward the designer and look-alike drugs was further strengthened by the Designer Drug Enforcement Act of 1986. In essence, the CSA is now applicable to all *analogs* of controlled substances that are intended for human use. ("Analogs are defined as chemicals that possess a structure that is substantially similar to that of a Schedule I or II drug, or have stimulant, depressant, or hallucinogenic properties, or are promoted for human use as having these properties." [14]) With all these measures, clandestine manufacturing of illegal substances was still increasing. It became necessary to add the Chemical Diversion and Trafficking Act of 1988 to regulate essential and precursor chemicals for the synthesis of abused drugs and machines for making the drugs into tablets and capsules.

Parallel to the legislative efforts, the federal government's enforcement measures also become broader and more diverse. For example, diplomatic and military interventions were implemented to suppress foreign supplies. Furthermore, the Department of Defense Authorization Act of 1982 defined the military's role in civilian drug enforcement activities.

A different line of approach designed to deter the "demand" end of the equation is also being emphasized, that is, drug-testing practices are extending from the criminal justice system and the military to the entire society. Thus, the issuance of the President's Executive Order No. 12564—Drug-Free Federal Workplace—on September 15, 1986, and the enactment of a new law, PL 100-7e, on July 7, 1987, permitted drug testing for employees of federal agencies. Mandatory drug testing has also been extended to employees of private-sector contractors who provide services or goods to governmental agencies, through the respective agencies' regulations. The Omnibus Transportation Employee Testing Act of 1991 requires drug testing for more than 7 million employees with safety-sensitive responsibilities in the commercial transportation industries. Furthermore, many state governments and private industries have also decided to implement drug-testing policies in parallel with federal regulations.

The Response and Contribution of the Scientific Communities: Drug Testing

A drug-testing policy of this magnitude cannot be implemented without a sound scientific and technical basis. Through the years, several federal agencies have been contributing to the development of technologies that are related to the analysis of abused drugs. For example, the National Institute of

Justice (NIJ) and its predecessor have been funding various law-enforcement agencies to enhance their ability to analyze abused drugs in dosage forms. The early emphasis of these agencies was on the acquisition of state-of-the-art instrumentation and the development of preliminary screen methodologies convenient for field applications. Recently, the NIJ has funded research projects that explore the use of hair as a specimen for drug testing. The DEA has also established its own laboratory system to assist in accomplishing the agency's enforcement missions.

The National Institute on Drug Abuse (NIDA), a federal agency delegated by the CSA of 1970 to conduct scientific research on drugs of abuse, has been conducting and supporting research efforts to study the behavioral and pharmacokinetic effects of drugs of abuse. It has also played a major role in helping the clinical and toxicological communities develop analytical methodologies for detecting and measuring drugs and their metabolites in biological fluids.

The concept and practice of workplace drug testing borrowed heavily from the nation's military experience. A significant reduction in drug use in the military population during recent years is substantiated by both survey and laboratory data. For example, past 30-day drug use among military personnel reported in surveys conducted in 1985 and 1988 showed a reduction from 8.9% to 4.8% [15], while military testing laboratories reported a decrease in the confirmed positive rate of almost 8% in 1983 to just under 2% in 1988 [16]. The decline was attributed directly to the testing program: "It's very effective in scaring the hell out of these guys"[16].

The U.S. Army first initiated mandatory drug testing in the early 1970s as part of an effort to control illicit drug use in Vietnam. The U.S. Navy soon developed an extensive drug-testing program. The program, conducted primarily on a random basis at a rate of almost three times per year per member of the Navy and Marine Corps, included the following drug categories: amphetamines, barbiturates, cannabinoids, cocaine, opiates, and phencyclidine [17]. "Prior to 1985, the military focused on identifying drug abusers for purposes of rehabilitation. In 1985, the military established forensic toxicology drug-testing laboratories in order to deter drug abuse and prevent drug abusers from entering the military by imposing disciplinary actions on identified abusers." [18]

The military drug-testing program established a two-step protocol in which urine specimens were first screened with immunoassay. Specimens that were found by immunoassay to include one of the following drug categories at or above a pre-set level (preliminary test cutoff) were tested by gas chromatography–mass spectrometry (GC–MS) for specific analytes. Only specimens that were found to include a targeted analyte at or above a pre-set level (confirmation cutoff) were reported as positive. During the most intense testing years, the Navy operated five major testing laboratories, in Norfolk, VA, Great Lakes, IL, Oakland, CA, San Diego, CA, and Jacksonville, FL; the Army operated three major testing laboratories, in Wiesbaden, Germany, Honolulu, HI, and Fort Meade, MD; and the Air Force operated one major testing laboratory, in San Antonio, TX. These laboratories are supported and monitored by the Armed Forces Institute of Pathology, which provides open and blind quality-control samples to evaluate the laboratories' performance [16].

With technical advances in testing methodologies, the military's experience, and the emphasis on deterrent efforts, NIDA was entrusted with, and it established, the final "Mandatory Guidelines for Federal Workplace Drug Testing Program" on April 11, 1988 (subsequently revised in 1994 [19]), following the issuance of the President's Drug-Free Federal Workplace Executive Order on September 15, 1986. The National Laboratory Certification Program, established under these Guidelines, oversees the laboratory's drug-testing activities on factors beyond the actual analytical method itself, including chain-of-custody, specimen and facility security, personnel qualification and training, test data interpretation, record management, and quality control and assurance matters such as method validation, standard and reference materials, and maintenance of laboratory instruments and facilities.

Drug testing is now accepted as an effective deterrent measure and a valuable investigative tool, and it is widely applied to different population groups for different reasons. Thus, the following drug-testing categories are commonly performed: pre-employment testing, postaccident testing, testing for suspicion of illegal drug use, voluntary employee requests for drug tests, random testing, and monitoring of rehabilitated employees. In a routine and high-volume testing environment, urine is considered the most effective and convenient specimen. Test methodologies normally include an initial immunoassay followed by confirmatory GC–MS tests of screened-positive samples for the following drug categories: amphetamines, barbiturates, benzodiazepines, cannabis, cocaine, lysergic acid diamide, methadone, methaqualone, opiates, phencyclidine, and propoxyphene. (No workplace drug urinalysis programs test for all drug categories.) Although detailed test methodologies are the subjects of individual chapters, brief principles and protocols are outlined here.

Immunoassay protocols utilize the competitive binding principle, in which a limited amount of antibody is allowed to react with a mixture containing a fixed amount of labeled (control) drugs and an unknown quantity of unlabeled (analyte) drugs. The presence and the quantity of a drug in the sample is evaluated on the basis of the amount of the labeled drug in the reacted or unreacted forms. The control drugs are labeled in different ways, each requiring different methods of detection and quantification. Radioactive isotopes, such as ^{125}I, are used for labeling in radioimmunoassay methods; an active enzyme, which is capable of converting (indirectly) nicotinamide adenine dinucleotide (NAD) to its reduced form (NADH), is coupled to the drug in enzyme-multiplied immunoassay techniques; and fluoresceins are coupled to the drug used in fluorescence polarization immunoassay [20].

Confirmatory procedures often include acid, base, or enzyme hydrolysis, followed by liquid- or solid-phase extraction, and often by derivatization. GC–MS protocols, operated under selected ion monitoring mode, are used almost universally. Criteria for conclusive drug identification include correct retention time and abundance ratios of the ions monitored. The internal standard method using deuterated analogs is most often employed for quantification. Specifically, the abundance ratio (of the ions selected for the analyte and the internal standard) in the test sample is divided by the same ion abundance ratio from the calibration standard. The result is then multiplied by the concentration of the analyte in the calibration standard.

Certification by the U.S. Department of Defense is required for a laboratory to conduct drug urinalysis for military personnel. Similarly, certification by the U.S. Department of Health and Human Services is required for a laboratory to engage in drug urinalysis for federal agencies and federal-government-regulated programs. These certification programs adopt strict administrative policies for reporting test results based on specific cutoff requirements [19] and regulate the laboratories' operations through on-site inspections and the provision and monitoring of proficiency test samples. These certification programs have not only assured the quality of drug urinalysis, they have set a higher standard for the operation of clinical and toxicological laboratories.

Making knowledge and appropriate analytical procedures readily available constitutes one essential component in formulating a comprehensive social, legal, and medical response to the widespread abuse of drugs. With this in mind, this book will address the following topics:

1. Characteristic features of drugs in their pre-used licit or illicit dosage forms or in complex biological matrices;

2. Sample pretreatment and separation techniques that are often applied prior to actual identification and quantification;

3. Methods suitable for preliminary screening and for conclusive qualitative and quantitative analysis of commonly abused drugs; and

4. Proper interpretation of these analytical findings.

Drug Classification

Drugs are, for the purpose of this book, limited to controlled substances and "dangerous drugs". Legal drugs that have a high index of toxicity [21], defined as the ratio of the number of deaths associated with a given drug to the number of prescriptions for that drug, are considered dangerous. For practical purposes, drugs that are commonly abused or frequently encountered in toxicological laboratories [2, 22–25] will be emphasized.

Drug classifications are based on criteria that are suitable for the intended use. For example, controlled substances are placed into five schedules for legal purposes. Classification systems are also established on the basis of chemical structures and biological activities. Because one of the major criteria for legal classification is based on the biological activity of a drug, and such activity is often related to its chemical structure, many drugs may fall in the same category under different classification systems. The table shown in Appendix I is an attempt to summarize these classifications as well as other information useful for convenient reference. The structural features of these drugs are designated in the last column of the table using the basic structural frameworks shown in Appendix III and the functional group designations shown in Appendix IV.

Chemical Structure

Spectrometric characteristics are typically reflections of functional groups embedded within the structural feature of the compounds. Thus, it has been proposed that drugs be grouped on the basis of [26–28]:

1. Ultraviolet absorption profile in the 200- to 340-nm range;

2. Effect of pH change;

3. Effect of decreased solvent polarity relative to water; and

4. Intensity of the absorption bands.

Drugs are often categorized on the basis of their acidity. The pK_a values of many commonly abused drugs are listed in the sixth column of Appendix I. The practical advantages of this classification include proper application of successful extraction and sample-pretreatment or cleanup procedures, and a better understanding of chromatographic and certain spectrometric characteristics, such as the pH and solvent effects mentioned above.

Except for a few groups of drugs, such as quaternary ammonium compounds [29], which are highly water-soluble and insoluble in organic solvents, most drugs can be extracted into organic solvents under a suitable pH. Because the major moieties of drugs and their metabolites are organic in nature, they tend to favorably partition into the organic phase when present in their neutral forms. Conversely, drugs in their salt forms tend to stay in aqueous media. Therefore, under acidic conditions, e.g., pH = 3, acidic drugs present in their neutral form can be easily extracted into the organic fraction.

$$A^- \; \underset{OH^-}{\overset{H^+}{\rightleftharpoons}} \; AH \qquad (1.1)$$

[Aqueous phase]   [Organic phase]

Similarly, basic drugs that exist in their neutral forms under alkaline conditions, that is, pH $\geq$ 10, are more likely to be extracted to the organic fraction under this condition.

$$BOH \; \underset{OH^-}{\overset{H^+}{\rightleftharpoons}} \; B^+ \qquad (1.2)$$

[Organic phase]   [Aqueous phase]

It should be noted, however, that when hydrochloride is added, some of these basic drugs form hydrochloride salts that may be chloroform-soluble and, therefore, chloroform-extractable.

Neutral drugs do not form salts under acidic or basic conditions. Thus, they can be extracted under either condition. Because the acidic fraction usually contains fewer drugs, neutral drugs are preferably removed with the acidic group in actual practice.

Legal Schedule

The Federal Controlled Substances Act, enacted in 1970, established a comprehensive pattern of control over the manufacture and distribution of drugs. The Act set up five schedules under which different criminal penalties are imposed for unlawful trafficking. Possession for one's own use of any controlled substance in any schedule is always a misdemeanor on the first offense.

The scheduling of a drug is based on [30]:

1. Its actual or relative potential for abuse;

2. Scientific evidence of its pharmacological effect, if known;

3. The state of current scientific knowledge regarding the drug or other substance;

4. Its history and current pattern of abuse;

5. The scope, duration, and significance of abuse;

6. What risk, if any, there is to public health;

7. Its potential for psychological or physiological dependence; and

8. Whether the substance is an immediate precursor of a substance already under control.

Only those that have no acceptable medical use in treatment in the United States can be classified as Schedule I drugs. The schedule of drugs listed in the Code of Federal Regulation [13] are indicated in the fourth column of Appendix I.

Pharmacological Effect

With the exception of a few peripherally acting drugs, such as carbon monoxide, arsenic, cyanide, insulin, and paraquat (all of which have been used as instruments of homicide or suicide), drugs that are frequently encountered in forensic and toxicological analyses are psychoactive compounds. These drugs are commonly categorized on the basis of their selective action on the brain and are grouped as stimulants, depressants, narcotics, and hallucinogens. Because it is the physiological and psychological effects of these drugs that result in their abuse, this classification system reveals significant information about the impact of these drugs on people. However, it should be noted that defining such broad categories of functions may not always encompass the broad spectrum of activities of a specific drug. The pharmacological effects of commonly encountered drugs are listed in the third column of Appendix I.

Samples in Criminal Investigations

A complete analytical process generally includes sampling, conversion of the analyte to a proper form for analysis (sample treatment), performance of one or more measurements, and interpretation of the data obtained [31]. Because samples handled by crime, toxicological, and clinical laboratories are seldom drugs in their pure forms, an understanding of sample characteristics will facilitate the selection of appropriate pretreatment and analytical procedures and the interpretation of analytical data.

Illicit Solid Dosage Forms

Drug samples encountered in crime laboratories are normally exhibited in dosage units and presented as their own entities or as counterfeits of brand-name pharmaceutical products. They often include the following characteristics:

1. Various forms of carriers or containers are used.

2. Various unintentional impurities, such as degradation products, synthesis by-products, and residual solvents, are present.

3. Intentional multiple-component drug combinations and the addition of adulterants and excipients are common.

4. The contents are often unexpected.

5. Many of these drugs may be sold as counterfeits of pharmaceutical products.

In addition to identifying the presence of a specific drug in an exhibit, laboratories are often requested to provide additional information that may be helpful to the investigation process. On the basis of the characteristics listed above, exhibits may be profiled and linked to common sources or routes of distribution (*see* Chapter 11).

Carriers

For purposes of facilitating illicit drug trade and appearing exotic to users, drugs of abuse are often distributed in different forms and are concealed in a wide variety of containers or carriers. Even in their pure forms, drugs may exhibit different appearances; for example, cocaine HCl may be distributed as rocks, flakes, or fine powder. Similarly, *d*-methamphetamine is available in clear, shiny crystals known as "ice" [33].

One of the most common forms of drug carriers in recent days is the impregnation of LSD in blotter paper [32, 34] that bears designs ranging from scenery, music notes, and cartoon characters to fantastic animals. Paper sheets are often perforated into dosage units. Although not as common, paper dosages of 4-bromo-2,5-dimethoxyamphetamine [35] and heroin [36] have also been encountered.

Unintended Impurities

Most "street drugs" are produced without proper quality control. Although highly sophisticated drugs and purified products have been encountered, most samples include impurities. Impurities may be synthesis by-products, residual chemicals and solvents, or degradation products. The exact source of an impurity is often not readily apparent. For example, there is quite a difference of opinion as to the origin of monoacetylmorphines present in heroin samples. It is generally agreed that 6-monoacetylmorphine (6-MAM) is a product of heroin hydrolysis [37, 38] and that 3-monoacetylmorphine (3-MAM) is produced by incomplete acetylation of morphine [39]. However, it has also been demonstrated that 6-MAM can be directly synthesized [40] and 3-MAM can be generated through hydrolysis of heroin [41].

Identification of impurities is important because of the potential harmful effects that they may have on users. The most profound example, as mentioned earlier in this chapter in the section "Drug Abuse and Drug Testing", is the unintended production of 1-methyl-4-phenyl-1,2,5,6-tetrahydropyridine (MPTP) in the process of manufacturing 1-methyl-4-phenyl-4-propionoxypiperidine (MPPP). This by-product (MPTP) has been linked to symptoms of Parkinson's disease and the death of users [42].

The identification of impurities in drug samples is also important from an investigative perspective. Knowledge of these impurities may provide a means of:

1. Determining the synthetic or natural origins of the sample;
2. Establishing the synthetic route employed [32]; and
3. Providing evidence of a link between seized samples or between material found at a laboratory and that circulating at the drug scene [32].

Impurities commonly associated with cocaine and amphetamine are listed in Tables 1.1 and 1.2, respectively. Some of these impurities may serve as indicators for determining whether the drug is derived from a synthetic or natural source (Table 1.1), and the particular route involved for the synthetic drug (Table 1.2). Similar information related to methamphetamine [42, 43] and heroin [44–48] samples have also been widely studied and reported. The information on component composition shown in Table 1.3 has been used as a basis for sample differentiation.

Solvents and other volatile compounds are another category of impurities that are commonly found in drug samples. Although residual solvents are also found in licit pharmaceutical products [58], they are more prominent in illicit preparations. Identification and quantification of solvents provide valuable information for sample comparison purposes. In cocaine, for example, solvents employed for extracting coca leaves become chemical residuals that are commonly detected in cocaine samples.

For the maximum conversion of cocaine base to cocaine hydrochloride [59–61], the solvent must [60]:

1. Be soluble to the base but insoluble to the hydrochloride;
2. Have the ability to retain water from the addition of hydrochloric acid;
3. Be volatile; and
4. Be readily available.

Table 1.1. Impurities commonly encountered in cocaine samples.

Compound name	Source and significance	Ref.
cis- and *trans*-Cinnamoylcocaine	a	[49, 50]
Cinnamic acid	b	[49]
Ecgonine	b,c	[50, 51]
Methylecgonine	b,c,d	[50, 51]
Benzoylecgonine	b,c	[50, 51]
Benzoic acid	b,c	[51]
Ecgonidine	b,c,e,f	[51]
Methylecgonidine	b,c,f	[51]
Ethylecgonidine	b,c,f	[51]
Methylpseudoecgonine	b,g	[51]
(+)-Cocaine	Synthetic	[54]
(+,−)-Pseudococaine	Synthetic	[54]
(+,−)-Allococaine	Synthetic	[54]
(+,−)-Pseudoallococaine	Synthetic	[54]
3-Aminomethyl-2-methoxycarbonyl-8-methyl-8-azabicyclo[3.2.1]oct-2-ene	Synthetic	[55]
3-Benzoyloxy-2-methoxycarbonyl-8-methyl-8-azabicyclo[3.2.1]octene	Synthetic	[55]
3-Benzoyloxy-8-methyl-8-azabicyclo[3.2.1]oct-2-ene	Synthetic	[55]

a Indication of "illicit preparation" from natural source: extraction of coca leaf followed by recrystallization.
b Indication of "pharmaceutical preparation" from natural source: hydrolysis of all alkaloids to ecgonine followed by methylation and benzoylation.
c Decomposition product.
d HCl–water hydrolysis product.
e This compound has also been reported in coca extract [52] and as a metabolite of cocaine [53].
f HCl hydrolysis product.
g NaOH hydrolysis product.

Table 1.2. Impurities commonly encountered in amphetamine samples.[a]

Compound name	Source and significance
N-Acetylamphetamine	Side reaction product of reductive amination route
Benzyl methyl ketone	Starting material for Leuckart reaction and reductive amination routes
Benzyl methyl ketoxime	Side reaction product of oxime and phenylnitropropene routes
α-Benzylphenethylamine	Side reaction product of reductive amination route
2-Benzyl-2-methyl-5-phenyl-2,3-dihydropyrid-4-one	Side reaction product of Leuckart reaction route
4-Benzylpyrimidine	Side reaction product of Leuckart reaction route
Dibenzyl ketone	Impurity in starting material (benzyl methyl ketone)
2,4-Dihydroxy-1,5-diphenyl-4-methylpent-1-ene	Side reaction product of reductive amination route
2,4-Dimethyl-3,5-diphenylpyridine	Side reaction product of Leuckart reaction route
2,6-Dimethyl-3,5-diphenylpyridine	Side reaction product of Leuckart reaction route
2,4-Dimethyl-3-phenyl-6-(phenylmethyl)pyridine	Side reaction product of Leuckart reaction route
N,N-Di(β-phenylisopropyl)amine[b]	Side reaction product of Leuckart reaction route, intermediate product of reductive amination route
N,N-Di(β-phenylisopropyl)formamide	Side reaction product of Leuckart reaction route
Formamide	Starting material for Leuckart reaction route
Formic acid	Starting material for Leuckart reaction route
N-Formylamphetamine	Intermediate product of Leuckart reaction route
2-Methyl-3-phenylaziridine	Side reaction product of oxime and phenylnitropropene routes
4-Methyl-5-phenyl-2-(phenylmethyl)pyridine	Side reaction product of Leuckart reaction route

Table 1.2. (Continued.)

Compound name	Source and significance
2-Methyl-3-phenyl-6-(phenylmethyl)pyridine	Side reaction product of Leuckart reaction route
4-Methyl-5-phenylpyrimidine	Side reaction product of Leuckart reaction route
1-Oxo-1-phenyl-2-(β-phenylisopropylimino)propane	Side reaction product of reductive amination route
N-(β-Phenylisopropyl)benzaldimine	Intermediate product of reductive amination route
N-(β-Phenylisopropyl)benzyl methyl ketimine[b]	Intermediate product of reductive amination route
2-Phenylmethylaziridine	Side reaction product of oxime and phenylnitropropene routes
Phenylnitropropene	Side reaction product of phenylnitropropene route

[a] Compounds listed in this table are reported by various sources as reviewed in Refs. [43] and [56].
[b] The simultaneous presence of these two compounds is indicative of a hydrogenation step in the synthesis process.

Table 1.3. Common narcotic contents of illicit heroin.

	SE Asia[a]		India		Pakistan Base		HCl		Turkey		Iran		Nigeria	
Compound[b]	1	2	1	2	1	2	1	2	1	2	1	2	1	2
Heroin	81	68	75	68	72	69	48	74	51	59	72	62	87	60
6-Monoacetylmorphine	2.7	2.4	4.4	2.1	4.6	1.7	17	2.8	2.2	1.8	1.8	0.9	4.9	1.7
Acetylcodeine	6.2	6.6	3.0	3.7	4.9	5.7	4.7	1.8	3.7	4.7	5.4	5.3	2.2	4.3

[a] "1" and "2" are duplicate measurements.
[b] Compositions (%) of these three compounds as listed here were used in Ref. [57] as a basis for the differentiation of samples from different geographic areas.

The most predominant solvents are acetone and ether. However, the use of methyl ethyl ketone as a replacement for ether is increasing [62, 63]. One study [50] reported the use of acetone–ether mixtures in 49% of 107 samples examined, acetone alone in 20%, and methyl ethyl ketone alone or in combination with acetone and/or ether in 33%. Approximately 75% of the clandestine samples contained traces (<0.05%) of ethanol [61].

Volatile components responsible for the "smell" of clandestine cocaine HCl are methyl benzoate and *cis-* and *trans-*methyl cinnamates. The decomposition of cocaine in the presence of excess HCl leads to the formation of benzoic acid and methanol, which in turn form methyl benzoate. The acid hydrolysis of *cis-* and *trans-*cinnamoylcocaine leads to the formation of *cis-* and *trans-*cinnamates. Quantification of these compounds by GC, performed using methyl caprate as an internal standard, shows that the combined amounts of odor-producing esters are less than 2% of the sample in most cases [61]. The methyl benzoate to methyl cinnamate ratio determines the smell of the cocaine: those samples with high ratios are described as "sweet" and those with low ratios are described as "fruity".

Data on sample solvents and volatile components obtained from 28 exhibits and a Merck standard are listed in Table 1.4 [61]. The feasibility of using this information for sample-differentiation purposes relies on the precision of measurement and the stability of volatile components in cocaine samples. The use of volatile-compound compositions in heroin samples as a basis for sample profiling has also been explored [64].

Intended Combinations

Multiple Active Ingredients. Intentional mixing of multiple active ingredients in a single dosage has also been reported. Popular combinations include heroin–phencyclidine (PCP) (Sunshine); heroin–cocaine (Speedball); cocaine–methamphetamine (Zoom); heroin–phentermine (Bam); pentazocine–tripelennamine (T's and Blues); and hydromorphone–codeine (Hits or Loads) [65–67]. The motivations [66] behind the abuse of the heroin–PCP combination are the complementary effects of the two substances and the production of higher profit. The anesthetic effects of PCP may ease the

Table 1.4. Solvents and volatile compounds in cocaine samples.[a]

Sample	Acetone	Ether	Methyl benzoate	Methyl cinnamates
1	0.40	0.051	0.101	0.026
2	0.37	0.128	0.027	0.023
3	0.50	0.164	0.009	0.016
4	0.28	0.049	0.030	0.016
5	0.50	0.134	0.141	0.044
6	0.92	0.214	0.007	< 0.005
.				
.				
.				
16	0.18	0.042	0.062	0.034
.				
.				
.				
25	0.97	0.277	0.010	0.017
26	0.54	0.118	< 0.005	< 0.005
27	0.29	0.087	0.585	< 0.005
28	0.84	0.204	0.005	< 0.005
Merck std	0.33	None	—	—

[a] Data (% composition) are taken from Ref. [61].

withdrawal pain of heroin use; on the other hand, heroin may ameliorate the undesirable effects of PCP. The addition of a small amount of PCP may boost the effect of low-purity heroin, which could then be sold as a high-quality product to yield larger profit. A primary reason for "speedballing" is allegedly the enhanced euphoric effect that cocaine brings to the heroin. In addition, the cocaine may retard the depressant effect that eventually comes when the heroin wears off [67]. Combined uses of marijuana, heroin, and alcohol; and cocaine and methamphetamine were reported [67] to produce an enhanced euphoria.

Table 1.5 summarizes secondary drugs and their frequencies of appearance in cocaine, heroin, marijuana, and phencyclidine specimens that were analyzed by the U.S. Drug Enforcement Administration's (DEA's) laboratories from 1990 to 1995. These data indicate the following:

1. Phencyclidine is very commonly used along with marijuana. For example, among the 300 phencyclidine specimens tested by the DEA's laboratories during 1993, marijuana was found in 169 (56.3%) cases.

2. Cocaine is occasionally a minor component in heroin samples. For example, among the 4,077 heroin samples tested during 1994, 210 (5.2%) were found to contain cocaine.

3. Cocaine–marijuana and heroin–depressants (such as phenobarbital, diazepam, and methaqualone) combinations have also been reported.

Excipients and Adulterants.   Similar to pharmaceutical preparations, inert materials called excipients are used to make up the bulk of drugs of abuse that are sold as solid dosage forms. Excipients include diluents, binders, lubricants, disintegrants, colorants, and flavoring agents [68, 69], each of which has a different function in the overall process of tablet manufacture and subsequent tablet disintegration in the body. Not every additive will necessarily be present in any given dosage, and the amount may vary. The most abundant excipient is usually the diluent, which is commonly found to constitute up to 80% of the bulk material. Examples of common diluents found in illicit tablets are listed in Table 1.6 [68].

It is not uncommon for adulterants to exclusively or partially replace the active ingredient in illicit drug trade. Thus, quinine is nearly universally used in heroin samples to give a bitter taste commonly believed to be associated with authentic heroin samples. Similarly, benzoic acid is used in fake cocaine samples to give an odor thought to be associated with authentic cocaine samples.

Table 1.5. Secondary drugs and their frequency as found in cocaine, heroin, marijuana, and phencyclidine cases tested by the Drug Enforcement Administration's laboratories from 1990 to 1995.[a,b]

| | Primary drug and number of annual cases in which secondary drugs were looked for | | | |
	Cocaine (12,095–17,568)[c]	Heroin (3,955–4,418)[c]	Marijuana (5,578–9,567)[c]	Phencyclidine (170–311)[c]
Cocaine	NA	56–210	12–18	NR–3
Heroin	23–46	NA	NR	NR
Marijuana	7–34	NR	NA	69–169
Phencyclidine	NR	NR	1–12	NA
Amphetamine	NR	NR	NR	NR
Methaqualone	NR	NR–39	NR	NR
Phenobarbital	NR	34–57	NR	NR
Diazepam	NR	6–34	NR	NR

[a] Based on statistical data provided by the Drug Enforcement Administration.
[b] Values are minimum and maximum number of annual cases in which subject secondary drugs were found (period: 1990–1995). Abbreviations: NA, not applicable; NR, maximum annual case < 3.
[c] Minimum and maximum number of annual cases in which secondary drugs were looked for (period: 1990–1995).

Table 1.6. Diluents used in illicit tablets.[a]

Lactose[b]	Sodium bicarbonate[b]
Glucose[b]	Calcium carbonate sucrose[b]
Sucrose[b]	Calcium sulfate (anhydrous)[b]
Maltose	Calcium sulfate (bihydrate)
Fructose	Calcium magnesium carbonate (dolomite)
Mannitol[b]	Calcium phosphate[b]
Cellulose	Calcium lactate methyl cellulose
Barium sulfate (Kaolin)[b]	Starches
Talc	Corn flour
Vitamin C (L-Ascorbic acid)	Dried powdered milks

[a] Data are taken from Ref. [68].
[b] These substances are also frequently used as pharmaceutical diluents.

From an investigative perspective, qualitative and quantitative comparisons of multiple active drug combinations and excipient–adulterant ingredients may provide information valuable for linking samples from different seizures. Thus, examples of drug combinations and excipient–adulterant ingredients are widely reported; examples are given in Table 1.7.

Unexpected Contents

The unexpected contents of certain illicit drugs are illustrated by a reported elaborate fake cocaine sample [75] that included ephedrine, fencamfamine, caffeine, benzoic acid, lidocaine, tetracaine, and an unidentified compound thought to be phenothiazine. The benzoic acid in the sample provided an odor that deceived one of the narcotics dogs.

Counterfeits

Bogus Quaalude tablets were very popular, partly because of the shortage of methaqualone powder resulting from the discontinuance of production and distribution of methaqualone by the Lemmon Pharmaceutical Company and the international control adopted by virtually all major producing and exporting countries. The contents of counterfeits of Quaalude, other methaqualone tablets, and several other pharmaceutical products are listed in Table 1.8.

Table 1.7. Common combination drugs and adulterant–excipients.

Drug name	Combination drug	Adulterant–excipient	Ref.
Heroin	Phenacetin, aspirin, caffeine	Lactose, mannitol, quinine, acetylquinine	[70]
Cocaine	Methaqualone	Starch	[71]
Methamphetamine	Fentanyl analogs	Procaine	[72]
Methaqualone	Pentobarbital, flurazepam	Acetaminophen	[73]
Cannabis	1-(1-Phenylcyclohexyl)pyrrolidine	1-Phenylcyclohexylmorphine	[74]

Table 1.8. Examples of common counterfeit products.

Brand name	Authentic content	Counterfeit content	Ref.
Quaalude (Lemmon)	Methaqualone base	Cyproheptadine	[76]
Quaalude (Rorer)	Methaqualone base	Diazepam, meprobamate	[77]
Mequin	Methaqualone	Diazepam	[78]
Mandrax	Methaqualone HCl	Secobarbital	[79]
Dilaudid	Hydromorphone	Propoxyphene HCl	[80]

Look-Alike and Designer Drugs

Look-alike and designer drugs are contemptible contributors to the modern drug-abuse problem. "The term look-alikes refers to tablets and capsules containing noncontrolled, over-the-counter ingredients, which are manufactured to resemble controlled products. Recently, this term has come to include products which, although not manufactured to resemble controlled products, contain the same ingredients and are promoted in the same manner as those resembling controlled products." [81] The stimulant look-alikes generally contain one or more of the following ingredients: caffeine, ephedrine, pseudoephedrine, and phenylpropanolamine. Depressant look-alikes may contain an antihistamine, such as doxylamine or chlorpheniramine, in combination with analgesics, such as acetaminophen or salicylamide.

A designer drug [6, 7, 82, 83] is commonly defined as "a substance other than a controlled substance that has a chemical structure substantially similar to that of a controlled substance in Schedules I or II or that was specially designed to produce an effect substantially similar to that of a controlled substance in Schedules I or II." They are normally analogs of compounds with proven pharmacological activity manufactured by underground chemists for sale on the street. The most common designer drugs are variations of fentanyl [7, 84] (Table 1.9), analogs of meperidine [85] (Table 1.10), and amphetamine- and methamphetamine-related 3,4-methylenedioxyamphetamine and 3,4-methylenedioxymethamphetamine [86–90] (Table 1.11).

Biological Samples

Samples received in toxicological and clinical laboratories for the analysis of abused drugs are normally biological fluids or tissues. In contrast to the dosage forms of drugs normally encountered in crime laboratories, the drugs presented in biological fluids or tissues are present in much lower concentrations. Often, the metabolites, instead of the abused drugs, are present. Basic knowledge of drug metabolic mechanisms is essential to effective handling of biological samples for the analysis of abused drugs and their metabolites.

Drug Metabolism

"Modern" approaches to drug analyses take advantage of the latest developments in analytical technologies and the accumulated knowledge derived from drug metabolism studies. Thus, drugs at low concentration levels that could not be detected before, and drug metabolites that were not known and

Table 1.9. Fentanyl and analogs — name, chemical structures, and year first identified as a drug of abuse. (Data are taken from Ref. [7].)

Name	Chemical structure	Year identified
Fentanyl		1982
α-Methylfentanyl		1979
p-Fluorofentanyl		1981
Benzylfentanyl		1982
α-Methylacetylfentanyl		1983
α-Methylacrylfentanyl		1983
3-Methylfentanyl		1983
Thienylfentanyl		1985
3-Methylthienylfentanyl		1985
β-Hydroxythienylfentanyl		1985
β-Hydroxy-(3-methyl)-thienylfentanyl		1985

previously neglected, are now frequently reported. As a result, positive results of all allegedly poisoned food and drink samples submitted for analysis to the London Metropolitan Police Forensic Science Laboratory have risen from less than 10% in the 1960s to over 90% in 1986 [91].

The applications of modern analytical methods to drug analysis are the subject of later chapters. This section provides a survey of the relevant metabolic principles and mechanisms of drugs, followed by a comprehensive list of the metabolites of commonly encountered drugs. Specialized references should be consulted for a more thorough understanding of metabolic mechanisms.

Administration, Absorption, Distribution, and Excretion of Drugs

The process of interaction between drugs and human organs is divided into three phases: the pharmaceutical (exposure) phase, the pharmacokinetic (toxicokinetic) phase, and the pharmacodynamic (toxicodynamic) phase. Pharmacodynamics is concerned with the study of interactions between the

Table 1.10. Meperidine and analogs — name, chemical structure, and year first identified as a drug of abuse. (Data are taken from Ref. [85].)

Name	Chemical structure	Year identified
Meperidine		—[a]
1-Methyl,4-propionoxy-4-phenylpyridine		1982
1-Methyl,4-phenyl,1,2,3,6-tetrahydropyridine		1982

[a] Data not available.

Table 1.11. Amphetamine and analogs — name, chemical structure, and year first identified as a drug of abuse.

Name	Chemical structure	Year identified	Ref.
Amphetamine (R = H)		—[a]	
Methamphetamine (R = CH₃)		—	
4-Methyl-2,5-dimethoxyphenethylamine		—	
p-Methylamphetamine		1974	[84]
3,4-Methylenedioxyamphetamine		1979	[87]
3,4-Methylenedioxymethamphetamine		1985	[86]
3,4-Methylenedioxyethamphetamine		1985	[86]
4-Bromo-2,5-dimethoxyphenethylamine		1985	[88]
N,p-Dimethylamphetamine		1989	[89]

drug (or its active metabolites) and the receptors in target tissues. While this subject is of great importance in understanding the mechanism of toxicological effects of drugs, it is normally not a major concern to drug analysts. On the other hand, many processes that take place during the pharmaceutical and pharmacokinetics phases are highly relevant to drug analysis.

The manner in which a drug is introduced into a biological system, that is, the route of administration, dictates the set of digestive and enzymatic conditions to which the drug is subjected and, therefore, the metabolic fate of the drug and its distribution in various tissues. Common routes of introduction include oral ingestion, intravenous and intramuscular injections, inhalation, and dermal application. Once a drug is introduced into a biological system, it is subjected to the processes of absorption, distribution, metabolism, and excretion, which are subjects of pharmacokinetics. Some basic principles governing the interactions between drugs and the human biological system with respect to drug introduction, absorption, distribution, and excretion will be addressed. Such knowledge facilitates the proper selection of specimens for analysis, sample pretreatment, and analytical data interpretation.

Exposure. Drug absorption is first controlled by the mechanisms available at the site of contact. Major absorption sites, site characteristics, and absorption preference related to various routes of administration are listed in Table 1.12. In general, oral ingestion of a drug often results in a high drug content in the stomach. However, certain drugs, such as propoxyphene and tricyclic antidepressants, which are rapidly absorbed in the gastrointestinal tract, may only leave a trace amount in the stomach. Weak bases, on the other hand, are favorably trapped in the acidic gastric environment and may exhibit substantial levels even if introduced intravenously. It is therefore suggested [92] that blood and liver concentration ratios are a better indication of drug administration routes.

Table 1.12. Route of administration and major characteristics affecting drug absorption.[a]

Route of administration	Absorption site	Absorption mechanism	Drug likely to be absorbed
Oral ingestion	Stomach Intestinal tract (cellular membrane)	Diffusion	Weak acid Weak base
Inhalation	Lung	Diffusion Filtration	Vapor with high solubility in blood Small particles (<1 μm)
Intravenous injection	Blood	Circulation	All
Intramuscular injection	Muscle	Diffusion Filtration	Lipophilic drug Small hydrophilic drug particle

[a] Based on information in Ref. [92].

Distribution. Once the drug finds its way through various absorption sites into the blood, it is distributed to various body tissues. The tissue-to-blood concentration ratios of a particular drug depend on the rate of blood perfusion to these tissues and the specific characteristics of the subject drugs, such as acidity, lipid solubility, and molecular size. Table 1.13 summarizes the major parameters affecting the retention of drugs and lists the categories of drugs that are favorably retained in various body tissues.

Concentrations of selected drugs, representing different drug classes, in various tissues found in fatal cases are listed in Table 1.14. It should be noted, however, that most of these entries are average reports of wide-ranging values, resulting from cases in which the subject drug may have been introduced through different routes. The survival time after the drug use may also vary, allowing different times for equilibrium between these tissues. The excretion rate of drugs is also affected by urinary pH (*see* the discussion later). Furthermore, many quantitative data can only be considered a "gross estimate" [93] when dealing with an "ill-defined sample of dubious origin". Thus, quantitative data obtained in postmortem cases should be interpreted with caution.

Table 1.13. Major mechanisms affecting distribution of drugs in tissues.[a]

Tissue	Major mechanism or characteristics	Drug likely to be deposited
Blood	Binding to proteins and erythrocytes (e.g., tricyclic antidepressants)	Acidic drug
Brain	pH gradient between plasma and cerebrospinal fluid	Weak basic drug
Liver	Diffusion, protein binding	All
Stomach	Low pH	Basic drug
Fat	Lipophilicity	Lipid-soluble compound

[a] Based on information in Ref. [92].

Table 1.14. Distribution of various categories of drugs in body tissues.

Drug	pK_a	Blood	Brain	Liver	Kidney	Urine	Ref.
Alkaline							
Cocaine	8.6	5.2 (1)	6.6 (1.3)	4.3 (0.83)	15 (2.9)	47 (9.0)	[94]
Amphetamine	9.9	8.6 (1)	2.9 (0.34)	30 (3.5)	17 (2.0)	237 (28)	[95]
Acidic							
Phenobarbital	7.2	64 (1)	63 (1.0)	138 (2.2)	84 (1.3)	38 (0.59)	[96]
Amobarbital	7.9	81 (1)	172 (2.1)	414 (5.1)	210 (2.6)	98 (1.2)	[97]
Neutral							
Methyprylon	12.0	59 (1)	—[a]	118 (2.0)	62 (1.1)	86 (1.5)	[98]
Paraldehyde	—	245 (1)	273 (1.1)	337 (1.4)	167 (0.68)	130 (0.53)	[97]

[a] Data not available.

Excretion. Drugs that are absorbed will most likely be excreted from the body in their original forms, as metabonates (nonenzymatic reaction products) and/or as metabolites. They are most commonly secreted in the bile and through the kidney, and they are eliminated in the feces and urine. Basically, drugs enter the liver, the major site of metabolism and excretion, and are then either secreted back into the blood or excreted into the bile.

Drugs or their metabolites that are hydrophilic and have higher molecular weights tend to be secreted into bile and found in the feces. Lipophilic drugs and metabolites tend to be returned to the blood. Because the blood perfusion rate to the kidney is high, drugs present in blood are likely to be excreted in the urine in ionized forms through glomerular filtration, or as lipid-soluble un-ionized molecules through tubular secretion.

Strongly acidic drugs are normally excreted in their conjugated forms and are not significantly affected the acidity of the urine. On the other hand, for weakly basic and weakly acidic drugs, the rates of excretion and the percentages that are excreted in their original forms are affected by the urinary acidity in two ways. As the urinary pH decreases, weakly basic drugs tend to exist in the ionic form, which is not favorable for renal tubular reabsorption (through renal tubule cells) back to the tissue. Thus, a higher percentage of these drugs is excreted as unconjugated forms, and their rate of excretion will increase. Conversely, a greater percentage of weakly acidic drugs will be excreted in their original forms at a higher rate under higher urinary pH conditions. Examples of these excretion phenomena are shown in Table 1.15.

Biotransformation of Drugs

Once a drug is exposed to a biological system, it may be converted to different forms as a result of solution chemistry, degradation, and enzymatic reactions. The biotransformation process is divided into two categories: metabolic reactions and conjugative reactions [101]. The metabolic reactions fall into three general classes: oxidation, reduction, and hydrolysis; the conjugation reactions combine products of metabolism with endogenous constituents such as amino acids (glycine and glutamine) and

Table 1.15. Rate and form of drug secretion as a function of urinary pH.

Drug	pK_a	pH of urine	% secreted	Ratio of original to conjugated form	Ref.
Methadone	8.6	5.1	24	10	[99]
		8.0	2.5	1.1	
Salicylate	3.0	<7.0	—[a]	5.7	[100]
		>7.0	—	9.0	

[a] Data not available.

glucuronic acid. Oxidation, reduction, and hydrolysis reactions are defined as phase I metabolism, while conjugation reactions are classified as phase II metabolism [102]. The phase I–phase II taxonomy helps explain the sequential nature of phase I processes, which often produce the functional groups needed for conjugation (phase II) reactions.

Biotransformation of drug molecules is facilitated by enzymatically catalyzed reactions in the blood, kidneys, intestinal mucosa, and, to the greatest extent, liver. The overall purpose of these transformations is to produce derivatives of higher solubility in the aqueous phase so that they may be readily excreted through the kidneys. Because these derivatives, in general, are less toxic than their parent compounds and more easily excreted, these biotransformation processes can be considered as detoxifying mechanisms [103] that have evolved as a protection mechanism against exogenous substances. (In cases where metabolites are more toxic, it may be inferred that the organism has not yet sufficiently evolved to develop a suitable detoxifying mechanism for this "new" substance encountered by the enzymatic system.)

Different enzyme systems in organs are responsible for various specific enzymatic reactions. From a drug analysis point of view, the ability to predict the potential reaction products of a drug with a set of given structural features is more important than an understanding of actual reaction processes and enzymatic systems. Thus, the biotransformation of several important functional groups commonly found in drugs is outlined first, and then a comprehensive list of metabolites of drugs frequently encountered in forensic analysis is given.

General Drug Biotransformation Patterns. Upon introduction into a biological system, a drug may be degraded through passive, nonenzymatic degradation reactions with the environment. Thus, some penicillins are unstable at acidic pH and are inactivated by gastric juices. These types of reactions may be predicted through solution chemistry. Of more significance are active enzymatic reactions. Phase I metabolism of compounds with major functional groups is summarized in Table 1.16. Phase II reactions [105] are summarized in Table 1.17.

Metabolites of Drugs Commonly Found in Forensic Analysis. Major metabolites derived from commonly encountered drugs are listed in Appendix II. Although metabolites in humans are of primary interest, metabolites found in experimental animals have been indicated where studies of drugs in humans cannot be found. The references cited, by no means exhaustive, are meant to provide a basis for looking up further information. The structural features of these metabolites are designated in Appendix III using basic structural frameworks, and the functional group designations are shown in Appendix IV.

Characteristics of Biological Samples

Biological samples commonly encountered include biological fluids that are traditionally used for clinical analysis, postmortem fluids and tissues, and unconventional body components, such as nails, hairs, and bone tissue. Biological fluids that are most commonly analyzed are blood, plasma (or serum), and urine. Mainly because of the noninvasive nature and the ease of sample collection, drug testing of saliva is gaining popularity and has been studied for its suitability in monitoring drug levels.

Table 1.16. Types of metabolic reactions and biotransformation of major functional groups.[a]

Type of reaction	Transformation	Reaction product
Oxidation		
ArH	Hydroxylation	Ar–OH
–CH$_3$		–COOH
–CH$_2$–R		–CH(OH)–R
–C–O[S]–R	*O[S]*–Dealkylation	–C–O[S]H
–CH$_2$–NH$_2$	Deamination	–CHO
–CHR–NH$_2$		–CRO
–NH–R	*N*–Dealkylation	–NH$_2$
R–C(=S)–R	Desulfuration	R–C(=O)–R
R–S–R	Sulfoxidation	R–S(=O)–R
Reduction		
–CH[R]=O	Aldehyde [ketone] to alcohol	–CH$_2$–OH[R]
–NO$_2$	Nitro to amine	–NH$_2$
R–S–S–R	Disulfide cleavage	R–SH
Ar–N=N–Ar	Azo reduction	Ar–NH$_2$
Hydrolysis		
COOR	Ester hydrolysis	COOH + HOR
CONR$_2$	Amide hydrolysis	COOH + HNR$_2$

[a] Based on information in Ref. [104].

Table 1.17. Major conjugation reactions of drugs with various functional groups.[a]

Functional group	Conjugation reaction	Conjugating agent
–O[S]H	Glucuronidation	Glucuronic acid
	Glucoside conjugation	Glucose
	Sulfation	Sulfate
	Methylation	*S*-Adenosylmethionine, 5-methyltetrahydrofolic acid
	Acetylation	Acetyl coenzyme A
–COOH	Glucuronidation	Glucuronic acid
	Glucoside reaction	Glucose
	Amino acid conjugation	Glycine, glutamine, ornithine, taurine
–NH$_2$	Glucuronidation	Glucuronic acid
	Sulfation	Sulfate
	Methylation	*S*-Adenosylmethionine, 5-methyltetrahydrofolic acid
	Acetylation	Acetyl coenzyme A
ArX	Glutathione conjugation	Glutathione

[a] Based on information in Ref. [105].

The ideal autopsy material [106] for evaluation of postmortem drug concentrations in fatal cases should be homogeneous and constantly available in adequate amounts. The possibility of contamination should be small, and the drug concentration in the material should be related to the toxic effect that might have contributed to death. The most frequently used autopsy blood is not without drawbacks, however. Blood is not homogeneous, and its composition may be influenced by postmortem processes like hemoconcentration or dilution. It is not always available, for example, in cases of excessive bleeding or in burned, putrefied, mummified, or exhumed bodies. For these and other reasons, postmortem tissues such as brain, liver, and muscle are also used for evaluation of drug contents.

The use of nails, hairs, and bone tissue for drug testing is rather unconventional. The convenience of obtaining nail and hair samples and the persistence of bone tissue after death are advantages that cannot be provided by other types of samples. Possibilities of detecting drugs and their metabolites in bloodstains, saliva stains, and perspiration stains provide information of investigative value. On rare occasions, insect larvae, found on human remains after disappearance of the usual toxicological specimens, may be analyzed.

The significance of selecting appropriate samples for the analysis of suspected drugs cannot be overemphasized. Thus, preferred samples and appropriate amounts for various toxicological examinations are listed in Table 1.18 [107] to serve as a guide for sample selection.

Table 1.18. Preferred samples for various toxicological examinations.[a]

Specimen	Amount	Toxic material
Bile	All	Narcotics
Blood	100 mL	All types
Bone	Portion of femur	Chronic metal poisoning
Brain	One-half	Ethanol, central nervous system depressants
Fat	100 g	Fat-soluble drugs (e.g., glutethimide), insecticides
Gastric contents	All	Ingested drugs or chemicals
Kidney	One kidney	Heavy metals
Liver	500 g	All types, especially organic bases
Lungs	200 g	Inhalants
Muscle	50 g	Any (in absence of other samples)
Urine	All	All types, metabolites
Vitreous humor	All	Any (in absence of blood)

[a] Data are taken from Ref. [107].

Biological Fluids

Special problems associated with the analysis of drugs in biological fluids are well addressed in a recent publication [108]. In general, "the ease with which samples can be analyzed increases with the degree of fluidity." [108] The order of fluidity of commonly encountered biological samples is listed in Table 1.19 [109]. Therefore, cerebrospinal fluid would be the easiest fluid to handle, whereas whole blood should probably be avoided if possible.

Table 1.19. Biological samples: order of fluidity and degree of difficulty of analysis.[a]

Liquids	Cerebrospinal fluid
	Tears
	Sweat
	Saliva
	Urine
	Bile
Mixed	Plasma–serum
	Blood
	Feces
	Heart–kidney–liver
	Lung–muscle
	Bone

[a] Data are taken from Ref. [109].

Artifacts derived from the use of various sample containers have been studied and reviewed [110–113]. Precautions should be exercised in data interpretation. For example, it was reported [110] that concentrations of tetrahydrocannabinol (Δ^9-THC) in whole-blood samples stored in plastic containers were generally lower than those stored in glass containers. Rubber septa were found [111] to contribute N-ethylbenzeneamine, which may interfere with the analysis of amphetamine. It was observed that waterproof labels incorporated inside sealed bottles to prevent sample tampering [113] caused reduced Δ^9-THC–carboxylic acid concentration.

Blood, Plasma, and Serum.   Blood is perhaps the most useful sample for the identification and quantification of drugs and for the interpretation of data of toxicological significance. Blood is a complex fluid containing solubilized proteins, dissolved fats and salts, and suspended cells. The major constituents, the red blood cells (erythrocytes), can be separated from the clear fluid (plasma) by centrifugation. If blood is allowed to stand without the addition of anticoagulating agents, the red cells will eventually clot and the resulting fluid, serum, can be decanted. Serum is thus, in most respects, similar to plasma except that it does not contain the soluble factors that lead to the clotting phenomenon.

In general, whole blood is not directly analyzed; plasma or serum is preferred. However, whole blood may be the sample of choice when trying to detect the presence of a drug, or when quantitative information is needed on a deteriorated blood sample where complete separation of plasma or serum from red cells is not possible. This is especially true if the drugs of interest have a high affinity for the erythrocytes. For example, the concentration of chlorthalidone in erythrocytes is about 40 times that in the plasma, and even 1% of hemolysis (which cannot be easily observed) will increase the apparent plasma concentration by 25% [112].

Because serum and plasma consist of a large amount of protein, and drugs often have a considerable affinity to protein, direct analysis of drugs in protein-removed serum or plasma may not reflect the total drug concentration. Therefore, protein is often denatured prior to extraction with an organic solvent. Alternatively, drugs strongly bound to protein may be freed with appropriate adjustment of pH.

Urine.   Compared to blood, plasma, or serum, urine is relatively free of protein, thus making it possible for direct extraction with an organic solvent. However, the interpretation of results obtained from urine samples is complicated by several factors, such as the amount of excretion, the variation of pH values, and the time lapse after the intake of the drug. It is often the total amount of excretion during a set period, rather than the concentration, that is of interest.

Diet often alters urinary pH, which greatly changes the patterns of drug excretion. Thus, weakly basic drugs, such as amphetamine, are more efficiently excreted in acidic urine, whereas weakly acidic drugs, such as barbiturates and salicylates, are more efficiently excreted in alkaline urine [114]. In an evaluation of urine samples for suitability in the preliminary drug screening of autopsy cases [115], it was reported that certain drugs were not found or were detected at low levels, whereas results obtained from liver and blood samples showed intoxicating concentrations. It was thus advocated that urine is not a sufficient material to be used alone for drug screening in systematic toxicological analysis.

Despite these concerns, urine is universally used for the now-widespread workplace drug-testing programs [116]. Compared to blood, urine has a longer detection time for most drugs. It also contains high concentrations of metabolites that are easily detectable. With cautious data interpretation [117–124] and regulation [125], it is possible to identify drug abusers in the workplace with minimal errors that are also correctable under most circumstances. If, however, the purpose of drug testing is to relate concentrations to impairment or toxicology, blood is the specimen of choice as it is most closely related to drug concentrations at effector sites.

Saliva.   Compared to urine collection, saliva sample-collection procedures are less invasive and cause less concern about violation of privacy and adulteration of samples. Because saliva can be used to estimate the actual, protein-unbound, circulating concentration of some drugs and their metabolites at the time of collection, test results may have potential for correlation with performance impairment [126]. Compared to blood, saliva samples can be directly analyzed without a prior extraction step. These advantages have promoted numerous investigations and several review articles on the suitability of saliva in therapeutic drug monitoring [127, 128] and in forensic applications [126, 129].

Saliva drug-analysis studies focus on distribution mechanisms, correlation of drug concentrations in saliva and other body fluids, and investigation of the validity of various sampling approaches. Only the unbound fraction of the drug in plasma is available for transfer to saliva, mainly through passive diffusion, which is governed by the saliva and plasma pH values and the structural features of the drugs and metabolites. Thus, the effect of salivary pH on the saliva-to-plasma ratios of cocaine ($pK_a = 8.6$), valproic acid ($pK_a = 4.6$), and benzoylecgonine (a zwitterion containing both acidic and basic functional groups, $pK_a = 2.25$ and 11.2) are shown in Figure 1.1 [126]. The ratios shown were calculated using the Henderson–Hasselbach equation on the basis of the net charge on the molecules at the indicated salivary pH and an assumed plasma pH of 7.4. Because the calculated values do not take into account the drug solubility in the lipid bilayer of the membranes of the salivary-gland acinar cells and potential active transfer mechanisms, they are not always consistent with actual observations.

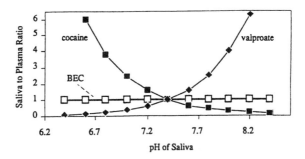

Figure 1.1. Effect of salivary pH on the saliva-to-plasma ratios of cocaine ($pK_a = 8.6$), valproic acid ($pK_a = 4.6$), and benzoylecgonine (BEC, a zwitterion containing both acidic and basic functional groups, $pK_a = 2.25$ and 11.2). The ratios shown were calculated using the Henderson–Hasselbach equation on the basis of the net charge on the molecules at the indicated salivary pH and an assumed plasma pH of 7.4. (Reproduced with permission from Ref. [126].)

The validity of sampling approaches and sample-treatment procedures has to be established before any data correlation can be conducted. It has been reported [130] that early saliva samples (<8 h) collected after oral or nasal ingestion of cocaine may show higher cocaine concentrations due to nasal cavity or mouth residues. Stimulation of saliva to near-maximum rates with citric acid candies causes an increase in salivary pH, resulting in reduced saliva drug concentrations and saliva-to-plasma ratio of basic drugs such as cocaine [131].

"Salivary vs. plasma" and "salivary drug concentration vs. dose" correlations were found to be highly significant in a large methadone-treatment population. The samples tested consisted of a clear liquid produced by freezing saliva samples at –40 °C for 24 h to effectively break the cellular components and suspended particles [132]. It should be noted, however, that current knowledge is generally not adequate for reliable prediction of the plasma levels of the drug based on salivary data. Thus, it has been suggested that "if both saliva and blood samples are available, qualitative analysis for drugs may be carried out using the saliva sample, while the blood sample is reserved for the quantitative estimation of the drugs detected in the saliva." [129]

Postmortem Tissues

Difficulties encountered in testing postmortem tissues include:

1. Certain materials produced during the putrefying process may interfere with the analysis of drugs of concern.

2. Drugs of interest may not survive the decaying or advanced state of putrefaction.

3. Pretreatment procedures for solid postmortem tissue may not always result in a consistent medium suitable for subsequent analysis.

4. Potential redistribution of drugs or metabolites in postmortem tissues may render data interpretation difficult.

Among several compounds reported, β-phenylethylamine was considered [133] the most frequently encountered putrefactive base that may interfere with the analysis of sympathomimetic amine drugs, such as amphetamine, methylhexaneamine, and tuaminoheptane. On the other hand, *p*-hydroxyphenylacetic acid and *p*-hydroxyphenylpropionic acid are common putrefactive acids that may interfere with ultraviolet analysis of 5,5-substituted barbiturates [134, 135]. The problematic nature of these putrefactive compounds is further complicated by the lack of predictive patterns of their production.

The stability of 56 drugs in putrefying blood [136, 137] and liver [136, 138] has been studied. It was concluded [138] that compounds with the following structural char.cteristics are prone to putrefactive decomposition:

1. Oxygen bonded to nitrogen but not to carbon or sulfur, as in nitro groups, *N*-oxides, and oximes;

2. Sulfur bonded as a "thiono" atom (C=S or P=S) or formed as part of a heterocyclic ring; and

3. The presence of an aminophenol structure, that is, OH and NH_2, on the same aryl nucleus.

Carbon bonds with oxygen and nitrogen, nitrogen bonds with hydrogen, and sulfur bonds with oxygen are generally stable. The destruction of labile structures is generally observed on the 3rd to 14th day of the putrefying process and is greatly enhanced by blowfly-borne bacteria. This helps explain the reports of the long-term stability of certain drugs in buried corpses [139] that had been insulated from blowfly contact by interment.

Postmortem tissues such as brain, liver, kidney, lung, and muscle are solid samples (Table 1.19) that require [107] a "destructive disruption" procedure and the use of a suitable liquid medium to render samples amenable to sampling and subsequent extraction and measurement methods. The end result of such procedures is the production of a heterogeneous suspension, often called a homogenate. Commonly used disruption procedures are mortar–pestle and blade–container homogenizing, sonication, and chemical dissolution. One of the major disadvantages of the mortar–pestle homogenizing approach is the lack of consistency, mainly due to the inadequate disruption force of this technique. Chemical dissolution, on the other hand, is likely to produce foaming, which may complicate the subsequent extraction process. Commonly used solvents and their general characteristics are listed in Table 1.20 [109].

Concentrations of drugs in various biological samples have been compared and are often of interpretive value [140–142]. However, it has been cautioned [143] that there is potential postmortem diffusion of drugs along a concentration gradient, from sites of high concentration in solid organs, into the blood, with resultant artifactual elevation of drug levels in blood.

The selection of a tissue for analysis depends on the drug of interest. For example, kidneys are often the choice if metal poisoning is suspected, whereas brain tissue is often used to investigate death due to solvent abuse or cyanide poisoning, and bile is selected if morphine-like compounds are suspected because of the high concentration of glucuronides from these drugs [112].

Vitreous Humor. Putrefaction is often less pronounced [144] in the vitreous humor than in blood, therefore making the removal of contaminants easier for the former type of sample. Because there is no direct blood supply to the vitreous humor, it is free from enzymes and the likelihood of microbial contamination. Analysis of blood and vitreous humor samples in cases of death caused by drug overdose [144–147] has indicated that the blood and vitreous humor concentration ratio, at equilibrium, depends on the following factors:

1. Solubility of the drug in the vitreous humor;

2. Lipid solubility of the drug, which affects the penetration of the blood–vitreous barrier; and

3. The percentage of the protein-bound drug in the blood.

The acidity of a drug affects its solubility in vitreous humor. Because the vitreous humor is acidic with respect to the plasma, weakly alkaline drugs will reach a higher concentration in vitreous humor, and may do so in spite of extensive protein-binding in the blood [146]. With the equilibrium ratio for each drug established, it is suggested [145] that the testing of drug levels in vitreous humor and blood may serve as a useful parameter for studying the survival time in drug-intoxication cases.

Table 1.20. General characteristics of homogenization media.[a]

Solvent	Advantages	Disadvantages
Distilled water	Relatively good solvent Does not destroy tissue constituents Final pH near 7.0	Degree of ionization may vary Does not denature enzymes consistently Final pH may vary with tissue
Dilute acid (0.5 N)	Relatively good solvent Denatures many enzymes Final pH 7.0	Considerable protein denaturation Compound(s) may be acid sensitive Foaming is minimal
Dilute alkali (0.5 N)	Relatively good solvent Denatures many enzymes Final pH 7.0	Considerable protein denaturation Compound(s) may be alkali sensitive May cause foaming (soaps)
Strong acid	Good solvent Denatures all enzymes Precipitates proteins Final pH 4.0	Clumping and aggregation may occur Compound(s) may be acid sensitive Tissue constituents may break down
Strong alkali (0.5 N)	Relatively good solvent Denatures all enzymes Precipitates proteins Final pH 10.0	Clumping and aggregation may occur Compound(s) may be alkali sensitive Tissue constituents may break down Foaming generally serious (soaps)
Organic solvent	Relatively chemically inert Denatures most enzymes Precipitates proteins	Clumping and aggregation generally severe May be poor solvents

[a] Data are taken from Ref. [109].

Liver.   The liver, as the primary drug-metabolizing organ, contains high drug levels and has traditionally been a favored postmortem tissue for the examination of drugs. The use of the liver is also favored because of its relatively large size. (The weight percentages [148] of muscle, fat, liver, brain, lungs, and kidneys in an average adult male are 40, 10, 2.5, 2, 0.7, and 0.4 percent, respectively). Because it is highly perfused with blood, drug concentrations in the liver tend to achieve instantaneous equilibrium with those in plasma. Thus, the liver is suitable for systematic qualitative drug-screening tests.

Brain.   The brain tends to resist postmortem putrefaction, and, therefore, it is useful for drug analysis if the body is found several days after death. With high solvent-retention capacity, the brain is often the sample of choice when solvent abuse is suspected.

Comparative analysis [149] of cocaine and its metabolite, benzoylecgonine, in the brain and blood has indicated that the brain-to-blood ratio of these drugs may be indicative of time lapse between drug intake and death and the drug-use pattern. Cocaine, as a lipophilic drug, is known to pass freely across the blood–brain barrier. At peak plasma cocaine concentration, the brain cocaine concentration is just over four times the plasma or serum concentration. Because the blood cocaine concentration falls slightly more rapidly than brain tissue concentration, the brain-to-blood ratio will increase, reaching a peak in the range of 8 to 10 between 1 and 2 h after administration. Thus, in cases of acute overdose of cocaine, death occurs from a few minutes to 1 to 3 h after administration, and the brain-to-blood ratios generally fall in the range of 4 to 10. On the other hand, chronic use of cocaine seems to alter the deposition pattern. After 14 days of chronic cocaine use, the brain-to-blood cocaine ratio following an acute dose was found to be 2.4.

The hydrolysis product of cocaine, benzoylecgonine, on the other hand, is restricted in its passage across the blood–brain barrier. The mean brain-to-blood ratio of 0.36 for benzoylecgonine in toxic cases [147] appears to be typical of the disposition ratio at 1 to 3 h after cocaine administration. On the other hand, the average brain-to-blood ratio of 1.4 for benzoylecgonine in an incidental case [149] was considered indicative of chronic abuse or use more than 8 to 10 h prior to death.

Stomach Contents. The analysis of stomach contents often reveals valuable information as to whether the drug was taken orally or through other routes. Absence of even a trace of drug in the stomach contents generally indicates administration of the drug through another route. It is not uncommon for whole capsules or tablets to be present, therefore providing definite evidence of oral consumption. Conversely, the presence of trace amounts of a drug cannot be interpreted as definite evidence of oral consumption. Some drugs, especially basic drugs, can be secreted from the blood into the acidic gastric juices after an injection.

Bile. Drugs removed from the liver, either as metabolites or in their unchanged forms, may be excreted in the bile. Morphine and acetaminophen are known to have high biliary secretion, morphine being mainly excreted as a glucuronide, whereas acetaminophen is secreted as a glucuronide and as glutathione conjugates [150].

Muscle. Muscle is not normally the preferred tissue for the analysis of drugs. Considering the possible drawbacks of preferred tissues, such as blood, the feasibility of using striated muscle from the extremities as the autopsy material for evaluation of postmortem drug concentration in fatal cases has been thoroughly evaluated. Results were compared with those obtained from blood and liver samples [106]. Drug concentrations in different muscles (legs and arms) were found to be as constant as those found in the corresponding blood samples.

The results summarized in Table 1.21 [106] indicate that, in fatal cases, drug concentrations of all acid and base classes are higher in the liver but occur at comparable levels in the blood and muscle. For nonacidic drugs, the percent standard deviation data for muscle is less variable than that for the blood. In all cases, drugs were detected in muscle when found in the blood or the liver. The detection [106] of drugs in muscle includes samples stored for more than one year (at –20 °C) in a body that was found several months after death, and in an exhumed body six months after burial.

Unconventional Samples

Samples such as hair, nails, bone tissue, and biological fluid stains are not commonly used in conventional toxicological and clinical analysis, but they may provide valuable information otherwise not available in forensic applications. Hair and nail sample-collection procedures are less invasive than those used for blood or urine. As postmortem specimens, hair, nails, and bone tissue can survive longer than other conventional biological tissues. The analysis of bloodstains, perspiration stains, seminal stains, and saliva stains is sometimes valuable from an investigative perspective.

The use of these unconventional samples is, however, often hindered by the lack of sufficient information in the literature to ensure the accuracy of analytical findings and indisputable data interpretation. A recent review article [151] was published to facilitate the evaluation of the database currently available.

Hair. The use of hair includes several advantages:

1. Sample collection is noninvasive.
2. Drugs are retained in hair much longer than in urine or serum, and they may therefore serve as a valuable indicator for the investigation of past drug exposure.
3. Sectional analysis of hair may provide information regarding the duration of drug exposure.
4. Certain drug metabolites that are of significant interpretative value are found in hair in higher concentration and retained for a long duration.

With the common practice of workplace drug urinalysis and the associated concerns about privacy-invading sample-collection procedures, several groups have recently advocated the use of hair as an alternate test specimen. Because studies on the analysis of drugs in hair are limited, and an understanding of the interpretation of analytical findings is generally lacking, numerous investigators have recently conducted studies aimed at improving the analytical procedure and understanding the underlying mechanisms that may facilitate data interpretation.

Table 1.21. Range, median, average, and percent standard deviation of drug concentrations (in μmol/kg) in blood, muscle, and liver from cases of suspected overdose.[a]

Drug	Sample	Range	Median	Average	Std dev (%)	No. of cases
Aprobarbital	Blood	190–440	340	317	30	7
	Muscle	170–640	320	368	40	7
	Liver	320–810	550	542	37	7
Pentobarbital	Blood	17–99	43	51	59	8
	Muscle	27–100	48	54	43	8
	Liver	68–500	225	272	62	8
Phenobarbital	Blood	20–570	310	346	35	10
	Muscle	250–1000	380	467	52	10
	Liver	310–1200	910	822	35	10
Amitriptyline	Blood	0.8–20	12	11	63	6
	Muscle	3.4–28	14	15.1	60	6
	Liver	96–200	115	138	36	6
Methadone	Blood	0.6–1.6	1.0	1.0	31	8
	Muscle	0.4–1.2	0.8	0.8	35	8
	Liver	1.4–8.6	4.3	4.3	56	8
Propoxyphene (without ethanol)	Blood	ND–64[b]	2.9	13	154	11
	Muscle	1.6–43	6.6	9.8	122	11
	Liver	3.4–120	49	49	80	11
Propoxyphene (with ethanol)	Blood	1.1–17	8.9	9.1	70	6
	Muscle	1.5–7.6	2.9	3.4	68	6
	Liver	5.6–210	52	71	106	6
Morphine (without ethanol)	Blood	0.1–44	0.8	4.8	233	15
	Muscle	0.1–14	0.6	2.0	180	15
	Liver	0.9–20	1.8	3.3	145	15
Morphine (with ethanol)	Blood	0.1–6.7	0.4	1.0	160	20
	Muscle	0.2–2.4	0.5	0.6	90	20
	Liver	0.4–4.1	1.4	1.5	63	20

[a] Data are taken from Ref. [106].
[b] ND: not detected.

Earlier exploratory works indicated the presence of opiate [152], amphetamine and related drugs [151–156], chloroquine and its metabolite [157], phencyclidine [158], phenobarbital [159], and cocaine [160] in hair. With the concern about using screening tests alone for reporting results, the U.S. Food and Drug Administration has issued a policy guideline warning against the use of radioimmunoassay alone to detect the presence of drugs of abuse [161]. Recent applications of mass spectrometric methods [162] have removed any doubt concerning the findings of drugs in hair specimens. More specific studies indicated the appearance of codeine in beard hair 6–8 days after administration [163]. Using methoxyphenamine as a model compound, it has been reported that the drug was detected in a band approximately 5 mm wide (equivalent to 1.7–2.4 periods of 7-day hair growth), and the amount decreased approximately 50% during a 5-month period [164]. It is not certain, however, whether the lack of "diffusion" within the hair shaft is true for other drug categories and will not be affected by the individual's physiological conditions [165].

In addition to the widely publicized potential of using data from segmental hair analysis to establish an expanded period of drug-exposure history, the special role of hair as a specimen for drug monitoring has been enhanced by recent findings that hair samples contain substantial quantities of parent drugs and other metabolites that are difficult to detect in urine specimens. For example, high concentrations of cocaine and cocaethylene (indicative of simultaneous consumption of cocaine and alcohol) [166], and heroin and 6-monoacetylmorphine (indicative of heroin use) [167], have been detected in hair samples. It is logical that these relatively lipophilic compounds are more likely to be detected in hair than in urine samples.

Because the biochemical processes involving the absorption of drugs and their metabolites into hair and the influence of external contamination and treatment are not adequately understood, hair still cannot be used as the routine sample for monitoring drug use. For example, it has been concluded that hair-washing procedures failed to completely remove environmental contamination; analysis of the methanolic wash of the hair of cocaine users also revealed the presence of cocaine metabolite, indicating the removal of cocaine from the interior as well as from the exterior surface of hair [163, 166–168]. It has also been proved that more than 90% of cocaine and benzoylecgonine can be removed from the drug user's head hair by a 20-h treatment with alkaline permanent wave [168].

Sweat.   Because drug compositions in sweat may partially contribute to the observation of drugs in hair [169], it becomes important to understand the drug-excretion patterns in sweat. "Sweat patches" used for collecting perspiration samples from human skin have improved from an earlier occlusive design [170] to the current dressing technology using very thin polyurethane and acrylate adhesives [171]. This new nonocclusive approach allows the moisture content of perspiration to evaporate and makes it possible to apply a single dressing to skin and maintain sterile attachment for several days with rare dermatological reactions. The final content of the collection pad is an accumulation or integration of that contributed to it by skin [172].

With the pH of sweat in the range of 4 to 6.8 [173] compared to a normal plasma pH of above 7.0, the concentration of a neutral basic drug in sweat is expected. It has been reported that cocaine in single doses as low as 50 mg can be detected for up to 7 days after use [172, 174]. The information given in Table 1.22 provides an interesting comparison of the distribution patterns of drugs and metabolites in urine, hair, and sweat following the consumption of cocaine and heroin [175].

Table 1.22. Relative concentrations of cocaine and heroin and their metabolites in urine, hair, and sweat.[a]

Sample	Cocaine	Heroin
Urine	Benzoylecgonine ≈ ecgonine methyl ester > cocaine	Morphine-3-glucuronide > morphine > 6-monoacetylmorphine
Hair	Cocaine > benzoylecgonine > ecgonine methyl ester	6-Monoacetylmorphine > heroin ≈ morphine
Sweat	Cocaine > ecgonine methyl ester > benzoylecgonine	Heroin ≈ 6-monoacetylmorphine > morphine

[a] Data are taken from Ref. [175].

Bone, Nails, and Biological Stains.   The resistance of bone tissues to postmortem putrefaction offers the advantage of prolonging the period of investigative capability of toxicological analysis. Although testing for metallic poisons such as arsenic, mercury, and lead in bone tissue is well known in classical toxicological analysis, the penetration of organic compounds into bone tissue remains practically unexplored. Noguchi and co-workers have stated: "Analysis of the bone was made on the basis that it represented a blood repository and drugs present in blood may be contained in the bone marrow. In the recently deceased, the bone marrow is heavily laden with blood, and compression yields oozing blood. In this case, the bone marrow was dry, and it was reasoned that drugs present in the blood would be deposited in the marrow as dehydration proceeded." [176] Indeed, in one study, amitriptyline was found in bone marrow extract.

Limited studies [177] have also indicated the possibility of detecting amobarbital and glutethimide in sternum and rib samples. These studies have further indicated that the amount of the drug sequestered in the sternum is relatively small if death occurs shortly after drug use. A higher drug concentration was found in a case in which death was thought to occur after a prolonged period following drug administration.

Nails presumably will retain undegraded drugs for a long time and thus have similar advantages to hair samples. The analysis of fingernail and toenail clippings [178] from methamphetamine users proved their potential in toxicological analysis. Methamphetamine levels in fingernails were found

[178] to be comparable to those in hair. The concentration in toenails was found to be higher, presumably because of its slower growth rate. Nails with their underside surface layer scraped off (to reduce the weight by about 20%) were found to contain similar drug levels, confirming deposition of drug in the true nail matrix.

Considering the investigative value of detecting drugs in the clothes worn by drug abusers, sweat, perspiration, semen, seminal stains, bloodstains, and menstrual bloodstains have also been reported. Among the drugs detected are diphenylhydantoin [179], morphine [180], phenobarbital [159], amphetamine [181], cocaine or its metabolite benzoylecgonine [160, 182], and methamphetamine [183]. Although everyday usage of these types of samples is unlikely, data obtained from them may provide unique information that is otherwise unavailable.

Meconium.   Concerns over the maternal use of illicit drugs, especially cocaine, during pregnancy have prompted the evaluation of newborn meconium as a specimen for the detection of intrauterine exposure to drugs [184]. Meconium offers the major advantage over urine of greater reliability of collection and an extended window for detecting drug use. As a nonhomogenous sample with layers formed at times of deposition in the intestine, meconium analysis is more cumbersome than urinary analysis, and it may also be harder to prepare standard and proficiency specimens for test-result assessment purposes. The current status of meconium testing has recently been reviewed [185]. It undoubtedly has great potential for use in the diagnosis of fetal drug exposure and the study of fetal metabolic pathways.

Extracts from meconium have been used successfully for the preliminary detection of amphetamine, opiates, phencyclidine, methadone, cocaine metabolites, and cannabinoid metabolites by EMIT [186, 187]. In a large-scale study, meconium was reported to be superior to maternal and newborn urines for the detection of cocaine metabolites, but not effective for cannabinoid metabolites [186]. In a study of 106 subjects for cocaine [188], the authors observed a higher positive rate from the meconium than from the corresponding urine samples—19 vs. 8 positives. This study also reported the detection of norcocaine and cocaethylene in meconium but not in urine samples. The greater efficiency for the detection of cocaine metabolites in meconium was attributed to the presence of 0.2 to 6.3 times more total m-hydroxybenzoylecgonine than benzoylecgonine in meconium [189]. Within the 50- to 1000-ng/mL concentration range studied, the cross-reactivities of m-hydroxybenzoylecgonine with TDx and EMIT II assays were found to be 131–150% and 0.63–1.10%, respectively [189]. Improvement in the confirmation rate of cannabinoid metabolites has also been reported recently [190].

Insect Larvae.   With remains in the advanced decomposition stage, insect larvae may serve as a suitable source for the identification of drugs present and the estimation of the postmortem intervals. Reported applications include the analysis of phenobarbital in *Cochliomyia macellaria* [191], opiates in *Calliphora vicina* [192] and *Boettcherisca peregrina* [193], cocaine in *Boettcherisca peregrina* [194], and methamphetamine in *Parasarcophaga ruficornis* [195]. Although correlations have been reported for opiate radioimmunoassay data obtained from larvae and controlled decomposing liver [192], it is highly uncertain whether quantitative data can be properly interpreted. It has been reported that the presence of drugs may affect the development rates of maggots and should be noted when estimating postmortem intervals [193–195].

References

1. Smith, R. M., *Newsweek,* June 16, 1986, 15.
2. Williams, A.; Peat, M.; Crouch, D.; Wells, J.; Finkle, B. *Public Health Rep.* **1985,** *100,* 19.
3. Taylor, R. A.; Weisman, A. P.; Gest, T., *U.S. News & World Report,* July 28, 1986, 48.
4. Johnston, L. D.; O'Malley, P. M.; Bachman, J. G. *Use of Licit and Illicit Drugs by American's High School Students 1975-1984;* U.S. Government Printing Office: Washington, DC, 1985.
5. *Drugs, Crime, and the Justice System;* Bureau of Justice Statistics. U.S. Government Printing Office: Washington, DC, 1992; p 126.
6. Baum, R. M. *Chem. Eng. News* **1985,** *63,* 7.

7. Henderson, G. L. *J. Forensic Sci.* **1988,** *33,* 569.
8. Sapienza, F. *DEA/Registrant Facts* **1982,** *8,* 3.
9. *Population Estimates 1992;* National Institute of Justice. U.S. Department of Justice. U.S. Government Printing Office: Washington, DC, 1993; SMA 93-2053.
10. *Data from the Drug Abuse Warning Network: Annual Emergency Room Data 1992*; U.S. Government Printing Office: Washington, DC, 1994; SMA 94-2080.
11. *Data from the Drug Abuse Warning Network: Medical Examiner Data 1992;* U.S. Government Printing Office: Washington, DC, 1994; SMA 94-2081.
12. *Drug Use Forecasting 1992 Annual Report;* National Institute of Justice. U.S. Department of Justice. U.S. Government Printing Office: Washington, DC; 1993.
13. 21 CFR 1300; U.S. Government Printing Office: Washington, DC, 1996; Chapter 2.
14. Shulgin, A. T. *The Controlled Substances Act: A Resource Manual of the Current Status of the Federal Drug Laws;* Lafayette, CA, 1988; p 234.
15. Bray, R. M.; Marsden, M. E.; Peterson, M. R. *Am. J. Public Health* **1991,** *81,* 865.
16. Marwick, C.; Gunby, P. *JAMA J. Am. Med. Assoc.* **1989,** *261,* 2784.
17. Willette, R. E. In *Urine Testing for Drugs of Abuse;* Hawks, R. L.; Chiang, C. N., Eds.; NIDA Research Monograph 73; National Institute on Drug Abuse: Rockville, MD, 1986; p 5.
18. Little, J. S.; Lukey, B. J.; Shimomura, E. T.; Fuhrmann, L. S. *J. Forensic Sci.* **1993,** *38,* 259.
19. *Fed. Regist.* **1994,** *59,* 29908.
20. Liu, R. J. H. *Forensic Sci. Rev.* **1994,** *6,* 19.
21. King, L.; Moffat, A. C. *Med. Sci. Law* **1983,** 23, 193.
22. Kalman, S. M.; Clark, D. R. *Drug Assay: The Strategy of Therapeutic Drug Monitoring;* Masson USA: New York, 1979; p 94.
23. Bailey, D. N.; Manoguerra, A. S. *J. Anal. Toxicol.* **1980,** *4,* 199.
24. Jones, D. W.; Adams, D.; Martel, P. A.; Rousseau, R. J. *J. Anal. Toxicol.* **1985,** *9,* 125.
25. Cimbura, G.; Lucus, D. M.; Bennett, R. C.; Warren, R. A.; Simpson, H. M. *J. Forensic Sci.* **1982,** *27,* 855.
26. Siek, T. J. *J. Forensic Sci.* **1974,** *19,* 193.
27. Siek, T. J.; Osiewicz, R. J. *J. Forensic Sci.* **1975,** *20,* 18.
28. Siek, T. J.; Osiewicz R. J.; Bath, R. J. *J. Forensic Sci.* **1976,** *21,* 525.
29. Nisikawa, M.; Tatsuno, M.; Suzuki, S.; Tsuchihashi, H. *Forensic Sci. Int.* **1991,** *51,* 131.
30. Vodra, W. W. *Drug Enforcement* **1975,** *2,* 37.
31. Sandell, E. B.; Elvin, P. J. In *Treatise on Analytical Chemistry: Part I Theory and Practice,* Vol. 1; Kolthoff, I. M.; Elving, P. J., Eds.; The Interscience Encyclopedia: New York, 1959.
32. Franzosa, E. S. *J. Forensic Sci.* **1985,** *30,* 1194.
33. *Microgram* **1990,** *23,* 66.
34. Humphreys, I. J. *Bull. Narc.* **1984,** *36,* 33.
35. *Microgram* **1982,** *15,* 188.
36. *Microgram* **1984,** *17,* 95
37. Nakamura, G. R.; Thornton, J. I.; Noguchi , T. T. *J. Chromatogr.* **1975,** *110,* 81.
38. Nakamura, G. R. *J. Forensic Sci.* **1960,** *5,* 259.
39. Moore, J. M.; Klein, M. *J. Chromatogr.* **1978,** *154,* 76.
40. Sy, W.-W.; By, A. W.; Neville, G. A.; Wilson, W. L. *Can. Soc. Forensic Sci. J.* **1985,** *18,* 86.
41. Davey, E. A.; Murray, J. B. *Pharm. J.* **1971,** *207,* 167.
42. Cantrell, T. S.; John, B.; Johnson, L.; Allen, A. C. *Forensic Sci. Int.* **1988,** *39,* 39.
43. Verweij, A. M. A. *Forensic Sci. Rev.* **1989,** *1,* 1.
44. Barnfiedl, C.; Burns, S.; Byrom, D. L.; Kemmenoe, A. V. *Forensic Sci. Int.* **1988,** *39,* 107.
45. Chiarotti, M.; Fucci, N. *Forensic Sci. Int.* **1988,** *47,* 47.
46. Clark, A. B.; Miller, M. D. *J. Forensic Sci.* **1978,** *23,* 21.
47. Moore, J. M.; Allen, A. C.; Cooper, D. A. *Anal. Chem.* **1984,** *56,* 642.
48. Kram, T. C.; Cooper, D. A.; Lurie, I. *Proceedings of the International Symposium on the Forensic Aspects of Controlled Substances;* FBI Academy: Quantico, VA, 1988; p 207.
49. Moore, A. M. *J. Assoc. Off. Anal. Chem.* **1973,** *56,* 1199.
50. Noggle, F. T., Jr.; Clark, R. *J. Assoc. Off. Anal. Chem.* **1982,** *65,* 756.
51. Lukaszewski, T.; Jeffery, W. K. *J. Forensic Sci.* **1980,** *25,* 499.
52. Toffoli, F.; Avico, U. *Bull. Narc.* **1965,** *17,* 27.
53. Lowry, W. T.; Lomonte, J. N.; Hatchett, D.; Garriott, J. C. *J. Anal. Toxicol.* **1979,** *3,* 91.
54. Allen, A. C.; Cooper, D. A.; Kiser, W. O.; Cottrell, R. C. *J. Forensic Sci.* **1981,** *26,* 12.
55. Cooper, D. A.; Allen, A. C. *J. Forensic Sci.* **1984,** *29,* 1045.
56. Sinnema, A.; Verweij, A. M. A. *Bull. Narc.* **1981,** *33,* 37.

57. O'Neil, P. J.; Gough, T. A. *J. Forensic Sci.* **1985,** *30,* 681.
58. Haky, J. E.; Stickney, T. M. *J. Chromatogr.* **1985,** *321,* 137.
59. Kiser, W. O. *Microgram* **1986,** *19,* 94.
60. Churchill, K. T. *Microgram* **1985,** *18,* 61.
61. Clark, C. C. *Microgram* **1981,** *14,* 136.
62. *Microgram* **1984,** *17,* 176.
63. *Microgram* **1985,** *18,* 57.
64. Chiarotti, M.; Fucci, N. *Forensic Sci. Int.* **1988,** *37,* 47.
65. *Microgram* **1984,** *17,* 15.
66. *Microgram* **1984,** *17,* 82.
67. *Microgram* **1983,** *16,* 169.
68. Gomm , P. J.; Humphreys, I. J. *J. Forensic Sci. Soc.* **1975,** *15,* 293.
69. Jackson, J. V. In *Clarke's Isolation and Identification of Drugs in Pharmaceuticals, Body Fluids, and Post-Mortem Material,* 2nd ed.; Moffat, A. C.; Jackson, J. V.; Moss, M. S.; Widdop, B.; Greenfield, E., Eds.; Pharmaceutical: London, 1986; p 50.
70. *Microgram* **1982,** *15,* 44.
71. *Microgram* **1981,** *14,* 75.
72. *Microgram* **1985,** *18,* 116.
73. *Microgram* **1983,** *16,* 88.
74. *Microgram* **1982,** *15,* 67.
75. *Microgram* **1982,** *15,* 25.
76. Navaratnam, V.; Rajananda, V.; Nair, N. K. *Microgram* **1985,** *18,* 7.
77. *Microgram* **1983,** *16,* 73.
78. *Microgram* **1981,** *14,* 22.
79. *Microgram* **1984,** *17,* 69.
80. *Microgram* **1985,** *18,* 145.
81. Sapienza, F. *Drug Enforcement* **1984,** *11,* 26.
82. Brewster, M.; Davis, F. T. *J. Forensic Sci.* **1991,** *36,* 587.
83. Wilson, W. L. *J. Forensic Sci. Soc.* **1991,** *31,* 233.
84. Moore, J. M.; Allen, A. C.; Cooper, D. A.; Carr, S. M. *Anal. Chem.* **1986,** *58,* 1656.
85. Weingarten, H. L. *J. Forensic Sci.* **1988,** *33,* 588.
86. Bost, R. O. *J. Forensic Sci.* **1988,** *33,* 576.
87. Bailey, K.; Beckstead, H. D.; Legault, D.; Werner, D. *J. Ass. Off. Anal. Chem.* **1974,** *57,* 1134.
88. Poklis, A.; Mackell, M. A.; Drake, W. K. *J. Forensic Sci.* **1979,** *24,* 70.
89. Ragan, F. A., Jr.; Hite, S. A.; Samuels, M. S.; Garey, R. E. *J. Anal. Toxicol.* **1985,** *9,* 91.
90. Bal, T. S.; Gutteridge, D. R.; Johnson, B. *Med. Sci. Law* **1989,** *29,* 186.
91. Jackson, J. V. In *Clarke's Isolation and Identification of Drugs,* 2nd ed.; Moffat, A. C.; Jackson, J. V.; Moss, M. S.; Widdop, B.; Greenfield, E., Eds.; Pharmaceutical: London, 1986; p 37.
92. Baselt, R. C.; Cravey, R. H. In *Introduction to Forensic Toxicology;* Cravey, R. H.; Baselt , R. C., Eds.; Biomedical: Davis, CA, 1981; p 243.
93. Chamberlain, J. *Analysis of Drugs in Biological Fluids;* CRC Press: Boca Raton, FL, 1985; p 2.
94. Baselt, R. C.; Cravey, R. H. *Disposition of Toxic Drugs and Chemicals in Man,* 3rd ed.; Year Book Medical: Chicago, IL, 1989; p 208.
95. Baselt, R. C.; Cravey, R. H. *Disposition of Toxic Drugs and Chemicals in Man,* 3rd ed.; Year Book Medical: Chicago, IL, 1989; p 49.
96. Bruce, A. M.; Smith, H. *Med. Sci. Law* **1977,** *17,* 195.
97. Rehling, C. J. In *Progress in Chemical Toxicology,* Vol. 3; Stolman, A., Ed.; Academic New York, 1967; p 363.
98. Baselt, R. C.; Cravey, R. H. *Disposition of Toxic Drugs and Chemicals in Man,* 3rd ed.; Year Book: Chicago, IL, 1989; p 561.
99. Baselt, R. C.; Casarett, L. J. *Clin. Pharm. Ther.* **1972,** *13,* 64.
100. Alpen, E. L.; Mandel, H. G.; Rodwell, V. W.; Smith, P. K. *J. Pharm. Exp. Ther.* **1951,** *102,* 150.
101. Gringauz, A. *Drugs: How They Act and Why;* Mosby: St. Louis, MO, 1978; p 305.
102. Williams, R. T. In *Symposium on Biological Approaches to Cancer Chemotherapy;* Academic: London, 1960; p 21.
103. Dutton, G. *Glucuronidation of Drugs and Other Compounds;* CRC Press: Boca Raton, FL, 1980; p 5.
104. Evans, M. A.; Baselt, R. C. In *Introduction to Forensic Toxicology;* Cravey, R. H.; Baselt, R. C., Eds.; Biomedical: Davis, CA, 1981; p 41.
105. Caldwell, J. In *Concepts in Drug Metabolism: Part A;* Jenner, P.; Testa, B., Eds.; Marcel Dekker: New York, 1980; p 211.

106. Christensen, H.; Steentoft, A.; Worm, K. *J. Forensic Sci. Soc.* **1983**, *25*, 191.
107. Rejent, T. A.; In *Introduction to Forensic Toxicology;* Cravey, R. H.; Baselt, R. C., Eds.; Biomedical: Davis, CA, 1981; p 140.
108. Chamberlain, J. *Analysis of Drugs in Biological Fluids,* 2nd. ed.; CRC Press: Boca Raton, FL, 1995; p 36.
109. Maickel, R. P. In *Drug Determination for Therapeutic and Forensic Contexts;* Reid, R.; Wilson, I. D., Eds.; Plenum: New York, 1984; p 3.
110. Christophersen, A. S. *J. Anal. Toxicol.* **1986**, *10*, 129.
111. Christophersen, A. S.; Bugge, A.; Dahlin, E.; Morland, J.; Wethe, G. *J. Anal. Toxicol.* **1988**, *12*, 147.
112. Toseland, P. A. In *Clarke's Isolation and Identification of Drugs in Pharmaceuticals, Body Fluids, and Post-Mortem Material,* 2nd ed; Moffat, A. C.; Jackson, J. V.; Moss, M. S.; Widdop, B.; Greenfield, E., Eds.; Pharmaceutical: London, 1986; p 111.
113. Bond, G. D., II; Chand, P.; Walia, A. S.; Liu, R. H. *J. Anal. Toxicol.* **1990**, *14*, 389.
114. Becket, A. H. In *Scitific Basis of Drug Dependence;* Steinberg, H., Ed.; Churchill: London, 1969; p 129.
115. Thelander, G.; Jonsson, J.; Schuberth, J. *Forensic Sci. Int.* **1983**, *22*, 189.
116. Willette, R. E. In *Urine Testing for Drugs of Abuse;* NIDA Research Monograph 73; Hawks, R. L.; Chiang, C. N., Eds.; National Institute on Drug Abuse: Rockville, MD, 1986; p 5.
117. Manno, J. E. In *Urine Testing for Drugs of Abuse;* NIDA Research Monograph 73; Hawks, R. L.; Chiang, C. N., Eds.; National Institute on Drug Abuse: Rockville, MD, 1986; p 54.
118. ElSohly, M. A.; Jones, A. B. *Forensic Sci. Rev.* **1989**, *1*, 13.
119. Cone, E. J.; Huestis, M. A. *Forensic Sci. Rev.* **1989**, *1*, 121.
120. Cody, J. T. *Forensic Sci. Rev.* **1990**, *2*, 63.
121. Liu, R. H. *Forensic Sci. Rev.* **1992**, *4*, 56.
122. Cody, J. T. *Forensic Sci. Rev.* **1993**, *5*, 109.
123. Cody, J. T. *Forensic Sci. Rev.* **1994**, *6*, 81.
124. Green, K. B.; Ischensmidt, D. *Forensic Sci. Rev.* **1995**, *7*, 1.
125. *Fed. Regist.* **1988**, *53*, 11979.
126. Schramm, W.; Smith, R. H.; Craig, P. A.; Kidwell, D. A. *J. Anal. Toxicol.* **1992**, *16*, 1.
127. Horning, M, G.; Brown, L.; Nowlin, J. *Clin. Chem. (Winston-Salem, N.C.)* **1977**, *23*, 157.
128. Mucklow, J. C. *Ther. Drug Monit.* **1982**, *4*, 229.
129. Idowu, O. R.; Caddy, B. *J. Forensic Sci. Soc.* **1982**, *22*, 123.
130. Schramm, W.; Craig, P. A.; Smith, R. H.; Berger, G. E. *Clin. Chem. (Winston-Salem, N.C.)* **1993**, *39*, 481.
131. Kato, K.; Hillsgrove, M.; Weinhold, L.; Gorelick, D. A.; Darwin, W. D.; Cone, E. J. *J. Anal. Toxicol.* **1993**, *17*, 338.
132. Wolff, K.; Hay, A.; Raistrick, D. *Clin. Chem. (Winston-Salem, N.C.)* **1991**, *37*, 1297.
133. Oliver, J, S.; Smith, H. *J. Forensic Sci. Soc.* **1973**, *13*, 47.
134. Batchelor, T. M.; Stevens, H. M. *J. Forensic Sci. Soc.* **1978**, *18*, 209.
135. Street, H. V. *J. Anal. Toxicol.* **1981**, *5*, 187.
136. Levine, B.; Blanke, R. V.; Valentour, J. C. *J. Forensic Sci.* **1983**, *28*, 102.
137. Levine, B. S.; Blanke, R. V.; Valentour, J. C. *J. Forensic Sci.* **1984**, *29*, 131.
138. Stevens, H. M. *J. Forensic Sci. Soc.* **1984**, *24*, 577.
139. Dunnett, N.; Ashton, P. G.; Osselton, M. D. *Vet. Hum. Toxicol.* **1979**, *21*, 199.
140. Suzuki, S.-I.; Inoue, T.; Hori, H.; Inayama, S. *J. Anal. Toxicol.* **1989**, *13*, 176.
141. Cone, E. J.; Weddington, W. W., Jr. *J. Anal. Toxicol.* **1989**, *13*, 65.
142. Cone, E. J.; Kumar, K.; Thompson, L. K.; Sherer, M. *J. Anal. Toxicol.* **1988**, *12*, 200.
143. Pounder, D. J.; Jones, G. R. *Forensic Sci. Int.* **1990**, *45*, 253.
144. Felby, S.; Olsen, J. *J. Forensic Sci.* **1969**, *14*, 507.
145. Sturner, W.; Garriott, J. C. *Forensic Sci.* **1975**, *6*, 31.
146. Sorensen, P. N. *Acta Pharmacol. Toxicol.* **1971**, *29*, 194.
147. Ziminski, K. R.; Wemyss, C. T.; Bidanset, J. H.; Manning, T. J.; Lukash, L. *J. Forensic Sci.* **1984**, *29*, 903.
148. Christophersen, A. S.; Tilstone, W. J.; Stead, A. H. In *Clarke's Isolation and Identification of Drugs in Pharmaceuticals, Body Dluids, and Post-Mortem Material,* 2nd ed.; Moffat, A. C.; Jackson, J. V.; Moss, M. S.; Widdop, B.; Greenfield, E., Eds.; Pharmaceutical: London, 1986; p 276.
149. Spiehler, V. R.; Reed, D. *J. Forensic Sci.* **1985**, *30*, 1003.
150. Wong, L. T.; Whitehouse, L. W.; Solomonraj, G.; Paul, C. J.; Thomas, B. H. *J. Anal. Toxicol.* **1979**, *3*, 260.
151. Inoue, T.; Seta, S. *Forensic Sci. Rev.* **1992**, *4*, 89.
152. Püschel, K.; Thomasch, P.; Arnold, W. *Forensic Sci. Int.* **1983**, *21*, 181.
153. Ishiyama, I.; Nagai, T.; Toshida, S. *J. Forensic Sci.* **1983**, *28*, 380.
154. Suzuki, O.; Hattori, H.; Asano, M. *J. Forensic Sci.* **1984**, *29*, 611.
155. Nakahara, Y.; Takahashi, K.; Tekeda, Y.; Fukui, S.; Tokui, T. *J. Forensic Sci.* **1991**, *36*, 70.

156. Nakahara, Y.; Takahashi, K.; Shimamine, M.; Tekeda, Y. *Forensic Sci. Int.* **1990**, *46*, 243.
157. Viala, A.; Deturmeny, E.; Aubert, C.; Estadieu, M.; Durand, A.; Cano, J. P.; Delmont, J. *J. Forensic Sci.* **1983**, *28*, 922.
158. Baumgartner, A. M.; Jones, P. F.; Black, C. T. *J. Forensic Sci.* **1981**, *26*, 576.
159. Smith, F. P.; Pomposini, D. A. *J. Forensic Sci.* **1981**, *26*, 582.
160. Smith, F. P.; Liu, R. J. H. *J. Forensic Sci.* **1986**, *31*, 1269.
161. *RIA Analysis of Hair to Detect the Presence of Drugs of Abuse;* U.S. Food and Drug Administration. National Technical Information Service: Springfield, VA, 1990; 7124.06.
162. Moeller, M. R. *J. Chromatogr.* **1992**, *580*, 125.
163. Cone, E. J. *J. Anal. Toxicol.* **1990**, *14*, 1.
164. Nakahara, Y.; Shimamine, M.; Takahashi, K. *J. Anal. Toxicol.* **1992**, *16*, 253.
165. Martz, R.; Donnelly, B.; Fetterolf, D.; Lasswell, L.; Hime, G. W.; Hearn, W. L. *J. Anal. Toxicol.* **1991**, *15*, 279.
166. Cone, E. J.; Yousefnejad, D.; Darwin, W. D.; Maguire, T. *J. Anal. Toxicol.* **1991**, *15*, 250.
167. Goldberger, B. A.; Caplan, Y. H.; Maguire, T.; Cone, E. J. *J. Anal. Toxicol.* **1991**, *15*, 226.
168. Welch, M. J.; Sniegoski, L. T.; Allgood, C. C.; Habram, M. *J. Anal. Toxicol.* **1993**, *17*, 389.
169. Wang, W. L.; Cone, E. J. *Forensic Sci. Int.* **1995**, *70*, 39.
170. Phillips, M. *Alcohol. Clin. Exp. Res.* **1984**, *8*, 51.
171. *Tegaderm Transparent Dressing Product Profile 70-2008-2115-8;* 3M Health Care Medical-Surgical Division, St. Paul, MN, 1988.
172. Burns, M.; Baselt, R. C. *J. Anal. Toxicol.* **1995**, *19*, 41.
173. Robinson, S.; Robinson, A. H. *Annu. Rev. Physiol.* **1954**, *34*, 202.
174. Spiehler, V.; Fay, J.; Fogerson, R.; Schoendorfer, D.; Niedbala, R. S. *Clin. Chem. (Winston-Salem, N.C.)* **1996**, *42*, 34.
175. Cone, E. J.; Hillsgrove, M. J.; Jenkins, A. J.; Keenan, R. M.; Darwin, W. D. *J. Anal. Toxicol.* **1994**, *18*, 298.
176. Noguchi, T. T.; Nakamura, G. R.; Griesemer, E. C. *J. Forensic Sci.* **1978**, *23*, 490.
177. Benko, A. *J. Forensic Sci.* **1985**, *30*, 708.
178. Suzuki, O.; Hattori, H.; Asano, M. *Forensic Sci. Int.* **1984**, *24*, 9.
179. Shaler, R. C.; Smith, F. P.; Mortimer, C. E. *J. Forensic Sci.* **1980**, *23*, 701.
180. Smith, F. P.; Shaler, R. C.; Mortimer, C. E.; Errichetto, L. T. *J. Forensic Sci.* **1980**, *25*, 369.
181. Smith, F. P. *Forensic Sci.* **1981**, *17*, 225.
182. Balabanova, S.; Schneider, E.; Buehler, G.; Krause, H. *Laboratoriumsmedizin* **1989**, *13*, 479.
183. Suzuki, S.-I.; Inoue, T.; Hori, H.; Inayama, S. *J. Anal. Toxicol.* **1989**, *13*, 176.
184. Ostrea, E.; Brady, M.; Gause, S.; Raymundo, A. L.; Stevens, M. *Pediatrics* **1992**, *89*, 107.
185. Moore, C.; Negrusz, A. *Forensic Sci. Rev.* **1995**, *7*, 103.
186. Wingert, W. E.; Feldman, M. S.; Kim, M. H.; Noble, L.; Hand, I.; Yoon, J. J. *J. Forensic Sci.* **1994**, *39*, 150.
187. Moriya, F.; Chan, K.-M.; Noguchi, T. T.; Wu, P. Y. K. *J. Anal. Toxicol.* **1994**, *18*, 41.
188. Browne, S.; Moore, C.; Negrusz, A.; Tebbett, I.; Covert, R.; Dusick, A. *J. Forensic Sci.* **1994**, *39*, 1515.
189. Steele, B. W.; Bandstra, E. S.; Wu, N.-C; Hime, G. W.; Hearn, W. L. *J. Anal. Toxicol.* **1993**, *17*, 348.
190. Moore, C. M.; Becker, J. W.; Lewis, D. E.; Leikin, J. B. *J. Anal. Toxicol.* **1996**, *20*, 50..
191. Beyer, J. C.; Enos, W. F.; Stajic, M. *J. Forensic Sci.* **1980**, *25*, 411.
192. Introna, F.; Lo Dico, C.; Caplan, Y. H.; Smialek, J. E. *J. Forensic Sci.* **1990**, *35*, 118.
193. Goff, M. L.; Brown, W. A.; Hewadikaram, K. A.; Omori, A. I. *J. Forensic Sci.* **1991**, *36*, 537.
194. Goff, M. L.; Omori, A. L.; Goodbrod, J. R. *J. Med. Entomol.* **1989**, *26*, 91.
195. Goff, M. L.; Brown, W. A.; Omori, A. I. *J. Forensic Sci.* **1992**, *37*, 867.

Chapter 2
Sample Pretreatment

Drug or drug-derived samples that are commonly encountered in forensic and clinical laboratories are present in three different categories of matrices:

1. Drugs that are in dosage forms, either prepared for pharmaceutical use or for illicit trade or consumption, are almost universally mixed with diluents and/or adulterants.

2. Plant-derived drugs, such as tetrahydrocannabinol, morphine, psilocin, and cathinone, are often minor constituents of gross plant materials.

3. Drugs and their metabolites in clinical and toxicological specimens are normally trace components in the presence of biological materials.

Since drugs in these matrices often exist as minor components in the presence of complex major constituents, some forms of sample pretreatment and/or concentration processes, ranging from simple filtration to heated distillation [1], are needed prior to the analytical step.

Extraction of Drugs in Dosage Forms

Prior to the application of an analytical methodology, drugs in dosage forms are normally extracted with an appropriate procedure to remove possible interfering materials. The most common extraction procedure involves the mixing of an organic solvent with the sample dissolved in water, followed by the removal of the solvent for further purification or analysis. The basic principle underlying this technique is that as organic compounds, drugs, when present in their free (un-ionized) forms, partition preferably into organic solvents. Thus, it is possible to separate drugs with different acidity through sequential extractions at different pH values (*see* the section "Drug Classification" in Chapter 1). In practice, the sample solution is usually adjusted to an appropriate pH prior to the addition of the organic solvent.

The completeness of removing a specific drug of interest into the organic phase depends on several factors. First, the percentage of the drug that is in its extractable un-ionized form depends on the pH of the medium and the pK_a value of the drug. Second, extraction efficiency varies with intrinsic factors, such as the drug's distribution ratio between the organic solvent and the aqueous phase, and extrinsic factors, such as the volume ratio of the two phases used and the number of extractions performed.

pH Adjustment

Proper adjustment of the sample acidity is essential for converting the drug of interest into a form suitable for partitioning into the organic solvent. The Henderson–Hasselbach equation [2] establishes the relationship between the pH of the medium, the pK_a of the drug, and the ratio of the drug in its ionized and un-ionized forms:

$$pH = pK_a + \log([A^-]/[HA])\tag{2.1}$$

where A^- is the base form and HA is the acid form of the drug of interest. With the known pK_a value of the drug and the desired level of the drug in its un-ionized form, one can calculate the pH to which the medium has to be adjusted. It is a common practice to extract basic and acidic drugs from the sample at a pH 2 to 3 units above and below, respectively, the pK_a values of the drugs [3]. It should be noted, however, that the medium pH can only affect the percentage of the drug that is present in its un-ionized form; the exact percentage of the drug that will be extracted depends on its partitioning characteristics between the selected aqueous and organic solvents.

Extraction Efficiency

The completeness of an extraction [4] depends not only on the intrinsic partition characteristics of the analyte between the selected solvent pair, but also on the volumes of the solvent phases and the number of extractions performed. The concentration, C_n, of the analyte left in the aqueous phase after a number, n, of extractions is given by:

$$C_n = C_0[V/(DV_0 + V)]^n = C_0[1/(1 + DV_0/V)]^n\tag{2.2}$$

where C_0 is the original concentration of the drug in the aqueous phase, V is the volume of the aqueous phase, V_0 is the volume of the organic phase, and D, the distribution ratio, is the ratio of the amount of analyte, in all chemical forms, in the organic phase to the amount in the aqueous phase. Thus, the amount of the analyte remaining after a single extraction depends on two factors: the intrinsic partition characteristics, D, and the volume ratio of the phases. This equation also shows that, when using a predetermined amount of the organic phase, the extraction is more efficient if it is carried out with small portions several times than with the total volume in one extraction.

The separation factor, α, is another parameter that deserves attention. Usually, an extraction involves more than one component. The preferential extraction of one component over another depends on the difference in the distribution ratios of these compounds and may be measured as follows:

$$\alpha = D_1/D_2\tag{2.3}$$

One can always strive to maximize the separation by selecting an appropriate solvent system and adjusting the phase volumes. The best possible separation through the adjustment of volume ratios follows the Bush–Densen equation:

$$V_0/V = 1/(D_1D_2)^{1/2}\tag{2.4}$$

Ideally, one desires the complete removal of analytes from the matrix and the complete separation of these analytes from each other. However, these two objectives are usually not complementary, especially when the analytes have similar structural characteristics. Only when one of the distribution ratios is relatively large and the other is small can a near-complete separation be quickly and easily achieved. If the separation factor is large but the smaller distribution ratio is of sufficient magnitude, it is necessary to resort to chemical parameters such as pH or masking agents to suppress the extraction of the unwanted component.

Specific Applications

The separation of noscapine and papaverine [5], both present in opium and normally extracted together, represents a typical application of selective extractions through the adjustment of medium acidity. When a mixture of papaverine (Chart 2.1a) and noscapine (Chart 2.1b) is dissolved in an organic solvent and the solution is treated with strong alkali (the alkaline solution is prepared in alcohol to ensure homogeneity of the reaction medium), noscapine is converted into noscapinate (Chart 2.1c), while papaverine remains unaffected. When the reaction mixture is mixed with an alkali aqueous solution, the noscapinate formed is dissolved in the aqueous layer, leaving papaverine in the organic layer. The aqueous layer is removed and treated with a strong acid in which relactonization takes place and noscapinate is reconverted back into noscapine.

A drug-extraction technique called alternate nonaqueous organic ratio (ANOR) has been applied to the extraction of various categories of drugs found in abused dosage forms [6, 7]. The procedure utilizes an organic solvent saturated with ammonium hydroxide or hydrochloric acid at a ratio of approximately one part acid (or base) to ten parts solvent. Drugs are extracted in their un-ionized or ion-pair forms. For example, when Darvon-N (a commercial preparation containing *d*-propoxyphene napsylate and acetaminophen) placed in filter paper is combined with ammonium hydroxide saturated hexane, the resulting *d*-propoxyphene base flows through the filter paper along with hexane. Amphoteric drugs, such as morphine, and acidic drugs, such as barbiturates, can be similarly extracted with hydrochloric acid saturated chloroform. Under the same conditions, drugs such as heroin, cocaine, phencyclidine, and pentazocine are extracted as ion-pairs. Drugs that have been successfully extracted with a combination of various solvents with acids or bases are listed in Table 2.1.

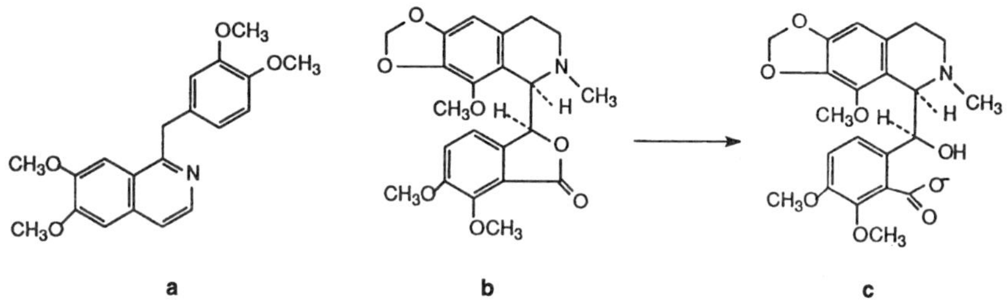

Chart 2.1. Chemical structures of papaverine (a), noscapine (b), and noscapinate (c).

Table 2.1. Extraction of drugs with acid- or base-saturated organic solvents.[a]

Extraction system	Drugs extracted	
NH$_4$OH-saturated CHCl$_3$	Hydromorphone	Morphine
	Benzphetamine (Didrex)	Lorazepam (Ativan)
	Flurazepam (Dalamne)	Benzodiazepines
	Phentermine (Ionamin)	Chlordiazepoxide (Librium)
	Cocaine	Pentazocine (Talwin)
	Methaqualone	
NH$_4$OH-saturated hexane	Phentermine (Fastin)	Propoxyphene (Darvon, Darvon-N)
	Codeine (Empirin #3)	Diethylpropion (Tenuate)
	Diazepam (Valium)	Chlorphentermine (Pre-Sate)
	Amphetamine	Ephedrine
	Phenylpropanolamine	Methamphetamine (Syndrox)
	Clortermine	3,4-Methylenedioxyamphetamine
	Meperidine (Demerol)	
HCl-saturated CHCl$_3$	Barbiturates	Diphenylhydantoin Na (Dilantin)
	Heroin	Phencyclidine
	Diazepam (Valium)	Pentazocine (Talwin)
	Cocaine	Meperidine
NH$_4$OH-saturated petroleum ether	Cocaine	
	Propoxyphene	
HCl-saturated diethyl ether/hexane (4:1)	Butalbital	
NH$_4$OH-saturated CHCl$_3$/hexane (4:1)	Chlordiazepoxide	

[a] Data are taken from Ref. [6].

Unique extraction techniques have been reported that utilize chemical reactions to facilitate the separation of structurally closely related compounds or to isolate a specific drug from common adulterants and diluents. Of particular interest is the separation of phentermine and methamphetamine based on the fact that a primary amine forms an insoluble salt with *d*-mandelic acid in dichloromethane, whereas a secondary amine does not. Thus, it is possible to identify phentermine and methamphetamine on the residue and the filtrate, respectively, after precipitating the phentermine–methamphetamine solution with *d*-mandelic acid [8]. On the other hand, it has been reported that the precipitate formed in the di-*p*-toluoyl-(−)-tartaric acid solution (in acetone), after cocaine samples are passed through activated alumina and the eluate is collected in the solution, is free of interference from 24 adulterants and diluents tested [9].

Extraction of Drugs in Plant Materials

Although direct mass spectrometric (MS) identification of psychotomimetic principles in plant materials have been reported [10, 11], commonly adopted procedures include an extraction step, which serves the purpose of removing interfering materials and concentrating the analytes. Marijuana, opium, psilocin- and psilocybin-containing mushrooms, and coca leaves are examples of plant materials that are commonly subjected to analysis in forensic laboratories. Prior to the extraction process, plant materials are often allowed to dry and are manicured by passing them through a sieve to remove seeds and stems [12]. Samples are also often ground to achieve homogeneity and to improve extraction efficiency [13–16]. Included in Table 2.2 are typical examples of plant materials, their principles, and extraction systems reported in the literature.

Pretreatment of Biological Samples

Drugs in biological samples may be present in their unchanged forms, as active metabolites, or as inactive metabolites, such as drug conjugates [23]. Many drugs and their metabolites may have a strong affinity to proteins, whereas most conjugates are highly hydrophilic and not amenable to extraction with

Table 2.2. Extraction of psychotomimetic principles from plant materials.

Plant material	Active component	Solvent system and procedure	Ref.
Cannabis	Cannabinoids	Chloroform	[12]
		Petroleum ether	[13]
		Ethanol (Soxhlet)	[17]
Opium	Morphine	Methanol (reflux)	[18]
		Sonicated under 5% acetic acid; then extracted with $CHCl_3$/methanol (3:1) at pH 8.5	[14]
		Dissolved in dimethyl sulfoxide; then eluted through Celite column with ether/light petroleum (7:3) under 2 N $NaCO_3$	[15]
Mushroom	Psilocin–psilocybin	Acetic acid (dilute)	[16]
		Methanol	[19]
Coca leaf	Cocaine	Chloroform/petroleum/methanol (2:1:1)	[20]
Anadenanthera	Bufotenine	Methanol	[21]
Khat	Cathinone	Evaporate methanol extract to near dryness; reconstitute with chloroform, followed by acidification and removal of neutral organic compounds by discarding chloroform; extract the alkaline aqueous phase with CH_2Cl_2	[22]

organic solvents. It becomes apparent that in order to achieve high recovery rates, drugs are often "released" from their binding with proteins and from their conjugated forms prior to the extraction step. In cases where the analysis of drug conjugates themselves is important [24], methods [25–28] suitable for the analysis of intact conjugates are used.

For most analytical purposes, sample-pretreatment procedures include three components: the hydrolysis of drug conjugates (if present), the removal of the binding proteins, and the extraction of the drugs from the resulting reaction mixture. Pre-extraction treatment procedures are valued on the basis of their effectiveness in releasing the targeted drugs intact (from the biological matrix) and in leaving a reaction mixture that is optimal for the subsequent extraction processes. It is possible, however, to employ a single process to achieve the drug-release and the drug-extraction goals. For example, zinc sulfate–sulfosalicylic acid reagent [29] (3.0% $ZnSO_4$ w/v and 0.5% sulfosalicylic acid w/v in 50% methanol water v/v) was found to be effective for the deproteinization of blood and homogenized liver or spleen tissue and the extraction of basic–amphoteric drugs into the supernatant. The pH of this deproteinization–extraction matrix mixture is about 6, under which the basic–amphoteric drugs are ionized and easily solvated by the supernatant phase. This mixture's pH is almost two units below the pK_as of the barbiturates and is not as effective for the extraction of these acidic drugs (see the section "pH Adjustment" earlier in this chapter). This process was found to be ineffective for the extraction of the acidic–neutral tetrahydrocannabinol carboxylic acid.

Acid and Enzyme Hydrolysis of Conjugated Drugs

During the biotransformation process, a drug may undergo an enzymatically catalyzed reaction to form a conjugate with an endogenous substance such as glucuronic acid (see the section "General Drug Biotransformation Patterns" in Chapter 1). Morphine glucuronide is the best-known example. Because the resulting conjugate is generally not amenable to extraction into the organic phase and cannot be easily handled by the gas-chromatographic method of analysis, it is normally hydrolyzed at the very early stage of the overall analytical scheme.

Hydrolysis is conventionally performed with HCl under rigorous conditions that often cause the decomposition of other concomitant compounds and generate a dirty reaction mixture. The use of enzymes to achieve the hydrolysis objective has been studied extensively and is the topic of a recent review [30]. The effectiveness of β-glucuronidase for the hydrolysis of morphine glucuronide (Chart 2.2) is compared with that of conventional acid-hydrolysis approaches. Listed below are a few general statements derived from the literature data [30]:

1. Conventional acid hydrolysis is more effective (it has a higher recovery and is less time-consuming), but it tends to cause decomposition of acid-labile compounds and leave a reaction matrix that can cause substantial noise in the analytical process.

2. The potency of β-glucuronidase from various sources and preparations varies. β-Glucuronidase from *Patella vulgate* appears to be very effective—it is active at 65 °C, producing a 90% (or higher) recovery with a 2-h incubation time.

3. With the implementation of appropriate quantitation and sensitive detection methodologies and the desire to detect acid-labile compounds, enzymatic hydrolysis may be preferred under most analytical circumstances.

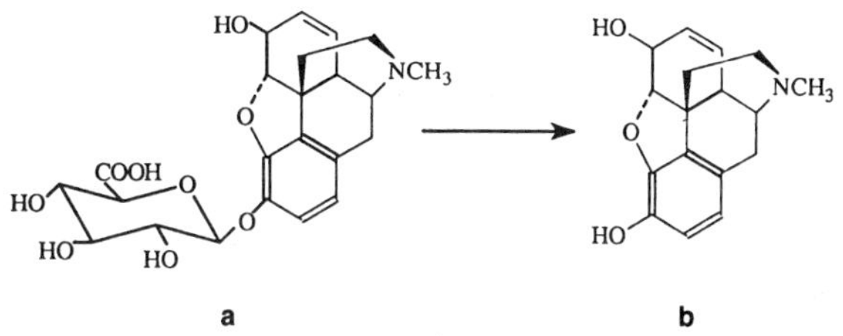

Chart 2.2. Chemical structures of morphine 3-β-D-glucuronide (a) and morphine (b).

Protein Removal

Protein in biological fluids or tissues is often removed from the sample matrix before a suitable extraction process is performed. If the analytes have high protein-binding capacities, the adopted protein-removal protocol should be able to release these analytes from protein so that they may not be lost. It has been reported [31] that phenytoin and valproate could not be detected following protein removal through ultrafiltration. Common protein-removal approaches include the use of:

1. Extreme temperatures, osmotic or ionic strength, or pH conditions;

2. Organic precipitation agents; or

3. Proteolytic enzymes.

Inorganic Acids and Salts

Common protein-precipitation procedures and their characteristics have been summarized and compared [32]. These procedures involve protein denaturation through acid hydrolysis, the use of extreme temperatures of 90–100 °C or freezing–thawing cycles, and co-precipitation with bases such as zinc hydroxide or barium sulfate–zinc hydroxide.

The recovery of alkaloids from blood and liver has been studied [33] using the following protein-precipitating reagents:

1. 5 N Hydrochloric acid;

2. 1 N Hydrochloric acid–aluminum chloride–citric acid;

3. Sodium tungstate–sodium bisulfate;

4. 1 N Hydrochloric acid–ammonium sulfate;

5. Trichloroacetic acid; and

6. Perchloric acid.

Results indicate rapid deproteination and good recoveries in the use of 5 N hydrochloric acid. However, this method causes hydrolysis of acid-labile alkaloids such as cocaine and orphenadrine. Hydrochloric acid (1 N)–aluminum chloride–citric acid results in slightly lower recoveries, but it offers the advantage of reasonable recoveries for acid-labile alkaloids. Poor recoveries were reported for other procedures. It was therefore recommended that aluminum chloride is the most appropriate agent for use in the general screening of tissues for alkaloids.

Organic Solvent Precipitation

The use of organic precipitation solvents in removing proteins from biological fluids or tissue homogenates has several advantages. The mild conditions used under the precipitation process minimize the possibility of decomposition of labile drugs. Furthermore, appropriate solvents may be selected so that they are compatible with the mobile phases used in HPLC analysis. Commonly used solvents include methanol, ethanol, and acetonitrile. Methanol is often preferred [34] because it produces a flocculent precipitate to give a clearer supernate suitable for direct injection. Acetonitrile precipitation often results in a gummy mass that may cause drug loss due to entrapment. The addition of sodium bisulfate–sodium chloride electrolyte [35] to a serum–acetonitrile mixture results in a cleaner upper phase and allows quantitative recoveries of acetaminophen, salicylic acid, and acetylsalicylic acid. Similarly, acetonitrile-containing acetic acid (35 mmol/L) is an effective protein-precipitating agent [36].

A mixture of ethanol and ethyl acetate (1:2) has been reported [37] to be an ideal solvent for the analysis of acetaminophen in homogenized liver. This solvent produces a single liquid phase containing a fine protein precipitate that can be easily filtered. Neither ethanol nor ethyl acetate alone is nearly as effective; ethanol does not precipitate the protein as well, while ethyl acetate forms two layers with the highly polar acetaminophen partitioning poorly into the ethyl acetate layer.

Proteolytic Enzymes

In parallel with the deconjugation of morphine glucuronide by β-glucuronidase in place of acid hydrolysis, proteolytic enzymes have been found to be valuable in releasing intact protein-bound drugs. For example, the harsh acid-hydrolysis conditions yield corresponding benzophenones (Chart 2.3) from the protein-bound benzodiazepines that contain a lactam ring. The resulting denatured tissue matrix often causes emulsion formation in the subsequent extraction step. Subtilisin Carlsberg, a non-group-specific proteolytic enzyme that cleaves most peptide bonds, can release intact benzodiazepines from tissue without the need for elaborate cleanup procedures [38].

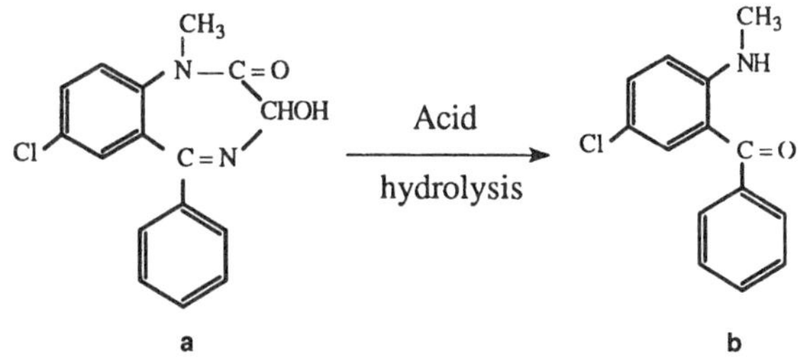

Chart 2.3. Enzymatic digestion (a) and acid hydrolysis (b) products of conjugated diazepam in biological samples.

Subtilisin Carlsberg has been effectively used [39] to liberate a variety of drugs from human tissues. Comparative studies [39] using subtilisin digestion and a conventional procedure indicated that the former procedure resulted in a higher recovery of drugs that were known to be tightly protein-bound or acid-labile, such as benzodiazepines or *d*-propoxyphene, and several other basic drugs studied. Combined use of subtilisin digestion and conventional sodium tungstate–sodium hydrogen sulfate precipitation [40] yields high recoveries of acidic drugs (barbiturates and salicylic acid), and prevents the co-extraction of endogenous tissue compounds. Similarly, the combined use of β-glucuronidase and aryl sulfatase [41] has been found to be an effective digestion method.

One important consideration is the comparability of the pH under which the drugs of interest remain stable and the enzyme utilized remains active. To illustrate this point, the stability of various drugs has been conducted [42] in the pH range of 7.0–11.0, within which subtilisin Carlsberg is active. Although drug recovery from liver tissues was found to be comparable at pH 7.4 and 10.3, it was found that drugs such as acetylsalicylic acid, chlorpropamide, cocaine, glutethimide, indomethacin, and nitrazepam are less stable at pH 10.4 while in control incubation experiments. Therefore, pH 7.4 is preferred for incubating biological tissues and subtilisin.

Empirical studies have shown that enzyme selection depends on the drug category and the major analytical concern. For example, among the four enzymes (subtilisin-A, papain, neutrase, and trypsin) studied, papain produced the highest recoveries (but with a longer time) for benzodiazepines and other basic drugs (chlordiazepoxide, chlorpromazine, diazepam, diphenhydramine, imipramine, nitrazepam, oxazepam, and promethazine). Neutrase, on the other hand, produced good results at a lower incubation temperature [43].

The exact mechanism of enzymatic digestion is often unknown. For example, β-glucuronidase was found [44] to produce higher recoveries of methaqualone than subtilisin Carlsberg and papain. The exact reason for this difference is not clear, but it is not due to the hydrolysis of the glucuronide linkage of the parent drugs or their hydroxyl metabolites, as papain and β-glucuronidase [44] produce the same results for chlorpromazine. Furthermore, β-glucuronidase from *E. coli* was found to be more effective in hydrolyzing tetrahydrocannabinol (THC) and 11-OH-THC glucuronides than was β-glucuronidase from *H. pomatia* [45].

Extraction

Drugs and drug metabolites in biological media are normally not suitable for direct qualitative or quantitative analyses for the following reasons:

1. The sample matrix, companion drugs, or drug metabolites may interfere with the analysis of the specific analyte.

2. The aqueous medium may not be compatible with the requirements of the analytical technique.

3. Concentrations of the analyte in the sample may be lower than the detection limit of the selected methods of analysis.

Thus, an extraction procedure is used to:

1. Remove interfering materials;

2. Convert the medium to a suitable solvent system; and

3. Concentrate the sample to a suitable volume.

The most suitable extraction process varies with the specific analytical objective and the type of analytical technique selected for the final measurement. For example, if chromatographic methods are to be used, co-extraction of multiple drug components may be desirable. If internal standards and a sensitive detection procedure are used, quantitative removal may not be required; speedy but inefficient procedures such as organic solvent precipitation [34–37] and microextraction [46] approaches may serve the purpose. If UV-visible or infrared spectrophotometric methods are selected for the final measurement, absorption characteristics of the solvent system will have to be considered.

Extraction procedures include the addition of a solvent system or an adsorption medium to the sample in a liquid state. The former approach is called liquid–liquid extraction, and the latter approach is called solid-phase extraction. Occasionally, samples may be lyophilized and mixed with a solvent to extract the sample in its solid form. Liquid–liquid extraction procedures often result in the transfer of the analyte from the aqueous phase into the organic solvent. The analyte may preferably partition into the organic phase in its un-ionized form or as an ion-pair; it may also be "salted out" of the aqueous phase with the high ionic strength of the medium. pH adjustment of the sample or the addition of a proper counterion may be used to facilitate a favorable partitioning of the drug in its un-ionized form or as an ion-pair. On the other hand, high concentrations of electrolytes are used to generate the ionic strength of the medium in salting-out approaches.

In a solid-phase extraction process, separation is performed with a solid additive or column material using hydrophilic, hydrophobic, or ionic groups attached to silica gels. Hydrophobic polystyrene resin, such as Amberlite XAD-2 (Rohm and Haas: Philadelphia, PA), has also been widely used as a weak absorbent for removing trace constituents that possess a hydrophobic moiety from aqueous medium.

Liquid–Liquid Extraction of Organic-Solvent-Soluble Drugs

The same principles underlying the extraction of drugs in dosage forms, as described in the first section of this chapter, apply to the extraction of drugs in biological matrices. On the basis of these principles, general extraction schemes [47] are developed for the extraction of drugs in biological fluids, such as urine and blood, involving extraction at acidic (pH 3) or basic (pH 8) conditions, followed by a sequential back-extraction with bicarbonate and sodium hydroxide to remove the strong acids and weak acids. Neutral drugs will remain in the organic solvent. Important parameters and specific requirements are discussed in the following sections.

Solvent Selection. Important parameters that have to be considered when selecting an appropriate solvent for the intended extraction system include density, volatility, polarity, selectivity, and solubility of the drugs. Obviously, solvents that are harmful to health or the environment should be avoided. The selected solvent also has to be compatible with the subsequent analytical step. For example, the solvent's optical window should be considered if UV spectrophotometry is to be used as the analytical tool.

A solvent heavier than water is preferred if a conventional separating funnel is used. On the other hand, it would be beneficial to use a lighter solvent if extraction is conducted in centrifuge tubes or if the phase separation is to be achieved by freezing the lower aqueous phase. The concentration process will be facilitated by using a solvent of high volatility.

Solvent polarity, P', determines, to a great extent, the solubility of the drug. On the other hand, the solvent selectivity parameters, x_e, x_d, and x_n, reflect the solvent's ability to function as a proton acceptor, a proton donor, and a strong dipole interactor [48]. The desired P' value can be obtained using a binary mixture [48]:

$$P' = \phi_a P_a + \phi_b P_b \qquad (2.5)$$

where ϕ_a and ϕ_b are the volume fractions of solvents with polarity values P_a and P_b, respectively.

Important parameters of commonly used solvents are listed in Table 2.3. The eight selectivity groups are graphically shown in Figure 2.1.

Even when these important solvent parameters are known, the efficiency of a specific solvent for an intended application requires empirical confirmation. For example, extraction recoveries of 28 acidic drugs were compared using hexane, toluene, diethyl ether, *n*-butyl chloride, and chloroform as extractants [52]. In the pH range of 1–2, the recoveries of most of these acidic drugs increase in the order of hexane > toluene > diethyl ether, following the increasing polarity of these solvents. Similarly,

Table 2.3. Solvents commonly used for liquid–liquid extraction.

Solvent	UV limit[a] (nm)	Boiling point[a] (°C)	Density[a] (g/cm³)	Eluotropic value (ε^o)[b]	Polarity[c] (P′)	Selectivity[c] x_e	x_d	x_n	Group
n-Pentane	210	36	0.626	0.00	0.0				—[d]
n-Hexane	210	69	0.660	0.00	0.1				—[d]
n-Heptane	210	98	0.684	0.00	0.2				—[d]
Cyclohexane	210	81	0.721	0.03	0.2				—[d]
Carbon tetrachloride	265	77	1.594	0.14	1.6				—[d]
Benzene	280	80	0.877	0.27	2.7	0.23	0.32	0.45	VII
Toluene	285	111	0.867	0.22	2.4	0.25	0.28	0.47	VII
Diisopropyl ether	220	68	0.726	—[f]	2.4	0.48	0.14	0.38	I
Diethyl ether[e]	220	35	0.713	0.29	2.8	0.34	0.13	0.34	I
Dichloromethane[e]	—[f]	40	1.327	0.32	3.1	0.29	0.18	0.53	V
Tetrahydrofuran[g]	220	66	0.881	0.35	4.0	0.38	0.20	0.42	III
Chloroform[e]	245	61	1.483	0.31	4.1	0.25	0.41	0.33	VIII
Acetone	330	56	0.790	0.43	5.1	0.35	0.23	0.42	VIa
Ethyl acetate	260	77	0.902	0.45	4.4	0.34	0.23	0.43	VIa
Cyclohexanol	—[f]	161	0.962	—[f]	4.7	0.36	0.22	0.42	VIa
n-Butanol	215	118	0.810	—[f]	4.1	0.56	0.28	0.24	II
Ethanol	210	79	0.789	—[f]	4.3	0.52	0.19	0.29	II
Dioxane[g]	215	101	1.033	—[f]	4.8	0.36	0.24	0.40	VIa
Methanol[g]	210	65	0.792	>0.73	5.1	0.48	0.22	0.31	II
Pyridine	330	115	0.980	0.55	5.3	0.41	0.22	0.36	III
Acetonitrile[g]	190	82	0.786	0.50	5.8	0.31	0.27	0.42	VIb
Acetic acid	260	118	1.049	>0.73	6.0	0.39	0.31	0.30	IV
Formamide[g]	270	211	1.133	—[f]	9.6	0.36	0.33	0.30	IV
Water	191	100	1.000	>0.73	10.2	0.37	0.37	0.25	VIII

[a] Most data are taken from Ref. [49].

[b] ε^o is the adsorption energy (on silica) per unit area of the solvent, thus indicating the solvent's eluting strength on silica. These values are taken from Ref. [50].

[c] P' is a measure of the solvent's ability to interact as a proton donor (x_d), proton acceptor (x_e), or dipole (x_n). Most data are obtained from Ref. [48].

[d] Selectivity group is irrelevant because of a low P' value.

[e] Common modifier used in normal-phase operation.

[f] Data not available.

[g] Common modifier used in reverse-phase operation.

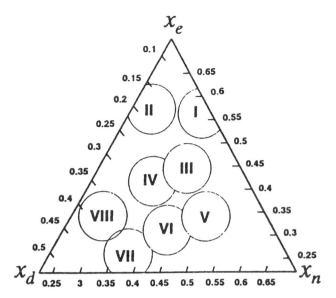

Figure 2.1. Classification of solvent selectivities—Groups I through VIII. The parameters x_e, x_d, and x_n represent the fraction of P′ (polarity index) contributed by interactions associated with ethanol (a proton acceptor), dioxane (a proton donor), and nitromethane (a dipole interactor). (Reproduced with permission from Ref. [51].)

the relatively polar methyl acetate was found [53] to be effective for extracting compounds that were sparingly soluble in ether. The recoveries observed for the chlorinated solvents cannot be readily explained.

Ether is commonly used for the extraction of basic drugs [54]. With the exception of caffeine, it appears that basic drugs of different structure types partition favorably in this solvent [55]. Solvents such as *n*-butyl acetate [56, 57] and *n*-butyl chloride [58] have also been found effective. However, *n*-butyl acetate tends to generate dirty extractions, while emulsion formation was observed with *n*-butyl chloride. Mixed solvent systems such as chloroform–ether [59], toluene–*n*-heptane–isoamyl alcohol [60], hexane–isoamyl alcohol [61, 62], chloroform–isoamyl alcohol [61], and toluene–isoamyl alcohol [62] have also been successfully used for the extraction of basic drugs.

Removal or Identification of Co-Extracted Materials. Many endogenous materials are inevitably co-extracted into the organic solvent with the drug of interest. Solvents vary in their ability to co-extract these materials. Extractants are cleaner when less-polar solvents are used. Thus, the cleanliness of extractants follows the order of ether > *n*-butyl chloride > *n*-butyl acetate [55].

The removal of these interfering materials is often needed to produce reliable results in the measurement step. A common practice involves the back-extraction of drugs into the aqueous phase followed by another extraction with the same or a different organic solvent system [61, 63]. It has also been reported that fatty acids or lipids can be effectively removed by partitioning the residue from a dried-down solvent extract between hexane–methanolic HCl [64] and petroleum ether–aqueous HCl [65] systems.

In cases where the removal of co-extracted endogenous materials is ineffective, these materials must be clearly identified in blank controls. Potential interference of these materials on the adopted chromatographic system must be understood. For example, it has been reported [66] that 2-phenethylamine, tryptamine, and indole were found in the alkaline extracts of autopsy blood and liver samples. The authors also reported that three additional peaks occurred regularly in acidic extracts, one in all extracts, and five in basic extracts. The appearance of these peaks may overlap with those of relevant analytes and render the identification of the analytes difficult.

Extraction-Derived Artifacts. Inorganic acids and bases used for pH adjustment and impurities in organic solvents pose a potential danger of chemically altering the drugs during the extraction process. It has been reported [67] that the use of diethyl ether containing peroxides as impurities resulted in rapid breakdown of methadone to form 1,5-dimethyl-3,3-diphenyl-2-ethylidene pyrrolidine, the major cyclic metabolite of methadone. Similarly, peroxides that formed in light-exposed HCl–MeOH, which was used to form a drug hydrochloride to prevent drug loss during solvent evaporation, were responsible [68] for lower yields in the extraction of chlorpromazine, amitriptyline, codeine, and methadone. Carbamate derivatives of tricyclic antidepressants were reportedly formed [69] due to the interactions of the analytes with phosgene, which may be present in chloroform not preserved with ethanol. The treatment of urine samples with ammonium hydroxide or sodium hydroxide and ammonium ions resulted in the formation of 2-(4-chlorophenoxy)-2-methylpropanamide from clofibrate [70]. It becomes apparent that artifacts of this nature have to be thoroughly considered in data interpretation.

Extraction of Water-Soluble Drugs

Because of the poor solubilities of polar drugs and drug metabolites in organic solvents, the removal of these compounds from biological fluids cannot be readily accomplished through the conventional liquid–liquid extraction procedure. Several approaches have been developed to resolve this difficulty.

Liquid–Solid Extraction of a Lyophilized Sample. In order to develop a universal approach capable of recovering all drug categories, a study was conducted [71] in which acetic acid was first added to urine samples, and then a standard lyophilization procedure was followed. Methanol was then used to extract both the acidic–neutral and basic drugs, and this was followed by simultaneous chromatographic analysis in a single solvent system. Acetone was added prior to the filtration to precipitate any inorganic salts, thus yielding a much cleaner extract.

A similar procedure was used specifically for the analysis of benzoylecgonine [72], a cocaine metabolite, and ritalinic acid [73], a metabolite of methylphenidate (a central nervous system stimulant of the amphetamine series that is used as a doping agent for sportsmen). In these two applications, the extracted benzoylecgonine and ritalinic acid were methylated with dimethylformamide [72] and diazomethane [73], respectively, prior to gas chromatographic (GC) analysis.

Ion-Pair Extraction. Ions of opposite charge tend to associate to form a neutral species called an ion-pair. The resulting ion-pair may be soluble in water to only a small extent, but relatively soluble in organic solvents, particularly when one of the ions is large and has an organic character [74]. The basic extraction principle [75] and application [76] of this technology have been reported.

This phenomenon has been successfully applied to the extraction of compounds such as benzoylecgonine [77], ritalinic acid [73], and quaternary ammonium compounds, as summarized in Table 2.4. To minimize the amounts of coextracted impurities, samples are commonly extracted with the same extraction solvent prior to the addition of the counterion. The efficiency of isolating a quaternary ammonium compound from an aqueous medium depends on [79]:

1. The nature of the counterion;
2. The concentration of the counterion;
3. The properties of the organic solvent; and
4. The pH of the aqueous medium.

A comparative study indicates [79] that dichloroethane gives cleaner extracts than those obtained with methylene chloride, while perchlorate is normally a more efficient counterion than iodide.

Ion-pairs thus obtained are normally not suitable for GC analysis. Thus, extractive alkylation processes, in which the ion-pairs are extracted and derivatized in a single step [62, 67], were used to form derivatives that can be readily analyzed by chromatographic methods. Alternatively, extracted ion-pairs were analyzed through pyrolysis in the injection port of a GC [69] or direct inlet probe MS analysis [84].

Salting-Out. A high concentration of electrolytes decreases the activity of "free" water. The addition of salting-out agents, such as polyvalent cations, results in a substantial "tying up" of water molecules [86]. This phenomenon allows the use of polar solvents, such as ethanol, as extractants and often lessens the emulsion-formation problem [87], a difficulty commonly encountered in conventional

Table 2.4. Examples of ion-pair extraction.

Compound extracted	Counterion agent	Extraction solvent	Ref.
Benzoylecgonine	Tetrahexylammonium hydrogen sulfate	Dichloromethane	[77, 78]
Ritalinic acid	Tetrabutylammonium hydrogen sulfate, tetrapentylammonium hydroxide	Dichloromethane	[73]
Quaternary ammonium compounds	Hexanitrodiphenylamine	Dichloromethane	[80]
	Picric acid	Dichloromethane	[81]
	Perchloric acid	Dichloromethane	[82]
	Bromthymol	Dichloromethane	[83]
	Potassium iodide–iodine	Dichloromethane	[84]
	Potassium iodide	Dichloromethane, dichloroethane	[79]
	Rose bengal	Chloroform	[85]

liquid–liquid extraction techniques. The general procedure involves the mixing of the polar solvent with the sample, followed by the salting-out of this solvent with the addition of a salt, thus allowing the use of ethanol to extract basic drugs [88] and cocaine metabolites [72, 89] from urine saturated with potassium carbonate.

Solvent–salt pairs, such as 2-propanol–ammonium sulfate [55, 87], ethyl acetate–ammonium carbonate [87], benzene–ammonium carbonate [87], and dichloromethane–ammonium carbonate [87], have been successfully used for the extraction of various drugs from biological media. Comparative studies [87] have indicated that the pairing of ethyl acetate with ammonium carbonate produces the best overall recovery (84–104%) of drugs, while the recoveries for the benzene–ammonium carbonate pair are excellent for some drugs but poor for others. The 2-propanol–ammonium carbonate combination produces excellent recoveries, but, unfortunately, it was found to coextract many interfering compounds in addition to the desired drugs.

Formation of Derivatives Favoring Organic Solvent Extraction. Conventional extraction approaches (using organic solvents or solid phases) may not work well for certain categories of drugs and metabolites that are relatively polar, amphoteric, and sensitive to oxidation. The formation of chloroformates under aqueous conditions prior to [90–92] (or at the same time as [93]) the extraction step has been found effective in overcoming these difficulties. Chloroformate derivatives are readily formed from compounds with phenolic, primary amine, and secondary amine functions at pH values ranging from 7.2 to 9.0 [91]. These derivatives are stable under aqueous conditions and show increasing lipophilic character with methyl, ethyl, and isobutyl chloroformates [90].

Since aliphatic hydroxy groups do not react with chloroformate-forming reagents, this method has been used in the blocking of phenol groups of catecholamine metabolites prior to derivatization of polar groups in the side-chain [90], and derivatizations of epinephrine and norepinephrine [91] and amphetamine-related compounds [92, 93].

Solid-Phase Extraction

Although effective and still widely used, conventional liquid–liquid extraction approaches have several disadvantages. In recent years, an alternate solid-phase extraction (SPE) approach, in which solid-phase adsorbents are used as the media for retaining the compounds of interest, followed by selective elution, has been intensely developed [50, 94]. The SPE approach offers the following advantages over conventional liquid–liquid procedures:

1. Less organic solvent usage;
2. No foaming problems;
3. Shorter sample-preparation time, and
4. Ease of incorporation into an automatic operation process.

Solid-phase extraction has now been widely adopted for the analysis of drugs of abuse in biological matrices [94–99] with four basic operation steps [100]: conditioning, loading, rinsing, and eluting. The first solvent for conditioning should be as strong as or stronger than the elution solvent. The second conditioning solvent should be the same as or as close to the strength of the loading solvent as possible. The solvent used for loading should be as weak as possible to result in the tightest or narrowest band of adsorbed analyte on the sorbent. A solvent that is slightly stronger or the same strength as the loading solvent is used as the rinse solvent. The rinse will elute unwanted sample components that are not as strongly retained as the analytes and also wash down small droplets of loading solvent adhering to the walls of the tube to ensure that all sample comes in contact with the sorbent. An ideal eluting solvent should elute the analytes within 5–10 bed volumes. The optimal amount of solvent to elute the analytes from a 500-mg cartridge is about 0.6–1.2 mL. Using a solvent that is too strong will result in the elution of unnecessary sample components that are more strongly retained than the analytes, whereas a solvent that is too weak will result in excessive elution solvent volumes, which negate the advantage of reducing solvent consumption with SPE cartridges. Sometimes, a desired solvent strength may have to be obtained by blending appropriate amounts of miscible solvents.

If a water-immiscible eluant is selected and the final analysis is to be performed with GC, a "drying" procedure should be applied between the rinsing and the eluting steps [101]. It has been reported [102] that the combination of vacuum and a small amount of methanol can produce a "dry" eluate without causing a substantial loss of drugs.

Solid-Phase Adsorbents and Intermolecular Forces.

Commonly used solid phases include conventional adsorbents, such as alumina, charcoal, Florisil (activated magnesium silicate), and unbounded silica gel; polymeric resins; and bonded silica gel. Silica is the most common conventional adsorbent used for the analysis of drugs. Because the silanol groups (Figure 2.2a) form hydrogen bonds with water molecules, these adsorbents inevitably contain different amounts of moisture. They require drying at an elevated temperature and storage in a desiccator to maintain effective and consistent performance.

Amberlite XAD-2, a cross-linked polystyrene polymer, represents another category of adsorbents commonly used for SPE. This polymeric resin adsorbs water-soluble organic substances through van der Waals interaction with the hydrophobic portion of the substrate. The following adsorption–elution characteristics are observed [103]:

1. Weak acids are more strongly adsorbed in the acidic form than in the salt form; consequently, weak acids can be eluted with caustic soda.

2. Weak bases are likewise strongly adsorbed in the free-base form and can be eluted with acid.

3. Polar solvents or polar solvent–water mixtures are effective eluants under these elution conditions.

Bonded silicas are now the most commonly used solid phases for the extraction of abused drugs from urine specimens. They are formed by the reaction of organosilanes with activated silica. The product is a sorbent with the functional group of the organosilane attached to the silica substrate through a silyl ether linkage (Figure 2.2b). Another reaction called endcapping may be performed subsequently; in this reaction some of the remaining silanol groups on the silica are deactivated (Figure 2.2c). The intent is to create a surface whose principal properties are due only to the functional group, with minimal interactions from the silica substrate. [94]

Depending on the functional groups attached, bonded silica gels are used as reversed-phase, normal-phase, or ion-exchange SPE media. Commonly used functional groups are listed in Table 2.5 together with their structures and primary interaction mechanisms [96].

The primary force of interaction is van der Waals attraction when nonpolar phases (in reverse-phase extraction) are used. Thus, nonpolar compounds in aqueous matrices are adsorbed and eluted with solvents of appropriate elutropic strength (Table 2.3). On the other hand, hydrogen bonding is the dominant interaction when polar phases (in normal-phase extraction) are used to remove polar compounds from organic matrices such as chloroform or diethyl ether. Aqueous buffers of high ionic strength are often required for the subsequent elution of the analytes from these sorbents.

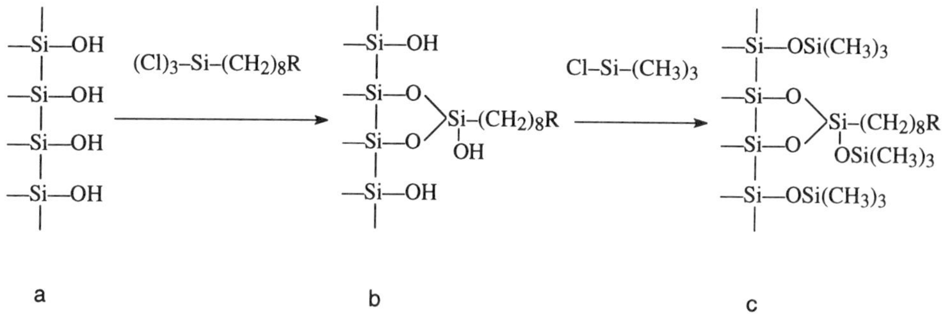

—Si—OH
|
—Si—OH   (Cl)₃–Si–(CH₂)₈R
| ———————————————→
—Si—OH
|
—Si—OH

—Si—OH
|
—Si—O
| \
—Si—O   >Si–(CH₂)₈R   Cl–Si–(CH₃)₃
| / | ———————————————→
—Si—OH   OH
|
—Si—OH

—Si—OSi(CH₃)₃
|
—Si—O
| \
—Si—O   >Si–(CH₂)₈R
| /   OSi(CH₃)₃
—Si—OSi(CH₃)₃
|

a   b   c

Figure 2.2. Structural characteristics of silica (a), a reaction product (b), and an "encapped" bonded silica sorbent (c) derived from a trifunctional derivatizing agent. (Based on figures in Ref. [50].)

Table 2.5. Bonded-phase silica properties.[a]

Phase	Structure	Primary mechanism
Ethyl (C-2)	$Si–CH_2–CH_3$	Nonpolar/polar
Octyl (C-8)	$Si–(CH_2)_7–CH_3$	Nonpolar
Octadecyl (C-18)	$Si–(CH_2)_{17}–CH_3$	Nonpolar
Cyclohexyl (CH)	$Si–C_6H_{12}$	Nonpolar phenyl
	$Si–C_6H_6$	Nonpolar cyanopropyl
	$Si–(CH_2)_3–CN$	Polar/nonpolar diol
	$Si–(CH_2)_3–O–CH_2–CH(OH)–CH_2(OH)$	Polar/nonpolar aminopropyl
	$Si–(CH_2)_3–NH_2$	Polar/anion exchange
Silica (Si)	$Si–OH$	Polar primary/secondary amine
	$Si–(CH_2)_3–N–(CH_3)_2–NH_2$	Polar/anion exchange
Benzenesulfonylpropyl (SCX)	$Si–(CH_2)_3–C_6H_6–SO_3{}^-$	Nonpolar/cation exchange
Sulfonylpropyl (PRS)	$Si–(CH_2)_3–SO_3{}^-$	Cation exchange
Carboxymethyl (CBA)	$Si–CH_2COO^-$	Cation exchange
Diethylaminopropyl (DEA)	$Si–(CH_2)_3NH(CH_2CH_3)_2{}^+$	Polar/anion exchange
Trimethylaminopropyl (SAX)	$Si–(CH_2)_3N(CH_3)_3{}^+$	Anion exchange

[a] Data are taken from Ref. [96].

The interaction of ions of opposite charges is the underlying mechanism for the ion-exchange mode of SPE. This method is often used to remove ionic compounds from aqueous media. For the extraction to be effective, the pH of the medium is adjusted to a level so that the compounds of interest are present in their ionic forms. Counterions and the ionic strength of the media also affect the ion-exchange-based extraction processes. For example, the presence of extraneous counterions with high affinity to the sorbent and high media ionic strength may cause undesired competition of these ions with the ionic analyte, thus hindering the retention of the analyte on the ion exchanger. On the other hand, these conditions may be desirable for the elution step.

Recent implementation of drug urinalysis programs in the workplace greatly accelerated the commercial development and production of disposable extraction cartridges that are prepacked with various bonded silica sorbents [99]. Bonded silica sorbents can also be fabricated into disc forms of various dimensions [98, 104]. In either form, many widely used commercial products employ mixed-mode bonded silica sorbents so that a broader range of drugs can be retained on a single cartridge or disc and then selectively eluted with various elution solvents separately. The most common combinations include hydrophobic and cation exchange groups.

Recent advances in manufacturing technology include the production of SPE discs with a rigid structure using a reduced bed mass and various silica pore sizes. This format eliminates the need for frits used in conventional packed-bed SPE columns and significantly reduces flow restriction and channeling. Advantages include reduction of void volume and high analyte recoveries with less interference from low-affinity materials, thus permitting the use of a smaller sample size, a smaller solvent volume, and a shorter processing time [105]. Sorbents in disc forms can also be stacked to achieve maximal mixed-mode operation.

Automation. To take full advantage of the approach, SPE procedures are usually implemented with some form of automation device. The most common and primitive approach makes use of vacuum manifolds, which can facilitate the holding of cartridges and controlled passage of various solvents through the cartridges. Several "medium" automation devices that are commercially available and promoted for SPE purposes include PREP (Du Pont: Wilmington, DE), ASPEC (Gilson: Middleton, WI), Sped-ed Wiz/Sped-ed Mate (Applied Separations: Allentown, PA), Tecan RSP ES (Tecan/SLT: Hillsborough, NC), Millilab Workstation (Waters: Milford, MA), and SpeedVac (Savant: Farmingdale, NY). These devices automate selected steps of the overall procedure and allow the operators to manually manipulate steps that they do not feel comfortable leaving to the machine. Total automation devices are also available. The Zymate modular systems (Zymark Corporation: Hopkinton, MA) have been fully tested and adopted by several laboratories in their production routine.

Application Examples. The SPE approach is now a common practice and is frequently reported in the literature. Application examples are cited here to illustrate the use of a conventional adsorbent, mixed-mode sorbents, and the newly developed disc-sorbent geometry. Extrelut (E. Merck: Darmstadt, Germany), a *conventional diatomaceous earth sorbent*, was used with three elution techniques [106]:

1. Differential elution at pH 2 for acidic and neutral drugs and pH 10 for basic drugs;

2. Single elution at pH 9 for all unconjugated drugs; and

3. Hydrolysis and elution at pH 9 for drugs in conjugated forms.

It was concluded that these column approaches produce eluates containing a smaller quantity of interfering materials than do eluates produced using the liquid–liquid extraction method, and such eluates are thus easier for chromatographic analysis. The differential-elution technique, although producing the same results as the single-elution technique, offered cleaner chromatograms.

Various studies [107–111] of Amberlite XAD-2, *a polymeric resin*, indicated that different categories of drugs can be effectively adsorbed at pH 8.5. Methanol was found to be an effective eluant, but the resulting extracts were impure [111–114]. Chloroform/2-propanol (3:1) and ethyl acetate/1,2-dichloroethane (3:2) [115] were found to be effective in extracting these drugs from the methanol eluate. Certain solvent systems that were found to be effective for special applications included 2-propanol/ethyl acetate/dichloroethane (25:45:30) for amphetamine [108], and ethyl acetate/dichloroethane (1:2) for phenobarbital [99] and morphine [109]. Sequential-elution techniques have also been widely applied. Thus, isopropyl ether, followed by chloroform/2-propanol (3:1) [116]; acetone followed by chloroform/methanol (3:1) [117]; and chloroform (pH 4.5) followed by chloroform/2-propanol (pH 11.6) [118] have also been applied.

Various *bonded silica sorbents* from different manufacturers are now widely used. Two studies are cited here to illustrate many important points that are pertinent to the pretreatment of biological samples, including enzymatic digestion, selection of an effective sorbent material, semiautomated operation, selection of a chemical-derivatization procedure, and the effects of cleanup procedures. In the first study [119], β-glucuronidase (as opposed to hydrochloric acid) was used to release intact oxazepam from urine, which was then extracted by three different bonded sorbents [Type W cartridge (Du Pont: Wilmington, DE), spe (J. T. Baker: Phillipsburg, NJ), and Bond Elut Certify (Analytichem International: Harbor City, CA)] and the use of the PREP Processor to automatically carry out the extraction steps. Methylation was found to be superior to trimethylsilylation for the stability of the derivatization products and the possibility of including a cleanup step after derivatization, thus resulting in a much cleaner final product (Figure 2.3) and greatly reducing the maintenance of the GC–MS system. The second study included a comparison of total selected ion chromatograms obtained from the final products of the SPE–derivatization process with and without using 1-chlorobutane to back-extract the alkalined eluate prior to its derivatization. Figure 2.4 indicates that a much cleaner chromatogram (lower) resulted from the additional cleanup step [120].

The newly available *bonded silica sorbent in disc form* has been evaluated for the isolation of tricyclic antidepressant drugs from serum [121], 9-THC-COOH from urine [122], and general drug screening [123, 124]. One obvious advantage is the ease of sample application and elution and the reduction of the required solvent volume. Thus, advances in chemical design of bonded phases and physical construction of sorbent materials enable high recovery and specific extraction of targeted analytes with a small eluant volume. Solid-phase extraction undoubtedly will be the preferred method for the isolation of abused drugs from various biological matrices.

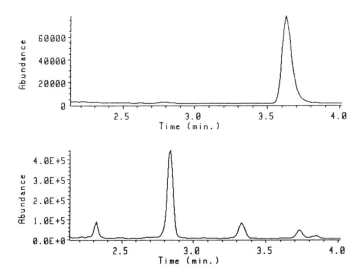

Figure 2.3. Total-ion (of the selected ions monitored) chromatograms showing that the methylation products (upper) are cleaner than the trimethylsilylation products (lower) of control samples containing added oxazepam. (Reproduced with permission from Ref. [119].)

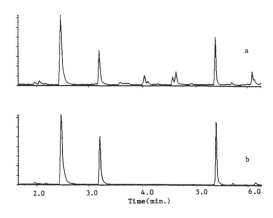

Figure 2.4. Total-ion (of the selected ions monitored) chromatograms resulting from solid-phase extraction without (upper) and with (lower) the additional cleanup step. (Reproduced with permission from Ref. [120].)

References

1. Majors, R. E. *LC-GC* **1989,** *7,* 91.
2. Christian, G. D. *Analytical Chemistry,* 2nd ed.; John Wiley & Sons: New York, 1977; p 212.
3. Smith, R. V.; Stewart, J. T. *Textbook of Biopharmaceutic Analysis;* Lee & Febiger: Philadelphia, 1981; p 43.
4. Dea, J. A. *Chemical Separation Method;* Van Nostrand Reinhold: New York, 1969.
5. Khanna, N.; Varshney, I. C.; Banerjee, S.; Singh, B. B. *Analyst (Cambridge, U. K.)* **1983,** *108,* 415.
6. Adair, A. R.; Noggle, F. T.; Odom, M. S.; Rhodes, M. A. *Microgram* **1983,** *16,* 220.
7. Saloom, J. M. *Microgram* **1983,** *16,* 229.
8. Ard, E. *Microgram* **1983,** *16,* 164.
9. Kessler, R. R. *Microgram* **1983,** *17,* 149.
10. Liu, J. R. H.; Fitzgerald, M. P.; Smith, G. V. *Anal. Chem.* **1979,** *51,* 1875.
11. Youssefi, M.; Cooks, R. G.; McLaughlin, J. L. *J. Am. Chem. Soc.* **1979,** *101,* 3400.
12. ElSohly, M. A.; Holley, J. H.; Lewis, G. S.; Russell, M. H.; Turner, C. E. *J. Forensic Sci.* **1984,** *29,* 500.
13. Turk, R. F.; Forney, R. B.; King, L. J. *J. Forensic Sci.* **1969,** *14,* 385.
14. Engelke, B. F.; Vincent, P. G. *J. Assoc. Off. Anal. Chem.* **1982,** *65,* 651.
15. Sperling, A. *J. Chromatogr.* **1984,** *294,* 297.
16. Casale, J. F. *J. Forensic Sci.* **1985,** *30,* 247.
17. Baker, P. B.; Bagon, K. R.; Gough, T. A. *Bull. Norcot.* **1980,** *32,* 48.
18. Vaidya, P. V.; Pundlik, M. D.; Meghal, S. K. *J. Assoc. Off. Anal. Chem.* **1980,** *63,* 685.
19. Lee, R. E. *J. Forensic Sci.* **1985,** *30,* 931.
20. Pe, U. W. *J. Forensic Sci.* **1977,** *10,* 261.
21. Chamakura, R. P. *Forensic Sci. Rev.* **1994,** *6,* 1.
22. Lee, M. M. *J. Forensic Sci.* **1995,** *40,* 116.
23. Tomasic, J. In *Drug Metabolite Isolation and Determination;* Reid, E.; Leppard, J. P., Eds.; Plenum: New York, 1983; p 149.
24. Chamberlain, J. *Analysis of Drugs in Biological Fluids;* CRC Press: Boca Raton, FL, 1985; p 45.
25. Caldwell, J.; Hutt, A. J.; Marsh, M. V.; Sinclair, K. A. In *Drug Metabolite Isolation and Determination;* Reid, E.; Leppard, J. P., Eds.; Plenum: New York, 1983; p 161.
26. Wilson, I. D.; Bhatti, A.; Illing, H. P. A.; Bryce, T. A.; Chamberlain, J. In *Drug Metabolite Isolation and Determination;* Reid, E.; Leppard, J. P., Eds.; Plenum: New York, 1983; p 181.
27. Harrison, M. P. In *Drug Metabolite Isolation and Determination;* Reid, E.; Leppard, J. P., Eds.; Plenum: New York, 1983; p 191.
28. Elledge, B. W.; Vharpentier, B. A. In *Analytical Methods in Forensic Chemistry;* Ho, M. M., Ed.; Ellis Horwood: New York, 1990; p 52.
29. Collins, C.; Muto, J.; Spiehler, V. *J. Anal. Toxicol.* **1992,** *16,* 340.
30. McCurdy, H. H. *Forensic Sci. Rev.* **1993,** *5,* 67.
31. Thormann, W.; Lienhard, S.; Wernly, P. *J. Chromatogr.* **1993,** *636,* 137.
32. Maickel, R. P. In *Drug Determination on Therapeutic and Forensic Context;* Reid, E.; Wilson, I. D., Eds.; Plenum Press: New York, 1984; p 3.
33. Stevens, H. M.; Owen, P.; Bunker, V. W. *J. Forensic Sci. Soc.* **1977,** *17,* 169.
34. Smith, R. V.; Stewart, J. T. *Textbook of Biopharmaceutic Analysis;* Lee & Febiger: Philadelphia, 1981.
35. Mathies, J. C.; Austin, M. A. *Clin. Chem.* 26, **1960.**
36. Smith, N. B. *Clin. Chim. Acta* **1984,** *144,* 259-272.
37. West, J. C. *J. Anal. Toxicol.* **1981,** *5,* 118.
38. Osselton, M. D.; Hammond, M. D.; Twichett, P. J. *J. Pharm. Pharmac.* **1977,** *29,* 460.
39. Ossleton, M. D. *J. Forensic Sci. Soc.* **1977,** *17,* 189.
40. Osselton, M. D.; Shaw, I. C.; Stevens, H. M. *Analyst* **1978,** *103,* 1160.
41. Predmore, D. B.; Christian, G. D.; Loomis, T. A. *J. Forensic Sci.* **1978,** *23,* 481.
42. Hammond, M. D.; Moffat, A. C. *J. Forensic Sci. Soc.* **1982,** *22,* 293.
43. Shankar, V.; Damodaran, C.; Sekharan, P. C. *J. Anal. Toxicol.* **1987,** *11,* 164.
44. Holzbecher, M.; Ellenberger, H. A. *J. Anal. Toxicol.* **1981,** *5,* 292.
45. Kemp, P. M.; Abukhalaf, I. K.; Manno, J. E.; Manno, B. R.; Alford, D. D.; McWilliams, M. E.; Nixon, F. E.; Fitzgerald, M. J.; Reeves, R. R.; Wood, M. J. *J. Anal. Toxicol.* **1995,** *19,* 62.
46. Ramsey, J.; Campbell, D. B. *J. Chromatogr.* **1971,** *63,* 303.
47. Widdop, B. In *Clarke's Isolation and Identification of Drugs,* 2nd ed.; Moffat, A. C.; Jackson, J. V.; Moss, M. S.; Widdop, B., Eds.; Pharmaceutical Press: London, 1986; p 3.

48. Snyder, L. R. *J. Chromatogr. Sci.* **1978,** *16,* 223.
49. Chamberlain, J. *Analysis of Drugs in Biological Fluids;* CRC Press: Boca Raton, FL, 1985; p 27.
50. Zief, M.; Kiser, R. *Solid Phase Extraction for Sample Preparation;* J T Baker: Phillipsburg, NJ, 1988.
51. Chen, X.-H. *Mixed-Mode Soild-Phase Extraction for the Screening of Drugs in Systematic Toxicological Analysis;* Ph.D. Thesis; State University Groningen: Groningen, the Netherlands, 1993.
52. Bailey, D. N.; Kelner, M. *J. Anal. Toxicol.* **1984,** *8,* 26.
53. Stevens, H. M. *J. Forensic Sci. Soc.* **1984,** *24,* 577.
54. Clifford, J. M.; Smyth, W. F. *Analyst* **1974,** *99,* 241.
55. Stevens, H. M. *J. Forensic Sci. Soc.* **1985,** *25,* 67.
56. Osselton, M. D.; Hammonol, M. D.; Moffat, A. C. *J. Anal. Toxicol.* **1980,** *4,* 105.
57. Rutherford, D. M. *J. Chromatogr.* **1977,** *137,* 439.
58. Dickson, S. J.; Cheary, W. T.; Missen, A. W.; Queree, E. A.; Shaw, S. M. *J. Anal. Toxicol.* **1980,** *4,* 74.
59. Parry, G. J. G.; Ferry, D. G. *J. Chromatogr.* **1976,** *128,* 166.
60. Peat, M. A.; Kopjuk, L. *J. Forensic Sci.* **1979,** *24,* 46.
61. Fretthold, D.; Jones, P.; Sebrosky, G.; Sunshine, I. *J. Anal. Toxicol.* **1986,** *10,* 10.
62. Balkon, J.; Donnelly, B. *J. Anal. Toxicol.* **1982,** *6,* 181.
63. Inoue, T.; Suzuki, S. *J. Forensic Sci.* **1986,** *31,* 1102.
64. Dudley, K. H. In *Trace Organic Sample Handling;* Reid, E., Ed.; Ellis Horwood: Chichester, U.K., 1981; p 189.
65. Tye, R.; Freitag, J. *J. Forensic Sci.* **1980,** *25,* 95.
66. Bogusz, M.; Erkens, M. *J. Anal. Toxicol.* **1995,** *19,* 49.
67. Osselton, M. D. *J. Forensic Sci. Soc.* **1976,** *16,* 299.
68. Swallow, W. H. *J. Forensic Sci. Soc.* **1981,** *22,* 297.
69. Wester, R.; Noonan, P.; Markos, C.; Bible, R., Jr.; Aksamit, W.; Hribar, J. *J. Chromatogr.* **1981,** *209,* 463.
70. Siek, T. J.; Rieders, E. F. *J. Anal. Toxicol.* **1981,** *5,* 194.
71. Broich, J. R.; Hoffman, D. B.; Goldner, S. J.; Ryauskas, S.; Umberger, C. J. *J. Chromatogr.* **1971,** *63,* 309.
72. Koontz, S.; Desemer, D.; Mackey, N.; Phillips, R. *J. Chromatogr.* **1973,** *85,* 75.
73. Wells, R.; Hammond, K. B.; Rodgerson, D. O. *Clin. Chem.* **1974,** *20,* 440.
74. Karger, B. L.; Snyder, L. R.; Horvath, C. *An Introduction to Separation Science;* John Wiley & Sons: New York, NY, 1973; p 251.
75. Cantwell, F. F.; Freiser, H. *Anal. Chem.* **1988,** *60,* 226.
76. Schill, G.; Borg, K. O.; Modin, R.; Persson, B. A. In *Progress in Drug Metabolism,* Vol. 2; Bridges, J. A.; Chasseaud, L. F., Eds.; John Willey & Sons: New York, NY, 1976.
77. Graas, J. E.; Watson, E. *J. Anal. Toxicol.* **1978,** *2,* 80.
78. Joern, W. A. *J. Anal. Toxicol.* **1987,** *11,* 110.
79. Lukaszewski, T. *J. Anal. Toxicol.* **1985,** *9,* 101.
80. Forney, R. B., Jr.; Carroll, F. T.; Nordgen, I. K.; Pettersson, B.-M.; Holmstedt, B. *J. Anal. Toxicol.* **1982,** *6,* 115.
81. De Ruyter, M. G. M.; Cronnelly, R.; Castagnoli, N., Jr.; *J. Chromatogr.* **1980,** *183,* 193.
82. Yakatan, G. J.; Tien, J.-Y. *J. Chromatogr.* **1979,** *164,* 399.
83. Balkon, J.; Donnelly, B.; Regent, T. A. *J. Anal. Toxicol.* **1983,** *7,* 237.
84. Nisikawa, M.; Tatsuno, M.; Suzuki, S.; Tsuchihashi, H. *Forensic Sci. Int.* **1991,** *51,* 131.
85. Poklis, A.; Melanson, E. G. *J. Anal. Toxicol.* **1980,** *4,* 275.
86. Dean, J. A. *Chemical Separation Methods;* van Nostrand Reinhold: New York, NY, 1969; p 41.
87. Horning, M. G.; Gregory, P.; Nowlin, J.; Stafford, M.; Lertratanangkoon, K.; Butler, C.; Stillwell, W. G.; Hill, R. M. *Clin. Chem.* **1974,** *20,* 282.
88. Gyllenhaal, O.; Johansson, L.; Vessman, J. *J. Chromatogr.* **1980,** *190,* 347.
89. De Jong, A. P. J. M.; Cramers, C. A. *J. Chromatogr.* **1983,** *276,* 267.
90. Meattherall, R. *J. Anal Toxicol.* **1995,** *19,* 316
91. Jonsson, J.; Kronstrand, R.; Hatanpää, M. *J. Forensic Sci.* **1996,** *41,* 148.
92. Bastos, M. L.; Kananen, G. E.; Young, R. M.; Monforte, J. R.; Sunshine, I. *Clin. Chem.* **1970,** *16,* 931.
93. Bastos, M. L.; Jukofsky, D.; Mule, S. J. *J. Chromatogr.* **1974,** *89,* 335.
94. Van Horne, K. C., Ed. *Sorbent Extraction Technology;* Analytichem International: Harbor City, CA, 1985.
95. Harkey, M. R.; Stolowitz, M. L. In *Advances in Analytical Toxicology,* Vol. 1; Baselt, R. C., Ed.; Biomedical Publications: Foster City, CA, 1984; Chapter 9.
96. Dimson, P.; Brocato, S.; Majors, R. E. *Amer. Lab.* **1986,** *18(10),* 86.
97. Majors, R. E. *LC-GC* **1986,** *4,* 972.
98. Majors, R. E. *LC-GC* **1991,** *9,* 332.
99. Platoff, G. E.; Gere, J. A. *Forensic Sci. Rev.* **1991,** *3,* 117.
100. *The Separation Times;* J&W Scientific: Folsom, CA, 1989; Vol. 3, p 3.
101. Chen, X.-H.; Franke, J.-P.; de Zeeuw, R. A. *Forensic Sci. Rev.* **1992,** *4,* 147.

102. Chen, X.-H.; Wijsbeek, J.; Franke, J.-P.; de Zeeuw, R. A. *J. Forensic Sci.* **1992,** *37,* 61.
103. *Amberlite XAD-2;* Rohm and Haas: Philadelphia, PA, 1975.
104. Markell, C.; Hagen, D. F.; Bunnelle, V. A. *LC-GC Int.* **1991,** *4,* 10.
105. Blevins, D. D.; Schultheis, S. K. *LC-GC* **1994,** *12,* 12.
106. Breiter, J.; Helger, R.; Lang, H. *Forensic Sci.* **1976,** *7,* 131.
107. Bastos, M. L.; Jukofsky, D.; Saffer, E.; Chedekel, M.; Mule, S. J. *J. Chromatogr.* **1972,** *71,* 549.
108. Kullberg, M. P.; Miller, W. L.; McGowan, F. Y.; Doctor, B. P. *Biochem. Medic.* **1973,** *7,* 323.
109. Miller, W. L.; Kullberg, M. P.; Banning, M. E.; Brown, L. D.; Doctor, B. P. *Biochem. Medic.* **1973,** *7,* 145.
110. Kullberg, M. P.; Gorodetzky, C. W. *Clin. Chem.* **1974,** *20,* 177.
111. Weissman, N.; Lowe, M. E.; Beattie, J. M.; Demetriou, J. A. *Clin. Chem.* **1971,** *17,* 875.
112. Fujimoto, J. J.; Wang, R. I. *Toxicol. Appl. Pharmocol.* **1970,** *16,* 186.
113. Hetland, L. B.; Knowlton, D. A.; Couri, D. *Clin. Chim. Acta.* **1972,** *36,* 473.
114. Wislocki, A.; Martel, P.; Ito, R.; Dunn, W. S.; McGuire, C. D. *Health Lab. Sci.* **1974,** *11,* 13.
115. Stajic, M.; Caplan, Y. H.; Backer, R. C. *J. Forensic Sci.* **1979,** *24,* 722.
116. Bastos, M. L.; Jukofsky, D.; Mule, S. J. *J. Chromatogr.* **1973,** *81,* 93.
117. Pranitis, P. A. F.; Milzoff, J. R.; Stolman, A. *J. Forensic Sci.* **1974,** *19,* 917.
118. Pranitis, P. A. F.; Stolman, A. *J. Forensic Sci.* **1975,** *20,* 726.
119. Langner, J. G.; Gan, B. K.; Liu, R. H.; Baugh, L. D.; Chand, P.; Weng, J.-L.; Edwards, C.; Walia, A. S. *Clin. Chem.* **1991,** *37,* 1595.
120. Gan, B. K.; Baugh, L. D.; Liu, R. H.; Walia, A. S. *J. Forensic Sci.* **1991,** *36,* 1331.
121. Lensmeyer, G. L.; Wiebe, D. A.; Darcey, B. A. *J. Chromatogr. Sci.* **1991,** *29,* 444.
122. Wu, A. H. B.; Liu, N.; Cho, Y.-J.; Johnson, K. G.; Wong, S. S. *J. Anal. Toxicol.* **1993,** *17,* 215.
123. Ensing, K.; Franke, J. P.; Termink, A.; Chen, X.-H.; de Zeeuw, R. A. *J. Forensic Sci.* **1992,** *37,* 460.
124. Blevins, D. D.; Henry, M. P. *Am. Lab.* **1995,** *27(8),* 32.

Section II

Preliminary Test and Chromatographic Methods

Definiteness of Analytical Findings

On the basis of the definiteness of the analytical finding that it yields, an analytical procedure can be considered a preliminary, conclusive, or "semi-conclusive" method. Preliminary methods of analysis are useful for screening purposes. They can be used in the field, where instrumentation needed for more definite methods of analysis cannot be conveniently applied. Typical examples of this category of methods include color tests and portable immunoassay kits. Preliminary methods of analysis that require major instrumentation and can only be performed under a *laboratory* environment may still find wide usage if they can be applied cost-effectively by rapidly processing a large number of samples. Typical examples of this category of methods are the various enzyme immunoassay, radioimmunoassay, and fluorescence polarization immunoassay methods that are now routinely used in workplace drug urinalysis and clinical laboratories. Preliminary analytical methods that are neither suitable for field use nor cost-effective for screening a large number of samples have limited applications.

Conclusive methods of analysis are required to provide the kinds of analytical results that can stand the stringent scrutiny processes that normally take place in a court of law. These methods, unfortunately, normally require major instrumentation, skilled personnel, and lengthy analytical procedures. MS is a good example of this type of method.

The operation procedure and instrumentation needed for "semiconclusive" methods of analysis generally fall between preliminary and conclusive methods of analysis. Microcrystal tests and chromatographic methods of analysis can be classified in this category. Because these methods do not provide the simplicity or high-throughput capacity of the preliminary methods, nor the definite results of the conclusive methods of analysis, their roles in the overall analytical scheme have been largely replaced by a wide variety of other methods that are now available. Thus, the use of once-popular microcrystal tests has significantly declined in recent years, as evidenced by the removal of the microcrystal test chapter from the second edition of *Clarke's Isolation and Identification of Drugs in Pharmaceuticals, Body Fluids, and Post-Mortem Material* [1]. The wide utilization of chromatographic methods, on the other hand, can be attributed to the development of automation, mixture-handling capacities, and convenient features for quantification, and their combined use with conclusive methods such as MS.

Underlying Test Principles and Cross-Reactivity

Methods based on several different principles are being used for preliminary analytical purposes. For example, chemical reactions are the underlying phenomena of color tests, physical interactions set the differentiating parameters in thin-layer chromatography, and antibody–antigen interactions are the basis of immunoassay methods.

Although the usefulness of an analytical method is often judged by its specificity, this is not always the case when evaluating a preliminary method of analysis. If the final goal of the preliminary test is to determine the presence of a specific drug, as in workplace drug-testing programs, specificity will be an important factor. On the other hand, if the purpose is to detect the presence of any drug or specific groups of drugs, a method that will respond to all drugs of interest will be advantageous. For example, cross-reactivity commonly associated with most immunoassays can be considered an advantage for preliminary screening of samples from traffic-offense cases in which drug use is suspected. Cross-reactivity to drugs with similar functional groups offers significant advantages in the following circumstances:

1. The exact structure of the drug is not known, as is often the case in designer-drug-related cases.

2. The biological specimens being tested contain new drugs, for which exact metabolic mechanisms are not yet established.

3. The concentration of the analytes is below the method's detection limit, while the cumulated responses derived from all structurally related compounds are detectable [2, 3].

Positive results derived from cross-reactivity avoid failure in detecting the presence of significant drugs without using multiple screen tests. Thus, the selection of a preliminary screen test very much depends on the final goal of the analytical process.

References

1. *Clarke's Isolation and Identification of Drugs in Pharmaceuticals, Body Fluids, and Post-Mortem Material,* 2nd ed.; Moffat, A. C.; Jackson, J. V.; Moss, M. S.; Widdop, B.; Greenfield, E. S., Eds.; Pharmaceutical: London, 1986.
2. Got, P.; Baud, F. J.; Sandouk, P.; Diamant-Berger, O.; Scherrmann, J. M. *J. Anal. Toxicol.* **1994,** *18,* 189.
3. Steele, B. W.; Bandstra, E. S.; Wu, N.-C.; Hime, G. W.; Hearn, W. L. *J. Anal. Toxicol.* **1993,** *17,* 348.

Chapter 3
Color Tests

Based on chemical reactions, color (spot) tests are often capable of characterizing certain groups of compounds, but they are less successful in detecting the presence of a specific compound. Because a vast number of homologs would react identically to a selected functional group of the reagent, color tests are unreliable for the identification of individual compounds. The names of reagents that are useful for the detection of common functional groups and reaction mechanisms are available in several general references [1–3].

In its simplest form, a color test involves the transfer, with a capillary tube, of a small amount of a control standard analyte (or the sample) to the well of a porcelain spot plate. The sample is allowed to dry, and then a drop of the selected reagent is added. When biological fluids, such as urine, are analyzed, an appropriate buffer may be added to maintain the reaction at a controlled pH [4]. To improve the effectiveness of the screen procedure, drugs in biological fluids are often first separated by thin-layer chromatographic techniques, and the spraying of color test reagents follows. This approach is best exemplified by the commercially available Toxi-Lab drug detection system now marketed by Ansys (Irvine, CA).

Color tests as applied to forensic drug analysis are used either to systematically screen for drugs that might be present or to identify the presence of a specific drug or drug group. Systematic screening is done through the use of a set of reagents and comparison of the resulting color formation with standard color charts [5, 6] or with those derived from testing controlled standards under identical conditions [7]. To improve test specificity, a systematic screen approach can be carried out through the sequential use of several reagents according to a predetermined test scheme [6, 8–10].

Comparison of Color Formation with a Color Chart Reference

With this approach, commonly used reagents (Table 3.1) are systematically tested against 25–100 µg of commonly abused drugs and other possible interfering compounds. Color formations during a specific time period are recorded and compared with the chart collection produced by the Inter-Society Color Council and the National Bureau of Standards (ISCC–NBS).

On the basis of this approach, the Law Enforcement Standards Laboratory at the National Bureau of Standards developed a system entitled the NIJ Standard for Color Test Reagents/Kits for Preliminary Identification of Drugs of Abuse [6]. "This standard is concerned with single reagents (or reagent combinations) that are used to give a preliminary confirmation of the identity of a suspected drug or class of drugs." [6] Colors produced by eleven reagents (1–11 in Table 3.1) with various drugs and other substances are recorded with the designations of the ISCC–NBS charts.

Table 3.1. Formulas of common color-test reagents.

Reagent	Content
1 Cobalt(II) thiocyanate	2.0 g cobalt(II) thiocyanate/100 mL distilled water
2 Dille-Koppanyi reagent (modified)	A: 0.1 g cobalt(II) acetate dihydrate/100 mL methanol. Add 0.2 mL glacial acid and mix. B: 5 mL isopropylamine/95 mL methanol To use: add 2 vol. Part A, then 1 vol. Part B.
3 Duquenois–Levine reagent	A: 2.5 mL acetaldehyde and 2.0 g vanillin/100 mL 95% ethanol B: Hydrochloric acid C: Chloroform To use: add 1 vol. of Parts A and B in order and determine the color produced. Add 3 vol. Part C and note if the color is extracted from the mixture of A and B.
4 Mandelin reagent	1.0 g ammonium vanadate/100 mL concentrated sulfuric acid
5 Marquis reagent	Carefully add 10 mL 40:60 (v:v) formaldehyde–water mixture to 100 mL concentrated sulfuric acid.
6 Nitric acid	Concentrated nitric acid
7 p-DMAB[a]	Add 2.0 g p-DMAB to 50 mL 95% ethanol and 50 mL concentrated hydrochloric acid
8 Ferric chloride	2.0 g anhydrous ferric chloride/100 mL distilled water
9 Froehde reagent	0.5 g molybdic acid or sodium molybdate/100 mL hot concentrated sulfuric acid
10 Mecke reagent	1.0 g selenious acid/100 mL concentrated sulfuric acid
11 Zwikker reagent	A: 0.5 g copper(II) sulfate pentahydrate/100 mL distilled water B: 5 mL pyridine/95 mL chloroform To use: add 1 vol. Part A, then 1 vol. Part B.
12 Ehrlich reagent	5 g p-DMAB/100 mL concentrated hydrochloric acid

[a] p-DMAB: p-dimethylaminobenzaldehyde.

Multiple Reagent Testing

The merit of using multiple reagents to increase the specificities of color tests has been advocated by several authors. In one approach [7], color formations resulting from the reaction of a series of nine spot tests with more than 200 compounds were compiled. To facilitate a manual or computer search process, a nine-digit color code was assigned to each drug, each digit representing the color formation resulting from the reaction of the subject drug with one of the nine reagents. Two directories based on the nine-digit color code were compiled. In one of these directories, drugs are alphabetically listed with their color code, whereas in the other directory, drugs are arranged numerically according to their color code. The latter directory is helpful for making an identification or for identifying possible alternates of an unknown sample on the basis of the observed color.

Alternately, multiple reagents are used in a flow-chart fashion [6, 9, 10], in which reagents are used sequentially to differentiate several possible drugs that may form similar colors with the reagent used in the previous step. A typical example of this approach is shown in Table 3.2 [6]. By following the specific order in performing these tests, it is possible to tentatively identify the drug being investigated or to narrow down the possibilities. Information obtained can be extremely informative in eliminating a suspected drug or determining the specific procedure needed for a conclusive identification.

Table 3.2. Flowchart for multiple and sequential use of reagents for color tests.[a]

First reagent	Second reagent		Third reagent		Fourth reagent	
Color	*Color*	*Drug*[c]	*Color*	*Drug*	*Color*	*Drug*
Marquis	**p-DMAB**		**Mandelin**		**HNO₃**	
No color	No color		Green		No color	Amphetamines
			Olive-brown		Red	Mescaline
					Co(SCN)₂	
					No color	Sugar
					Greenish-blue/blue	Demerol
Orange-brown			No color			
	LSD					
Purple	Purple					
	Co(SCN)₂					
Yellow-green	No color	STP				
	Greenish-blue/blue	Marezine, Ritalin				
	Co(SCN)₂					
Pink-red	No color	Aspirin, Contac, Dristan, Excedrin				
	Greenish-blue/blue	Methadone, phencyclidine				
	Co(SCN)₂		**Duquenois–Levine**			
No color	No color		No color	Methaqualone		
			Purple	Marijuana (CHCl₃ extract)		
			Mandelin			
			No color	Cocaine, quinine		
			Orange	Brompheniramine, Librium		
			Co(SCN)₂			
			Greenish-blue/blue	Darvon, heroin•HCl		
			No color	Mace, morphine		
	Mandelin					
	Greenish-blue/blue					
	Red-brown					
	Mandelin					
Purple-violet-	Purple-red	Methapyrilene•HCl				
black	Olive	Codeine, opium				
	Pale blue	MDA				
	HNO₃					
	Red-brown	Mace				
	Pink-red	Methapyrilene•HCl				
	Orange	Codeine, morphine				
	Yellow	Heroin•HCl				
	Greenish-yellow	Chlorpromazine, MDA				
	Dille–Koppanyi					
Tan	Purple	Pento-, pheno-, secobarbital				

[a] Data are taken from Ref. [9].

[b] See Table 3.1 for reagent formulations.

[c] Abbreviations: *p*-DMAB: *p*-dimethylaminobenzaldehyde; LSD: lysergic acid diethylamide; MDA: 3,4-methylenedioxyamphetamine; STP: 2,5-dimethoxy-4-methylamphetamine.

References

1. Feigl, F.; Anger, V. *Spot Tests in Organic Analysis,* 7th ed.; Elsevier: Amsterdam, Netherlands, 1966.
2. Jungreis, E.; Ben-Dor, L. In *Organic Spot Test Analysis;* Svehla, G., Ed.; Elsevier: Amsterdam, Netherlands, 1980; Vol. 10, p 1.
3. Jungreis, E. *Spot Test Analysis: Clinical, Environmental, Forensic, and Geochemical Applications;* John Wiley & Sons: New York, 1985.
4. Saker, E. G.; Solomons, E. T. *J. Anal. Toxicol.* **1979,** *3,* 220.
5. *NIJ Standard for Color Test Reagents Kits for Preliminary Identification of Drugs of Abuse;* U.S. Department of Justice. U. S. Government Printing Office: Washington, DC, 1981.
6. Valapoldi, R. A.; Wicks, S. A. *J. Forensic Sci.* **1974,** *19,* 636.
7. Johns, S. H.; Wist, A. A.; Najam, A. R. *J. Forensic Sci.* **1979,** *24,* 631.
8. Jungreis, E. *Spot Test Analysis: Clinical, Environmental, Forensic, and Geochemical Applications;* John Wiley & Sons: New York, 1985; p 76.
9. Masoud, A. N. *J. Pharm. Sci.* **1981,** *64,* 841.
10. Hider, C. L. *J. Forensic Sci. Soc.* **1971,** *11,* 257.

Chapter 4
Thin-Layer Chromatography

Thin-layer chromatography (TLC) represents one of the simplest forms of chromatographic technology. This technique is widely used in drug analysis, mainly because of its simplicity, low cost, capability of simultaneous drug detection, and the possibility of using various detection approaches that can provide a certain degree of specificity. The recent development of "high performance" TLC, through the use of stationary phases with a smaller particle–pore size and a narrower particle–pore size distribution, has further improved analysis time, detection limits, and reproducibility in quantification and retention characteristics. Commercially available procedures such as Toxi-Lab (Ansys: Irvine, CA) [1] also facilitate the standardization of operation and the data-interpretation process.

As a chromatographic technique, the selected stationary phases can be coated on glass plates and on plastic or aluminum sheets. Although the major mode of separation in TLC analysis is based on physical interaction between the analyte and adsorbents such as silica gel and alumina, stationary phases that facilitate separation through partition and ion-exchange processes have also been successfully applied to drug analysis. Both normal-phase (where the stationary phase is more polar than the mobile phase) and reversed-phase [2, 3] (where the stationary phase is less polar than the mobile phase) systems have been used.

TLC Systems

In contrast to gas chromatography (and, to a lesser extent, high-performance liquid chromatography), in TLC the variation of the mobile-phase composition is the major operating parameter that may be used to maximize resolution improvement. In gas chromatography, the choice of the mobile phase is quite limited, and the availability of various stationary phases renders the resolution of analytes with diverse retention characteristics possible, whereas in TLC, silica gel is the only stationary phase that has found broad application.

The basic mechanisms underlying the separation of different drug categories under various solvent conditions are different. For example, acidic drugs migrate as undissociated acids when neutral solvents, such as chloroform/ether (75:25), are used, but they migrate as anions with alkaline solvents, such as chloroform/isopropanol/ammonia (45:45:10). This principle does not apply to basic drugs. With alkaline or neutral solvents, basic drugs migrate as undissociated bases, whereas if an acidic solvent is used, the resulting protonated base BH^+ usually becomes too polar to migrate on silica gel. Thus, bromides were introduced to convert the protonated base to an ion-pair, BH^+Br^- [4]. It was reported that the values of the retardation factor, R_f, thus obtained are more reproducible without the presence of basic components, such as NaOH, KOH, ammonia, or diethylamine, in the mobile phase.

Parameters [5] that should be considered in the selection of a TLC system include:

1. Sensitivity;
2. Time taken for analysis;
3. Distribution of the R_f (or $R_f \times 100$) values across the plate;
4. Reproducibility of the measurement of R_f values; and
5. Correlation of chromatographic properties between systems.

It has been concluded [5] that there is little practical difference between systems in terms of sensitivity and time taken for analysis, whereas the distribution of R_f values for the drugs of interest and their reproducibility vary among systems and are the most important considerations. Correlation of chromatographic parameters obtained becomes important when multiple systems are used in the same or different laboratories.

Reproducibility

A good TLC system should generate reproducible R_f values at different times and different laboratories. The minimal distance obtainable from a system, the so-called error factor [5] or discriminating distance [6], that will distinguish two compounds, depends on the precision of R_f values obtained from a system. Factors that affect reproducibility may be associated with experimental techniques or with sample characteristics. The former factors include [5]:

1. Solvent purity;

2. Accurate makeup of solvent composition;

3. Completeness in removing the solvent used to apply the test substances;

4. Equilibrium of solvent components in the developing tank;

5. Temperature;

6. Substrate mass; and

7. Appropriate conditioning of the plate.

Interfering material coextracted from biological samples may affect the retention characteristics of different TLC systems to a different extent. It has been reported [7] that intra- and interlaboratory standard deviations of R_f values obtained from extracted drugs are at least twice those obtained from pure drugs. With the three systems studied, the precision of the results followed the order of methanol > methanol/butanol (60:40) containing 0.1 M NaBr > chloroform/methanol (90:10) with the plate dipped in 0.1 M KOH before use. The same study further reported that the precision and accuracy of the R_f values were independent of the extraction method used and the type of biological matrix from which the drugs were extracted.

Correction of Retention Data

To ensure the efficient use of reference databases compiled by other laboratories, raw retention data are often corrected on the basis of standard substances that show comparable retention behavior with respect to the concerned analytes. Using two standard substances, the corrected retention data, R_f^c, of the analyte is obtained using the Galanos–Kaponlas equation [8]:

$$R_f^c = aR_f + b \tag{4.1}$$

$$a = [(R_f^c)_t - (R_f^c)_u]/[(R_f)_t - (R_f)_u] \tag{4.2}$$

$$b = (R_f^c)_t - a(R_f)_t \tag{4.3}$$

where $(R_f^c)_t$, $(R_f^c)_u$, $(R_f)_t$, and $(R_f)_u$ are the retention data of standard substances t and u in the reference database and those obtained from the test laboratory, respectively. This approach was applied to a nonpolar single-component solvent [9] and a polar multicomponent solvent [10] system to provide substantial improvements in reproducing retention data. The improvement in data reproducibility was less profound for the latter solvent system.

Other reported correction approaches include the use of three reference substances and a nonlinear graphic correction procedure [11], four reference substances [12], and starting point and three reference substances [6]. The solvent systems and reference substances used in these studies are listed in Table 4.1 for reference. It was further suggested [13] that the reference substances should be spotted at the center and at the edges of the plate so that an erratic plate composition or development procedure may be detected.

Table 4.1. TLC systems and R_f correction reference substances.

Adsorbent	Solvent	Reference compounds	Ref.
Silica gel	Benzene	2,4-Dinitrophenylhydrazone of methanal, *n*-dodecanal	[9]
Silica/Kieselguhr	*n*-Butanol/acetic acid/water (5:1:2)	Dimethyl yellow, cochineal red	[10]
Silica gel; alkaline silica	Methanol/ammonia (100:1.5); methanol; methanol/ammonia (100:1.5)	Mescaline, amphetamine, benzphetamine; *N*-methyl-, *N,N*-dimethyl-, *N,N*-dibutyltryptamine; ergometrine, 8β-, 8α-ergotamine	[11]
Silica gel/NaOH	Cyclohexane/toluene/diethylamine (75:15:10)	Codeine, desipramine, pethidine, dipipanone	[12]
	Chloroform/methanol (9:1)	Desipramine, dipipanone, caffeine, meclozine	[12]
	Acetone	Amitriptyline, procaine, mepivacaine, meclozine	[12]
Silica	Methanol; methanol/butanol (6:4); chloroform/methanol (9:1); ethyl acetate/cyclohexane/methanol/ conc. NH$_4$OH (70:15:10:5)	Codeine, nikethamide, benzocaine	[6]
Silica (ion-pair adsorption)	Methanol; methanol/butanol (6:4); chloroform/methanol (9:1); ethyl acetate/cyclohexane/methanol (70:15:15); all with NaBr	Same as above	[6]

Evaluation of System Separation Power

Various approaches have been explored to evaluate the separation power of the TLC systems for an intended analysis. Parameters used for these evaluations include identification power (IP) [6], discriminating power (DP) [5, 14], and information content (IC) [15]. The IP of a system is defined as the number of compounds that can be unequivocally distinguished from each other based on the R_f values generated by that system. Two compounds are considered distinguishable if their difference in R_f values is larger than a minimal distance, the discriminating distance. If more than one system is used, two compounds are considered distinguishable if and only if there exists at least one system in which the difference of the R_f values of these two compounds is more than the discriminating distance.

DP is defined as the probability that two drugs selected at random can be separated by a TLC system. It is calculated by comparing the measured R_f value for each drug with the R_f value for each of the other drugs. If the difference between these R_f values is less than a predetermined error window, the two drugs are considered undiscriminated. The DP of a system in which N compounds are investigated with a total number, M, of pairs of undiscriminated R_f values is defined as follows:

$$DP = 1 - [2M/N(N - 1)] \qquad (4.4)$$

If more than one system is used, M is defined as the number of pairs unresolved in all systems concerned.

For the calculation of IC, R_f values in a data bank are divided into m classes of a given class width, for instance $0.05R_f$ units. For each of the m classes, the probability that an unknown compound will appear with an R_f value within the limit of this class equals r/n, where r is the number in the class and n is the total number of entries of the data bank. IC is expressed in bits by

$$IC = \sum_{k=1}^{m} r/n \log_2(r/n) \qquad (4.5)$$

Selection of Developing Systems

The selection of a suitable mobile phase for a specific analysis depends on the analytical need. In cases where the drug category in the specimen is completely unknown, a solvent system capable of eluting a broad spectrum of drug categories but limited in resolving drugs within a category may be useful [16, 17]. On the other hand, if the specific drug within a drug category is to be identified, a solvent system that may fine-tune the resolution of the drugs with similar structural features should be used. Thus, selected systems are recommended for the resolution of neutral drugs [18-20], acidic drugs [5, 20, 21], and basic drugs [5, 22–26], and for specific drug groups such as amphetamines [27], barbiturates [3], benzodiazepines [2, 28, 29], cannabinoids [30–34], cocaine and its metabolites [35], lysergic acid diethylamide related compounds [36, 37], opiates [38], and benzodiazepines [28, 39, 40]. Table 4.2 includes some systems that have been recommended for the resolution of various categories of drugs. Each system should be valuable for screening for the presence of a specific drug. However, the combined use of more than one system is required for more specific identification.

Table 4.2. Selected TLC systems recommended for the resolution of various drug categories.

Drug category	*Adsorbent*	*Solvent system*	*Ref.*
All categories	Silica	Ethyl acetate/methanol/ammonia (17:2:1)	[41]
		Ethyl acetate/methanol/ammonia/water (150:18:0.5:10 and 150:22.5:0.8:4)	[16]
		Ethyl acetate/1-propanol/ammonia (40:30:3); chloroform/ether/methanol/ammonia (75:25:5:1)	[42]
Basic	Silica/KOH	Methanol/ammonia (100:1.5); cyclohexane/toluene/diethylamine (75:15:10); chloroform/methanol (9:1); acetone	[5]
		Ethyl acetate/methanol/ammonia (85:10:5)	[23]
	Silica	Toluene/methanol/ammonia (50:50:1); 2-propanol/*n*-heptane/ammonia (50:50:1)	[25]
		Chloroform/ethanol (5:1)	[24]
		Ethyl acetate/cyclohexane/ammonia/methanol/water (70:15:2:8:0.5)	[45]
		Methanol; methanol/butanol (with NaBr) (6:4)	[6]
Neutral/acidic	Silica	Chloroform/acetone (4:1); ethyl acetate/methanol/ammonia (85:10:5); ethyl acetate; chloroform/methanol (9:1)	[5]
Neutral	Silica	Ether/acetone (3:1); chloroform/acetone (22:3)	[19]
Amphetamine	Silica	Ethyl acetate/methanol/water/ammonia (95:3.5:1.5:0.75)	[27]
		Ethyl acetate/cyclohexane/*p*-dioxane/methanol/water/ammonia (50:50:10:10:0.5:1.5)	[45]
Opium	Silica/$NaCO_3$	Chloroform/ethanol (8:2)	[34]
	Silica	Ethyl acetate/methanol/ammonia (85:10:5)	[38]
		Benzene/methanol (8:2)	[43]
		Benzene/dioxane/ammonia/ethanol (50:40:5:5)	[44]
		Ethyl acetate/cyclohexane/*p*-dioxane/methanol/water/ammonia (50:50:10:10:1.5:0.5)	[45]
Methadone/cocaine	Silica	Ethyl acetate/cyclohexane/methanol/ammonia (56:40:0.8:0.4); ethyl acetate/cyclohexane/ammonia (50:40:0.1)	[17]
Barbiturates	Silica	Ethyl acetate/cyclohexane/ammonia/methanol/water (70:15:2:8:0.5); ethyl acetate/cyclohexane/methanol/ammonia (56:40:0.8:0.4)	[17]
		2-Propanol/chloroform/ammonia (9:9:2)	[46]
Cannabinoids	Silica	Benzene	[30, 32]
		Hexane/ether/acetic acid (87:12:1)	[32]
		Ether/hexane (1:4); benzene/diethylamine (95:5)	[33]
		Benzene/hexane/diethylamine (25:10:1)	[34]
		Chloroform/1,1-dichloroethane (15:10); xylene/1,4-dioxane (19:1)	[31]
Benzodiazepines	Aluminum oxide	Chloroform/toluene/ethanol (40:60:2)	[37]
	Silica	Chloroform/acetone (9:1)	[28]
		Ether/methylene chloride/diethylamine/triethylamine (90:10:2:1)	[39]

Although the specificity of an identification can always be improved by using as many solvent systems as possible, the use of more than two systems does not normally improve the discrimination significantly and is not realistic in practical applications. Thus, only two-system recommendations are listed in Table 4.3. These selections are based on purely empirical data [17], the correlation of R_f values [47], or calculations derived from IP [6], DP [5, 14], and IC [15]. It was recommended that the combined use of two solvent systems for the resolution of basic drugs should include a system that will separate the drugs in their ion-pair forms [6].

Table 4.3. Recommended combination of two solvent systems for TLC.

Drug category	Adsorbent	Solvent system	Ref.
All categories	Silica[a]	Ethyl acetate/cyclohexane/methanol/ammonia (70:15:10:5) and ethyl acetate/cyclohexane/ammonia (50:40:0.1); ethyl acetate/cyclohexane/ammonia/methanol/water (70:15:2:8:0.5) and ethyl acetate/cyclohexane/ammonia (50:40:0.1)	[17]
Basic	Silica; Cellulose F	Methanol and 1-butanol/water/citric acid (870:310:4.8)	[47]
	Silica/KOH	Cyclohexane/benzene/diethylamine (75:15:10) and acetone buffer, pH 4.58 (95 °C)	[14]
	Silica	Methanol and methanol/1-butanol (with NaBr) (6:4)	[6]
	Silica/KOH	Cyclohexane/benzene/diethylamine (75:15:10) and acetate buffer, pH 4.58 (90 °C)	[15]
	Silica/KOH	Methanol/ammonia (100:1.5) and cyclohexane/toluene/diethylamine (75:15:10)	[5]
Neutral	Kieselguhr[b]	Ethyl acetate and dioxane/methylene chloride/water (1:2:1)	[8]
Acidic/neutral	Silica	Ethyl acetate/methanol/ammonia (85:10:5) and ethyl acetate	[5]

[a] The same plate is developed consecutively using the two solvent systems listed in the next column.

[b] The same plate is developed consecutively (in two dimensions) using the two solvent systems listed in the next column.

Detection

In addition to its simplicity, TLC technology may also provide somewhat selective detection through the use of various reagents for color formation with the resolved analytes. The availability of this approach offsets, to some extent, problems associated with the limited resolution that can be achieved in the limited length of TLC plates, and difficulties in precise R_f value measurements. The use of more-specific detection techniques is especially helpful in general screen cases, where the presence of various drugs is possible. Thus, multiple detection reagents are recommended [17, 41, 42] for achieving the desired degree of discrimination. Detections are essentially conducted through the use of various reagents that form products displaying characteristic color under white or ultraviolet radiation.

To allow for a systematic detection of various drugs in a general screen, several reagents are sequentially applied to the same plate, where each spray characterizes additional compounds that were not revealed or discriminated in the previous step. Thus, the selection of detection reagents, their application sequence, and the resulting color formations were evaluated in several comprehensive studies [17, 41, 42]. Some of the reported results are summarized in Table 4.4.

Alternately, if specific compounds are to be identified, various reagents known to produce characteristic colors may be applied. Reagents recommended for the detection of commonly abused drugs, and resulting color formations, are listed in Table 4.5.

Standardized Operation

The chromatographic behavior of various drugs under different solvent systems has been well-characterized in several comprehensive studies [5, 16–26]. Although these references are valuable for

Table 4.4. Sequential application of reagents for systematic identification of drugs of abuse on TLC plates.[a]

First reagent (Ehrlich) Color	Second reagent (iodoplatinate) Color change	Drug	Ref.
Purple	Disappear	Lysergide	[42][b]
Yellow	Dark brown	Procaine•HCl, benzocaine	
Blue	Violet purple	Psilocin	
None	Purple	Caffeine	
None	Brown	Ephedrine sulfate, psilocybin, methaqualone	
None	Orange	Mescaline•HCl	
None	Pink	Secobarbital	
None	Deep blue	Methapyrilene•HCl	

First reagent (ninhydrin) Color	Second reagent (diphenylcarbazone/HgSO₄) Color change	Third reagent (iodoplatinate/Dragendorff) Color change	Drug	Ref.
Pink	No change	Brown	Amphetamine	[41][c]
Violet	No change	Disappear	Phenylpropanolamine	
—	Pink	Disappear	Amo-, buta-, hexo-, pheno-, secobarbital, diallylbarbituric acid	
—	Violet	Disappear	Heptabarbital	
—	Purple	Disappear	Pentobarbital	
—	—	Violet	Diphenhydramine, ethylmorphine, methylphenidate, diazepam, methamphetamine	
—	—	Red/orange	Cocaine	
—	—	Red/violet	Codeine, heroin, papaverine, procaine, meperidine, quinine, quinidine, strychnine	
—	—	Purple	Dihydrocodeine, dihydrocodeinone	
—	—	Dark blue	Morphine	
—	—	Orange	Propoxyphene	
—	—	Orange/purple	Noscopine	
—	—	Orange/violet	Methadone	
—	—	Brown	Phenacetin, caffeine	
—	—	Blue	Nicotine	

First reagent (ninhydrin) UV	Color 90 °C	Second reagent (diphenylcarbazone/HgSO₄)	Third reagent (H₂SO₄/UV)	Fourth reagent (iodoplatinate)	Fifth reagent (iodine-KI/ammoniacal silver nitrate)	Drug	Ref.
			Color change				
Lt purple	Purple	Pink	Orange	Tan orange/green	Tan	LSD	[17][d]
Purple/lt purple	Lt purple/lt purple	Pink/red orange	Disappear	—	—	Amphetamine sulfate	
Ft purple	Purple/pk purple	Pink	Lt orange	Tan	Tan	Ephedrine	
Lt purple	Purple/dk purple	Pink/red	Lt orange	—	—	Methamphetamine	
Gray/ft purple	Purple/gy purple	Pink	Green/orange green	Tan	Brown	N,N-Dimethyltryptamine	

Drug					
Oxazepam	Lt purple	Disappear	Lt blue	Lt blue	—
Mephentermine	Gy purple	Purple/ft purple	Lt orange	Brown gray/lt brown	Tan
Phenethylamine	Dk purple	Dk purple	Orange	Brown	Brown
Phentermine	Ft purple	Lt purple/gray	Lt yellow	Orange	Tan
2,5-Dimethyoxymethamphetamine	—	Lt purple/gy purple	Brown	Tan/pink	Tan
Methadone	—	Gy purple	—	Brown	Brown
Pseudoephedrine	—	Lt purple	—	—	—
Codeine phosphate	Gray	Gray	Orange, then disappear	Lt brown/blue	Lt brown/brown
Meperidine	—	Gray	Orange	Gy brown/lt brown	Dk brown/brown
Papaverine	Decolorized	Disappear	Yellow	Brown	Lt brown
Quinine	—	—	Blue purple	Brown	Dk brown
Caffeine	—	—	—	—	Brown/gy brown
Cocaine•HCl	—	—	—	Brown/gy brown	Gy brown
Heroin	—	—	—	Brown/blue	Tan/blue
Hydrocodone	—	—	—	Blue/brown	Dk brown
Methaqualone	—	—	—	Brown	Brown
Amobarbital, butabarbital	Purple	—	Tan	—	—
Barbital	Lt purple	—	Disappear	—	—
Pentobarbital	Purple	—	Brown	—	—
Phenobarbital	Purple	—	Disappear	—	—
Secobarbital	Purple	—	Brown	—	—
Diazepam & metabolite	—	—	Yellow/green	Tan	Brown
Propoxyphene	—	—	Tan	Brown	Brown
Morphine	—	—	Green/yellow	Navy blue	Brown/blue
Hydromorphone	—	—	Green	Brown blue/navy blue	Lt brown/dk brown
Mescaline	—	—	Yellow/lt orange	Brown	Brown/dk brown
Phencyclidine	Pink orange	—	—	Brown	Brown

a Abbreviations: dk: dark; gy: grey; ft: faint; lt: light.

b Solvent system: for a complete unknown, two plates are prepared and developed in chloroform/ether/methanol/conc. NH_4OH (75:25:5:1) and ethyl acetate/1-propanol/conc. NH_4OH (40:30:3). Reagents: Ehrlich: 0.2% p-dimethylaminobenzaldehyde in 10% HCl; iodoplatinate in 10% HCl; water (A: 10% hexachloroplatinic(IV) acid solution; B: 6% KI solution).

c Solvent system: ethyl acetate/methanol/conc. ammonium hydroxide (170:20:10). Reagents: ninhydrin: 0.1% ninhydrin in acetone; diphenylcarbazone: 0.01% diphenylcarbazone in equal parts of acetone and water; $HgSO_4$: 0.5 g mercuric oxide in conc. H_2SO_4, slowly added to 200 mL water; iodoplatinate: mix A and B, then dilute to 250 mL with water (A: 1 g platinic chloride in 10 mL water; B: 60 g KI in 200 mL water), stored in refrigerator; Dragendorff: mix A, B, and C, then dilute with 100 mL water and 25 mL glacial acetic acid (A: 1.3 g bismuth subnitrate in 60 mL water; B: 15 mL glacial acetic acid; C: 12 g KI in 30 mL water). Special spray and observation procedures: air-dried plate is heated at 75 °C for 10 min prior to the ninhydrin spray, then placed under UV light; $HgSO_4$ spray is applied heavily immediately after the diphenylcarbazone spray.

d Solvent systems and development procedure: first develop with ethyl acetate/cyclohexane/methanol/ammonia (75:15:10:5) or ethyl acetate/cyclohexane/ammonia/methanol/water (70:15:2:8:0.5); the air- then oven-dried (85–90 °C) plate is developed with ethyl acetate/cyclohexane/methanol/ammonia (50:40:0.1) until the solvent reaches 14.5 cm (45–50 min). Reagents: ninhydrin: 0.5% ninhydrin in 1-butanol; diphenylcarbazone, $HgSO_4$: see note c above; silver nitrate: 1% in water; H_2SO_4: 0.5% in water; iodoplatinate: mix equal volumes of stock solution and 2 N HCl, with the stock solution made out of 5% platinum trichloride/10% KI/water (5:45:50); iodide–KI: mix and shake A and B to form a clear solution, then add 33.8 mL conc. HCl (A: 2 g iodine in 50 mL 95% ethanol; B: 2 g iodine in 16.2 mL water); ammoniacal silver nitrate: mix 15 mL 5 N NH_4OH and 50% $AgNO_3$, then add more (about 20 mL) 5 N NH_4OH until clear. Sequence and regions of plate sprayed: ninhydrin: 0–4.5 cm; diphenylcarbazone followed by silver acetate: 0–9.5 cm, then $HgSO_4$: 4.5–14.5 cm; H_2SO_4: 0–9.0 cm; iodoplatinate: whole plate; iodine–KI: 4.5–14.5 cm; ammoniacal silver nitrate: 0–4.5 cm. Plate treatment and observation of color and color changes: only abbreviated descriptions are provided here; original reference should be consulted for further details.

Table 4.5. Recommended detection reagents for selected drug groups.[a]

Drug	Reagent, treatment, and color formation				Det. limit	Ref.
	Formaldehyde vapor Mandelin's	Water	366 nm	Dragen-dorff		[27][b]
Amphetamine	Yellow to brown	Pale olive	Blue	Brown	2.5	
2,5-Dimethoxyamphetamine (DMA)	Yellowish green	Bt green to or yellow	Dp orange	Brown	5	
2,5-Dimethoxy-4-ethylamphetamine	Greenish yellow	Yellow	Dull blue	Brown	10	
2,5-Dimethoxy-4-methamphetamine	Greenish yellow	Yellow	Dull blue	Brown	10	
N-Ethyl-3,4-methylenedioxyamphetamine	Blue–green	Light olive	Neg.	Brown	7.5	
Methamphetamine	Yellow to brown	Pale olive	Blue	Brown	2.5	
4-Methoxyamphetamine	Blue–purple	Fades	Dull	Brown	2.5	
2-Methoxy-4,5-methylenedioxyamphetamine	Greenish yellow	Greenish yellow	Faint/neg.	Brown	7.5	
N-Methyl-3,4-methylenedioxyamphetamine	Blue–green	Light olive	Neg.	Brown	7.5	
3,4-Methylenedioxyamphetamine	Blue–green	Gray–tan	Bt blue	Brown	5	
Phendimetrazine	Neg., yellow to green	Neg. or pale green	Neg. or blue	Brown	5	
Phenmetrazine	Neg., yellow to green	Neg. or pale green	Neg. or blue	Brown	5	
Phentermine	Yellow to green	Pale green	Blue	Brown	15	
Phenylpropanolamine	Yellow to brown	Pale olive	Blue	Brown	15	
Pseudoephedrine	Yellow to green	Pale green	Blue	Brown	15	
2,4,5-Trimethoxyamphetamine	Yellow	Fades	Neg.	Brown	10	
2,4,6-Trimethoxyamphetamine	Rose	Pink–tan	Faint or neg.	Brown	5	
3,4,5-Trimethoxyamphetamine	Yellow	Fades	Dull blue	Brown	5	
3,4,5-Trimethoxyphenethylamine (mescaline)	Orange	Fades	Dull green	Brown	5	
	Iodoplatinate					[48][c]
d-Amphetamine	Purple–yellow				5	
Ephedrine	Light purple				5	
Methamphetamine	Purple–gray				5	
Phenmetrazine	Purple				5	
	1% Silver acetate, UV irradiation, then 0.02 M potassium permanganate					[48][c]
Amobarbital	Light yellow				5	
Phenobarbital	Light pink				5	
Pentobarbital	Light pink				5	
Secobarbital	Bright yellow				5	
Sodium barbital	White				5	
	0.5% 2,6-Dichloroquinone chlorimide (in 2-propanol)	1:1 Ammonia solution				[49][d]
Codeine	Dark gray	Black			15	
Morphine	Green yellow	Green–yellow			5	
Narcotine	Brown–green	Green with brown periphery			5	
Papaverine	Light orange	Light orange			10	
Thebaine	Bluish gray	Deep green			10	
	Fast Blue BB					[31][e]
Cannabinoid acids	Red–orange					
Cannabichromene	Purple (becomes orange–brown)					
Cannabigerol	Orange					
Tetrahydrocannabivarinol	Red					
Cannabivarinol	Purple					
Tetrahydrocannabinol	Red					
Cannabinol	Purple					
Cannabidiol	Yellow					

Table 4.5. (Continued.)

Drug	Reagent, treatment, and color formation				Ref.
	Potassium iodoplatinate		Exposed to HCl fumes		[28]ᶠ
	254 nm	350 nm	254 nm	350 nm	
Chlordiazepoxide	Absorbs	Yellow	Absorbs	Yellow	
Diazepam	Absorbs	—	Green	Green	
Nitrazepam	Absorbs	Absorbs	—	Blue	
Oxazepam	Absorbs	—	Green	Green	
Medazepam	Green	Blue	Orange	Orange	
Bromazepam	Absorbs	—	Yellow	Yellow	
Dibenzepin	Greenish blue	Greenish blue	Greenish blue	Greenish blue	

NOTE: Detection limit: in ng/mL.

ᵃ Abbreviations: bt: bright; dp: deep; or: orange.

ᵇ Solvent systems: ethyl acetate/methanol/water (95:3.5:1.5) plus 7.5 μL conc. NH₄OH per mL of solvent; acetone plus 5 μL conc. NH₄OH per mL of solvent. Reagent: Mandelin's: heat 200 mg ammonium metavanadate with 250 mL conc. H₂SO₄ until completely dissolved; modified (iodinated) Dragendorff's: mix A and B, then dilute to 250 mL with water (A: heat 200 mg bismuth subnitrate, 5 mL water, and 10 mL glacial acetic acid until completely dissolved; B: 5 g KI and 2 g iodine in 100 mL dist. water).

ᶜ Solvent systems: ethanol/dioxane/benzene/NH₄OH (5:40:50:5); ethyl acetate/methanol/NH₄OH (85:10:5). Reagent: iodoplatinate (see footnote b of Table 4.4).

ᵈ Solvent systems: chloroform/ethanol (8:2). The 1:1 ammonia solution is subsequently sprayed on the plate after the application of the 2,6-dichloroquinone chlorimide spray.

ᵉ Solvent systems and development procedure: first develop with chloroform (ethanol-free)/1,1-dichloroethane (15:10), spray with diethylamine, then develop at a right angle with xylene (mixed isomers)/1,4-dioxane (9:1). Reagent: Fast Blue BB: dissolve 0.5 g Fast Blue BB salt in 50 mL water, then dilute to 100 mL with methanol; add 2 drops of 2 N HCl to improve stability.

ᶠ Solvent system: chloroform/toluene/ethanol (40:60:2). Reagent: iodoplatinate (see footnote b of Table 4.4).

system selection and for comparison of results, maximal standardization of operation steps, including extraction–concentration, developing, and detection, is made possible through the use of commercial Toxi-Lab kits, which include extraction tubes, evaporation discs, silica gel embedded fiberglass plates, and color-developing solutions. The Toxi-Lab system has been described in a recent book chapter [50], and training programs and regular newsletters are available from the manufacturer [51, 52]. A typical operation [53] of a Toxi-Lab kit involves the addition of 5 mL of urine or 2 mL of serum to an appropriate Toxi-Tube, followed by mixing for 1 min. After centrifugation, the extract is evaporated onto a small disc of chromatographic media, and the dried disc (now impregnated with the extracted drugs) is inserted into an open hole of a Toxi-Gram, which includes various drug standards. The Toxi-Gram is then developed in a small jar containing the recommended solvent system. Following development, the chromatogram is sequentially dipped into a series of reagents that produce color changes for numerous drugs. The availability of Photo-Grams (photographs of drug and metabolite detection characteristics) and standard drugs on discs for parallel development with the analyte minimizes subjective data interpretation.

Two general analytical schemes are used in the Toxi-Lab system: A and B. System A is designed for basic and neutral drugs, whereas System B is designed for acidic and neutral drugs. While these general procedures are effective for many drugs, special procedures are needed for correctly detecting drugs and metabolites that are:

1. Too polar for effective extraction;

2. Present in low concentration;

3. Present mainly as conjugates and require prior hydrolysis; and

4. Present with other drugs having similar structural features and retention characteristics.

Table 4.6 summarizes some approaches used to overcome these problems.

Table 4.6. Special approaches for improving performance of the Toxi-Lab system.

Special procedure	Problem to overcome	Drug to be tested
Hydrolysis	Present mainly as conjugates	Benzodiazepines, morphine, marijuana metabolites
Special extraction procedure	Too polar for normal extraction	Marijuana and cocaine metabolites
Chemical derivatization [54]	Inadequate separation	Structurally closely related opiates
Development of solvent system	Inadequate separation	Sympathomimetic amines
Detection system	Ambiguous identification	Morphine

It should be noted that this standardized procedure cannot improve the inherent low sensitivity associated with TLC technology. With some exceptions, it is suitable for detecting toxic concentrations of most commonly used and abused drugs. Many drugs in their therapeutic concentrations can also be detected. Just as in most standardized procedures, modifications are commonly needed for specific applications. Since coextracted lipids may cause interference, revised extraction procedures are recommended [55] when Toxi-Lab is adopted for the analysis of liver specimens.

References

1. Michaud, J. D.; Jones, D. W. *Am. Lab.* **1980,** *12,* 104.
2. Sohr, C. J.; Buechel, A. T. *J. Anal. Toxicol.* **1982,** *6,* 286.
3. Cserháti, T.; Bojarski, J.; Fenyvesi, E.; Szejtli, J. *J. Chromatogr.* **1986,** *351,* 356.
4. De Zeeuw, R. A.; Van Mansvelt, F. J. W.; Greving, J. E. *J. Chromatogr.* **1978,** *148,* 255.
5. Stead, A. H.; Gill, R.; Wright, T.; Gibbs, J. P.; Moffat, A. C. *Analyst (Cambridge, U. K.)* **1982,** *107,* 1106.
6. De Zeeuw, R. A.; Schepers, P.; Greving, J. E.; Franke, J.-P. In *Instrumental Applications in Forensic Drug Chemistry;* Klein, M.; Kruegel, A. V.; Sobol, S. P., Eds.; U.S. Government Printing Office: Washington, DC, 1978; p 167.
7. Bogusz, M.; Klys, M.; Wijsbeek, J.; Franke, J.-P.; de Zeeuw, R. A. *J. Anal. Toxicol.* **1984,** *8,* 149.
8. Galanos, D. S.; Kapoulas, V. M. *J. Chromatogr.* **1964,** *13,* 128.
9. Dhont, J. H.; Vinkenborg, C.; Compaan, H.; Ritter, F. J.; Labadie, R. P.; Verweij, A.; De Zeeuw, R. A. *J. Chromatogr.* **1970,** *47,* 376.
10. Dhont, J. H.; Vinkenborg, C.; Compaan, H.; Ritter, F. J.; Labadie, R. P.; Verweij, A.; De Zeeuw, R. A. *J. Chromatogr.* **1972,** *71,* 283.
11. Phillips, G. F.; Gardiner, J. *J. Pharm. Pharmacol.* **1969,** *21,* 793.
12. Moffat, A. C. *J. Chromatogr.* **1975,** *110,* 341.
13. Bogusz, M.; Bialka, J.; De Zeeuw, R. A.; Franke, J.-P. *J. Anal. Toxicol.* **1985,** *9,* 139.
14. Moffat, A. C.; Smalldon, K. W. *J. Chromatogr.* **1974,** *90,* 9.
15. De Cleraq, H.; Massart, D. L. *J. Chromatogr.* **1975,** *11,* 1.
16. Wahl, K.; Rejent, T. *J. Anal. Toxicol.* **1979,** *3,* 216.
17. Kaistha, K. K.; Tadrus, R.; Janda, R. *J. Chromatogr.* **1975,** *107,* 359.
18. Haywood, P. E.; Horner, M. W.; Rylance, H. J. *Analyst (Cambridge, U. K.)* **1967,** *92,* 711.
19. McElwee, D. J. *J. Anal. Toxicol.* **1979,** *3,* 266.
20. Baselt, R. C. *Analytical Procedures for Therapeutic Drug Monitoring and Emergency Toxicology;* Biomedical: Davis, CA, 1980; p 10.
21. Owen, P.; Pendlebury, A.; Moffat, A. C. *J. Chromatogr.* **1978,** *161,* 195.
22. Moffat, A. C.; Clare, B. *J. Pharm. Pharmocol.* **1974,** *16,* 665.
23. Cole, R. K. *Microgram* **1981,** *14,* 144.
24. Dutt, M. C.; Poh, T. T. *J. Chromatogr.* **1981,** *206,* 267.
25. Gubitz, G.; Wintersteiger, R. *J. Anal. Toxicol.* **1980,** *4,* 141.
26. Kalman, S. M.; Clark, D. R. *Drug Assay: The Strategy of Therapeutic Drug Monitoring;* Masson: New York, 1979; p 92.
27. O'Brien, B. A.; Bonicamp, J. M.; Jones, D. W. *J. Anal. Toxicol.* **1982,** *6,* 143.
28. Stevens, H. M.; Jenkins, R. W. *J. Forensic Sci. Soc.* **1971,** *11,* 183.
29. Schuetz, H. *J. Anal. Toxicol.* **1978,** *2,* 147.

30. Turk, R. F.; Dharir, H. I.; Forney, R. B. *J. Forensic Sci.* **1969**, *14*, 389.
31. Fowler, R.; Gilhooley, R. A.; Baker, P. B. *J. Chromatogr.* **1971**, *171*, 509.
32. Masound, A. N.; Doorenbos, N. J. *J. Pharm. Sci.* **1973**, *62*, 313.
33. Hughes, R. B.; Kessler, R. R. *J. Forensic Sci.* **1979**, *24*, 842.
34. Baggi, T. R. *J. Forensic Sci.* **1980**, *25*, 691.
35. Rafla, F. K.; Epstein, R. L. *J. Anal. Toxicol.* **1979**, *3*, 59.
36. Kraus, L.; Stahl, E.; Thies, W. *Bull. Narc.* **1980**, *32*, 67.
37. Ardrey, R. E.; Moffat, A. C. *J. Forensic Sci. Soc.* **1979**, *19*, 253.
38. Pe, W. *J. Forensic Sci. Soc.* **1983**, *23*, 221.
39. Schuetz, H. *J. Anal. Toxicol.* **1978**, *2*, 147.
40. Klein, C. M.; Lau-cam, C. A. *J. Chromatogr.* **1985**, *350*, 273.
41. Davidow, B.; Petri, N. L.; Quame, B. *Am. J. Clin. Pathol.* **1968**, *50*, 714.
42. Masound, A. N. *J. Pharm. Sci.* **1976**, *65*, 1585.
43. Rao, N. V. R.; Tandon, S. N. *Forensic Sci.* **1977**, *9*, 103.
44. Thornton, J. I.; Dillon, D. J. *J. Forensic Sci. Soc.* **1966**, *6*, 42.
45. Kaistha, K. K.; Jaffe, J. H. *J. Pharm. Sci.* **1972**, *61*, 679.
46. Jackson, J. V.; Clatworthy, A. J. In *Chromatographic and Electrophoretic Techniques;* Smith, I.; Seakinsm J. W. T., Eds.; Heinemann: London, U.K., 1976; Vol. 1, p 406.
47. Smalldon, K. W. *J. Forensic Sci. Soc.* **1971**, *11*, 171.
48. Mulé, S. J. *J. Chromatogr.* **1969**, *39*, 302.
49. Baggi, T. R.; Rao, N. V. R.; Murty, R. K. *Forensic Sci.* **1976**, *8*, 265.
50. Brunk, S. D. In *Analysis of Addictive and Misused Drugs;* Adamovics, J. A., Ed.; Marcel Dekker: New York, 1995; p 41.
51. *Training Programs 1 and 2, Cat No. 496 and 497;* ANSYS: Irvine, CA.
52. *Toxi-News;* ANSYS: Irvine, CA.
53. Jones, D. W.; Rousseau, R. *J. Clin. Lab Products* **1973**, *12*, 23.
54. Dietzen, D. J.; Koenig, J.; Turk, J. *J. Anal. Toxicol.* **1995**, *19*, 299.
55. Whitter, P. D.; Cary, P. L. *J. Anal. Toxicol.* **1986**, *10*, 68.

Chapter 5
Immunoassay

Immunoassay and Workplace Drug Urinalysis

The Basis of Immunoassays

The most important aspects of immunoassay technologies are:

1. The production of an antibody possessing the desired specificity, affinity, and sensitivity;

2. The development of a mechanism for the reaction of the specific antibody with the analyte; and

3. The design of a detection system suitable for measuring the occurrence and extent of the specific reaction.

The antibody is produced in an animal (polyclonal antibodies) or in an organic culture (monoclonal antibodies) in response to an antigenic complex composed of the drug of interest coupled to a carrier protein.

Most immunoassay procedures adopt the *competitive binding principle,* in which the antibody is allowed to react with a mixture of labeled (control) and unlabeled (sample) drugs. The presence and the quantity of a drug in the sample is evaluated on the basis of the quantities of the labeled drug in the reacted or unreacted forms. The control drugs are labeled in different ways, each requiring different methods of detection and quantification. For example, radioactive isotopes, such as ^{125}I, are used for labeling in radioimmunoassay (RIA) methods; active enzymes, which are capable of converting (indirectly) nicotinamide adenine dinucleotide (NAD) to its reduced form (NADH), are coupled to the drug in enzyme multiplied immunoassay techniques (EMIT); fluoresceins are coupled to the drug used in fluorescence polarization immunoassay (FPIA); and particles are attached to the drug in particle immunoassay (PIA). (The basic principles of operation and the applications of these methods are discussed later in sections addressing each of these methodologies.)

Several immunological assay systems have been developed for the detection of drugs in biological specimens. Commercial kits are available for routine use in clinical, toxicological, and forensic laboratories. Major immunoassays that have found significant applications in preliminary testing of commonly abused drugs are listed in Table 5.1.

The methodologies listed in Table 5.1 are considered *heterogeneous* if a phase-separation step is needed prior to detection. The detection process is designed to measure the extent of the labeled antigen linked (directly or indirectly) to the antibody, thus reflecting the amount of the test drug (unlabeled antigen) in the sample. Methodologies based on the measurement of radioactivity are heterogeneous because the detecting device cannot differentiate the source of the radioactivity (free or bound labeled antigen). Therefore, a separation step is required.

Methodologies based on a change in optical intensities do not require the separation step if these properties are modified through the substrate's linkage (directly or indirectly) to the antibody. They are considered *homogeneous* immunoassays. Both heterogeneous and homogeneous immunoassays are used for workplace drug urinalysis. Basic methodologies and cross-reactivities of the most common commercial products are discussed in this chapter.

Modified with permission from *Forensic Science Review,* 1994, Vol. 6, pages 19–57. Copyright 1994 Central Police University, Taiwan.

Table 5.1. Major classification of immunoassays and commercial products for workplace drug urinalysis.

Immunoassay class	Commercial source and trade name[a]	Drug groups with a commercially available assay kit[b]
Radioimmunoassay (RIA)	Roche: Abuscreen	Amphetamine, barbiturates, benzodiazepines, cannabinoids, cocaine metabolite, lysergic acid diethylamide, methamphetamine, methaqualone, morphine, phencyclidine
	DPC: Double antibody Coat-A-Count	Amphetamine, benzodiazepines, cannabinoids Barbiturates, cocaine metabolite, fentanyl, lysergic acid diethylamide, methadone, methamphetamine, morphine, phencyclidine
	Immunalysis[c]	Cannabinoids, phencyclidine
Enzyme immunoassay[d] (EIA)	Syva: EMIT	Amphetamine, barbiturate, cannabinoid, cocaine metabolite, phencyclidine
	Boehringer Mannheim: CEDIA	Amphetamine, barbiturate, benzodiazepine, cannabinoid, cocaine metabolite, methadone, methaqualone, phencyclidine, propoxyphene
Fluorescence polarization immunoassay (FPIA)	Abbott: TDx/ADx	Amphetamine–methamphetamine, barbiturates, benzodiazepines, cannabinoids, cocaine metabolite, methadone, opiates, phencyclidine, propoxyphene
Particle immunoassay[e] (PIA)	Roche: Abuscreen Online	Amphetamines, barbiturates, benzodiazepines, cannabinoids, cocaine metabolite, methadone, opiates, phencyclidine

[a] Commercial manufacturers: Roche: Roche Diagnostic Systems (Branchburg, NJ); DPC: Diagnostic Products Corporation (Los Angeles, CA); Immunalysis: Immunalysis Corporation (Glendale, CA); Syva: Syva Company (Palo Alto, CA); Boehringer Mannheim: Boehringer Mannheim Corporation (Indianapolis, IN); Abbott: Abbott Laboratories (Irving, TX).

[b] As of August 1994.

[c] Several other assay kits that were made available by this manufacturer in the past (see Table 5.3) are currently being re-introduced and evaluated.

[d] Enzyme immunoassay reagents from numerous sources have recently become available. These reagents are not included mainly because of the lack of literature data resulting from evaluations performed by independent laboratories.

[e] In parallel with the nomenclature convention applied to RIA, EIA, and FPIA, the term *particle immunoassay* (PIA) is hereby adopted for immunoassays that use particles of various natures and sizes as the label and the basis of the detection mechanism.

On-Site versus Laboratory-Based Immunoassay Kits

Two categories of immunoassay kits are widely available for preliminary testing of drugs of abuse. Laboratory-based immunoassays require major instrumentation and can only be performed in a laboratory environment, but they can be cost-effective through rapid processing of a large number of samples. With batch-mode operation, quality-control practices can be conveniently applied. This is normally the method of choice in a large-scale testing setting.

On-site immunoassay kits, on the other hand, offer rapid access to test results, are self-contained, and can be performed in the field by individuals with little analytical chemistry experience. Because tests are performed on a one-at-a-time basis and results are obtained mainly through visual observations, uniform quality-control practices and objectivity are harder to achieve. These kits are uniquely suitable for preliminary sample screening at remote sites. The recent development, applications, and evaluations of these kits are included in a last section of this chapter.

Role of Immunoassay in Workplace Drug Urinalysis

The adoption of immunoassays as the "official" preliminary test methods in workplace drug urinalysis programs reflects the important role of this methodology in drug testing. In clinical laboratories, immunological methods are not conventionally evaluated on the basis of their specificity (see the section "Underlying Test Principles and Cross-Reactivity" prior to Chapter 3). Specificity, however, is the primary concern when an immunoassay is used in a workplace drug urinalysis program.

Current practices in workplace drug urinalysis require a two-step test protocol that is based on different underlying principles of the pairing techniques [1]. An immunoassay is used as the screen-test methodology for identifying presumptive positive samples that generate responses equivalent to (or above) those generated by a targeted analyte at the cutoff concentration. These presumptive positive samples are further tested for specific drugs and metabolites by gas chromatography–mass spectrometry (GC–MS) procedures. Only samples that are confirmed to include the targeted drugs at (or above) the cutoff concentrations can be reported as positive. Table 5.2 lists the analytes targeted and the cutoff concentrations of immunoassays and GC–MS tests adopted by the drug urinalysis laboratory certification programs under the auspices of the Department of Health and Human Services and the Department of Defense [2, 3].

Table 5.2. Cutoff levels[a] of immunoassays and GC–MS tests adopted by the U.S. laboratory certification programs.

Drug category	Immunoassay cutoff (ng/mL)			GC–MS test cutoff (ng/mL)		
	Analyte targeted	DoD[b]	HHS[b]	Analyte targeted	DoD	HHS
Amphetamine	Amphetamines	—	1,000			
	Amphetamine	500	—[e]	Amphetamine	500	500
	Methamphetamine	500	—[e]	Methamphetamine	500[c]	500[d]
Barbiturates	Secobarbital	200	—[e]	Butalbital	200	—[e]
				Amobarbital	200	—[e]
				Pentobarbital	200	—[e]
				Secobarbital	200	—[e]
Cocaine	Cocaine metabolites	150	300	Benzoylecgonine	100	150
Opiates	Opiates	300	300	Morphine	4,000	300
				Codeine	2,000	300
				6-Monoacetylmorphine	10	—[e]
Phencyclidine	Phencyclidine	25	25	Phencyclidine	25	25
Cannabinoids	Cannabinoids	50	50	9-THC-COOH[f]	15	15
LSD[g]	LSD	0.5	—[e]	LSD	0.2	—[e]

[a] As of September 1, 1994.

[b] DoD: U.S. Department of Defense; HHS: U.S. Department of Health and Human Services.

[c] The sample should also contain 200 ng/mL of amphetamine to report positive for methamphetamine [4]. Additional testing to differentiate *d*- and *l*-methamphetamine may be required.

[d] The sample must also contain 200 ng/mL of amphetamine to report positive for methamphetamine [4].

[e] Testing not included by the HHS program.

[f] 9-THC-COOH: 11-nor-Δ^9-tetrahydrocannabinol-9-carboxylic acid.

[g] LSD: lysergic acid diethylamide.

Correlation of Preliminary-Test and Confirmation-Test Results

With the aforementioned two-step test protocol and reporting policy, the cutoff concentration adopted for an immunoassay should correspond well with the cutoff concentration of the compound targeted by GC–MS. If an inappropriately low immunoassay cutoff value is adopted, an excess number of negative GC–MS results will be reported, causing the overall analytical procedure to be *financially inefficient*. On the other hand, if the preliminary test cutoff value is set too high, an excessive number of positive samples may be rejected as negatives in the preliminary screening process without being submitted to the GC–MS test step, causing the testing program to be *technically inefficient*.

A test protocol based on an immunoassay reagent with broad cross-reactivity, and which is therefore sensitive to those drugs and metabolites not targeted by the corresponding GC–MS procedure, may require the adoption of a higher cutoff value to avoid generating an excessive number of presumptive positive results that are not confirmed by GC–MS. Thus, the relationship between the immunoassay and GC–MS tests and the underlying immunoassay *cross-reacting characteristics* and *interference susceptibility* are important issues in workplace drug urinalysis and will be emphasized.

To a much lesser extent, immunoassays are also used for analyzing drugs in blood, serum, and other biological specimens. Characteristics of these applications will also be included in this chapter.

Methodologies and Cross-Reactivities

Radioimmunoassay

Basic Methodology

RIA represents one of the most powerful and earliest immunoassay technologies used for high-volume drug screening. Most commonly used RIA methods are based on competitive binding, in which the free unlabeled antigen (the analyte in the sample) competes with the ^{125}I-labeled antigen, in proportion to their respective concentrations, to a limited amount of antibody in the reaction mixture, as shown in the first part of Figure 5.1.

To assess the drug's presence and its concentration in a sample, the radioactivity measured from the sample's antibody-bound fraction is compared with data established by a series of standards. Because the detection mechanism cannot differentiate the sources of the radioactivity, it is necessary to separate the antibody-bound radiolabeled drug (the reaction product) from the antibody-free radio-labeled drug (unreacted) prior to radioactivity counting.

Among various processes used for phase separation, Abuscreen and Immunalysis kits use a precipitating second antibody, Ab_2, to remove the antibody-bound fraction from the reaction mixture, as shown by the second part of Figure 5.1. Newer reagents provided by these manufacturers allow for the addition of Ab_1 and Ab_2 at the same time, thus greatly simplifying the test protocol. Coat-A-Count kits use immobilized antibodies on the wall of a polypropylene tube, thus making the separation step automatic.

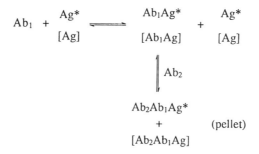

Figure 5.1. Schematic of the competitive immunoassay techniques. Substances inside brackets are present only if the specimen tested contains the analyte. Ag*: the labeled form of the drug; Ag: the drug (analyte) to be measured; Ab_1: the antibody capable of binding the drug (the limiting factor in the reaction); Ab_2: the second antibody capable of binding Ab_1 to form aggregates. (Reproduced with permission from Ref. [63].)

Various dose–response curves [5] established by control samples with known amounts of the drug can be used for the attempted quantification of the drug in an unknown sample. A typical standard curve is obtained by plotting B/B_0 (counts obtained from the test sample related to counts obtained from the zero-dose control) against increasing concentrations of the analyte in the set of control samples (Figure 5.2). Another popular dose–response curve plots logit(B/B_0) against $\log_e[Ag]$, where $[Ag]$ is the concentration of the analyte (antigen).

$$\log_e[(B/B_0)/(1 - B/B_0)] = a + b \log_e[Ag] \tag{5.1}$$

The latter plot is advantageous in that a linear relationship of the two parameters can be obtained within a limited concentration range.

Cross-Reactivities

Cross-reactivity data of common commercial kits are summarized in Table 5.3. Most of these data are taken directly from the respective reagent package inserts, while others are reported by independent laboratories. The cross-reacting compounds listed in this table are by no means complete. It is highly possible that some cross-reacting compounds may not have been tested and are therefore not listed.

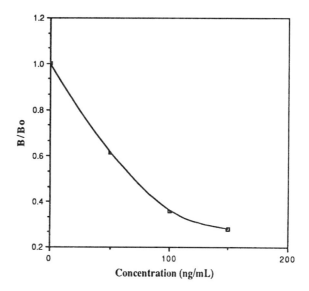

Figure 5.2. Dose–response curve for RIA. (Data taken from Abuscreen Radioimmunoassay for Phencyclidine package insert dated Oct. 1988. Reproduced with permission from Ref. [63].)

Table 5.3. Reported cross-reactivities of commercial RIA kits.[a]

Manufacturer, assay name, and assay specifics	Cross-reacting compound	% cross-reactivity	Concentration tested (ng/mL)
Roche Abuscreen RIA for Amphetamine (High Specificity)	3,4-Methylenedioxyamphetamine	95; 17	1,000; 10,000
Calibrator: *d*-amphetamine	*d,l*-Amphetamine	87; 17	1,000; 10,000
Control range: 0–1,500 ng/mL	Hydroxyamphetamine HCl	35; 18	1,000; 10,000
Date: Nov. 1991	*l*-Amphetamine	27; 7.7	1,000; 10,000
	β-Phenylethylamine HCl	6.4; 2.6	1,000; 10,000
	Tyramine HCl	3.1; 1.5	1,000; 10,000
	3,4-Methylenedioxymethamphetamine HCl	1.8; 0.9	1,000; 10,000
	Phenylpropanolamine HCl	1.1; 0.13	1,000; 10,000
	3,4-Methylenedioxyamphetamine[b]	158–107	100–10,000
	3,4-Methylenedioxymethamphetamine[b]	1.6–0.26	1,000–100,000
	2-Methoxyamphetamine[b]	1.6–0.54	1,000–100,000
	2,5-Dimethoxyamphetamine[b]	1.4–0.23	1,000–100,000
	2,5-Dimethoxy-4-methylamphetamine[b]	1.4–0.12	1,000–100,000
DPC Double Antibody Amphetamine	Hydroxyamphetamine	52–29	50–1,000
Calibrator: amphetamine	Phenylethylamine	2.8	10,000
Control range: 0–1,000 ng/mL	Tyramine	2.3; 1.6	1,000; 10,000
Date: April 1992	3-Methoxy-3,4-methylenedioxyamphetamine	1.4–1.1	500–10,000
	3,4-Methylenedioxyamphetamine[c]	173–108	100–10,000
	l-Amphetamine[c]	9.2–7.6	1,000–100,000
	2,5-Dimethoxyamphetamine[c]	4.2; 2.1	1,000; 10,000
	2-Methoxyamphetamine[c]	4.1–0.66	1,000–100,000
	N-Hydroxy-3,4-methylenedioxyamphetamine[c]	1.5–0.37	1,000–100,000

Table 5.3. (Continued.)

Manufacturer, assay name, and assay specifics	Cross-reacting compound	% cross-reactivity	Concentration tested (ng/mL)
Roche Abuscreen RIA for Methamphetamine (High Specificity)	3,4-Methylenedioxymethamphetamine	108	—
	l-Methamphetamine	2.8	—
Calibrator: d-methamphetamine	l-Ephedrine[d]	<1; ≥0.1	
Control range: 0–1,500 ng/mL	d-Pseudoephedrine[d]	<1; ≥0.1	
Date: April 1992			
DPC Coat-A-Count Methamphetamine	3,4-Methylenedioxymethamphetamine	438; 289	100; 500
Calibrator: methamphetamine	d,l-Methamphetamine	76; 67	500; 1,000
Control range: 0–1,000 ng/mL	l-Methamphetamine	3.9; 3.3	10,000; 100,000
Date: Oct. 1992	Propylhexadrine	3.9–1.3	1,000–100,000
	Ranitidine	2.3; 1.2	10,000; 100,000
	l-Ephedrine	0.9	100,000
	3,4-Methylenedioxymethamphetamine[e]	>100->100	100–10,000
	Phenethylamine[e]	14–0.1	100–10,000
	N,N-Dimethylamphetamine[e]	10–3.1	100–10,000
	Amphetamine[e]	6.5–0.3	100–10,000
	Phenylephrine[e]	2.4–0.2	100–10,000
	N-Ethylamphetamine[e]	2.4–0.5	100–10,000
	Ephedrine[e]	1.7–0.5	100–10,000
	Phenmetrazine[e]	1.6–0.02	100–10,000
	Diphenhydramine[e]	1.3–0.02	100–10,000
	l-Ephedrine[d]	<1; ≥0.1	
	d-Pseudoephedrine[d]	<1; ≥0.1	
	Trimethobenzamide[f]	<1; ≥0.1	
Roche Abuscreen RIA for Barbiturates	Aprobarbital	70	—
Calibrator: secobarbital	Allylcyclopentylbarbituric acid	44	—
Control range: 0–400 ng/mL	Allylisobutylbarbituric acid	27	—
Date: Nov. 1991	Butabarbital	22	—
	Pentobarbital	21	—
	Diallylbarbituric acid	21	—
	p-Hydroxyphenobarbital	16	—
	Amobarbital	7	—
	Phenobarbital	6	—
	Barbital	3	—
DPC Coat-A-Count Barbiturates	Phenobarbital	1,820; 5,335	10; 100
Calibrator: secobarbital	Phenobarbital	182; 5,335	10; 100
Control range: 0–10,000 ng/mL	Butabarbital	256	1,000
Date: Jan. 1992	Amobarbital	228	1,000
	Pentobarbital	124; 153	1,000; 10,000
	Allylcyclopentylbarbituric acid	121; 93	1,000; 10,000
	Allobarbital	69; 67	1,000; 10,000
	Aprobarbital	39; 49	1,000; 10,000
	Butalbital	35; 25	1,000; 10,000
	Thiopental	32; 37	1,000; 10,000
	Barbital	22; 44	1,000; 10,000
	Mephobarbital	5.7–8.2	1,000–100,000
Roche Abuscreen RIA for Benzodiazepines	Diazepam[g]	284	—
Calibrator: oxazepam	N Methyloxazepam (temazepam)	212	—
Control range: 0–200 ng/mL	N-Desmethyldiazepam	202	—
Date: Oct. 1990	Alprazolam	144	—
	α-Hydroxyalprazolam	131	—

Table 5.3. (Continued.)

Manufacturer, assay name, and assay specifics	Cross-reacting compound	% cross-reactivity	Concentration tested (ng/mL)
Roche Abuscreen RIA for Benzodiazepines (Continued)	Pinazepam[g]	131	—
	Midazolam	55	—
	4-Hydroxyalprazolam	37	—
	Nitrazepam	31	—
	Flunitrazepam	31	—
	Desmethylchlordiazepoxide	30	—
	Hydroxyethylflurazepam	30	—
	Medazepam[g]	30	—
	Demoxepam	26	—
	Halazepam[g]	25	—
	Desalkylflurazepam	23	—
	Clonazepate[g]	22	—
	Prazepam[g]	21	—
	3-Hydroxyflunitrazepam	21	—
	Triazolam	20	—
	Didesethylflurazepam	19	—
	4-Hydroxytriazolam	13	—
	Desmethylflunitrazepam	10	—
	α-Hydroxytriazolam	5	—
	Chlordiazepoxide[g]	5	—
	Lorazepam	3	—
	Desmethylmedazepam	3	—
	Flurazepam	2	—
	Clonazepam	2	—
DPC Double Antibody Benzodiazepines Calibrator: oxazepam Control range: 0–1,000 ng/mL Date: March 1992	Alprazepam	354; 330	50; 100
	Tempazepam	352; 420	50; 100
	Diazepam	302; 390	50; 100
	Nitrazepam	96–32	50–1,000
	Demoxepam	94–47	50–1,000
	α-Hydroxyalprazepam	90–54	50–1,000
	Flunitrazepam	43–32	50–1,000
	Midazolam	21–11	50–1,000
	Desmethyldiazepam	20–58	50–1,000
	Bromazepam	16–20	50–1,000
	Chlordiazepoxide	6–15	50–1,000
	Halazepam	14–4	50–10,000
	Clorazepate	8–8	50–10,000
	Prazepam	8–3	50–10,000
	Triazolam	6–2	50–10,000
	Medazepam	4–9	50–10,000
	Lorazepam	4–1	50–10,000
	Flurazepam	3–1	50–10,000
	Clonazepam	2–1	50–10,000
DPC Double Antibody Buprenorphine Calibrator: buprenorphine Control range: 0.5–10 ng/mL Date: June 1992	*N*-Dealkylbuprenorphine	234–22	1–100
	Etorphine	1.8–0.07	10–100,000
Roche Abuscreen RIA for Cannabinoids Calibrator: 9-THC-COOH Control range: 0–150 ng/mL Date: Feb. 1992	11-nor-Δ^8-THC-COOH	76	—

Table 5.3. (Continued.)

Manufacturer, assay name, and assay specifics	Cross-reacting compound	% cross-reactivity	Concentration tested (ng/mL)

DPC Double Antibody Cannabinoids
  Calibrator: 9-THC-COOH
  Control range: 0–100 ng/mL
  Date: April 1993
	11-nor-Δ^8-THC-COOH
	11-nor-Δ^8-THC
		97; 125
		>100; >100
			10; 100
			100; 1,000

Immunalysis Urine Cannabinoids
Direct RIA Kit
  Calibrator: 9-THC-COOH
  Control range: 0–100 ng/mL
  Date: Feb. 1993
	11-nor-Δ^9-THC
	11-Hydroxy-Δ^9-THC
	Cannabinol
	Cannabidiol
	8 β-11-Dihydroxy-Δ^9-THC
		<5
		<5
		<5
		<5
		<5
			—
			—
			—
			—
			—

Roche Abuscreen RIA for Cocaine Metabolite
  Calibrator: benzoylecgonine
  Control range: 0–600 ng/mL
  Date: Nov. 1991
	Ecgonine
	Cocaine
		2.9–0.88
		1.2–0.72
			1,000; 100,000
			1,000–100,000

DPC Coat-A-Count Cocaine Metabolite
  Calibrator: benzoylecgonine
  Control range: 0–5,400 ng/mL
  Date: June 1993
	Cocaine
	Cocaethylene
	Ecgonine methyl ester
	l-Cocaine[h]
	l-Benzoylecgonine[h]
	l-Norcocaine[h]
	d-Cocaine[h]
	l-Ecgonine[h]
	l-Benzoylnorecgonine[h]
	l-Ecgonine methyl ester[h]
	d-Pseudococaine[h]
		4200–3900
		4300
		408; 5.7
		7,259
		104
		64
		7.4
		5.6
		1.9
		1.3
		1.0
			1–100
			100
			5000; 10,000
			50
			300
			50
			5,000
			5,000
			5,000
			5,000
			5,000

Immunalysis Cocaine Metabolite
Direct RIA Kit
  Calibrator: benzoylecgonine
  Control range: 0–2,000 ng/mL
  Date: Sept. 1987
	Cocaine
	Ecgonine
		>100
		50
			—
			—

DPC Coat-A-Count Fentanyl
  Calibrator: fentanyl
  Control range: 0–7.5 ng/mL
  Date: April 1993
	$trans$-3-Methylfentanyl
	p-Fluorofentanyl
	cis-3-Methylfentanyl
	Thienylfentanyl
	3-Methylfentanyl
	α-Methylfentanyl
	α-Methylthiofentanyl
	2-Hydroxyfentanyl
	Norfentanyl
	p-Fluorofentanyl[i]
	Thienylfentanyl[i]
	3-Methylfentanyl[i]
	α-Methylfentanyl[i]
		32–58
		32–28
		4.8–27
		26–16
		22–15
		12–5
		7.2–9.7
		8.2–8.4
		1–7.6
		28
		16
		15
		5
			5–10
			5–50
			5–10
			5–50
			5–50
			5–50
			5–100
			5–100
			5–100
			50
			50
			50
			50

Roche Abuscreen RIA for LSD
  Calibrator: LSD
  Control range: 0–1 ng/mL
  Date: April 1993
	Lysergic acid N-(methylpropyl)amide	24	—

Table 5.3. (Continued.)

Manufacturer, assay name, and assay specifics	Cross-reacting compound	% cross-reactivity	Concentration tested (ng/mL)
DPC Coat-A-Count LSD Calibrator: LSD Control range: 0–3 ng/mL Date: March 1992	Lysergic acid methylpropylamide 2-Oxo-LSD[j] Lysergic acid methylpropylamide[j] Lysergic acid monoethylamide[j] Nor-LSD[j]	5.6; 1.5 11 ≈ 6 ≈ 2 ≈ 1	100; 1,000 — — — —
DPC Coat-A-Count Methadone Calibrator: *d*-methadone Control range: 0–500 ng/mL Date: March 1992	*l*-Methadone 1-α-Acetylmethadol (LAAM)	101–64 50–60	10–300 200–1,000
Roche Abuscreen RIA for Methaqualone Calibrator: methaqualone Control range: 0–750 ng/mL Date: Nov. 1989	(None listed)		
DPC Coat-A-Count Methaqualone Calibrator: methaqualone Control range: 0–500 ng/mL Date: Nov. 1987	4-Hydroxymethaqualone	250	100
Roche Abuscreen RIA for Morphine Calibrator: morphine Control range: 0–600 ng/mL Date: Oct. 1991	Ethyl morphine Codeine Morphine-3-glucuronide Dihydrocodeine Hydrocodone Dihydromorphine Hydromorphone 6-Monoacetylmorphine *N*-Norcodeine Thebaine Oxycodone Dihydrocodeine[k] Dihydromorphine[k] Norcodeine[k] Hordenine[l]	159 156 26 20 12 14; 7.3 12; 6.1 9.0; 5.7 3.4; 2.1 2.4; 1.1 1.3; 0.44 51 8.6 4.8	— — — — — 1,000; 10,000 1,000; 10,000 1,000; 10,000 1,000; 10,000 1,000; 10,000 1,000; 10,000 10,000 10,000 10,000
DPC Coat-A-Count Morphine Calibrator: morphine Control range: 0–500 ng/mL Date: May 1992	Nalorphine Normorphine Hydromorphone Nalorphine[j] Dihydrocodeine[j] Normorphine[j]	24–27 9.0–9.1 1.5–0.75 22 4.7 8.7	10–500 100–1,000 100–10,000 10,000 10,000 10,000
Immunalysis Urine Heroin/Morphine Direct RIA Kit Calibrator: morphine Control range: 0–500 ng/mL Date: Sept. 1987	Morphine-3-glucuronide Codeine *N*-Allylnormorphine	75 6 2	— — —
Roche Abuscreen RIA for Phencyclidine Calibrator: phencyclidine Control range: 0–50 ng/mL Date: Nov. 1991	1-[1-(2-Thienyl)cyclohexyl]piperidine HCl 1-(1-Phenylcyclohexyl)pyrrolidine HCl *N,N*-Diethyl-1-phenylcyclohexylamine HCl	10 7.0 2.4; 0.8	1,000 1,000 1,000; 10,000

Table 5.3. (Continued.)

Manufacturer, assay name, and assay specifics	Cross-reacting compound	% cross-reactivity	Concentration tested (ng/mL)
DPC Coat-A-Count Phencyclidine	1-[1-(2-Thienyl)cyclohexyl]piperidine	95–97	10–100
Calibrator: phencyclidine	1-(1-Phenylcyclohexyl)-4-hydroxypiperidine	2.2–1.1	500–5,000
Control range: 0–250 ng/mL			
Date: April 1993			
Immunalysis Phencyclidine Direct RIA Kit	1-[1-(2-Thienyl)cyclohexyl]piperidine	50	—
Calibrator: phencyclidine	1-[1-(2-Thienyl)cyclohexyl]morpholine	20	—
Control range: 0–50 ng/mL	1-(1-Phenylcyclohexyl)pyrrolidine	10	—
Date: Jan. 1992	1-(1-Phenylcyclohexyl)morpholine	8	—

[a] With the exceptions of data that are footnoted, data listed in this table are taken from the respective reagent package inserts as specified in the first column of the table. Only those compounds that were reported to show ≥1% cross-reactivity are listed. Compounds are listed in descending order of their reported cross-reactivities. Data listed in the "% cross-reactivity" and the "Concentration tested" columns may be separated by ";" or "–" meaning that two data values (for ";") or a range of data (for "–") were reported in the original literature.

[b] Data are taken from Ref. [6].

[c] Data are taken from Ref. [7].

[d] Data were reported in Ref. [8]. Using 1,000 ng/mL as the cutoff, a negative result and a positive result were observed in controls with 100,000 ng/mL and 1,000,000 ng/mL, respectively, of these compounds. Thus, the cross-reactivities of these compounds are <1% but ≥0.1%.

[e] Data are taken from Ref. [9].

[f] Data are taken from Ref. [10]. Using 1,000 ng/mL as the cutoff, a negative result and a positive result were observed in controls with 100,000 ng/mL and 1,000,000 ng/mL, respectively, of these compounds. Thus, the cross-reactivities of these compounds are < 1% but ≥ 0.1%.

[g] Oxazepam is a metabolite of these compounds.

[h] Data are taken from Ref. [11].

[i] Data are taken from Ref. [12].

[j] Data are taken from Ref. [13].

[k] Data are taken from Ref. [14].

[l] Data are taken from Ref. [15].

Despite the fact that cross-reactivity with structurally related drugs may serve other purposes well, the two-step test protocol and the test-result reporting policy adopted by workplace drug urinalysis programs have motivated RIA reagent manufacturers to develop and produce assay kits that will generate test results closely related to the GC–MS test results of a limited number of targeted drugs. This trend is clearly demonstrated by the differences in reagent specificity (as shown in Table 5.3) for drugs that are and are not adopted for monitoring in workplace drug urinalysis programs. Thus, highly specific kits for amphetamine, methamphetamine, benzoylecgonine, 11-nor-Δ^9-tetrahydrocannabinol-9-carboxylic acid (9-THC-COOH), morphine–codeine, and phencyclidine are available. To demonstrate this point further, reagent specificity data, as summarized in Table 5.4, indicate that earlier reagents from the same manufacturers were not as specific as current ones.

Reagents that are highly specific toward one enantiomeric form have also been developed [16] and marketed. For example, Roche's d-methamphetamine kit, which reportedly cross-reacts with l-methamphetamine by a mere 2.8%, may exclude a substantial percentage of methamphetamine-containing urine samples derived from people who use Vicks nasal inhaler [17].

Application and Sample Characteristics

When applying an immunoassay for testing different biological fluids or tissues, the following sample characteristics are to be considered:

1. The concentration of a specific drug or metabolite is different in different biological fluids or tissues.

2. The concentration of a drug or metabolite present in a specific biological fluid or tissue varies with the time lapse following the drug exposure.

3. The sample matrix may contain different cross-reacting and interfering components.

Table 5.4. Variation of specificities of Abuscreen Radioimmunoassay for Cannabinoid kits.

Cross-reacting compound	Approximate % cross-reactivity[a]						
	Mar. 1987	*Mar. 1988*	*Nov. 1988*	*June 1989*	*May 1990*	*Oct. 1991*	*Feb. 1992*
11-nor-Δ^8-THC-COOH	244	76	76	76	76	49.5	76
11-Hydroxy-Δ^9-THC	38	20	20	<5	<5	2.8	<5
Δ^9-THC	5	3	3	<5	<5	<1	<5
8 β-11-Dihydroxy-Δ^9-THC	11	7	7	<5	<5	1.9	<5
8 α–Hydroxycannabinol	<5	13	13	<5	<5	1.4	<5
11-Hydroxycannabinol	<5	7	7	<5	<5	<1	<5
Cannabinol	<5	<5	<5	<5	<5	<1	<5
Cannabinol	<5	<5	<5	<5	<5	<1	<5

[a] Data are taken from Abuscreen Radioimmunoassay for Cannabinoid reagent package inserts dated March 1987, March 1988, Nov. 1988, June 1989, May 1990, Oct. 1991, and Feb. 1992.

The qualitative and quantitative examination of methamphetamine and amphetamine (as a metabolite of methamphetamine) in urine and blood illustrates a typical example of differential distribution patterns that exist in different biological fluids. Qualitatively, it was reported [18] that methamphetamine alone was found in 148 blood samples and methamphetamine plus amphetamine in only 9 blood samples, while a much higher proportion of urine samples was reported [18] to contain both methamphetamine and amphetamine. Quantitatively, it was concluded [19] that 1000 ng/mL and 50 ng/mL of methamphetamine should be used, respectively, for screening urine and blood samples.

Data shown in Figures 5.3 and 5.4 [20] clearly illustrate the difference in 9-THC-COOH concentrations (at various time intervals) in plasma and urine collected from heavy and light users.

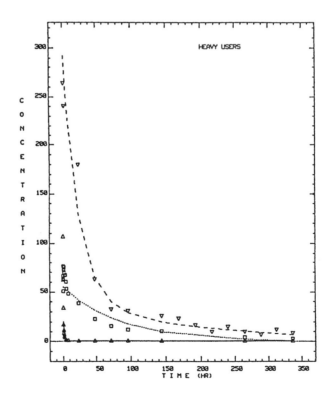

Figure 5.3 Curves of concentration vs. time after smoking one marijuana cigarette containing 18 mg of THC. The experimental subjects were heavy users. Shown are THC in plasma (triangles), and 9-COOH-THC in plasma (squares) and in urine (inverted triangles). (Reprinted with permission from Ref. [20].)

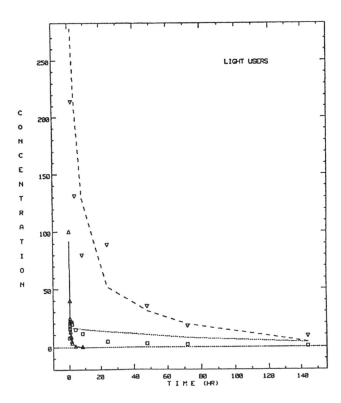

Figure 5.4. Curves of concentration vs. time after smoking one marijuana cigarette containing 18 mg of THC. The experimental subjects were light users. Shown are THC in plasma (triangles), and 9-COOH-THC in plasma (squares) and in urine (inverted triangles). (Reprinted with permission from Ref. [20].)

Matrix variations of different biological fluids will also affect test results and require different sample-preparation procedures. Although one study concluded [21] that, in general, RIA reagents designed for screening urine samples can be adopted for screening blood samples without significant observable matrix effects, the same study stated that a false negative phencyclidine was observed, perhaps because of the high nonspecific binding of the radiolabeled drug by the blood.

In order to minimize matrix effects, several studies recommended methanol extraction of blood for screening cannabinoid metabolites [22–26], while chloroform/isopropyl alcohol (3:1) elution of blood samples adsorbed on XAD-2 columns was recommended [27] for screening morphine-related drugs. Nevertheless, the precision of data derived from methanol extracts of blood samples is still poorer than that reported for urine samples. A recent study [28] revealed that methanol precipitation of interfering substances (possibly proteins) is useful for testing urine samples that showed aberrant RIA results for cannabinoids.

Extraction procedures are definitely required when samples such as bloodstains are tested. It has been reported [29] that physiological saline elutes about 43% of amphetamine from bloodstains, while the use of detergents causes a detrimental effect on the elution and/or detection process. On the other hand, it was recommended [29] that 0.1% of ionic sodium dodecyl sulfate (SDS) in physiological saline should be used as the eluant for phenobarbital. Similarly, SDS solution was more effective than un-ionic Triton X-100 in eluting phenytoin from bloodstains [29].

Enzyme Immunoassay

Basic Methodology

The enzyme-multiplied immunoassay technique represents the most widely used enzyme immunoassay (EIA) technology applied to drug screening. The enzyme-multiplied immunoassay technique is based on the absorbance change (at 340 nm) caused by the reduction of NAD to NADH. This reaction is coupled by the oxidation of glucose-6-phosphate to 6-phosphogluconolactone, as shown in Figure 5.5. The latter oxidation is catalyzed by glucose-6-phosphate dehydrogenase (G6P-DH) attached to the free, but not the bound, antigen.

In practice, a standard amount of the enzyme-labeled antigen and a constant limiting quantity of the antibody are used in every standard and test sample. Under the competitive reaction process, the concentration level of the analyte in a test sample determines the amount of the enzyme-labeled antigen that remains unbound. The amount of the enzyme-labeled antigen that remains unbound will determine the oxidation rate of glucose-6-phosphate to 6-phosphogluconolactone, thus indirectly determining the absorbance change caused by the reduction of NAD to NADH. Because physical separation of the bound from the unbound antigen is unnecessary for the measurement of the absorption change, this is a homogeneous immunoassay. A typical dose–response curve obtained from calibrators is shown in Figure 5.6.

Recently, Boehringer Mannheim Corporation marketed a new homogeneous assay system, CEDIA, based on genetically engineering β-galactosidase into a large polypeptide (an enzyme acceptor, EA) and a small polypeptide (an enzyme donor, ED). Both the ED and EA are enzymatically inactive but spontaneously associate to form enzymatically active tetramers that hydrolyze chlorophenol red β-galactopyranoside to provide a spectrophotometrically measurable change at 420 nm. In the assay, the analyte is labeled with an ED, and it competes with the analyte in the test specimen (if present) for the limited analyte-specific antibody. The ED on the antibody-bound analyte–ED conjugate will not assemble with the EA to form the active enzyme. Thus, the quantity of the analyte in the test specimen modulates the amount of β-galactosidase formed (Figure 5.7) and indirectly determines the signal change generated by the enzyme–substrate system incorporated in the assay process [30].

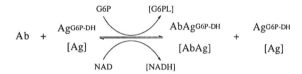

Figure 5.5. Schematic of the EMIT enzyme immunoassay. Substances inside brackets are present only if the specimen tested contains the analyte. Ag$^{G6P\text{-}DH}$: the enzyme-labeled form of the drug; Ag: the drug (analyte) to be measured; Ab: the antibody capable of binding the drug (the limiting factor in the reaction); G6P: glucose-6-phosphate; G6PL: 6-phosphogluconolactone; NAD: nicotinamide adenine dinucleotide; NADH: reduced form of NAD.

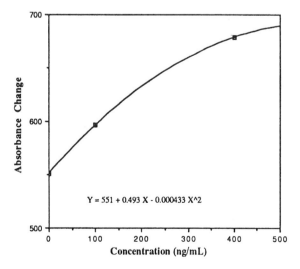

Figure 5.6. Example of a dose–response curve for EMIT d.a.u. fit with a second-order polynomial model. (Reproduced with permission from Ref. [63].)

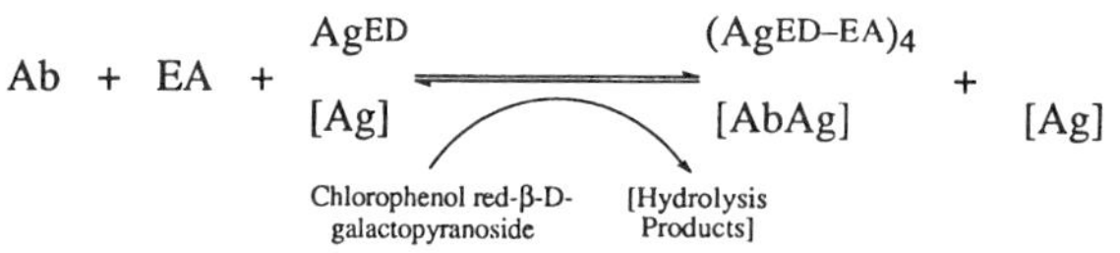

$$
\text{Ab} + \text{EA} + \begin{array}{c} \text{AgED} \\ [\text{Ag}] \end{array} \rightleftharpoons \begin{array}{c} (\text{AgED--EA})_4 \\ [\text{AbAg}] \end{array} + [\text{Ag}]
$$

Chlorophenol red-β-D-   [Hydrolysis
galactopyranoside   Products]

Figure 5.7. Schematic of the CEDIA enzyme immunosorbent assay. Substances inside brackets are present only if the specimen tested contains the analyte. Ag^{ED}: the labeled form of the drug; Ag: the drug (analyte) to be measured; Ab: the antibody capable of binding the drug (the limiting factor in the reaction); EA: a large polypeptide (enzyme acceptor); ED: a small polypeptide (enzyme donor). Both EA and ED are enzymatically inactive but spontaneously associate to form active tetramers.

EIA has also been applied to screening for fentanyl [31], cocaine [32], and other drugs [33] and metabolites using a heterogeneous approach called enzyme-linked immunosorbent assay (ELISA). In two applications [31, 32], a fixed amount of the analyte coated on the solid-phase medium competes with the analyte in the test sample to react with the first antibody added. The solid-phase medium is then washed, and an enzyme-labeled second antibody is added. The solid-phase medium is washed again, and an enzyme substrate is then added. The extent of the substrate reaction is determined by the amount of the second antibody-bound enzyme present (bound indirectly to the solid-phase medium). The amount of the solid-phase-bound, second antibody–enzyme complex is determined by the amount of the first antibody present (bound to the analyte coated on the solid-phase medium), which, in turn, is determined by the concentration of the analyte present in the test sample (Figure 5.8). Therefore, the extent of the substrate reaction indicates the concentration of the analyte in the test sample.

$$
\text{Ab}_1 + \begin{array}{c} \text{Ag-}| \\ [\text{Ag}] \end{array} \longrightarrow \begin{array}{c} \text{Ab}_1\text{Ag-}| \\ [\text{Ab}_1\text{Ag}] \end{array} + \begin{array}{c} \text{Ag-}| \\ [\text{Ag}] \end{array} \xrightarrow{\text{Washing}} \text{Ab}_1\text{Ag-}| + [\text{Ag-}|]
$$

1. Enzyme-Ab₂
2. Washing

Enzyme-Ab₂Ab₁Ag-|
+
[Ag-|]

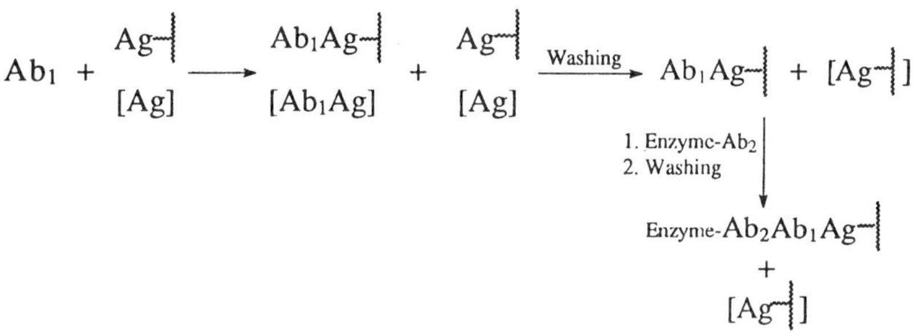

Figure 5.8. Schematic of the enzyme-linked immunosorbent assay. Substances inside brackets are present only if the specimen tested contains the analyte. Ag-|: drug coated on the solid-phase medium; Ag: the drug (analyte) to be measured; Ab_1: the antibody capable of binding the drug (the limiting factor in the reaction); Ab_2: the second antibody capable of binding Ab_1 to form aggregates. (Reproduced with permission from Ref. [63].)

In a third application [32], the antibody is immobilized on polystyrene beads. After exposure to the test sample and a fixed amount of enzyme-labeled antigen, the polystyrene beads are washed, and the enzyme substrate is then added. The substrate reaction, as determined by the amount of the bound enzyme-labeled antigen, is monitored by the absorbance change. Because the concentration of the analyte in the test sample determines how much of the enzyme will be bound (indirectly) to the solid phase, and available for catalyzing the substrate reaction, it can therefore be related to the absorbance change monitored.

Because the measured phenomenon is based on the activity of an enzyme capable of catalyzing multiple reactions for the conversion of an unlimited quantity of reaction products, EIA can, in theory, be very sensitive. In practice, however, the amplification resulting from the accumulation of the reaction products is more than canceled out by the relatively insensitive photometric method used for the detection and quantification of the reaction products [34].

Cross-Reactivities

Because EMIT kits are the most widely used immunoassay, their cross-reacting characteristics are frequently reported by users. The fundamental principle of EIA, relying on the enzyme activity to produce the monitored event, renders a decreased specificity and high level of background [34] that are commonly observed for EMIT. Because the enzyme activity is also modified by any nonspecific binding to the antigen–enzyme complex, the measured reaction product is responsive to this untargeted reaction with reduced specificity and a high noise level.

Some users have identified the exact cross-reacting compounds and quantitative cross-reactivity data, whereas others have reported only the observed phenomena. The former category of user-reported data (cross-reacting compound identified) is listed in Table 5.5 along with data provided in the manufacturers' reagent package inserts. Literature reports, in which the exact cross-reacting compounds are not identified, are included in the section "Interference" in this chapter.

Because the list of cross-reacting compounds can never be complete, lists of compounds with negative cross-reactivity [35] are very informative and should be checked when doubt arises.

Cross-reactivity data of the newly available CEDIA kits, as reported by the manufacturer, are summarized in Table 5.6.

Table 5.5. Compounds reported cross-reacting to Emit drug abuse urine assays.[a]

Manufacturer, assay name, and assay specifics	Cross-reacting compound[a]	Positive response concentration (ng/mL)[a]
Syva EMIT d.a.u. Amphetamine Class	d,l-Amphetamine	300
Calibrator and cutoff: d-amphetamine;	Mephentermine	400
300 ng/mL	Phentermine	400
Control range: 0–2,000 ng/mL	Tranylcypromine	500
Date: Jan. 1993	Isometheptene	500
	d-Methamphetamine	1,000
	d,l-Ephedrine	1,000
	Phenmetrazine	1,000
	Phenylpropanolamine	1,000
	Nylidrin	2,000
	Isoxsuprine	6,000
	d,l-Pseudoephedrine[b]	10,000
	Pseudoephedrine[b]	15,000
Syva EMIT d.a.u. Monoclonal	d-Amphetamine[c]	≤400
Amphetamine/Methamphetamine	3,4-Methylenedioxyamphetamine[d]	1,000
Calibrator and cutoff: d-methamphetamine;	d,l-Amphetamine	1,000
1,000 ng/mL	3,4-Methylenedioxymethamphetamine[d]	3,000
Control range: 0–3,000 ng/mL	l-Amphetamine	10,000
Date: Nov. 1989	l-Methamphetamine	12,000
	Phentermine[b]	300
	Chloroquine[b]	3,500
	Methoxyphenamine[b]	17,000
	Ranitidine[b,e]	62,000
	N-Acetylprocainamide[b]	215,000
	Procainamide[b]	855,000
	d-Ephedrine[f]	100,000
	l-Ephedrine[f]	1,000,000
	d-Pseudoephedrine[f]	1,000,000
	l-Pseudoephedrine[f]	1,000,000

Table 5.5. (Continued.)

Manufacturer, assay name, and assay specifics	Cross-reacting compound[a]	Positive response concentration (ng/mL)[a]
Syva EMIT d.a.u. Monoclonal Amphetamine/Methamphetamine (Continued)	d,l-Norephedrine[f]	1,000,000
	d,l-Norpseudoephedrine[f]	1,000,000
	Phentermine[g]	10,000
	Chlorpromazine[g]	200,000
	Chloroquine[g]	200,000
	l-Ephedrine[g]	200,000
	N-Acetyl procainamide[g]	200,000
	Phenmetrazine[g]	200,000
	Phenylpropanolamine[g]	200,000
	Quinicrine[g]	200,000
	Ranitidine[g]	200,000
	Tyramine[g]	200,000
	Tolmetin[h]	—
Syva EMIT II Monoclonal Amphetamine/Methamphetamine	d-Amphetamine	1,000
Calibrator and cutoff: d-methamphetamine; 1,000 ng/mL	d,l-Methamphetamine	1,200
Control range: 0–3,000 ng/mL	d,l-Amphetamine	1,500
Date: Nov. 1993	Benzphetamine[i]	1,500
	l-Methamphetamine[j]	2,000
	Phentermine	2,000
	3,4-Methylenedioxyamphetamine	3,000
	l-Amphetamine	6,000
	3,4-Methylenedioxymethamphetamine	6,000
	Phenmetrazine	6,000
	Mephentermine	10,000
	Methoxyphenamine	25,000
	Fenfluramine	36,000
	Tranylcypromine	65,000
	Propranolol	160,000
	l-Ephedrine	180,000
	Tyramine	200,000
	Phenylpropanolamine	290,000
	Chloroquine	380,000
	Norpseudoephedrine	380,000
	Quinacrine	400,000
	Pseudoephedrine	670,000
	Selegiline[i]	—
	Benzphetamine[g]	10,000
	Phentermine[g]	10,000
	l-Ephedrine[g]	200,000
	Mephentermine[g]	200,000
	Phenmetrazine[g]	200,000
	Bupropion[k]	—
Syva EMIT d.a.u. Barbiturate	Butalbital	150
Calibrator and cutoff: secobarbital; 200 ng/mL	Aprobarbital	180
Control range: 0–1,000 ng/mL	Talbutal	200
Date: May 1993	Cyclopentobarbital	200
	Alphenal	300
	Amobarbital	300
	Butabarbital	300
	Pentobarbital	300
	5-Ethyl-5-(4-hydrophenyl)barbituric acid	600

Table 5.5. (Continued.)

Manufacturer, assay name, and assay specifics	Cross-reacting compound[a]	Positive response concentration (ng/mL)[a]
Syva EMIT d.a.u. Barbiturate	Phenobarbital	700
(Continued)	Barbital	1,000
	Thiopental	10,000
	Mephobarbital[b]	750
	Heptabarbital[b]	3,900
	Butyvinal[b]	5,000
	Allobarbital[b]	10,000
	Hexobarbital[b]	100,000
	Methohexital[b]	367,000
	p-Hydroxyphenytoin[b]	460,000
Syva EMIT II Barbiturate	Talbutal	150
Calibrator and cutoff: secobarbital; 200 ng/mL	Aprobarbital	200
Control range: 0–1,000 ng/mL	Cyclopentobarbital	200
Date: Jan. 1993	Butabarbital	200
	Pentobarbital	200
	Alphenal	400
	Amobarbital	450
	Barbital	1,500
	5-Ethyl-5-(4-hydrophenyl)barbituric acid	1,500
	Phenobarbital	1,500
	Thiopental	12,000
Syva EMIT d.a.u. Benzodiazepine	Clonazepam	2,000
Calibrator and cutoff: oxazepam; 300 ng/mL	Demoxepam	2,000
Control range: 0–1,000 ng/mL	Desalkylflurazepam	2,000
Date: Nov. 1987	N-Desmethyldiazepam	2,000
	Diazepam	2,000
	Flunitrazepam	2,000
	Lurazepam	2,000
	Nitrazepam	2,000
	Chlordiazepoxide	3,000
	Lorazepam	3,000
	Alprazolam[b]	100
	Prazepam[b]	100
	Medazepam[b]	145
	Halazepam[b]	155
	Triazolam[b]	170
	Clobazam[b]	230
	Temazepam[b]	260
	Lormetrazepam[b]	310
	N-1-Desalkylflurazepam[b]	322
	Bromazepam[b]	380
	Camazepam[b]	2,400
	Tetrazolam[b]	2,700
	Oxazolam[b]	4,000
	Clotiazepam[b]	4,500
	Ketazolam[b]	4,500
	Clorazepate[b]	6,150
	Midazolam[b]	185,000
	Nordiazepam[l]	500
	3-Hydroxydesalkylflurazepam[l]	1,000
	Hydroxyethylflurazepam[l]	75,000

Table 5.5. (Continued.)

Manufacturer, assay name, and assay specifics	Cross-reacting compound[a]	Positive response concentration (ng/mL)[a]
Syva EMIT II Benzodiazepine	Alprazolam	100
Calibrator and cutoff: oxazepam; 200 ng/mL	N-Desmethyldiazepam	100
Control range: 0–1,000 ng/mL	Midazolam	100
Date: Jan. 1993	Flurazepam	110
	Prazepam	110
	Diazepam	110
	Triazolam	120
	α-Hydroxyalprazolam	120
	α-Hydroxytriazolam	120
	1-N-Hydroxyethylflurazepam	130
	Medazepam	130
	Halazepam	140
	Tertazepam	150
	Clobazam	180
	Temazepam	190
	Clorazepate	200
	Nitrazepam	200
	Ketazolam	210
	Flunitrazepam	220
	N-Desalkylflurazepam	230
	Lormetazepam	230
	Clonazepam	250
	Bromazepam	340
	Clotiazepam	400
	Demoxepam	500
	Norchlordiazepoxide	670
	Lorazepam	750
	Chlordiazepoxide	800
	Oxaprozin [m]	—
Syva EMIT d.a.u. Cannabinoid 100 ng	8 β-Hydroxy-Δ^9-THC	200
Calibrator and cutoff: 9-THC-COOH;	11-Hydroxy-Δ^9-THC	200
100 ng/mL	11-Hydroxy-Δ^8-THC	200
Control range: 0–200 ng/mL	8 β-11-Dihydroxy-Δ^9-THC	300
Date: Oct. 1992		
Syva EMIT II Cannabinoid 100 ng	8 β-Hydroxy-Δ^9-THC	200
Calibrator and cutoff: 9-THC-COOH;	11-Hydroxy-Δ^9-THC	200
100 ng/mL	11-Hydroxy-Δ^8-THC	200
Control range: 0–200 ng/mL	8 β-11-Dihydroxy-Δ^9-THC	200
Date: Feb. 1993		
Syva EMIT d.a.u. Cocaine Metabolite	m-Hydroxybenzoylecgonine[n]	300
Calibrator and cutoff: benzoylecgonine;		
300 ng/mL		
Control range: 0–3,000 ng/mL		
Date: Jan. 1992		
Syva EMIT II Cocaine Metabolite	(None listed)	
Calibrator and cutoff: benzoylecgonine;		
300 ng/mL		
Control range: 0–3,000 ng/mL		
Date: July 1993		

Table 5.5. (Continued.)

Manufacturer, assay name, and assay specifics	Cross-reacting compound[a]	Positive response concentration (ng/mL)[a]
Syva EMIT d.a.u. Methadone 　Calibrator and cutoff: methadone; 　　300 ng/mL 　Control range: 0–1,000 ng/mL 　Date: Jan. 1992	(None listed)	
Syva EMIT II Methadone 　Calibrator and cutoff: methadone; 　　300 ng/mL 　Control range: 0–1,000 ng/mL 　Date: Jan. 1993	(None listed)	
Syva EMIT d.a.u. Methaqualone 　Calibrator and cutoff: methaqualone; 　　300 ng/mL 　Control range: 0–1,500 ng/mL 　Date: Oct. 1992	Mecloqualone 3′-Hydroxy methaqualone 4′-Hydroxy methaqualone 2′-Hydroxymethylmethaqualone Mecloqualone[o] 2-Methyl-3-*o*-(4′-hydroxy-2′-methylphenyl)-4(3*H*)-quinazolinone[j] 2-Methyl-3-*o*-(3′-hydroxy-2′-methylphenyl)-4(3*H*)-quinazolinone[j] 2-Methyl-3-*o*-(2′-hydroxy-2′-methylphenyl)-4(3*H*)-quinazolinone[j] 2-Methyl-3-*o*-tolyl-3-hydroxy-4(3*H*)-quinazolinone[j] 2-Methyl-3-*o*-tolyl-6-hydroxy-4(3*H*)-quinazolinone[j]	1,000 1,000 1,000 5,000 300 400 500 1,200 1,500 30,000
Syva EMIT II Methaqualone 　Calibrator and cutoff: methaqualone; 　　300 ng/mL 　Control range: 0–1,500 ng/mL 　Date: Jan. 1993	Mecloqualone 4′-Hydroxymethaqualone 3′-Hydroxymethaqualone 2′-Hydroxymethyl methaqualone	300 300 550 2,000
Syva EMIT d.a.u. Opiate 　Calibrator and cutoff: morphine; 　　300 ng/mL 　Control range: 0–1,000 ng/mL 　Date: June 1992	Codeine Hydrocodone Hydromorphone Levorphanol Morphine-3-glucuronide Oxycodone Dihydrocodeine[b] Monoacetylmorphine[b] Levallorphan[b] Norlevorphanol[b] Oxymorphone HCl[b] Dihydrocodeine[p] Dihydromorphine[p] Levorphanol[p] Norcodeine[p]	1,000 1,000 3,000 3,000 3,000 50,000 260 460 1,000 23,000 82,000 2,2 10,000 1,6 10,000 1,4 10,000 1,0 10,000
Syva EMIT II Opiate 　Calibrator and cutoff: morphine; 　　300 ng/mL 　Control range: 0–1,000 ng/mL 　Date: Jan. 1993	Codeine Hydrocodone Hydromorphone Levorphanol Morphine-3-glucuronide Oxycodone	1,000 1,000 3,000 3,000 3,000 50,000

Table 5.5. (Continued.)

Manufacturer, assay name, and assay specifics	Cross-reacting compound[a]	Positive response concentration (ng/mL)[a]
Syva EMIT d.a.u. Phencyclidine	1-(1-Phenylcyclohexyl)morpholine (PCM)	1,000
Calibrator and cutoff: phencyclidine;	1-(1-Phenylcyclohexyl)pyrrolidine (PCPy)	1,000
75 ng/mL	1-[1-(2-Thienyl)cyclohexyl]piperidine (TCP)	1,000
Control range: 0–400 ng/mL	1-[1-(2-Thienyl)cyclohexyl]pyrrolidine (TCPy)	1,000
Date: May 1988	4-Phenyl-4-piperidinocyclohexanol	2,000
	N,N-Diethyl-1-phenylcyclohexylamine (PCDE)	3,000
	1-(4-Hydroxypiperidino)phenylcyclohexane	3,000
	1-[1-(2-Thienyl) cyclohexyl]morpholine (TCM)	5,000
Syva EMIT II Phencyclidine	1-(1-Phenylcyclohexyl)morpholine (PCM)	1,000
Calibrator and cutoff: phencyclidine; 25 ng/mL	1-(1-Phenylcyclohexyl)pyrrolidine (PCPy)	1,000
Control range: 0–75 ng/mL	1-[1-(2-Thienyl)cyclohexyl]piperidine (TCP)	1,000
Date: Dec. 1992	1-[1-(2-Thienyl)cyclohexyl]pyrrolidine (TCPy)	1,000
	4-Phenyl-4-piperidinocyclohexanol	2,000
	N,N-Diethyl-1-phenylcyclohexylamine (PCDE)	3,000
	1-(4-Hydroxypiperidino)phenylcyclohexane	3,000
	1-[1-(2-Thienyl) cyclohexyl]morpholine (TCM)	5,000
Syva EMIT d.a.u. Propoxyphene	Norpropoxyphene	10,000
Calibrator and cutoff: propoxyphene; 300 ng/mL		
Control range: 0–1,000 ng/mL		
Date: Sept. 1992		
Syva EMIT II Propoxyphene	Norpropoxyphene	4,200
Calibrator and cutoff: propoxyphene; 300 ng/mL		
Control range: 0–1,000 ng/mL		
Date: Dec. 1992		

[a] With the exceptions of data that are footnoted, data listed in this table are taken from the respective reagent package inserts as specified in the first column of the table. When present at the concentration level listed in the "Positive response concentration" column, the cross-reacting compounds were reported to show an equal or greater response than that observed for the calibrator at the cutoff concentration.

[b] These compounds are listed in *Syva EMIT Drug Abuse Urine Assays Cross-Reactivity List* [35] as cross-reacting compounds and show positive results if present at concentrations equal to or higher than the concentrations listed in the "Positive response concentration" column.

[c] Poklis et al. [36] also reported a similar cross-reactivity of *d*-amphetamine.

[d] Poklis et al. [37] reported positive results with approximately 800 ng/mL and 3,000 ng/mL of methylenedioxyamphetamine and methylenedioxymethamphetamine, respectively. The $S(+)$-isomer of the former compound is approximately two to three times more responsive than the $R(-)$-isomer, while the $S(+)$-isomer of the latter compound is approximately 15 to 20 times more responsive than the $R(-)$-isomer.

[e] Poklis et al. [38] reported that approximately 100,000 ng/mL of ranitidine generated an absorbance change equivalent to 1,000 ng/mL of *d*-methamphetamine. This study also reported that ranitidine did not interfere with the EMIT d.a.u. polyclonal amphetamine assay.

[f] Data are taken from Ref. [8].

[g] Data are taken from Ref. [39]

[h] Data are taken from Ref. [40].

[i] Amphetamine and methamphetamine are metabolites of these drugs.

[j] Urine samples (from subjects exposed to Vicks inhaler) containing 1,390, 1,290, and 740 ng/mL of *l*-methamphetamine (determined by GC–MS) did not test positive by this reagent [41].

[k] Bupropion and its three metabolites (threo and erythro amino alcohol metabolites and a morpholinol metabolite) were all cross-reactive [42].

[l] Data are taken from Ref. [43].

[m] Data are taken from Ref. [44].

[n] Steele et al. [45] reported a cross-reactivity ranging from 0.63% to 1.10% in the 50- to 600 ng/mL concentration range studied.

[o] Data are taken from Ref. [46].

[p] Data are taken from Ref. [14].

Table 5.6. Compounds reported cross-reacting to CEDIA drug abuse urine assays.[a]

Manufacturer, assay name, and assay specifics	Positive response cross-reacting compound[a]	% cross-reactivity[a]
CEDIA Amphetamines	d-Amphetamine	101
Calibrator and cutoff: d-methamphetamine;	3,4-Methylenedioxymethamphetamine	70
1,000 ng/mL	d,l-Methamphetamine	67
Control range: 0–5,000 ng/mL	d,l-Amphetamine	52
Date: Jan. 1993	3,4-Methylenedioxyamphetamine	2.2
	Phentermine	1.9
CEDIA Barbiturate	Talbutal	163
Calibrator and cutoff: secobarbital;	Cyclopentobarbital	115
200 (or 300) ng/mL	Amobarbital	108
Control range: 0–3,000 ng/mL	Butalbital	97
Date: 1994	Phenobarbital	88
	Butabarbital	85
	Aprobarbital	75
	Pentobarbital	66
	Barbital	18
CEDIA Benzodiazepine	Diazepam	311
Calibrator and cutoff: nitrazepam;	Estazolam	306
200 (or 300) ng/mL	Nordiazepam	245
Control range: 0–3,000 ng/mL	Clobazam	239
Date: 1994	Alprazolam	209
	Flurazepam	204
	Prazepam	194
	Desalkylflurazepam	161
	α-OH-Alprazolam	150
	Temazepam	139
	Medazepam	122
	Halazepam	109
	Flunitrazepam	107
	Nitrazepam	96
	Demoxepam	70
	Clorazepate	68
	Bromazepam	54
	Chlordiazepoxide	21
	Triazolam	16
	Lormetazepam	11
	α-OH-Triazolam	11
	Delorazepam	9.5
	Lorazepam	3.2
	Clonazepam	1.7
CEDIA Cocaine	Cocaethylene	60
Calibrator and cutoff: benzoylecgonine; 300 ng/mL	Cocaine	60
Control range: 0–5,000 ng/mL	Ecgonine methyl ester	3
Date: 1994		
CEDIA Multi-Level THC	11-nor-Δ^8-THC-COOH	106
Calibrator and cutoff: 9-THC-COOH;	11-OH-Δ^9-THC	56.8
50 (or 25 or 100) ng/mL	1-Δ^9-THC-glucuronide	43.3
Control range: 0–150 ng/mL	Δ^9-THC	10.7
Date: 1994	8,11-di-OH-Δ^9-THC	6.4
	Cannabinol	5.5
	8 β-OH-Δ^9-THC	4.3

Table 5.6. (Continued.)

Manufacturer, assay name, and assay specifics	Positive response cross-reacting compound[a]	% cross-reactivity[a]
CEDIA Opiate	Codeine	125
Calibrator and cutoff: morphine;	Morphine-3-glucuronide	94
300 ng/mL	6-Monoacetylmorphine	81
Control range: 0–2,000 ng/mL	Diacetylmorphine	67
Date: 1994	Hydromorphone	66
	Dihydrocodeine	62
	Hydrocodone	59
	Morphine-6-glucuronide	57
	Oxycodone	3.1
	Oxymorphone	1.9
CEDIA Phencyclidine	1-(1-Phenylcyclohexyl)-4-hydroxypiperidine (PCHP)	106
Calibrator and cutoff: phencyclidine;	Phenylcyclohexylpyrrolidine (PHP)	68
25 ng/mL	Phenylcyclohexamine (PCE)	37
Control range: 0–150 ng/mL	1-[1-(2-Thienyl)cyclohexyl]piperidine (TCP)	31
Date: 1994	4-Phenyl-4-piperidinocyclohexanol (4-OH-PCP)	2.5

[a] All data listed in this table are taken from the respective reagent package inserts as specified in the first column of the table. Only those compounds that were reported to show ≥1% cross-reactivity are listed. Compounds are listed in descending order of their reported cross-reactivities.

Application and Sample Characteristics

Urine samples are typically tested directly. Vitreous humor can be satisfactorily tested following centrifugation [47] or simple dilution with water [48]. Blood samples are typically pretreated prior to testing. Pretreatment procedures range from one-step addition of chloroform [49], methanol [49–53], acetone [54], or N,N-dimethylformamide [55] to full-fledged extraction and cleanup procedures [56–59], which presumably generate extracts that are suitable for chromatographic methods of analysis. Although the latter approach may generate excellent test data, the extensive pretreatment procedure precludes it from being adopted in a high-volume testing environment. A compromise procedure involving the use of acetone, vortexing, centrifugation, evaporation, and reconstitution with a 1:1 mixture of EMIT buffer and methanol was reported [60] to yield good results on postmortem blood specimens.

EIA is not intended for the quantitative determination of a single component. However, dose–response curves can be constructed using data generated from carefully prepared standards [47, 61–63]. A typical concentration vs. absorbance change plot is shown in Figure 5.6 [63]. This plot was obtained using data from a set of standards that are fitted with polynomial equations of various degrees. The resulting best-fit equation is then used to estimate the apparent concentration of the analyte in test samples.

The correlation of the chemical structural features with the assay responses has also been explored [64]. Specifically, EMITtox Serum Tricyclic Antidepressant Assay was used for the study of methanolic extract of blood samples spiked with 12 tricyclic antidepressants (Chart 5.1). The relative importance of different structural features was determined by comparing pairs of drugs that differed structurally from each other by only one feature. Five separate molecular features were compared in this manner: the endocyclic double bond of the alicyclic (middle seven-membered) ring, the ring oxygen, the branched side chain, the secondary amino side chain, and chlorine on the aromatic ring. The relative importance of four additional structural features was indirectly compared using the information obtained from the five structural features directly obtained. It was concluded that the relationship between the molecular structure and the EMIT assay response on a molecular basis is not only quantitative but also additive.

Fluorescence Polarization Immunoassay

Basic Methodology

In parallel with the RIA and EIA technologies, the common test protocol of FPIA is also based on the competitive binding of labeled antigen and unlabeled antigen, in proportion to their concentrations, to a limited quantity of antibody in the reaction mixture. A fluorophore is used as the label. The antibody-bound fluorophore emits at the same plane as the exciting polarized radiation, while the fluorophore on the free antigen emits at a different plane. The extent to which the labeled antigen is bound to the antibody can be monitored through the measurement of the extent of polarization. Thus, the separation step as shown for the heterogeneous RIA (second part of Figure 5.1) is not needed.

The fast dissociation rate constant and the stability of the antibody–fluorescein-labeled drug complex have a significant impact on the test procedure [65–69]. With the fast dissociation rate constant, the antibody and the fluorescein-labeled drug can be premixed as a single reagent. Thus, the test procedure will only involve the addition of the test sample to the premixed reagent, followed by an incubation period (to allow the displacement of a proportional amount of fluorescein-labeled drug by the analyte in the test sample) and polarization measurement. Because the premixed reagent has proven to be stable and the reaction equilibrium, once attained, remains stable, frequent calibration of the assay is unnecessary [67]. It should be noted, however, that degradation of the commercial reagent has been reported [70].

It has also been reported [65, 67] that the combination of several antisera and fluorescein-labeled antigens allows the detection of multiple drugs with a single reagent. However, because each antiserum may have different cross-reacting characteristics [67], the correlation of polarization readings with the identities and concentrations of individual analytes present may require a sophisticated experimental design.

Cross-Reactivity

Compared to absorption detection methods, measurement of the fluorescence polarization signal is less likely to be interfered with by the sample matrix. FPIA also has an advantage over fluorescence methods because any potential intrinsic fluorescence derived from the sample matrix is unlikely to cause detection interference in polarization measurements. It has been reported, however, that interference derived from fluorescence of the sample matrix may occur if the measuring device lacks adequate optical sophistication [28].

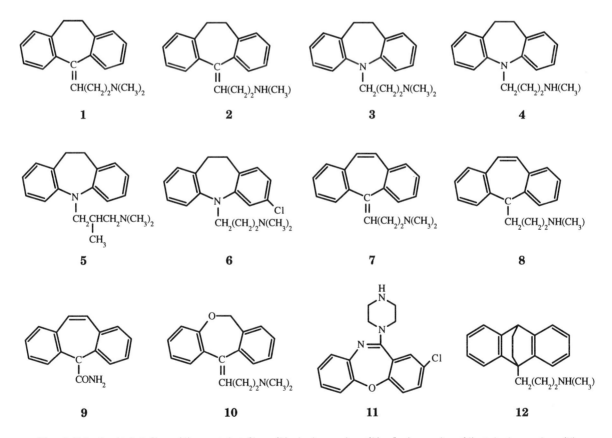

*Chart 5.1. Amitriptyline (**1**), nortriptyline (**2**), imipramine (**3**), desipramine (**4**), trimipramine (**5**), clomipramine (**6**), cyclobenzaprine (**7**), protriptyline (**8**), carbamazepine (**9**), doxepin (**10**), amoxapine (**11**), and maprotyline (**12**).*

Compounds that were reported to cross-react with Abbott TDx/ADx reagents are summarized in Table 5.7.

Application and Sample Characteristics

Problems that may arise when testing blood samples include:

1. The high blank intensity caused by high turbidity; and

2. Low optical intensity due to background chromophores that strongly attenuate the excitation beam [81].

It has been reported [82] that putrefactive bases may also interfere with the testing of amphetamine, as is often observed [50] in EIA-based methodologies.

In general, the pretreatment procedure required for FPIA testing of blood samples is not as extensive as those required for EIA-based methodologies. Satisfactory results have been reported for testing blood samples without any pretreatment [83], with a simple dilution of saline [81], or after the precipitation of proteins with acetone [54] or 10% trichloroacetic acid [84].

Table 5.7. Reported cross-reactivity of Abbott TDx/ADx assays.[a]

Manufacturer, assay name, and assay specifics	Compound	% cross-reactivity[a]	Concentration tested (ng/mL)[a]
Abbott TDx Amphetamine Class	d-Methamphetamine	260–56	100–100,000
Calibrator: d,l-amphetamine	d,l-Methamphetamine	230–198	100–10,000
Range: 0–6,000 ng/mL	Propylhexedrine	140–51	100–10,000
Date: 1993	d-Amphetamine	110–44	100–100,000
	l-Methamphetamine	92; 85	500; 1,000
	l-Amphetamine	51–40	150–3,000
	d,l-Ephedrine	46–18	500–10,000
	l-Ephedrine	39–4.2	1,000–100,000
	Cyclopentamine	36–28	1,000–10,000
	Mephentermine	35–20	1,000–10,000
	Labetalol	34; 23	1,000; 10,000
	3,4-Methylenedioxyethylamphetamine	30–2.5	5,000–1,000
	3,4-Methylenedioxymethamphetamine	28–3.1	1,000–100,000
	Phentermine	25–20	1,000; 10,000
	Benzphetamine	24–18	1,000–10,000
	Tranylcypromine	22–20	1,000–10,000
	Nylidrin	21–15	1,000–10,000
	Chlorphentermine	20–13	1,000–10,000
	Phenylpropanolamine	17–11	1,000–10,000
	Phenmetrazine	15–3.6	1,000–100,000
	l-Pseudoephedrine	12–9.6	1,000–10,000
	Benzathine	12–5.6	1,000–100,000
	Fenfluramine	12–4.0	1,000–100,000
	Isoxsuprine	12–4.0	1,000–100,000
	Isometheptene	12	1,000–10,000
	Tauaminoheptene	10–6.9	1,000–10,000
	d-Pseudoephedrine	5.9–4.5	10,000–100,000
	Phenethylamine	5.2; 5.5	10,000; 20,000
	Haloperidol	3.8; 1.8	10,000; 100,000
	Phenelzine	3.7; 4.7	10,000; 100,000
	l-Norpseudoephedrine	3.1–2.9	1,000–100,000
	Monoethylpropionaldehyde	1.0 3.0	10,000–100,000
	Aminopropylphenone	2.7–1.6	1,000–100,000

Table 5.7. (Continued.)

Manufacturer, assay name, and assay specifics	Compound	% cross-reactivity[a]	Concentration tested (ng/mL)[a]
Abbott TDx Amphetamine Class (Continued)	Methoxyphenamine	2.6–1.4	1,000–100,000
	N-Methylephedrine	1.7–0.9	10,000–100,000
	Dezocine	1.5	10,000
	Amantadine	1.3	10,000
	Fenfluramine[b]	14	10,000
	Phentermine[b]	5	10,000
	Phenmetrazine[b]	4	10,000
	Phenethylamine[b]	2	10,000
	3,4-Methylenedioxymethamphetamine[c]	118–18	150–10,000
	3,4-Methylenedioxyethylamphetamine[c]	47–12	150–10,000
	3,4-Methylenedioxyamphetamine[d]	465–503	1,000–10,000
	Ritodrine[e]	0.6–1.4	125,000–350,000
	d-Methamphetamine[f]	207–66	200–5,000
	d-Amphetamine[f]	116–53	200–10,000
	3,4-Methylenedioxyethylamphetamine[f]	73–4.5	200–100,000
	l-Methamphetamine[f]	70–116	200–2,000
	l-Amphetamine[f]	66–58	200–5,000
	3,4-Methylenedioxymethamphetamine[f]	47–7.4	200–10,000
	4-Hydromethamphetamine[f]	27–4.7	200–5,000
	3,4-Methylenedioxyamphetamine[f]	26–4.8	200–10,000
	Ephedrine[f]	26–3.9	1,000–100,000
	Phenylpropanolamine[f]	13–7.4	1,000–50,000
Abbott ADx Amphetamine/ Methamphetamine II Calibrator: d-amphetamine Range: 0–8,000 ng/mL Date: 1993	d,l-Amphetamine	80–217	150–3,000
	4-Chloroamphetamine	73–122	300–5,000
	d-Methamphetamine	100–60	150–8,000
	3,4-Methylenedioxyethylamphetamine	70–96	300–3,000
	3,4-Methylenedioxymethamphetamine	63–51	300–8,000
	d,l-Methamphetamine	57–43	150–8,000
	3,4-Methylenedioxy-N-ethylamphetamine	47–17	300–8,000
	l-Amphetamine	37–29	300–8,000
	Propylhexedrine	34; 19	1,000; 10,000
	p-Hydroxyamphetamine	27–31	1,000–10,000
	Fenfluramine	13–5.5	1,000–50,000
	Isometheptene	11–10	2,000–50,000
	Mephentermine	4.7; 5.3	10,000; 100,000
	l-Methamphetamine	4.3; 5.1	3,000; 8,000
	4-Methyl-2,5-dimethoxyamphetamine	4.7–2.4	3,000–100,000
	4-Ethyl-2,5-dimethoxyamphetamine	4.7–1.9	3,000–100,000
	Phenethylamine	2.4; 3.1	10,000; 100,000
	Methoxyphenamine	2.0; 2.4	10,000; 100,000
	Labetalol	0.6–1.9	25,000–150,000
	Phenmetrazine	1.0; 0.8	10,000; 100,000
	d,l-Amphetamine[g]	120–210	200–2,000
	3,4-Methylenedioxyamphetamine[g]	136–170	200-2,000
	d-Methamphetamine[g]	108–86	200–5,000
	3,4-Methylenedioxymethamphetamine[g]	92–104	200–5,000
	4-Hydromethamphetamine[g]	79–73	500–10,000
	3,4-Methylenedioxyethylamphetamine[g]	67–31	200–10,000
	l-Amphetamine[g]	61–66	200–10,000
	d,l-Methamphetamine[g]	58–66	200–10,000

Table 5.7. (Continued.)

Manufacturer, assay name, and assay specifics	Compound	% cross-reactivity[a]	Concentration tested (ng/mL)[a]
Abbott ADx Amphetamine/ Methamphetamine II (Continued)	2-Methoxyamphetamine[g]	27–35	500–10,000
	l-Methamphetamine[g,h]	7.2–10	1,000–10,000
	2,5-Dimethoxyamphetamine[g]	6.5–4.1	5,000–50,000
	Phenmetrazine	1.0; 0.8	10,000; 100,000
	Labetalol	0.6–1.9	25,000–150,000
	Phentermine[i]	Negative	10,000
	Mephentermine[i]	Negative	200,000
	Phenmetrazine[i]	Negative	200,000
	Phenylpropanolamine[i]	Negative	200,000
	Tyramine[i]	Negative	200,000
Abbott TDx Barbiturates Calibrator: secobarbital Control range: 0–2,000 ng/mL Date: 12/01/1987	Butobarbital	160–129	200–1,200
	Amobarbital	155–133	200–1,200
	Phenobarbital	105–155	105–1,200
	Cyclopentobarbital	145–115	200–1,200
	Alphenal	110–137	200–1,200
	Butalbital	100–101	200–2,000
	Pentobarbital	105–72	200–2,000
	Brallobarbital	90–91	200–2,000
	Talbutal	90–84	200–2,000
	Butabarbital	85–60	200–2,000
	5-Ethyl-5-(4-hydroxyphenyl)barbituric acid	60–54	200–2,000
	Aprobarbital	45–36	200–2,000
	Allobarbital	25–29	200–2,000
	p-Hydroxyphenytoin	3.4	10,000
	Glutethimide	1.8	10,000
	Phenytoin	1.0–1.0	1,000–100,000
	Barbital	1.0	2,000
	Glutethimide[j]	15	2,000
	Phenytoin[j]	10	2,000
	Primidone[j]	2	2,000
	Phenytoin[k]	—	1,800–22,000
	5-(p-Hydroxyphenyl)-5-phenylhydantoin[k]	—	12,300–934,000
Abbott TDx Barbiturates II U Calibrator: secobarbital Range: 0–2,000 ng/mL Date: 1992	Cyclopentobarbital	867	200
	Talbutal	266–250	200–700
	Butabarbital	245–236	200–1,200
	Butalbital	114–106	200–1,200
	Alphenal	108–82	200–2,000
	Brallobarbital	94–83	200–2,000
	Cyclobarbital	80–70	200–2,000
	Phenobarbital	71–51	200–2,000
	Pentobarbital	65–68	200–2,000
	Aprobarbital	62–65	200–2,000
	Metharbital	53–46	200–2,000
	Butobarbital	46–52	200–2,000
	Amobarbital	36–34	200–2,000
	Allobarbital	32–29	200–2,000
	Thiopental	12–7.0	700–2,000
	5-Ethyl-5-(4-hydroxyphenyl)barbituric acid	9.7–6.4	700–2,000
	Glutethimide	9.0; 4.8	1,000; 10,000
	Barbital	6.4; 6.0	1,200; 2,000

Table 5.7. (Continued.)

Manufacturer, assay name, and assay specifics	Compound	% cross-reactivity[a]	Concentration tested (ng/mL)[a]
Abbott TDx Benzodiazepines	Estazolam	132; 88	100; 1,000
Calibrator: nordiazepam	Diazepam	123–144	200–2,400
Range: 0–2,400 ng/mL	Prazepam	119–70	200–2,400
Date: 1993	Alprazolam	117–61	200–2,400
	Medazepam	99–47	200–2,400
	1-*N*-Hydroxyethylflurazepam	90–52	200–2,400
	Midazolam HCl	99–45	200–2,400
	Nimetazepam	90–19	100–10,000
	Nitrazepam	89–31	200–2,400
	Triazolam	83–23	200–2,400
	Oxazepam	76–36	200–2,400
	Flurazepam	75–27	200–2,400
	Temazepam	74–48	200–2,400
	Flunitrazepam	70–31	200–2,400
	Desalkylflurazepam	59–37	200–2,400
	Lorazepam	50–17	200–2,400
	Clonazepam	48–15	200–2,400
	Bromazepam	42–14	200–2,400
	Demoxepam	34–13	200–2,400
	Clobazam	27; 8.1	1,000; 10,000
	Chlordiazepoxide	23–6.7	200–2,400
	Norchlordiazepoxide	22–7.4	200–2,400
	Lormetazepam	18–0.9	1,000–100,000
	Oxaprozin	1.8	10,000
	Zomepirac	1.1	10,000
Abbott TDx Cannabinoids	11-Nor-Δ^8-THC-9-COOH	111–109	25–100
Calibrator: 9-THC-COOH	8 β-11-Hydroxy-Δ^9-THC	108–29	25–200
Range: 0–135 ng/mL	8 β-11-Dihydroxy-Δ^9-THC	107–29	25–200
Date: 1993	11-Hydroxy-Δ^9-THC	72–60	25–200
	Cannabinol	56–35	25–200
Abbott TDx Cocaine Metabolite	Cocaine	0.4; 0.8	10,000; 100,000
Calibrator: benzoylecgonine	*m*-Hydroxybenzoylecgonine[l]	150–13	50–1,000
Range: 0–5,000 ng/mL	Ecgonine methyl ester[m]	1.2	10,000
Date: 1986			
Abbott TDx Methadone	*l*-α-Methadol	56–20	250–4,000
Calibrator: methadone	*p*-Hydroxymethadone	48–26	250,000–4,000,000
Range: 0–4,000 ng/mL	*l*-α-Acetylmethadol	26–11	500–4,000
Date: 1989	*d*-β-Acetylmethadol	14; 7.8	1,000; 4,000
	l-β-Acetylmethadol	13; 6.5	1,000; 4,000
	d-α-Methadol	4.5	4,000,000
	l-α-Acetyl-*N*-normethadol	2.8	4,000,000
Abbott TDx Opiates	Codeine	120–114	50–500
Calibrator: morphine	Hydrocodone	120–47	50–1,000
Range: 0–1,000 ng/mL	Hydromorphone	114–37	50–1,000
Date: 1986	Dihydromorphine	108–47	50–1,000
	6-Monoacetylmorphine	96–45	50–1,000
	Levorphanol	79–7.9	100–10,000
	Ethylmorphine	77–95	200–1,000
	Dihydrocodeine	68–53	200–1,000

Table 5.7. (Continued.)

Manufacturer, assay name, and assay specifics	Compound	% cross-reactivity[a]	Concentration tested (ng/mL)[a]
Abbott TDx Opiates (Continued)	Diacetylmorphine	69–40	200–1,000
	Thebaine (dimethylmorphine)	63–6.9	100–10,000
	Morphine-3β-D-glucuronide	58–36	50–1,000
	Levallorphan	36–0.3	100–100,000
	Promethazine	35–0.1	100–100,000
	Oxycodone	24–0.5	200–100,000
	Oxymorphone	18–0.4	200–100,000
	Nalorphine	14–0.7	1,000–100,000
	N-Norcodeine	6.7–0.5	1,000–100,000
	Cyclazocine	4.1–0.8	1,000–10,000
	N-Normorphine	4.1–0.3	1,000–100,000
	Meperidine	3.6–0.1	1,000–250,000
	Alphaprodine	3.4–0.2	1,000–100,000
	Naloxone	3.3–0.2	1,000–100,000
	Naltrexone	2.7–0.1	1,000–100,000
	Doxepin	1.3–0.1	100–50,000
	Dihydrocodeine[n]	67	10,000
	Dihydromorphine[n]	46	10,000
	Levorphanol[n]	8.1	10,000
	Nalorphine[n]	3.3	10,000
	Norcodeine[n]	2.8	10,000
	Normorphine[n]	1.2	10,000
	Fenoprofen[o]	—	—
	Flurbiprofen[o]	—	—
	Indomethacin[o]	—	—
	Ketoprofen[o]	—	—
	Tolmetin[o]	—	—
Abbott TDx Phencyclidine II Calibrator: phencyclidine Range: 0–500 ng/mL Date: 1994	1-[1-(2-Thienyl)cyclohexyl]piperidine	>5	100
	4-Hydroxypiperidine phencyclidine	>0.5	1,000
	Dextromethorphan	Not specified	
Abbott TDx Propoxyphene Calibrator: propoxyphene Range: 0–1,500 ng/mL Date: 1991	N-Norpropoxyphene	80–30	200–1,500
	Norpropoxyphene[p]	93–29	100–1,500
TDx Tricyclic Antidepressants Calibrator: imipramine Range: 0–1,000 ng/mL Date: 1993	Opipramol	620–62	10–1,000
	Nortriptyline	97–81	100–500
	Imipramine N-oxide	89–92	75–1,000
	Fluphenazine	85–0.8	100–100,000
	Amitriptyline	80–91	100–500
	Desipramine	90–87	100–500
	Dothiepin	69–59	100–500
	Trimipramine	67–55	100–500
	Protriptyline	63–54	100–500
	Cyclobenzaprine	53; 43	300; 1,000
	Clomipramine	51–41	100–500
	Doxepin	42–32	100–500
	Cyprohcptadinc	30	1,000
	Prochlorperazine	30; 18	100; 1,000
	Perphenazine	28; 23	300; 1,000
	Nordoxepin	27	500

Table 5.7. (Continued.)

Manufacturer, assay name, and assay specifics	Compound	% cross-reactivity[a]	Concentration tested (ng/mL)[a]
TDx Tricyclic Antidepressants (Continued)	Chlorpromazine	20; 14	300; 1,000
	2-Hydroxydesipramine	18; 12	300; 1,000
	2-Hydroxyimipramine	16; 9.7	300; 1,000
	cis-10-Hydroxyamitriptyline	13	300
	Promethazine	13; 9.0	300; 1,000
	cis-10-Nortriptyline	8.7	300
	Maprotiline	8.2	1,000
	Mianserin	5.2	1,000
	Phenothiazine	4.3	10,000
	Thioridazine	2.8–0.4	1,000–100,000
	Pimozide	2.7–0.3	1,000–100,000
	Orphenadrine	2.6–0.5	1,000–100,000
	Amoxapine	2.0	1,000

[a] With the exceptions of data that are footnoted, data listed in this table are taken from the respective reagent package inserts as specified in the first column of the table. Only those compounds that were reported to show ≥1% cross-reactivity are listed. Compounds are listed in descending order of their reported cross-reactivities. Data listed in the "% cross-reactivity" and the "Concentration tested" columns may be separated by ";" or "–", meaning that two data values (for ";") or a range of data (for "–") were reported in the original literature.

[b] Data are taken from Ref. [71].

[c] Data are taken from Ref. [72].

[d] Data are taken from Ref. [73].

[e] Data are taken from Ref. [74].

[f] Data are taken from Ref. [75] using Amphetamine Class reagent.

[g] Data are taken from Ref. [75] using Amphetamine/Methamphetamine II reagent.

[h] Simonick and Watts [76] reported an 8% cross-reactivity.

[i] Data are taken from Ref. [39]. Positive results were observed with the levels of the compounds tested.

[j] Data are taken from Ref. [77].

[k] Data are taken from Ref. [78].

[l] Data are taken from Ref. [45].

[m] Data are taken from Ref. [79].

[n] Data are taken from Ref. [14].

[o] Data are taken from Ref. [44].

[p] Data are taken from Ref. [80].

Particle Immunoassay

Basic Methodology

In parallel with the use of radioactive isotopes, fluorescein, and enzymes, particles of appropriate size can also be used as labels [85] to serve as the basis for detecting whether a targeted antibody–antigen reaction has occurred. Thus, latex particles were used for the development of immunoassay test methodologies for morphine [86], barbiturates [87], and methamphetamine [88, 89].

The test protocol is based on the competitive binding of latex-particle-labeled drugs with the analyte (if present) in the test sample for a limited amount of antibody available (*see* the first part of Figure 5.1, where the asterisk represents latex particles in this case). For a negative sample, the antibody cross-links to sufficient latex-particle-labeled drug molecules to produce agglutination particles that are large enough for visual detection. A highly positive sample will result in a smooth milky appearance of the original reaction medium.

This technology has recently been commercialized by Roche Diagnostic Systems and marketed as Abuscreen Ontrak assay kits, which are convenient for on-site applications. Because visual inspection is the basis of detection, the differentiation of samples containing the analyte at or near the cutoff level will always be somewhat subjective. Although this line of products may be useful for field applications, an objective detecting mechanism and automation process are more effective for applications in a high-volume test environment.

Indeed, the same manufacturer has marketed a different line of products (Abuscreen Online), in which mechanisms for objective detection and automation are featured. Working under the same principle, this test methodology adopts a microparticle label and a photometric detection device [90]. On the basis of the competitive-binding principle used for all immunoassays addressed earlier, a negative sample results in the formation of a microparticle lattice, which blocks light transmission and increases absorbance. The presence of the analyte in a positive sample inhibits lattice formation in a degree proportional to its concentration. Thus, final absorbance after a given reaction time decreases proportionally to the analyte concentration in the specimen.

Because this is a relatively new product, independent literature data are generally lacking. The cross-reactivity data as provided in the product-package inserts for the currently available assay kits are shown in Table 5.8.

Interference

Interference can be broadly defined as the observation of a test result that does not provide the intended diagnostic finding reflecting the true status of the specimen. Interferences that have been widely studied and reported are the false positive responses (on the initial test) resulting from the presence of cross-reacting compounds listed in Tables 5.3 and 5.5–5.8. (These positive initial tests are then identified and eliminated by GC–MS procedures.)

Table 5.8. Reported cross-reactivity of Abuscreen OnLine assays.[a]

Manufacturer, assay name, and assay specifics	Cross-reacting compound[a]	% cross-reactivity[a]
Roche Abuscreen OnLine for Amphetamines	d,l-Amphetamine	51
Calibrator: d-amphetamine and d-methamphetamine	3,4-Methylenedioxyamphetamine	32
Control range: 0–2,000 ng/mL	p-Hydroxyamphetamine	14
Date: April 1992	l-Amphetamine	2
	β-Phenethylamine	2
Roche Abuscreen OnLine for Barbiturates	Cyclopentobarbital	95
Calibrator: secobarbital	Aprobarbital	68
Control range: 0–400 ng/mL	Allobarbital	61
Date: Feb. 1992	Butabarbital	41
	Butalbital	40
	Pentobarbital	35
	Phenobarbital	32
	Amobarbital	28
	p-Hydroxyphenobarbital	27
	Barbital	21
Roche Abuscreen OnLine for Benzodiazepines	Alprazolam	96
Calibrator: nordiazepam	(α-Hydroxyalprazolam)[b]	112
Control range: 0–200 ng/mL	(4-Hydroxyalprazolam)[b]	146
Date: Sept. 1992	Bromazepam	75
	Chlordiazepoxide	55
	(Desmethylchlordiazepoxide)[b]	60
	Clonazepam	56
	Clorazepate K+ salt	43
	Demoxepam	96
	Diazepam	105
	(Oxazepam)[b]	98
	(N-Methyloxazepam)[b]	95
	Flunitrazepam	52
	(Desmethylflunitrazepam)[b]	56
	(3-Hydroxyflunitrazepam)[b]	24

Table 5.8. (Continued.)

Manufacturer, assay name, and assay specifics	Cross-reacting compound[a]	% cross-reactivity[a]
Roche Abuscreen OnLine for Benzodiazepines (Continued)	Flurazepam	61
	(Desalkyflurazepam)[b]	49
	(Didesethylflurazepam)[b]	84
	(Hydroxyethylflurazepam)[b]	88
	Lorazepam	59
	Medazepam	40
	(Desmethylmedazepam)[b]	38
	Midazolam	96
	Nitrazepam	81
	(7-Aminonitrazepam)[b]	52
	Pinazepam	106
	Prazepam	84
	Triazolam	96
	(α-Hydroxytriazolam)[b]	98
	(4-Hydroxytriazolam)[b]	52
Roche Abuscreen OnLine for Cannabinoids	8 α-Hydroxy-Δ^9-THC	22
Calibrator: 9-THC-COOH	11-Hydroxy-Δ^9-THC	18
Control range: 0–150 ng/mL	Δ^9-THC	11
Date: Sept. 1992	8 β-11-Dihydroxy-Δ^9-THC	10
	11-Hydroxycannabinol	5
	Cannabinol	2
Roche Abuscreen OnLine for Cocaine Metabolite	Ecgonine	2.3
Calibrator: benzoylecgonine		
Control range: 0–600 ng/mL		
Date: Jan. 1991		
Roche Abuscreen OnLine for Methadone	Methadol	130
Calibrator: methadone	Hydroxymethadone	49
Control range: 0–600 ng/mL	L-α-Acetylmethadol HCl	45
Date: July 1993	Promethazine	2.5
Roche Abuscreen OnLine for Opiates	Codeine	199
Calibrator: morphine sulfate	Dihydromorphine	178
Control range: 0–600 ng/mL	6-Monoacetylmorphine	80
Date: Jan. 1991	Thebaine	79
	Hydrocodone	77
	Dihydromorphine	73
	Hydromorphone	73
	Morphine-3-glucuronide	62
	Ethylmorphine	39
	Oxycodone	6
	Meperidine	3
	N-Norcodeine	2
Roche Abuscreen® OnLine for Phencyclidine	Thienylcyclohexylpiperidine	64
Calibrator: phencyclidine		
Control range: 0–50 ng/mL		
Date: April 1992		

[a] Data listed in this table are taken from the respective reagent package inserts as specified in the first column of the table. Only those compounds that were reported to show ≥1% cross-reactivity are listed. Compounds are listed in descending order of their reported cross-reactivities.

[b] Compounds listed with indentation and inside parentheses are metabolites of the preceding drugs.

In addition to interferences caused by known cross-reacting compounds, the following sample conditions may generate a test result leading to an incorrect interpretation of the sample status:

1. The presence of the targeted analyte derived from sources other than the targeted drugs of abuse;

2. The presence of cross-reacting compounds with unknown structures;

3. Specimen conditions that cause nonspecific binding;

4. Specimen conditions that interfere with the assay's detection mechanism; and

5. The presence of adulterants that degrade the analytes or alter their interacting characteristics.

One area that has attracted much attention in the drug-testing communities is the response of various immunoassays toward the intentional addition of adulterants. Some of these adulterants may actually destroy the targeted drugs, thus rendering the specimen truly negative. Under this circumstance, immunoassays (and other test methodologies) are expected to respond negatively. However, other adulterants may just cause nonspecific binding or create interference in the detection mechanism. Immunoassays based on antibodies with desired specificities and a detection mechanism that is robust toward interfering conditions are most desirable.

Targeted Analytes Derived from Unintended Exposure, Food Consumption, and Licit Medication

Some of the analytes targeted as indicators of drug abuse may be derived from unintended exposure, food consumption, or licit medication. *Low* quantities of marijuana and cocaine-related metabolites have been detected in urine samples from individuals who were subjected to passive inhalation [91] or skin absorption [92]. A recent RIA analysis of hair samples collected from undercover narcotics officers revealed low-level (<5 ng/10 mg), presumably external, cocaine contamination (removable by isopropyl alcohol and phosphate buffer washing) [93].

It is well known [94] that morphine and codeine may be observed in urine samples collected from individuals consuming food items with poppy seeds or prescriptions containing morphine or codeine. Methamphetamine detection may also be caused by using Vicks nasal inhaler [95] and other medications. Methamphetamine and amphetamine have been reported as the metabolites of a substantial number of licit drugs. A list recently compiled [96, 97] by a drug urinalysis expert included amphetaminil, benzphetamine, clobenzorex [98], deprenyl, dimethylamphetamine, ethylamphetamine, famprofazone, fencamine, fenethylline, fenproporex, Furfenorex, mefenorex, prenylamine, and mesocarb. Therefore, interpretation of test results must always be conducted carefully by those with a thorough knowledge of the subject.

Unknown Cross-Reacting Compounds

It has been reported that unknown metabolites of chlorpromazine [99], brompheniramine [99], and labetalol [100] have caused EMIT d.a.u. Monoclonal Amphetamine/Methamphetamine Assay to generate false positive results. Metabolites, not the parent drugs, are believed to be the responsible cross-reacting compounds because:

1. These drugs were prescribed for the urine specimen donors.

2. These parent drugs were present in the urine specimens.

3. Studies on control samples with various concentrations of the parent drugs *alone* failed to generate a positive result.

Because reference metabolites of these drugs are not available, the exact cross-reacting metabolites cannot be identified.

Urine specimens from patients using pipothiazine or fluspirilene prescriptions also generated false positive results when tested by the EMIT d.a.u. Monoclonal Amphetamine/Methamphetamine Assay [101]. Because no study was conducted on the parent drugs alone, it is not known whether the parent drugs or their metabolites caused the false results.

Other interferences reported include:

1. False EMIT positives for amphetamine by benzathine [102];
2. False EMIT positives for cannabinoids by an acute dose of ibuprofen and a chronic dose of naproxyn [103]; and
3. False TDx positives for barbiturates by chronic doses of ibuprofen and naproxyn [103].

The false positives from the studies of cannabinoids and barbiturates [103] were observed only in a small fraction of specimens studied—the vast majority of specimens did not generate false positive results.

False EMIT negatives of cannabinoid assays [104, 105] and other drugs [106] have also been reported. In one study, an alarming six false negatives (out of 41 patient samples tested) were reported [104]. Improved performance has been reported [107] with the new calibration formulation that does not use a surfactant and uses 9-THC-COOH (instead of 11-nor-Δ^8-THC-COOH) as the calibrator.

Nonspecific Binding

The causes for many observed false negative results are often unknown. On the basis of the observation that the average absorbance change of completely negative postmortem urine samples is lower than that generated by samples collected from healthy persons, it has been postulated [108] that some inhibitors to the EMIT reactions may be present in the postmortem urine samples. It is possible that the observed lower absorbance changes are due to the presence of nonspecific interacting materials that cause a higher initial absorbance value. Indeed, it has been reported [109] that many postmortem urine specimens had values of absorbance change lower than those produced by the negative calibrators.

In an RIA study [26], nonextracted negative blood specimens exhibit a significant degree of nonspecific binding, and they therefore yield lower B/B_0 values (left section in Figure 5.9) as compared to negative reference urine. When both the blood and reference urine specimens are extracted, a suitable distribution around the reference is observed (right section in Figure 5.9).

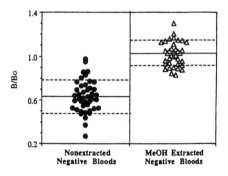

Figure 5.9. The distribution of B/B_0 for nonextracted negative blood samples referenced against the manufacturer's supplied urine-based negative (left), and negative methanol-extracted blood samples referenced against methanol-extracted blank bovine blood (right). (Reproduced with permission from Ref. [26].)

Detection Mechanism

EIA also suffers a potential spectrometric interference caused by substances that are present in the sample. Here are a few reported examples in which spectrometric detection has been identified as the source of interference:

1. *p*-Nitrophenol (a metabolite of methyl parathion) [110] and tolmetin (a nonsteroidal anti-inflammatory drug) and its metabolites [111] can absorb strongly in the 340-nm region at pH 8.0 and thus cause interference.

2. The presence of metronidazole [112] or mefenamic acid [113] causes excessively high initial absorbance values, thus preventing the assessment of EMIT test data.

3. Salicyluric acid (the principal metabolite of aspirin) interferes with the EMIT test methodology by reducing the molar absorptivity of NADH at 340 nm [114].

The availability of excess reagent antibody has also been attributed [115] as the cause of EMIT false negative test results for samples containing benzoylecgonine. It was reasoned that when excess antibody is added, the amount of enzyme-labeled drug bound by the antibody is increased. This increase results in a greater amount of enzyme being inhibited by the antibody. The resulting decreased signal (decreased conversion of NAD to NADH) would decrease the sensitivity of the EMIT assay near the threshold concentration. When the total antibody concentration approaches the total amount of drug (the sum of enzyme-labeled and non-enzyme-labeled species), the amount of free enzyme-labeled drug decreases toward zero, and the EMIT absorbance signal is markedly reduced, causing the observed false negative result. The use of a "high tech" adulterant that may tamper with the underlying assay detection mechanism is intellectually challenging and can only be used by those who have ready access to the specific antibody.

Adulterants

Many common accessible adulterants have been found to affect the responses of common immunoassays. The effects of adulterants vary with the drug categories tested and the immunoassay methodologies used. Adulterants that were reported to cause significant interference have been reviewed recently [116]. Information included in this review and newer data are summarized in Table 5.9.

Table 5.9. Effects of adulterants on immunoassays.

Substance	Concentration	Method	Drug category[a]	Sample[b]	Effect[c]	Ref.
Acetic acid	0.41 M	EMIT	OPI	+	– – – –	[117]
		RIA	OPI	+	–	[117]
		OnLine	OPI	+	–	[117]
Ammonia	5%, 10%	RIA	COC	–	– –	[118]
	5%, 10%, 14%	RIA	THC	+ / –	+++	[118, 119]
	14%	RIA	PCP	+ / –	+++ / ++	[119]
	10%	FPIA	BAR	+ / –	+++ / ++	[120]
	14%	FPIA	PCP	+	– – –	[119]
	14%	FPIA	THC	+ / –	+++ / +	[119]
L-Ascorbic acid	10%	RIA	AMP, OPI, THC	+	–	[118]
	10%	RIA	THC	–	–	[118]
	10%	FPIA	THC	+	–	[119]
Bleach, liquid[d]	12 μL/mL	EMIT	THC	+	– – – –	[121]
	23 μL/mL	EMIT	AMP, BAR, OPI	+	– – – –	[121]
	42 μL/mL	EMIT	COC	+	– – – –	[121]
	125 μL/mL	EMIT	BEN	+	– – – –	[121]
	10%	RIA	AMP	+	– – – –	[118]
	10%, 5%	RIA	OPI , THC	+	– – – –	[118]
	50%	RIA	PCP	+	– – – –	[119]
	5%	RIA	THC	–	++	[119]
	50%	RIA	THC	+ / –	++++	[119]
	5%, 50%, 10%	FPIA	THC	+	– – – –[e]	[119, 120]
	10%	FPIA	OPI	+	– – – –	[120]
	2.5% (v/v)	EMIT	OPI	+	–	[117]
		RIA	OPI	+	–	[117]
		OnLine	OPI	+	–	[117]

Table 5.9. (Continued.)

Substance	Concentration	Method	Drug category[a]	Sample[b]	Effect[c]	Ref.
Bleach, liquid[d]	1 drop/10 mL	EMIT	THC	+	−[f]	[122]
(Continued)	10.5 μL/mL	EMIT	AMP, BEN, OPI, PCP, THC	+	− − − −	[123]
	10.5 μL/mL	RIA	AMP, OPI, PCP	+	− − − −	[123]
	10.5 μL/mL	RIA	BAR	+	++	[123]
	10.5 μL/mL	FPIA	AMP, OPI, PCP, THC	+	− − − −	[123]
	10.5 μL/mL	FPIA	BEN	−	+++	[123]
Blood	1 drop/10 mL	EMIT	THC	+	−[f]	[122]
	10%	FPIA	THC	+ / −	− −	[120]
Detergent, ionic	10%	RIA	COC	+	− −	[118]
	10%, 5%	RIA	THC	+ / −	+	[118]
	10%	FPIA	BAR	+ / −	+	[120]
	10%	FPIA	THC	−	+	[120]
Detergent, liquid	1 drop/10 mL	EMIT	THC	+	−[f]	[122]
Drano (NaOH +	12 μL/mL	EMIT	THC	+	− − − −	[121]
NaHClO$_4$)	23 μL/mL	EMIT	AMP, BAR, OPI	+	− − − −	[121]
	42 μL/mL	EMIT	COC	+	− − − −	[121]
	125 μL/mL	EMIT	BEN	+	− − − −	[121]
	10%	RIA	AMP, BAR, COC, OPI, PCP, THC	+ / −	++++	[118]
	1%	RIA	COC	+	− − − −[e]	[118]
	10%	FPIA	COC	+	− − − −	[120]
	10%	FPIA	PCP	+	− − −	[120]
Ethanol	20%	RIA	THC	+	+	[123]
	20%	FPIA	THC	+	+	[123]
Ethylene glycol	20%	RIA	THC	+	+	[123]
	20%	FPIA	THC	+	+	[123]
Goldenseal	0.9%	RIA	THC	+	− −	[118]
	0.9%	FPIA	THC	+	− −	[120]
	0.9%	FPIA	BAR	+ / −	− − −	[120]
	0.9%	FPIA	AMP	−	++	[120]
	30 mg/mL	EMIT	THC	+	− − − −	[121]
Hydrogen peroxide	6 μL/mL	EMIT	BEN	+	− −	[123]
	6 μL/mL	RIA	THC	+	++	[123]
	6 μL/mL	FPIA	BEN	+ / −	++	[123]
	6 μL/mL	FPIA	THC	+	++	[123]
Lime solvent	0%	RIA	AMP, OPI	+	− −	[118]
	10%	RIA	THC	−	++++	[118]
Phosphate	5%, 10%	RIA	COC	+	− − − −	[118]
	5%, 10%	RIA	THC	+ / −	+++	[118]
	10%	RIA	AMP, PCP	+ / −	++	[118]
	10%	RIA	BAR	+ / -	+++	[118]
	10%	FPIA	OPI	+	− − − −	[120]
	10%	FPIA	PCP	+	− − − −	[120]
Potassium hydroxide	0.5 M, 5 M	RIA	THC	+ / −	++++	[119]
	0.5 M, 5 M	RIA	PCP	+ / −	++++	[119]
	5 M	FPIA	THC	+ / −	+++	[119]
	0.5 M, 5 M	FPIA	PCP	+	− − −	[119]
2-Propanol	20%	RIA	THC	+	+	[123]
	50%	RIA	THC	+ / −	+++	[119]
	20%	FPIA	THC	+	+	[123]
	50%	FPIA	THC	+ / −	++	[119]

Table 5.9. (Continued.)

Substance	Concentration	Method	Drug category[a]	Sample[b]	Effect[c]	Ref.
Salicyluric acid[g]	≈ 7 g/L	EMIT	All	+	– – – –	[108, 124]
Salt	0.25 g/mL	EMIT	THC	+	–[f]	[122]
	160 mg/mL	EMIT	MED	+	– – – –	[125]
	50 mg/mL	EMIT	BAR, OPI, MED	+	– – – –	[126]
	50 mg/mL	EMIT	AMP, BEN, COC, OPI, PCP, THC	+ / –	– – – –	[123]
	50 mg/mL	EMIT	BAR	+	– – – –	[123]
	75 mg/mL	EMIT	AMP, BAR, COC	+	– – – –	[121]
	50 mg/mL	EMIT	OPI, THC	+	– – – –	[121]
	10%	RIA	THC	+	– –	[118]
	50 mg/mL	FPIA	BEN	+	– –	[123]
	10%	FPIA	THC	+	– –	[120]
Soap, liquid[h]	1 drop/5 mL	EMIT	THC	+	–[f]	[122]
	10 μL/mL	EMIT	MED	+	– – –	[122]
	10%	RIA	THC	+ / –	+++	[118]
		RIA	PCP	+	– – –	[119]
	5%, 10%	RIA	THC	+	– – – –	[119]
	5%, 10%	FPIA	PCP	–	+	[119]
	5%, 10%	FPIA	THC	–	++	[119, 120]
	10%	FPIA	BAR	+ / –	++++	[120]
	10%	FPIA	AMP	+ / –	+++	[120]
	12 μL/mL	EMIT	THC	+	– – – –	[121]
	23 μL/mL	EMIT	BAR	+	– – – –	[121]
	42 μL/mL	EMIT	BEN	+	– – – –	[121]
	2%	EMIT	BEN, THC	+ / –	– – – –	[123]
	2%	EMIT	PCP	+	– – – –	[123]
	2%	EMIT	BAR	+	++	[123]
	2%	RIA	BEN, THC	+ / –	++++	[123]
	2%	FPIA	AMP, BAR, BEN, THC	+ / –	+++	[123]
Sodium bicarbonate	40 mg/mL	EMIT	BAR	+	++	[123]
	40 mg/mL	EMIT	PCP	+	– – – –	[123]
	40 mg/mL	RIA	AMP, BAR, THC	+ / –	++	[123]
	40 mg/mL	FPIA	PCP	+	– – – –	[123]
UrinAid (glutaraldehyde)[i]	≈6.3% (v/v)	EMIT	AMP, COC, OPI, PCP, THC	+	– – – –	[127]
		FPIA	AMP, COC, OPI, PCP, THC	+	– –	[127]
		FPIA	PCP	+	– – – –	[127]
		OnLine	PCP	–	++++	[127]
		OnLine	AMP	–	++++	[127]
		OnLine	THC	–	++	[127]
		RIA	AMP, COC, OPI, PCP, THC	–	++++	[127]
Vanish	1-10%	RIA	AMP, OPI	+	– –	[118]
	1%, 10%	RIA	THC	+	– –	[118]
	5%	RIA	THC	–	+++	[118]
	10%	FPIA	THC	–	+	[120]
Vinegar	125 μL/mL	EMIT	THC	+	– –	[121]
	0.5 drops/mL	EMIT	THC	+	–[f]	[122]
	50%	RIA	THC	–	++	[119]
	50%	FPIA	PCP	+	– –	[119]
	10%	FPIA	THC	+	– – –	[120]
Visine	125 μL/mL	EMIT	THC	+	– – – –	[121]
	107 μL/mL	EMIT	BEN	+	– – –	[121]

Table 5.9. (Continued.)

Substance	Concentration	Method	Drug category[a]	Sample[b]	Effect[c]	Ref.
Visine (Continued)	10%	RIA	THC	+	− −	[118]
	10%	FPIA	THC	+	− −	[120]

[a] Abbreviations for drug categories: AMP: amphetamine; BAR: barbiturate; BEN: benzodiazepine; COC: cocaine metabolite; MED: methadone; OPI: opiate; PCP: phencyclidine; THC: cannabinoid metabolite.

[b] "+" and "−" designate that samples tested were with and without, respectively, the targeted analyte.

[c] "+" and "−" designate enhanced and reduced response. One, two, three, and four symbols indicate a slight, moderate, significant, and very significant effect, respectively. Since different measures were used for reporting interferences, the extents of interference shown in this column were estimated by this author. Original articles should be consulted for more precise information.

[d] $NaHClO_4$ is the main ingredient used in bleach preparations. Different brands were used by different investigators: Clorox in Refs. [121–123]; Cabbco in Ref. [118]; and Giant Food in Ref. [119].

[e] The targeted analyte was believed to have been degraded [120].

[f] A negative result was obtained from a known positive sample. No information concerning the magnitude of the change in responses was given.

[g] Salicyluric acid, the principal metabolite of salicylate (aspirin), was reported to interfere with the measurement of NADH formed in the assay by reducing the molar absorptivity of NADH at 340 nm. Measuring the EMIT assay signal at 376 nm, where salicyluric acid has no absorbance, eliminates the interference.

[h] Different brands of liquid soap were used by different investigators: Joy in Ref. [123]; Ivory in Ref. [119]; and Derma Cidol 2000 in Refs. [119] and [120]. The identities of the four brands used in Ref. [125] were not reported.

[i] Abnormal test results in EMIT, FPIA, and RIA were reportedly caused by the depression of absorbance rate change, increase in background fluorescence, and interference with pellet formation, respectively.

Most of the studies included in Table 5.9 did not compare the effects of the adulterants on various immunoassays under the same conditions. It is therefore difficult to make general statements concerning the robustness of one methodology over the others. It seems to be clear, however, that cannabinoid assays are more susceptible to the interfering effects of adulterants.

Numerous mechanisms have been proposed to account for the observed interference [118, 120, 121, 123], but the exact cause of these interferences is generally unknown. It has been proved, however, that bleach actually causes the degradation of 9-THC-COOH [120]. Visine was believed to increase the adhesion of 9-THC-COOH to the borosilicate glass specimen containers, thereby reducing the availability of 9-THC-COOH in antibody-based assays [128].

Simultaneous Multiple-Drug Detection

Simultaneous multiple-drug detection based on the proven EIA, FPIA, and RIA technologies has also been explored. The approaches involve either the simple mixing of antisera to single drugs [129–134] or the use of polyvalent antibodies [135]. The former approach is represented by recent FPIA [129, 130], EIA [131], and RIA [134] studies. In the FPIA study, fluorescein-labeled amphetamine, secobarbital, and benzoylecgonine were mixed with the following antisera: anti-amphetamine, anti-secobarbital, anti-phenobarbital, and anti-benzoylecgonine. It was reported that the cross-reactivity characteristics of the combination reagent were comparable to those of individual antisera. The detection limits (1–2.5 µg/mL) reported by this study, however, are less than desirable.

Similar approaches were used in the EIA study [131], in which the antibody reagents were mixed into a simple dilute solution. The enzyme-labeled drugs were combined similarly for use in testing urine samples for the presence of as many as 11 drugs: amphetamine, barbiturates, benzodiazepine, cocaine metabolite, methadone, methaqualone, opiate, phencyclidine, phenytoin, propoxyphene, and tricyclic antidepressants. Test results of 325 samples were compared with those obtained by thin-layer chromatography (TLC). The most significant difference was that 16 samples were found to be negative by TLC but positive by the mixed-reagent approach. Of the 16 samples, 12 were found to contain drugs, some of which were not the targeted drugs. Using the absorbance change between a blank and a 300-ng/mL amphetamine standard as the cutoff value, a range of 36 untargeted drugs generated positive responses at the 10-µg/mL concentration level. It appears that this approach is effective for differentiating samples that are free of drugs from those containing drugs. The resulting positive samples will need to be rescreened by a more specific assay for respective target drugs.

The RIA reagent [134] combined two highly specific single-antigen assays into one system that was found to exhibit (1) a nearly equivalent response to *d*-amphetamine and *d*-methamphetamine, (2) high cross-reactivity to MDA (3,4-methylenedioxyamphetamine) (<100%) and MDMA (3,4-methylenedioxymethamphetamine) (77%), (3) low cross-reactivity to *l*-amphetamine and *l*-methamphetamine, and (4) little cross-reactivity to commonly encountered over-the-counter drugs.

The alternate approach for simultaneous multiple-drug detection involves the use of a polyvalent antibody prepared by sequential or simultaneous immunization of rabbits with protein conjugates of compounds related to morphine, barbiturates, cocaine, and diazepam [145]. Data provided by this RIA study indicated good sensitivities, allowing the use of 75, 100, and 40 ng/mL as the cutoffs for barbiturate, benzoylecgonine, and morphine, respectively. No false negatives and a 4.3% false positive rate were reported when compared with single-drug RIA testing on 1000 urine samples. This approach has the advantages of providing more-uniform antisera and preventing the further reduction in potency (especially for low-titered drugs) that may result from the in vitro mixing approach.

Comparison of Immunoassay Performance Characteristics

When drug testing is used for clinical purposes to determine the presence of a specific drug or drug category in the sample, test results are evaluated together with other diagnostic information to make appropriate medical decisions. For the reasons stated earlier (see the section "Underlying Test Principles and Cross-Reactivity" prior to Chapter 3), it may be advantageous to use an immunoassay reagent possessing broad cross-reactivities that will respond to compounds with similar structures. On the other hand, when an immunoassay is used in a workplace drug urinalysis program, positive and negative results depend on the adopted apparent analyte cutoff concentration. Thus, technologies and issues pertaining to the selection of an appropriate cutoff level are crucial, and the following factors should be carefully considered.

1. Immunoassays are generally responsive to compounds with similar structural features and may also be vulnerable to detection-related interference.
2. GC–MS is specific for the identification and quantification of individual compounds.
3. Current practice in workplace drug-testing programs requires reporting positive results only if GC–MS results (for a targeted drug or metabolite) are at or above a cutoff concentration, and prior immunoassay screen test results are also at or above an identical or higher cutoff level.

In light of the workplace urinalysis requirements, manufacturers often strive to reduce their reagents' cross-reacting characteristics and improve the detection methodology to achieve a better correlation with GC–MS, thereby increasing overall test efficiency. The ideal situation is that all samples that have screened positive are confirmed to contain the target drug or metabolite at or above the GC–MS cutoff level, while those that have screened negative contain no targeted drugs or metabolites (or contain them at levels below the GC–MS cutoff).

In addition to selecting appropriate cutoff-concentration levels for both immunoassay and GC–MS tests, the following immunoassay performance characteristics greatly affect an immunoassay's ability to generate test results that best meet the above-mentioned ideal situation:

1. Susceptibility to *interferences* caused by cross-reactivity or detection-related mechanisms; and
2. The immunoassay's *sensitivity* as reflected by the assay's measurement precision and the magnitude of separation between signals generated by a specimen slightly above the cutoff and a specimen slightly below the cutoff.

The performance characteristics of various immunoassays are commonly compared on a qualitative level by grossly examining the consistency of the positive and negative results generated by the immunoassay and the GC–MS tests. They can also be compared on a more "refined" level by looking into the assay's sensitivity parameters or the quantitative correlation between the immunoassay's apparent analyte concentration and the GC–MS test data.

One factor that has great commercial significance is the assay's "overall ease and speed of analysis", which are not addressed further here except to quote Abercrombie and Jewel [136] on the matter: "We prefer the fully automated EMIT II and OnLine assays for high-volume urine testing, in comparison with our laboratory's semi-automated RIA tests and the limited-throughput TDx system.".

Consistency of Immunoassay and GC–MS Test Results

Numerous immunoassay comparison studies have been reported [136–150]. Most of these studies emphasize evaluating the ability of various immunoassays to provide positive and negative results that are consistent with those obtained by the GC–MS confirmatory test.

A series of similar publications [137–139] compared the ability of various immunoassays to identify true positives for GC–MS tests. Data summarized in Table 5.10 indicate the following:

1. For assays of cannabinoids, the effectiveness of identifying true positives is in the order of Abuscreen RIA ≥ Abuscreen OnLine > TDx > EMIT II.

2. For assays of barbiturates, the effectiveness of identifying true positives is in the order of Abuscreen OnLine ≈ TDx > EMIT II > Abuscreen RIA.

3. For assays of cocaine and opiates, all immunoassays show comparable results.

These studies also concluded that only a very small number of false positive were produced by each immunoassay, and these cannot be used as the basis for differentiating the overall efficiency of the immunoassays.

Table 5.10. Immunoassays' effectiveness in selecting specimens for GC–MS testing.[a,b]

Drug Category	Abuscreen RIA				EMIT II			EMIT 700	Abuscreen	OnLine	TDx
Marijuana	99.1;	99[c];	100[c];	99.2[d]	88.4;	88.7[c];	90.3[d]	90.6	99.1	99.4[d]	94.6
Cocaine	99.6;	99.6[c];	100[c];	99.6[d]	99.3;	97.4[c];	98.2[d]	96	99.6	98.9[d]	98.9
Opiates	100;	100[c];	99.1[c];	100[d]	96.9;	100[c];	100[d]	98.3	100	100[d]	100
Barbiturates	78.0;	75[c];	84[c];	78.8[d]	88.0;	86[c];	84.6[d]	97.5	100	96[d]	100

[a] Entries are the numbers of positives detected by respective immunoassays divided by the total number of GC–MS-confirmed samples (detected by any immunoassay) multiplied by 100. Except for the data noted, data presented in this table are taken from Ref. [137].

[b] Immunoassay cutoffs for marijuana, cocaine, opiates, barbiturates, and phencyclidine are 50, 150, 300, 200, and 25 ng/mL, respectively. GC–MS cutoffs for 9-THC-COOH, benzoylecgonine, codeine, morphine, butalbital (or phenobarbital), and phencyclidine are 15, 100, 300, 300, 200, and 25 ng/mL, respectively.

[c] Data are taken from Ref. [138].

[d] Data are taken from Ref. [139].

The effects of immunoassay cutoffs on the consistency of immunoassay and GC–MS test results are demonstrated in a recent study [143] comparing several commercial cannabinoid immunoassay kits. Specimens adopted for this study were collected over an extended time period from clinically controlled marijuana smokers; the study thus reflects distribution patterns of different cannabinoid metabolites. The data summarized in Table 5.11 clearly indicate increases in the number of true positives when the immunoassays' cutoff is changed from 100 ng/mL to 50 ng/mL while keeping the GC–MS test's cutoff at 15 ng/mL. With variation in magnitude, all immunoassays show an overwhelming increase in assay sensitivity as compared to the deterioration in assay specificity, thus improving overall test efficiency.

Assay Precision and Sensitivity

Measurement precision and sensitivity (as reflected by the magnitude of the separation between signals generated by specimens slightly above and specimens slightly below the cutoff) determine an assay's effectiveness in serving as a preliminary test method in drug urinalysis. Data summarized in Table 5.12 [151] allow a comparison of these parameters, which are generated by three formulations derived from EMIT phencyclidine reagents. The EMIT d.a.u. data imply difficulties in consistently (95.5% certainty) giving negative test results for specimens with phencyclidine present at 18.8 ng/mL. Similarly, specimens with phencyclidine at 31.3 ng/mL may not consistently (95.5% certainty) test positive.

More comprehensive studies [137–139] compared assay precision and sensitivity (in terms of the slope calculated from the data of the negative and the cutoff calibrators) for several commercially available immunoassay kits (Table 5.13). It appears that the ability of these immunoassays to differentiate a specimen's response that is slightly below the cutoff from that of the calibration standard are in the order of Abuscreen RIA > TDx ≈ Abuscreen OnLine > EMIT II.

A recent study [117] reported that test-precision data might be affected by the presence of adulterants (baking soda, table salt, Clorox bleach, and acetic acid). Clorox (in 1.5% v/v) was found to affect precision for all three opiate assays tested (EMIT d.a.u., Abuscreen RIA, and OnLine), whereas only EMIT d.a.u. was affected by baking soda and table salt.

Table 5.11. Percent change in cannabinoid immunoassays' effectiveness in selecting specimens for GC–MS testing when the cutoff concentration is reduced from 100 to 50 ng/mL.[a]

Assay Name	Sensitivity[b]	Specificity[b]	Efficiency[b]
EMIT d.a.u. 100	+29.2	−2.6	+2.4
EMIT II 100	+28.5	−1.5	+3.2
Abuscreen OnLine	+43.0	−1.0	+6.0
Abuscreen RIA	+35.8	−1.5	+4.4
DRI[c]	+30.5	−1.6	+3.4
ADx	+53.6	−1.9	+6.9
Mean	+36.8	−1.7	+4.4
Std dev	15.0	0.5	1.7

[a] Data are taken from Ref. [143].

[b] Definitions: sensitivity: [true positives/(true positives + false negatives)] × 100; specificity: [true negatives/(true negatives + false positives)] × 100; efficiency: [(true negatives + true positives)/total no.] × 100; true positives: specimens with immunoassay results equal to or greater than the specified cutoff concentration and ≥ 15 ng/mL 9-THC-COOH by GC–MS; true negatives: specimens with results less than the cutoff concentrations for both the immunoassay and the GC–MS tests; false positives: specimens with immunoassay results equal to or greater than the specified cutoff concentration and <15 ng/mL 9-THC-COOH by GC–MS; false negatives: specimens with immunoassay results less than the specified cutoff concentration and ≥15 ng/mL 9-THC-COOH by GC–MS.

[c] DRI: Diagnostic Reagents (Mountain View, CA).

Table 5.12. Assay precision and separation data (at and near cutoff) derived from EMIT phencyclidine reagents.[a]

Reagent formulation	18.8 ng/mL			25.0 ng/mL				31.3 ng/mL		
	Mean	SD	[Mean + 2SD]	[Mean − 2SD]	Mean	SD	[Mean + 2SD]	[Mean − 2SD]	Mean	SD
EMIT d.a.u.	221	1.96	225	220	224	2.20	228	227	230	1.31
EMIT II	184	1.86	188	191	194	1.46	197	202	206	1.91
EMIT mixed	236	1.18	238	240	242	1.20	244	248	250	1.05

[a] Data are taken from Ref. [151].

Table 5.13. Comparison of assay precision and sensitivity.[a]

Drug category	Parameter[b]	EMIT II		Abuscreen OnLine		TDx[c]	Abuscreen RIA
		Negative	Cutoff	Negative	Cutoff		
Marijuana	Mean	0.292	0.347	1.129	0.057		
	Std dev	0.007	0.018	0.723	0.051		
	CV[d]	2.4	5.6	5.0	5.7		
	Slope	11		81.2		81.0	514
Cocaine	Mean	0.203	0.228	0.614	0.305		
	Std dev	0.003	0.003	0.032	0.018		
	CV	1.5	1.2	5.3	5.8		
	Slope	1.6		20.6		11.0	223
Opiates	Mean	0.127	0.183	0.565	0.148		
	Std dev	0.004	0.003	0.022	0.009		
	CV	2.0	1.4	3.9	6.3		
	Slope	1.9		13.9		29.7	177
Barbiturates	Mean	0.228	0.291	0.819	0.283		
	Std dev	0.007	0.008	0.030	0.014		
	CV	3.2	2.8	3.7	4.9		
	Slope	3.2		26.8		13.5	141
Phencyclidine	Mean	0.190	0.258	0.500	0.221		
	Std dev	0.002	0.007	0.030	0.006		
	CV	1.1	2.6	5.9	2.8		
	Slope	27.2		112		—	178

[a] With the exceptions of data that are footnoted, data presented in this table are taken from Ref. [139].

[b] Slopes were calculated on the basis of the curves between the negative and cutoff calibrators. EMIT II and Abuscreen OnLine values are multiplied by 10^4 and TDx values by 10^2. Abuscreen OnLine and TDx slopes are listed as positive for easy comparison with EMIT II values.

[c] Data are taken from Ref. [137].

[d] Coefficient of variance (in %).

Quantitative Data Correlation Between GC–MS and Different Immunoassays

The data shown in Table 5.10 [143] clearly demonstrate that reasonable agreement between preliminary-test and confirmatory-test results can be obtained only if the immunoassay cutoff concentration that is adopted corresponds well with the cutoff level of the GC–MS test. The optimal immunoassay cutoff can be estimated by correlating the immunoassay's apparent "quantitative" results with GC–MS results. Although no perfect correlation can be expected, a more specific immunoassay will produce a better quantitative correlation with the GC–MS result, thus allowing the selection of a cutoff concentration that will correspond well with an adopted GC–MS cutoff-concentration level.

Selection of Samples for Correlation Studies

The nature of samples selected for the correlation studies is crucial. First, samples have to be true clinical samples, with the distribution of metabolites representative of users' excretion patterns. Using controls that are spiked with the analyte alone does not take into account the effect of the immunoassays' cross-reacting characteristics on the parameters evaluated. Second, samples with no analyte or high concentration levels of the analyte are not good candidates either. The inclusion of true negative samples (0 ng/mL of analyte) in the study will generate much better correlations but will reduce the potential differentiation of test methods compared. Samples with high concentration levels of the targeted analyte (and associated metabolites or parent drugs) will generate immunoassay results that are out of the methods' dynamic ranges. Thus, ideal samples used for correlation studies should be true clinical samples in which the concentrations of the targeted analyte are near the GC–MS cutoff level. Data generated from these samples are most suitable for the correlation of immunoassay apparent analyte concentration with the concentration of a specific analyte (determined by a GC–MS protocol).

Cannabinoids

Correlation Studies of Commercial Immunoassays. A study [63] was conducted to compare the commercial cannabinoid immunoassay kit Abuscreen RIA with EMIT d.a.u. and TDx in the correlation of their respective immunoassay's apparent 9-THC-COOH concentrations with the 9-THC-COOH concentration determined by GC–MS. Considering the sample-selection criteria outlined in the last paragraph, out of a total sample population of approximately 1360 initially tested by Abuscreen RIA, only 26 specimens that generated apparent 9-THC-COOH concentrations in the 50–200-ng/mL range were selected for this study and analyzed by TDx and GC–MS. The same study [63] also compared the same correlation characteristics for the commercial cannabinoid immunoassay kits Abuscreen RIA and EMIT d.a.u. Similarly, out of a sample population of approximately 5070 initially tested by Abuscreen RIA, only 47 specimens that generated apparent 9-THC-COOH concentrations in the 50–150-ng/mL range were selected and analyzed by EMIT d.a.u. and GC–MS. Because samples thus selected are highly relevant, potential differences in these assay kits' abilities to generate results that can be correlated with the GC–MS results are enhanced.

A linear model was used to correlate the immunoassay and the GC–MS results. A linear model, undoubtedly, does not fully describe the relationship between the immunoassay and the GC–MS results for the following reasons:

1. The distributions of other immunoassay-responding cannabinoids and metabolites (that are not targeted by the GC–MS analysis) in the sample population are not necessarily proportional to the concentration of 9-THC-COOH.

2. The cross-reacting characteristics of immunoassays toward these cannabinoids and metabolites may vary at different concentration levels.

Because the concentrations of the targeted analyte, and presumably those of other cannabinoids and metabolites, are all within a narrow concentration range and centered at a 9-THC-COOH concentration (15 ng/mL), on which a corresponding immunoassay concentration is to be projected, a linear model should represent a reasonable approximation.

Results from the aforementioned linear-model correlations of various immunoassay and GC–MS data are graphically compared in Figure 5.10 (Abuscreen RIA vs. TDx) and in Figure 5.11 (Abuscreen RIA vs. EMIT d.a.u.). The resulting correlation parameters are summarized in Table 5.14. Because the same samples were used for the immunoassays evaluated, the correlation parameters are highly indicative of the performance characteristics of these immunoassay reagent kits manufactured at the time of the study.

A later study [152] adopted the same approach to compare the Abuscreen RIA and Immunalysis RIA immunoassay kits. Samples used for this comparison were drawn, however, from two different sample populations but with similar sample-selection criteria. The resulting correlation results are graphically presented in Figure 5.12, while the corresponding correlation parameters are included in Table 5.14 (rows 5 and 6). Again the resulting correlation parameters are highly indicative of the performance characteristics of these immunoassay reagent kits manufactured at the time of the study.

The last two sets of data in Table 5.14 are included for reference only and cannot be compared with the others. Sample selection criteria adopted for the DPC RIA study [153] are different from those of the other three sets. The significantly different data obtained from methanol extract of blood and nonextracted blood specimens [154] might have resulted from the differences in matrix effect and variations in the distribution of immunoassay-responding components in the methanol extract and the original blood specimens.

Interpretation of Correlation Parameters. Because immunoassays are responsive to 9-THC-COOH and other marijuana metabolites normally present in a urine sample collected from a person with marijuana exposure, and GC–MS analysis is specific for 9-THC-COOH only, the apparent 9-THC-COOH concentration reported by an immunoassay is higher than that obtained from a GC–MS procedure. The extent of this inflated apparent 9-THC-COOH concentration depends on:

1. The cross-reacting characteristics of the reagent used; and

2. The concentrations of 9-THC-COOH and other cross-reacting marijuana metabolites present in the sample.

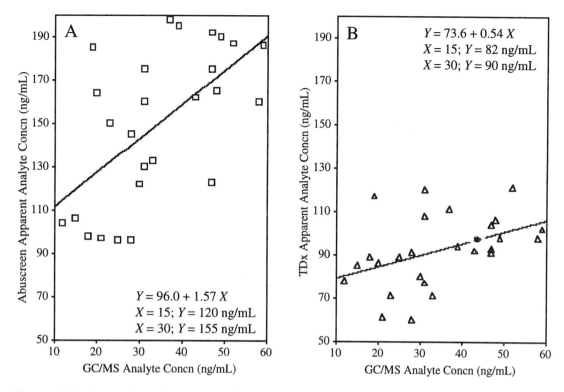

Figure 5.10. Correlation of apparent immunoassay and GC–MS 9-THC-COOH concentrations: Abuscreen RIA (reagent with package insert dated March 1987) vs. GC–MS (A); and TDx (reagent with operation manual dated 12/01/1987) vs. GC–MS (B). The same 26 samples (selected from a total population of approximately 1360) were used for this comparative study. (Data taken from Ref. [63].)

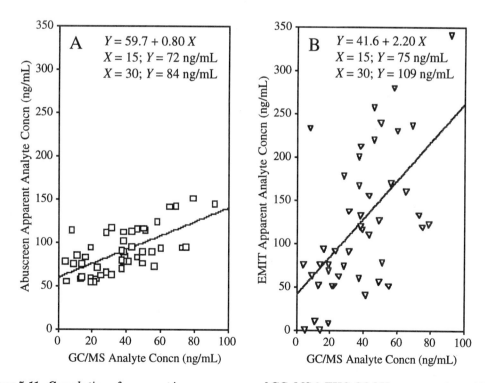

Figure 5.11. Correlation of apparent immunoassay and GC–MS 9-THC-COOH concentrations: Abuscreen RIA (reagent with package insert dated June 1989) vs. GC–MS (A); and EMIT d.a.u. 100 ng Assay Kit for Cannabinoids (reagent with package insert dated Nov. 1987) vs. GC–MS (B). The same 47 samples (selected from a total population of approximately 5070) were used for this comparative study. (Data taken from Ref. [63].)

Table 5.14. Correlation of immunoassay and GC–MS test data for cannabinoids.

Immunoassay	No. of data points	Total sample population	Correlation coefficient	Correlation equation[a]	Immunoassay apparent 9-THC-COOH equivalent to 15 ng/mL	Ref.
Abuscreen RIA[b]	26	1,360	0.601	$Y = 96.0 + 1.57X$	120 ng/mL	[63]
TDx[b,c]	26	1,360	0.438	$Y = 73.6 + 0.54X$	82 ng/mL	[63]
Abuscreen RIA[d]	47	5,070	0.658	$Y = 59.7 + 0.80X$	72 ng/mL	[63]
EMIT d.a.u.[e]	47	5,070	0.575	$Y = 41.6 + 2.20X$	75 ng/mL	[63]
Abuscreen RIA[f]	24	26,400	0.379	$Y = 56.2 + 0.86X$	69 ng/mL	[152]
Immunanalysis[g]	24	10,300	0.542	$Y = 45.8 + 1.54X$	69 ng/mL	[152]
DPC[h]	10	—	0.638	$Y = 9.5 + 2.96X$	54 ng/mL	[153]
Abuscreen RIA	122[i]	—	0.681	$Y = 6.5 + 0.31X$	11 ng/mL	[154]
	92[j]	—	0.228	$Y = 8.2 + 0.078X$	9.4 ng/mL	[154]

[a] X: GC–MS 9-THC-COOH concentration; Y: immunoassay *Cannabis* metabolites concentration (expressed in 9-THC-COOH equivalents).
[b] The reagent used for this study is best described by the package insert dated March 1987.
[c] The reagent used for this study is best described by the operation manual dated 12/01/1987.
[d] The reagent used for this study is best described by the package insert dated June 1989.
[e] The reagent used for this study is best described by the package insert dated Nov. 1987.
[f] The reagent used for this study is best described by the package insert dated Feb. 1992.
[g] The reagent used for this study is best described by the package insert dated Feb. 1992.
[h] Data are taken from Ref. [153] with correlation performed by this author (R. H. Liu).
[i] Methanol extracts of blood specimens.
[j] Unextracted blood specimens.

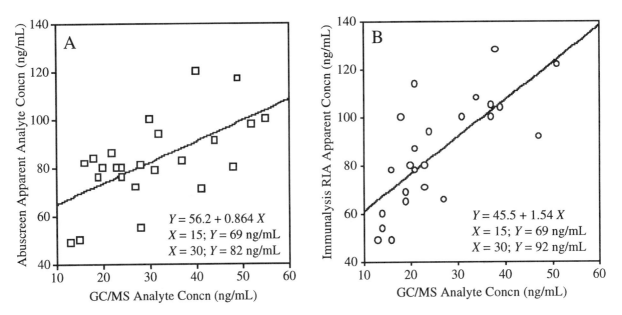

Figure 5.12. Correlation of apparent immunoassay and GC–MS 9-THC-COOH concentration: Abuscreen RIA (reagent with package insert dated Feb. 1992) vs. GC–MS (A); Immunalysis RIA for Cannabinoids (reagent with package insert dated Feb. 1992) vs. GC–MS (B). Different samples (selected from total sample populations of approximately 26, 330, and 12,870, respectively) were used for this comparative study. (Reproduced with permission from Ref. [152].)

Whether these inflated apparent RIA concentrations can be reasonably correlated with their respective GC–MS results depends primarily on:

1. The consistency of the relative concentrations of 9-THC-COOH and other cross-reacting metabolites in the samples; and
2. The specificity of the reagent used.

Because the distribution of 9-THC-COOH and other cross-reacting metabolites in a sample is controlled by the metabolic mechanism, with secondary variations (from individual to individual) randomly distributed in the sample population, the "goodness" of the aforementioned correlation will primarily depend on the reagent specificity. The *correlation coefficients* listed in Table 5.14 suggests that Abuscreen RIA may be more reliable in predicting the concentration of 9-THC-COOH in urine in the presence of common cannabinoid metabolites. This statement is valid only for comparing those reagent lots (from Roche, Abbott, Syva, and Immunalysis) used in the subject study.

It is interesting to explore the possibility of using the resulting *regression equations* for estimating appropriate cutoff values for an immunoassay (inflated immunoassay apparent 9-THC-COOH concentrations) that best correspond to a selected GC–MS concentration. Thus, using the regression equations shown in Table 5.14, the calculated Abuscreen RIA, EMIT d.a.u., TDx, and Immunalysis RIA apparent 9-THC-COOH concentrations that correspond to 15 ng/mL of 9-THC-COOH are shown in the next-to-last column of the table.

The equivalent apparent 9-THC-COOH concentrations of the immunoassay kits shown in the next-to-last column of Table 5.14 strongly suggest that the cross-reacting characteristics of these assays toward other metabolites associated with marijuana exposure are significantly different. If the same cutoff concentration is used for all immunoassays compared, these immunoassays would generate different numbers of false positives and false negatives. Or a sample with a 9-THC-COOH concentration in the neighborhood of 15 ng/mL may test positive by one immunoassay but test negative by the other. With the reagent lots used to generate data shown in Table 5.14, if the same cutoff value is used for all immunoassay kits, the likelihood of producing a preliminary test positive result for a sample with a borderline 9-THC-COOH concentration (15 ng/mL) will be in the order of Abuscreen RIA (March 1987) > TDx (12/01/1987) > EMIT d.a.u. (Nov. 1987) > Abuscreen RIA (June 1989) ≈ Immunalysis RIA (Feb. 1992) ≈ Abuscreen RIA (Feb. 1992). (The difference observed for the three Abuscreen RIA sets will be examined and commented on in the section "Quantitative Data Correlation of GC–MS and Immunoassays Performed with Reagents Manufactured at Different Time Periods".) This is the equivalent of saying that the likelihood of producing a false positive preliminary test result for a sample with a 9-THC-COOH concentration slightly below the GC–MS cutoff value is in the order of Abuscreen RIA (March 1987) > TDx (12/01/1987) > EMIT d.a.u. (Nov. 1987) ≈ Abuscreen RIA (June 1989) ≈ Immunalysis RIA (Feb. 1992) ≈ Abuscreen RIA (Feb. 1992). Conversely, the likelihood of producing a false negative preliminary test result for a sample with a 9-THC-COOH concentration slightly above the GC–MS cutoff value is in the order of Abuscreen RIA (Feb. 1992) ≈ Immunalysis RIA (Feb. 1992) ≈ Abuscreen RIA (June 1989) ≈ EMIT d.a.u. (Nov. 1987) > TDx (12/01/1987) > Abuscreen RIA (March 1987).

Cocaine

A similar correlation study was conducted [155] using TDx and Abuscreen RIA reagents for cocaine metabolites. Using the sample-selection criteria adopted for the cannabinoids assay studies [63], 62 samples were selected (out of a total sample population of 3300) for the correlation study. The resulting correlation equations and apparent immunoassay analyte concentrations equivalent to 150 ng/mL of benzoylecgonine are summarized in the first two rows of Table 5.15.

Table 5.15. Correlation of immunoassay and GC–MS test data for cocaine.

Immunoassay	No. of data points	Total sample population	Correlation coefficient	Correlation equation[a]	Immunoassay benzoylecgonine equivalent to 150 ng/mL
TDx[b]	62	3,300	0.766	$Y = 76 + 0.72X$	190 ng/mL
Abuscreen RIA[b]	62	3,300	0.467	$Y = 299 + 0.532X$	380 ng/mL
Abuscreen RIA[c]	—	—	0.925	$Y = 0.046 + 0.958X$	144 ng/mL

[a] X: GC–MS benzoylecgonine concentration; Y: immunoassay cocaine metabolites concentration (expressed in benzoylecgonine equivalents).

[b] Data are taken from Ref. [155].

[c] Data are taken from Ref. [156]. These data were obtained from eluates of blood spotted onto a filter-paper matrix. Apparent Abuscreen RIA benzoylecgonine concentration equivalent to 150 ng/mL of benzoylecgonine was calculated by this author (R. H. Liu) using the equation established by the original authors.

To demonstrate the significance of selecting an appropriate cutoff for immunoassays to improve the efficiency of workplace drug urinalysis, the agreements of positive and negative results between the immunoassays and GC–MS data were evaluated using different immunoassay cutoff concentrations (Table 5.16). If 300 ng/mL and 150 ng/mL are adopted as the immunoassay and GC–MS cutoffs, respectively, Abuscreen RIA generated three false negatives and 21 false positives and TDx generated 16 false negatives and no false positives. Thus, it appears that the 300-ng/mL cutoff is too low for Abuscreen RIA but too high for TDx, for those reagents provided by the manufacturers at that time. If the corresponding apparent immunoassay cutoffs estimated from the correlation equations were adopted as the immunoassay cutoffs (190 ng/mL for TDx and 380 ng/mL for Abuscreen RIA), Abuscreen RIA would have generated 5 false negatives and 16 false positives, while TDx would have generated 3 false negatives and 6 false positives. Because these data are highly dependent on reagent specificities, there is no doubt that reagents provided at different time periods by these same manufacturers will perform differently, as demonstrated in the next paragraph and in the next section.

Table 5.16. Dependency of false positive and false negative results on the selection of immunoassay cutoff concentrations.[a]

Immunoassay	Cutoff concentration (ng/mL)[b]		No. of false negatives	No. of false positives
	GC–MS	Immunoassay		
TDx	150	300	16	0
Abuscreen RIA	150	300	3	21
TDx	150	190	3	6
Abuscreen RIA	150	380	5	16

[a] Data are taken from Ref. [155].

[b] The GC–MS concentration is expressed in ng/mL of benzoylecgonine, whereas those for immunoassays are expressed in benzoylecgonine equivalents, based on the calibration curves generated at the time of analysis.

Correlation parameters of data generated by eluates of blood spotted onto a filter-paper matrix [156] are included in the last row of Table 5.16. Much better correlation between the Abuscreen RIA and GC–MS results was observed in this blood study—the RIA and GC–MS data are statistically indistinguishable. This might have resulted from (1) the more specific reagent produced by the manufacturer at the time this study was conducted, or (2) the absence of cross-reacting compounds in the eluates resulting from the sample-preparation procedure.

Opiates

Table 5.17 summarizes the data from a study of opiates [157] that was performed using TDx, EMIT, and Abuscreen RIA reagents with morphine as the targeted analyte. DPC's Morphine Coat-A-Count reagent was also included in this study. However, because of its high specificity toward free morphine, results are not comparable with those of the other three immunoassays and are not included in the table.

Because urine samples used in this study were collected from subjects administered intramuscularly with heroin and morphine, variations in opiate distributions in these samples were minimized, resulting in the good correlations shown in Table 5.17. The sources of opiates in random opiate-containing samples include exposure to various opiate-contributing sources (such as poppy seeds and opiate-containing prescriptions) with different opiate compositions.

Because immunoassay reagents cross-react significantly with other opiates, such as codeine, poorer correlations are expected for data generated by random clinical samples with diverse opiate distributions. However, this was not observed for vitreous humor specimens that were hydrolyzed prior to Abuscreen RIA analysis [158]. Apparent RIA morphine concentration correlated well with GC–MS results, as shown in the least row of Table 5.17.

Table 5.17. Correlation of immunoassay and GC–MS test data for opiates.

Immunoassay	No. of data points	Correlation coefficient	Correlation equation[a]	Immunoassay apparent morphine equivalent to 300 ng/mL	Specimen	Ref.
TDx	233	0.994	$Y = -22.0 + 0.918X$	350.8 ng/mL	Urine	[157][b]
Abuscreen RIA	233	0.995	$Y = 27.1 + 0.907X$	360.6 ng/mL	Urine	[157]
EMIT	194	0.973	$Y = 6.2 + 0.692X$	424.6 ng/mL	Urine	[157]
Abuscreen RIA	10	0.995	$Y = 0.0093 + 0.924X$	277.2 ng/mL	Vitreous humor	[158][c]

[a] X: GC–MS morphine concentration; Y: immunoassay opiate metabolites concentration (expressed in morphine equivalents).

[b] Opiates were derived from intramuscular administration of heroin and morphine.

[c] Vitreous samples were hydrolyzed with β-glucuronidase–arylsulfatase prior to RIA analysis.

Quantitative Data Correlation of GC–MS and Immunoassays Performed with Reagents Manufactured at Different Time Periods

The cannabinoid study [63] concluded that the immunoassay apparent 9-THC-COOH concentrations corresponding to a specific GC–MS concentration vary with the immunoassay kits marketed by different manufacturers. A further study [158] was conducted to evaluate correlations between immunoassay and GC–MS results using a single immunoassay (Abuscreen RIA for Cannabinoids) with reagents provided by the same manufacturer at different times. Resulting correlation data are shown in Figure 5.13. Respective correlation coefficients and linear-regression equations are summarized in Table 5.18. It is clear that the correlation between the Abuscreen RIA and GC–MS results improves in parallel with the reagents' increasing specificity data as shown in the package inserts (Table 5.4).

The regression equations shown in Table 5.18 are used to estimate appropriate cutoff values (for these reagents manufactured at different time periods) that correspond to a specific GC–MS 9-THC-COOH concentration (15 ng/mL). Results are listed in the last column. These empirical data suggest that if the 9-THC-COOH concentration determined by GC–MS is the primary concern and the immunoassay is to be performed efficiently and statistically correlate well with the GC–MS results, the corresponding Abuscreen RIA cutoff levels should be approximately 115, 100, 70, and 55 ng/mL using the reagents described in the package inserts dated March 1987, March 1988/November 1988, June 1989/May 1990, and October 1991, respectively. This kind of information prompted the U.S. Department of Defense and the Department of Health and Human Services laboratory certification programs to alter the cannabinoid cutoff level to 50 ng/mL.

Selection of Immunoassay Cutoff in Workplace Drug Urinalysis

Empirical data shown in Tables 5.14, 5.15, 5.17, and 5.18 demonstrate that correlations of immunoassay and GC–MS results vary with the methodologies and manufacturers of the immunoassay kits. The correlations also vary with the lots of those kits provided by the same manufacturer. However, significant correlations (correlation coefficients are 1.0 for perfect correlation and 0.0 for no correlation) do exist in all cases. The immunoassay apparent analyte concentrations that correspond to a specific GC–MS analyte concentration may be approximated (calculated) from the regression equations.

Reagent specificities determine the immunoassay apparent analyte concentrations that correspond to a specific GC–MS analyte concentration. The scientific validity of the calculated apparent analyte concentrations varies with the degree of correlation between the immunoassay and GC–MS results. The numerical values of the immunoassay apparent 9-THC-COOH concentration that correspond to 15 ng/mL of 9-THC-COOH is in the order of TDx > EMIT d.a.u. > Immunalysis RIA > Abuscreen RIA. These comparisons are based on data generated by reagents dated October 1991/February 1992, February 1992, December 1987, and November 1987 for Abuscreen RIA, Immunalysis RIA, TDx, and EMIT d.a.u., respectively. Newer TDx and EMIT II reagents may perform differently.

The calculated immunoassay apparent analyte concentration may be adopted as the cutoff for routine use after some adjustments. The adjustments are dictated by the testing policy: the balance between the number of false negatives and the number of unconfirmed presumptive positives that are considered acceptable.

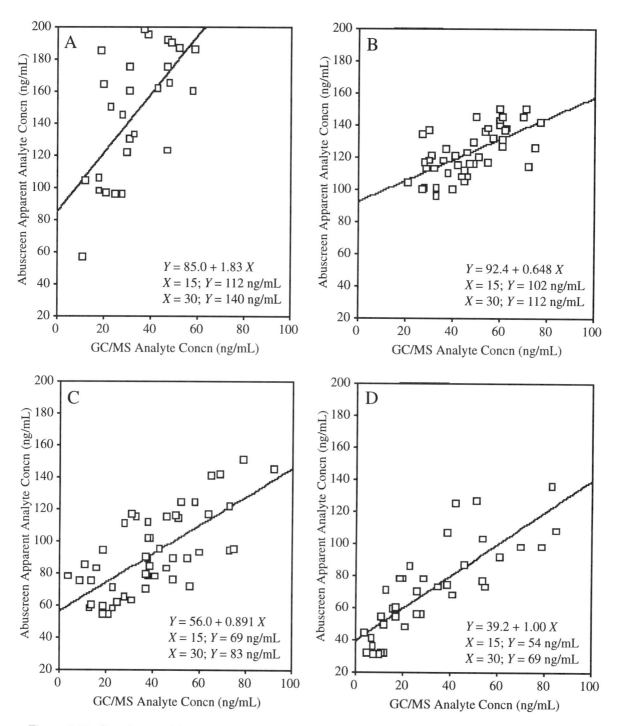

Figure 5.13. Correlation of GC–MS 9-THC-COOH concentration with Abuscreen RIA for Cannabinoids apparent 9-THC-COOH concentration, using the reagent provided by the same manufacturer at different dates: reagent package inserts dated March 1987 (A); March 1988 and Nov. 1988 (B); May 1990 (C); and March 1988 and Oct. 1991 (D). (Reproduced with permission from Ref. [159].)

Because the apparent immunoassay analyte concentration corresponding to a specific analyte concentration varies with the immunoassay methodologies, scientifically speaking, different cutoff values should be used for different immunoassay methodologies and manufacture lots. For the cannabinoid immunoassay kits mentioned above, appropriate cutoff values are in the order of TDx > EMIT > Immunalysis RIA > Abuscreen RIA, ranging from approximately 82 to 54 ng/mL. Thus, the immunoassay reagent manufacturers should carefully study specificity characteristics of each manufacturing lot and provide these correlation data for users' evaluation and adaptation.

Table 5.18. Correlation of GC–MS 9-THC-COOH concentration with apparent 9-THC-COOH concentration generated by Abuscreen RIA reagents manufactured at different time periods.[a]

Reagent package insert date	Test date	Correlation coefficient	Correlation equation[b]	RIA apparent 9-THC-COOH equivalent to 15 ng/mL
March 1987	11/09/87–1/21/88	0.426	$Y = 85.6 + 1.82X$	113 ng/mL
March 1988/Nov. 1988	11/29/88–4/24/89	0.405	$Y = 92.4 + 0.65X$	102 ng/mL
June 1989/May 1990	6/08/89–2/25/90	0.500	$Y = 56.0 + 0.89X$	69.4 ng/mL
Oct. 1991	10/04/91–2/19/92	0.665	$Y = 39.2 + 1.00X$	54.0 ng/mL

[a] Data are taken from Ref. [158].

[b] *X*: GC–MS 9-THC-COOH concentration; *Y*: Abuscreen RIA *Cannabis* metabolites concentration (expressed in 9-THC-COOH equivalents).

Development and Comparison of "Quick Test" Immunoassay Kits

The recent development of immunoassay kits that are uniquely suitable for testing a few samples at a remote site and offer rapid access to test results include Abuscreen Ontrak (Roche Diagnostic Systems: Branchburg, NJ), EZ-SCREEN (Environmental Diagnostics: Burlington, NC), accuPINCH (HYCOR Biomedical: Garden Grove, CA), and Triage panel (Biosite Diagnostics: San Diego, CA).

Abuscreen Ontrak Assay kits are available for amphetamines, barbiturates, benzodiazepines, cocaine (metabolites), morphine, phencylidine, and THC [160]. These assay kits are based on the principles of latex agglutination–inhibition, relying on the competition for binding to antibody between latex–drug conjugate and drug that may be present in the test sample. Typically, a urine sample is placed in the mixing well with antibody reagent, buffer, and latex reagent. In the absence of drug in the sample, the latex–drug conjugate forms large particles (agglutination) by binding to the antibody. When a sufficient amount of drug is present, it competes with latex conjugate for the limited antibody present. Therefore, a positive urine will not change the smooth milky appearance of the mixture [160]. True negative (zero drug concentration) and highly positive samples can be easily distinguished, whereas consistent recognition of positive and negative appearance resulting from samples with drug concentrations near the cutoff level may not always be attainable by all test performers. Thus, an evaluation study reported discrepancies in cutoff verifications with a 200-ng/mL morphine standard, a 250-ng/mL morphine control, and a 20% above-the-cutoff THC standard [161].

Cross-reacting compounds that are reported in one manufacturer's reagent inserts are summarized in Table 5.19.

Triage panel is a newly marketed competitive immunoassay designed for simultaneously detecting seven abused drug classes in urine [162]. The panel includes a reaction well containing three lyophilized beads: the colloidal gold–drug conjugates with representative drugs of the seven classes (conjugates); the antibodies for the seven targeted drug classes; and the buffer intended to maintain the reaction medium at pH 7.5–8.5. The panel also includes discrete drug-class-specific zones, each containing a solid phase (nylon membrane) with immobilized antibody for a specific drug class. A wash solution is also provided with the panel.

The assay protocol includes the addition of the sample to the reaction well for a 10-min incubation; transferring the reaction mixture to the seven discrete drug-class-specific zones; allowing for drainage; and the addition of three drops of the wash solution. The appearance of a red bar in a specific zone indicates the presence of the respective drug at or above the cutoff level. On the basis of the competitive immunoassay principle, drugs present in the sample compete with the conjugates for the limited amount of antibody in the reaction well. Conjugates left unbound in the reaction mixture, when transferred to the discrete drug-class-specific zones, will bind to the respective antibodies immobilized in these specific zones and, upon drainage of the reaction mixture and the wash solution, show the color bar. In the absence of the drug or the presence of the drug below a preset cutoff level, no free conjugates remain in the reaction mixture. Therefore, no conjugate will be retained by the antibodies immobilized on the membrane, and no color bar will be observed.

Table 5.19. Reported cross-reactivity of Abuscreen Ontrak assays.[a]

Manufacturer, assay name	Cross-reacting compound[a]	% cross-reactivity[a]
Roche Abuscreen Ontrak for Amphetamines	p-Hydroxyamphetamine	50
Calibrator: amphetamine	β-Phenethylamine	4
Cutoff: 1000 ng/mL	Tyramine HCl	1
Date: April 1992		
Roche Abuscreen Ontrak for Barbiturates	Allylcyclopentylbarbituric acid	800
Calibrator: secobarbital	Barbital	200
Cutoff: 200 ng/mL	Aprobarbital	100
Date: April 1992	Allobarbital	100
	Amobarbital	100
	Butabarbital	80
	Butalbital	80
	Pentobarbital	40
	Phenobarbital	29
	p-Hydroxyphenobarbital	27
Roche Abuscreen Ontrak for Benzodiazepines	Flunitrazepam	80
Calibrator: nordiazepam	(Desmethylflunitrazepam)[b]	59
Cutoff: 100 ng/mL	(3-Hydroxyflunitrazepam)[b]	27
Date: April 1992	Nitrazepam	67
	(7-Aminonitrazepam)[b]	40
	Diazepam	59
	(Oxazepam)[b]	50
	(n-Methyloxazepam)[b]	53
	Alprazolam	53
	(α-Hydroxyalprazolam)[b]	67
	(4-Hydroxyalprazolam)[b]	67
	Clonazepam	53
	Pinazepam	44
	Prazepam	40
	Lorazepam	40
	Midazolam	40
	Clorazepate	33
	Chlordiazepoxide	27
	(Desmethylchlordiazepoxide)[b]	27
	Demoxepam	27
	Flurazepam	27
	(Desalkyflurazepam)[b]	40
	(Didesethylflurazepam)[b]	53
	(Hydroxyethylflurazepam)[b]	–
	Medazepam	27
	(Desmethylmedazepam)[b]	27
	Triazolam	–
	(α-Hydroxytriazolam)[b]	53
	(4-Hydroxytriazolam)[b]	27
	Halazepam	20
Roche Abuscreen Ontrak for THC	8 α-Hydroxy-Δ^9-THC	>100
Calibrator: 9-THC-COOH	11-Hydroxy-Δ^9-THC	>100
Cutoff: 50 ng/mL	Δ^9-THC	10
Date: April 1992	8 β-11-Hydroxy-Δ^9-THC	>100
	11-Hydroxycannabinol	>100
	Cannabinol	10

Table 5.19. (Continued.)

Manufacturer, assay name	Cross-reacting compound[a]	% cross-reactivity[a]

Roche Abuscreen Ontrak for Cocaine (Metabolites)
    Calibrator: benzoylecgonine
    Cutoff: 300 ng/mL
    Date: April 1992
	Cocaine HCl	20

Roche Abuscreen Ontrak for Morphine
    Calibrator: morphine
    Cutoff: 300 ng/mL
    Date: April 1992
	Codeine
	Dihydromorphine
	Ethylmorphine HCl
	Dihydrocodeine bitartrate
	Hydromorphone HCl
	Morphine-3-glucuronide
	Hydrocodone bitartrate
	Thebaine
		120
		75
		75
		60
		38
		38
		38
		19

Roche Abuscreen Ontrak for Phencyclidine
    Calibrator: phencyclidine
    Cutoff: 25 ng/mL
    Date: April 1992
	1-[1-(2-Thienyl)cyclohexyl]piperidine
	1-(1-Phenylcyclohexyl)pyrrolidine
	3-*N,N*-Diethyl-1-phenylcyclohexylamine
	3-*N,N*-Dimethyl-1-phenylcyclohexylamine
		100
		50
		17
		1

[a] Data listed in this table are taken from the respective reagent package inserts as specified in the first column of the table. Only those compounds that were reported to show ≥1% cross-reactivity are listed. Compounds are listed in descending order of their reported cross-reactivities.

[b] Compounds listed with indentation and inside parentheses are metabolites of the preceding drugs.

The Triage panel manufacturer suggested [162] that the technology is inherently more sensitive than latex-agglutination-based assays, in that it is based on a response function that is practically zero below the cutoff level and increases rapidly once the cutoff level has been exceeded. Nevertheless, the slope of the dose–response curve is not infinitely steep, and it is affected by (1) the relative affinity of the antibody for the drug conjugate and the drug [163], and (2) matrix effects in individual urine samples producing a certain degree of fluctuation in the threshold concentration [163]. Thus, the color intensity of the red bars may depend on the drug concentration. Table 5.20 includes the drug-concentration ranges within which positive results may be expected to decrease from 100% to 0%. The information provided by the manufacturer is compared with data reported in an independent study [165].

Table 5.20. Drug concentration ranges producing positive Triage panel test results.

Drug class	Cutoff (ng/mL)	Targeted drugs	Range (ng/mL) that may result in positive	
			Manufacturer's data[a]	Independent report[b]
Amphetamines	1,000	*d*-Amphetamine	900–500	
		d-Methamphetamine	1,100–700	900 (30%)[c]
Barbiturates	300	Amo-, buta-, pento-, and secobarbital	360–240	270 (70%)[c]
Benzodiazepines	300	Oxazepam glucuronide, temazepam glucuronide,	350–100	
		α-Hydroxyalprazolam glucuronide	350–240	
Cocaine	300	Benzoylecgonine	400–240	240 (100%)
Opiates	300	Codeine, morphine	400–300	300 (70%)[c]
Phencyclidine	25	Phencyclidine	30–20	22.5 (60%)
THC	100	9-THC-COOH	100–75	80 (80%)

[a] 100% and 0% of positive results were observed for samples with the drug concentrations at the higher and the lower ends, respectively. Data are taken from Ref. [163].

[b] Percentage of positive results observed for samples with designated drug concentration. Data are taken from Ref. [164].

[c] *d*-Methamphetamine, secobarbital, and morphine, respectively, were used as the targeted analytes in these studies.

A large-scale comparative study [165] indicated that results provided by the Triage panel are essentially equivalent to the instrument-based EMIT d.a.u. assays. The Triage panel was reportedly more specific for *d*-methamphetamine [165]; however, it produced many more false amphetamine results for postmortem urine samples [163] because of cross-reactivity with tyramine, a typical putrefaction product.

The Triage panel utilizes antibodies designed to detect the glucuronide conjugates of two broad classes of benzodiazepines: benzodiazepines that are conjugated to glucuronic acid or contain a large functional group at the 1-position on the heterocyclic seven-membered ring, and benzodiazepines that are conjugated with glucuronic acid via the hydroxyl group at the 3-position on the heterocyclic seven-membered ring. The former group includes flurazepam (α-hydroxyethylflurazepam), triazolam, alprazolam, midazolam, and nordiazepam. The latter group includes chlordiazepoxide, chlorazepate, oxazepam, temazepam, lorazepam, desalkyflurazepam, and diazepam [166]. Benzodiazepine screening results produced by the Triage panel were reported to have a high degree of concordance with GC–MS tests and to exhibit markedly better diagnostic accuracy than EMIT and ADx [166]. It has been reported [163], however, that the detection limit of the Triage panel for 7-aminoflunitrazepam, the major urinary metabolite of flunitrazepam, is in the range of 500–1000 ng/mL.

Substantial evaluation studies on EZ-SCREEN Cannabinoid and accuPINCH THC have also been reported. The former kit was reportedly highly sensitive to 9-THC-COOH and had low cross-reactivity to THC and other cannabinoids. However, test-result reading—which is based on comparing the reduction in color with the gray–blue at the negative control port—is apparently not an easy task; the reported overall agreement between the three readers involved in the study was approximately 80% [167]. *accu*PINCH THC reportedly produced valid positive *B* readings (9-THC-COOH ≥ 100 ng/mL) with relatively low cross-reactivity to THC and other cannabinoids, and it was relatively insensitive to changes in sample temperature. The test, however, is highly sensitive to sample turbidity [168].

References

1. Ferrara, S. D.; Tedeschi, L.; Frison, G.; Brusini, G.; Castagna, F.; Bernardelli, B.; Soregaroli, D. *J. Anal. Toxicol.* **1994,** *18,* 278.
2. *Directive No. 1010.1;* U.S. Department of Defense. 1984.
   U.S. Department of Defense memorandum, 1991.
3. *Fed. Regist.* **1994,** *59,* 29908.
4. *Notice to All DHHS/NIDA Certified Laboratories;* U.S. Department of Health and Human Services. 1990.
5. Rodbard, D. *Clin. Chem.* **1974,** *20,* 1255.
6. Cody, J. T. *J. Anal. Toxicol.* **1990,** *14,* 50.
7. Cody, J. T. *J. Anal. Toxicol.* **1990,** *14,* 321.
8. D'Nicuola, J.; Jones, R.; Levine, B.; Smith, M. L. *J. Anal. Toxicol.* **1992,** *16,* 211.
9. Appel, T.; Wade, N. A. *J. Anal. Toxicol.* **1989,** *13,* 274.
10. Jones, R.; Klette, K.; Kuhlman, J. J.; Levine, B.; Smith, M. L.; Watson, C. V.; Selavka, C. M. *Clin. Chem.* **1993,** *39,* 699.
11. Cone, E. J.; Mitchell, J. *J. Forensic Sci.* **1989,** *34,* 32.
12. Henderson, G. L.; Harkey, M. R.; Jones, A. D. *J. Anal. Toxicol.* **1990,** *14,* 172.
13. Altunkaya, D.; Smith, R. N. *Forensic Sci. Int.* **1990,** *47,* 113.
14. Cone, E. J.; Dickerson, S.; Paul, B. D.; Mitchell, J. M. *J. Anal. Toxicol.* **1992,** *16,* 72.
15. Singh, A. K.; Granley, K.; Misrha, U.; Naeem, K.; White, T.; Yin, J. *Forensic Sci. Int.* **1992,** *54,* 9.
16. Niwaguchi, T.; Kanda, Y.; Kishi, T.; Inoue, T. *J. Forensic Sci.* **1982,** *27,* 592.
17. Fitzgerald, R. L.; Ramos, J. M., Jr.; Bogema, S. C.; Poklis, A. *J. Anal. Toxicol.* **1988,** *12,* 255.
18. Rasmussen, S.; Cole, R.; Spiehler, V. *J. Anal. Toxicol.* **1987,** *13,* 263.
19. Bailey, D. *Clin. Toxicol.* **1987,** *25,* 399.
20. Barnett, G.; Willette, R. E. In *Advances in Analytical Toxicology;* Baselt, R. C., Ed.; Year Book Medical: Chicago, IL, 1989; Vol. 2, p 232.
21. Altunkaya, D.; Smith, R. N. *Forensic Sci. Int.* **1990,** *47,* 195.
22. Bergman, R. A.; Lukaszewski, T.; Wang, S. Y. S. *J. Anal. Toxicol.* **1981,** *5,* 85.
23. Childs, P. S.; McCurdy, H. H. *J. Anal. Toxicol.* **1984,** *8,* 220.

24. Clatworthy, A. J.; Oon, M. C. H.; Smith, R. N.; Whitehouse, M. J. *Forensic Sci. Int.* **1990,** *46,* 219.
25. McCurdy, H. H.; Callahan, L. S.; Williams, R. D. *J. Forensic Sci.* **1989,** *34,* 858.
26. Moody, D. E.; Rittenhouse, L. F.; Monti, K. M. *J. Anal. Toxicol.* **1992,** *16,* 297.
27. Cimbura, G.; Koves, E. *J. Anal. Toxicol.* **1981,** *5,* 296.
28. Spiehler, V. R.; Sedgwick, P. *J. Anal. Toxicol.* **1985,** *9,* 63.
29. Smith, F. P. In *Advances in Forensic Science;* Lee, H. C.; Gaensslen, R. E., Eds.; Biomedical: Foster City, CA, 1985; Vol 1.
30. Henderson, D. R.; Friedman, S. B.; Harris, S. B.; Manning, W. B.; Zoccoli, M. A. *Clin. Chem.* **1986,** *32,* 1637.
31. Ruangyuttikarn, W.; Law, M. Y.; Rollins, D. E.; Moody, D. E. *J. Anal. Toxicol.* **1990,** *14,* 160.
32. Cone, E. J. *J. Forensic Sci.* **1989,** *34,* 991.
33. Niedbala, R. In *Face Off with the American Disease;* Roche Diagnostic Systems: Nutley, NJ, 1986; p 27.
34. Schmidt, D. E.; Ebert, M. H. In *Neuromethods 10: Analysis of Psychiatric Drugs;* Boulton, A. A.; Baker, G. B.; Coutts, R. T., Eds.; Hymana: Clifton, NJ, 1988; p 241.
35. *Syva Emit® Drug Abuse Urine Assays Cross-Reactivity List;* Syva: Palo Alto, CA, 1991.
36. Poklis, A.; Hall, K. V.; Eddleton, R. A.; Fitzgerald, R. L.; Saady, J. J.; Bogema, S. C. *Forensic Sci. Int.* **1993,** *59,* 49.
37. Poklis, A.; Fitzgerald, R. L.; Hall, K. V.; Saady, J. J. *Forensic Sci. Int.* **1993,** *59,* 63.
38. Poklis, A.; Hall, K. V.; Still, J.; Binder, S. R. *J. Anal. Toxicol.* **1991,** *15,* 101.
39. Dasgupta, A.; Saldana, S.; Kinnaman, G.; Smith, M.; Johansen, K. *Clin. Chem.* **1993,** *39,* 104.
40. Joseph, R.; Dickerson, S.; Willis, R.; Frankenfield, D.; Cone, E. J.; Smith D. R. *J. Anal. Toxicol.* **1995,** *19,* 13.
41. Poklis, A.; Jortani, S. A.; Brown, C. S.; Crooks, C. R. *J. Anal. Toxicol.* **1993,** *17,* 284.
42. Nixon, A. L.; Long, W. H.; Puopolo, P. R.; Flood, J. G. *Clin. Chem.* **1995,** *41,* 955.
43. Poklis, A. *J. Anal. Toxicol.* **1981,** *5,* 174.
44. Matuch-Hite, T.; Jones, P., Jr.; Moriarity, J. *J. Anal. Toxicol.* **1995,** *19,* 130.
45. Steele, B. W.; Bandstra, E. S.; Wu, N.-C.; Hime, G. W.; Hearn, W. L. *J. Anal. Toxicol.* **1993,** *17,* 348.
46. Sutheimer, C. A.; Hepler, B. R.; Sunshine, I. *J. Anal. Toxicol.* **1983,** *7,* 83.
47. Vogel, J.; Hodnett, C. N. *J. Anal. Toxicol.* **1981,** *5,* 307.
48. Ojanpera, I.; Vuori, E.; Nieminen, R.; Penttila, A. *J. Forensic Sci.* **1986,** *31,* 707.
49. Slightom, E. L.; Cagle, J. C.; McCurdy, H. H.; Castagna, F. *J. Anal. Toxicol.* **1982,** *6,* 22.
50. Gjerde, H.; Christophersen, A. S.; Skuterud, B.; Klenmetsen, K.; Morland, J. *Forensic Sci. Int.* **1990,** *44,* 179.
51. Asselin, W. M.; Leslie, J. M.; McKinley, B. *J. Anal. Toxicol.* **1988,** *12,* 207.
52. Perrigo, B. J.; Joynt, B. P. *J. Anal. Toxicol.* **1989,** *13,* 235.
53. Peel, H. W.; Perrigo, B. J. *J. Anal. Toxicol.* **1981,** *5,* 165.
54. Bogusz, M.; Aderjan, R.; Schmitt, G.; Nadler, E.; Neureither, B. *Forensic Sci. Int.* **1990,** *48,* 27.
55. Blum, L. M.; Klinger, R. A.; Rieders, F. *J. Anal. Toxicol.* **1989,** *13,* 285.
56. Weingarten, H. L.; Trevias, E. L. *J. Anal. Toxicol.* **1982,** *6,* 88.
57. Slightom, E. L. *J. Forensic Sci.* **1978,** *23,* 292.
58. Hepler, B. R.; Sutheimer, C.; Sunshine, I.; Sebrosky, G. F. *J. Anal. Toxicol.* **1984,** *8,* 78.
59. Bress, W. C.; Bidanset, J. H.; Lukash, L. *J. Anal. Toxicol.* **1982,** *6,* 264.
60. Lewellen, L. J.; McCurdy, H. H. *J. Anal. Toxicol.* **1988,** *12,* 2608.
61. Walberg, C. B.; McMarron, M. M.; Schulze, B. W. *J. Anal. Toxicol.* **1983,** *7,* 106.
62. Sarandis, S.; Pichon, R.; Miyada, D.; Pirkle, H. *J. Anal. Toxicol.* **1984,** *8,* 59.
63. Weaver, M. L.; Gan, B. K.; Allen, E.; Baugh, L. D.; Liao, F. Y.; Liu, R. H.; Langner, J. G.; Walia, A. S.; Cook, L. F. *Forensic Sci. Int.* **1991,** *49,* 43.
64. Asselin, W. M.; Leslie, J. M. *J. Anal. Toxicol.* **1991,** *15,* 167.
65. Colbert, D. L.; Childerstone, M. *Clin. Chem.* **1987,** *33,* 1921.
66. Colbert, D. L.; Gallacher, G.; Mainwaring-Burton, R. W. *Clin. Chem.* **1985,** *31,* 1193.
67. Colbert, D. L.; Smith, D. S.; Landon, J.; Sidki, A. M. *Clin. Chem.* **1984,** *30,* 1765.
68. Colbert, D. L.; Smith, D. S.; Landan, J.; Sidki, A. M. *Ann. Clin. Biochem.* **1986,** *23,* 37.
69. Eremin, S. A.; Gallacher, G.; Lotey, H.; Smith, D. S.; Landon, J. *Clin. Chem.* **1987,** *33,* 1903.
70. Bacaj, P. J.; Boyd, J. C.; Herold, D. A. *Clin. Chem.* **1990,** *36,* 818.
71. Caplan, Y. H.; Levine, B.; Goldberger, B. *Clin. Chem.* **1987,** *33,* 1200.
72. Kunsman, G. W.; Manno, J. E.; Cockerham, K. R.; Manno, B. R. *J. Anal. Toxicol.* **1990,** *14,* 149.
73. Ruangyuttikarn, W.; Moody, D. E. *J. Anal. Toxicol.* **1988,** *12,* 229.
74. Nice, A.; Maturen, A. *Clin. Chem.* **1989,** *35,* 1542.
75. Cody, J. T.; Schwarzhoff, R. *J. Anal. Toxicol.* **1993,** *17,* 26.
76. Simonick, T. F.; Watts, V. W. *J. Anal. Toxicol.* **1992,** *16,* 11.
77. Caplan, Y. H.; Levine, B. *J. Anal. Toxicol.* **1989,** *13,* 289.
78. Siff, K. S.; Finkler, A. E. *Clin. Chem.* **1988,** *34,* 1359.

79. Poklis, A. *J. Anal. Toxicol.* **1987,** *11,* 228.
80. Kintz, P.; Mangin, P. *J. Anal. Toxicol.* **1993,** *17,* 222.
81. Lee, C.-W.; Lee, H.-M. *J. Anal. Toxicol.* **1989,** *13,* 50.
82. Kintz, P.; Tracqui, A.; Mangin, P.; Lugnier, A.; Chaumont, A. *Clin. Chem.* **1988,** *34,* 2374.
83. Caplan, Y. H.; Levine, B. *J. Anal. Toxicol.* **1988,** *12,* 265.
84. McCord, C. E.; McCutcheon, J. R. *J. Anal. Toxicol.* **1988,** *12,* 295.
85. Craine, J. E. In *Analytical Methods in Forensic Chemistry;* Ho, H. H., Ed.; Ellis Horwood: New York, 1990; p 371.
86. Ross, R.; Horwitz, C. A.; Hager, H.; Usategui, M.; Burke, M. D.; Ward, P. C. J. *Clin. Chem.* **1975,** *21,* 139.
87. Heveran, J. E.; Cox, M.; Tonchen, A.; Bergamini, J. A. *J. Forensic Sci.* **1978,** *23,* 470.
88. Aoki, K.; Kuroiwa, Y. *Forensic Sci. Int.* **1985,** *27,* 49.
89. Niwaguchi, T.; Inoue, T.; Kishi, T.; Kanda, Y.; Niwase, T.; Nakadate, T.; Inayama, S. *J. Forensic Sci.* **1979,** *24,* 319.
90. Looney, C. E. *J. Clin. Immunoassay* **1984,** *7,* 90.
91. Cone, E. J.; Huestis, M. A. *Forensic Sci. Rev.* **1989,** *1,* 121.
92. Baselt, R. C.; Chang, J. Y.; Yoshikawa, D. M. *J. Anal. Toxicol.* **1990,** *14,* 383.
93. Mieczkowski, T. *Microgram* **1995,** *28,* 193.
94. ElSohly, M. A.; Jones, A. B. *Forensic Sci. Rev.* **1989,** *1,* 13.
95. Fitzgerald, R. L.; Ramos, J. M., Jr.; Bogema, S. C.; Poklis, A. *J. Anal. Toxicol.* **1988,** *12,* 255.
96. Cody, J. T. *J. Chromatogr.* **1992,** *580,* 77.
97. Cody, J. T. *Forensic Sci. Rev.* **1993,** *5,* 111.
98. Tarver, J. A. *J. Anal. Toxicol.* **1994,** *18,* 183.
99. Olsen, K. M.; Gulliksen, M.; Christophersen, A. S. *Clin. Chem.* **1992,** *38,* 611.
100. Poklis, A. *Clin. Chem.* **1992,** *38,* 2560.
101. Crane, T.; Dawson, C. M.; Tickner, T. R. *Clin. Chem.* **1993,** *39,* 549.
102. Badcock, N. R.; Zoanetti, G. D. *Clin. Chem.* **1987,** *33,* 1080.
103. Rollins, D. E.; Jennison, T. A.; Jones, G. *Clin. Chem.* **1990,** *36,* 602.
104. Berkabile, D. R.; Meyers, A. *J. Anal. Toxicol.* **1989,** *13,* 63.
105. Joern, W. A. *J. Anal. Toxicol.* **1989,** *13,* 126.
106. Fenton, J.; Schaffer, M.; Chen, N. W.; Bermes, E. W. *J. Forensic Sci.* **1980,** *25,* 314.
107. Lehrer, M.; Meenan, G. M. *J. Anal. Toxicol.* **1990,** *14,* 62.
108. Taylor, J. *J. Forensic Sci.* **1989,** *34,* 1055.
109. Isenschmid, D.; Caplan, Y. H. *J. Forensic Sci.* **1989,** *34,* 1056.
110. Giblin, V. R.; Hite, S. A.; Samuels, M. S.; Ragan, F. A., Jr. *J. Anal. Toxicol.* **1983,** *7,* 297.
111. Joseph, R.; Dickerson, S.; Willis, R.; Frankenfield, D.; Cone, E. J.; Smith, D. R. *J. Anal. Toxicol.* **1995,** *9,* 13.
112. Tamayo, C. L.; Tena, T. *J. Anal. Toxicol.* **1991,** *15,* 159.
113. Crane, T.; Badminton, M. N.; Dawson, C. M.; Rainbow, S. J. *Clin. Chem.* **1993,** *39,* 549.
114. Linder, M. W.; Valdes, R., Jr. *Clin. Chem.* **1994,** *40,* 1512.
115. Critchfield, G. C.; Wilkins, D. G.; Loughmiller, D. L.; Davis, B. W.; Rollins, D. E. *J. Anal. Toxicol.* **1993,** *17,* 69.
116. Cody, J. T. *Forensic Sci. Rev.* **1990,** *2,* 63.
117. Smith, F. P.; Reuschel, S. A.; Jenkins, K. C. *Science & Justice* **1995,** *35,* 65.
118. Cody, J. T.; Schwarzhoff, R. *J. Anal. Toxicol.* **1989,** *13,* 277.
119. Bronner, W.; Nyman, P.; von Minden, D. *J. Anal. Toxicol.* **1990,** *14,* 368.
120. Schwarzhoff, R.; Cody, J. T. *J. Anal. Toxicol.* **1993,** *17,* 14.
121. Mikkelsen, S. L.; Ash, K. O. *Clin. Chem.* **1988,** *34,* 2333.
122. Schwartz, R. H.; Hayden, G. F.; Riddile, M. *Am. J. Dis. Child.* **1985,** *139,* 1093.
123. Warner, M. *Clin. Chem.* **1989,** *35,* 648.
124. Wagener, R. E.; Linder, M. W.; Valdes, R., Jr. *Clin. Chem.* **1994,** *40,* 608.
125. Vu Duc, T. *Clin. Chem.* **1985,** *31,* 658.
126. Kim, H. J.; Cerceo, E. *Clin. Chem.* **1976,** *22,* 1935.
127. Goldberger, B. A.; Caplan, Y. H. *Clin. Chem.* **1994,** *40,* 1605.
128. Pearson, S. D.; Ash, K. O.; Urry, F. M. *Clin. Chem.* **1990,** *35,* 636.
129. Colbert, D. L.; Childerstone, M. *Clin. Chem.* **1987,** *33,* 1921.
130. Ruangyuttikarn, W.; Moody, D. E. *J. Anal. Toxicol.* **1988,** *12,* 229.
131. Black, D. L.; Goldberger, B. A.; Caplan, Y. H. *Clin. Chem.* **1987,** *33,* 367.
132. Ward, C.; McNally, A. J.; Rusyniak, D.; Salamone, S. J. *J. Forensic Sci.* **1994,** *39,* 1486.
133. Mule, S. J.; Witlock, E.; Jukofsky, D. *Clin. Chem.* **1975,** *21,* 81.

134. Usategui-Gomez, M.; Heveran, J. E.; Cleeland, R.; McGhee, B.; Telischak, Z.; Awdziej, T.; Grunberg, E. *Clin. Chem.* **1975,** *21,* 1378.

135. Kaul, B.; Davidow, B.; Millian, S. J. *J. Anal. Toxicol.* **1977,** *1,* 14.

136. Abercrombie, M.; Jewell, J. *J. Anal. Toxicol.* **1986,** *10,* 178.

137. Armbruster, D. A.; Schwarzhoff, R. H.; Pierce, B. L.; Hubster, E. C. *Clin. Chem.* **1993,** *39,* 2137.

138. Armbruster, D. A.; Schwarzhoff, R. H.; Pierce, B. L.; Hubster, E. C. *J. Anal. Toxicol.* **1994,** *18,* 110.

139. Armbruster, D. A.; Schwarzhoff, R. H.; Pierce, B. L.; Hubster, E. C. *J. Forensic Sci.* **1993,** *38,* 1326.

140. Budgett, W. T.; Levine, B.; Xu, A.; Smith, M. L. *J. Forensic Sci.* **1992,** *37,* 632.

141. Ferrara, S. D.; Tedeschi, L.; Frison, G.; Brusini, G.; Castagna, F.; Bernardelli, B.; Soregaroli, D. *J. Anal. Toxicol.* **1994,** *18,* 278.

142. Frederick, D. L.; Green, J.; Fowler, M. W. *J. Anal. Toxicol.* **1985,** *9,* 116.

143. Huestis, M. A.; Mitchell, J. M.; Cone, E. J. *Clin. Chem.* **1994,** *40,* 729.

144. Irving, J.; Leeb, B.; Foltz, R. L.; Cook, C. E.; Bursey, J. T.; Willette, R. E. *J. Anal. Toxicol.* **1984,** *8,* 192.

145. Jones, A. B.; ElSohly, H. N.; Arafat, E. S.; ElSohly, M. A. *J. Anal. Toxicol.* **1986,** *10,* 178.

146. Kogan, M. J.; Razi, J. A.; Pierson, D. J.; Willson, N. J. *J. Forensic Sci.* **1986,** *31,* 494.

147. Swartz, J. *Clin. Chem.* **1992,** *38,* 1000.

148. Wells, D. J.; Barnhill, M. T. *Clin. Chem.* **1989,** *35,* 2241.

149. Armbruster, D. A.; Hubster, E. C.; Kaufman, M. S.; Ramon, M. K. *Clin. Chem.* **1995,** *41,* 92.

150. Baker, D. P.; Murphy, M. S.; Shepp, P. F.; Royo, V. R.; Caldarone, M. E.; Escoto, B.; Salamone, S. J. *J. Forensic Sci.* **1995,** *40,* 108.

151. Fyfe, M. J.; Chand, P.; McCutchen, C.; Long, J. S.; Walia, A. S.; Edwards, C.; Liu, R. H. *J. Forensic Sci.* **1993,** *38,* 156.

152. Brendler, J.; Liu, R. H. **1997,** *43,* 688.

153. Altunkaya, D.; Clatworthy, A. J.; Smith, R. N.; Start, I. J. *Forensic Sci. Int.* **1991,** *50,* 15.

154. Moody, D. E.; Rittenhouse, L. F.; Monti, K. M. *J. Anal. Toxicol.* **1992,** *16,* 297.

155. Baugh, L. D.; Allen, E. E.; Liu, R. H.; Langner, J. G.; Fentress, J. C.; Chadha, S. C.; Cook, L. F.; Walia, A. S. *J. Forensic Sci.* **1991,** *36,* 79.

156. Henderson, L. O.; Powell, M. K.; Hannon, W. H.; Miller, B. B.; Martin, M. L.; Hanzlick, R. L.; Vroon, D.; Sexson, W. R. *J. Anal. Toxicol.* **1993,** *17,* 42.

157. Cone, E. J.; Dickerson, S.; Paul, B. D.; Mitchell, J. M. *J. Anal. Toxicol.* **1993,** *17,* 156.

158. Liu, R. H.; Edwards, C.; Baugh L. D.; Weng, J.-L.; Fyfe, M. J.; Walia, A. S. *J. Anal. Toxicol.* **1994,** *18,* 65.

159. *Abuscreen ONTRAK® Assay for Amphetamines, Barbiturates, Benzodiazepines, Cocaine Metabolites, Morphine, PCP, and THC;* Roche Diagnostic Systems: Branchburg, NJ; 1992.

160. Armbruster, D. A.; Krolak, J. M. *J. Anal. Toxicol.* **1992,** *16,* 172.

161. Buechler, K. F.; Moi, S.; Noar, B.; McGrath, D.; Villela, J.; Clancy, M.; Shenhav, A.; Colleymore, A.; Valkirs, G.; Lee, T.; Bruni, J. F.; Walsh, M.; Hoffman, R.; Ahmuty, F.; Nowakowski, M.; Buechler, J.; Mitchell, M.; Boyd, D.; Stiso, N.; Anderson, R. *Clin. Chem.* **1992,** *38,* 1678.

162. Röhrich, J.; Schmidt, K.; Bratzke, H. *J. Anal. Toxicol.* **1994,** *18,* 407.

163. *Triage™ Panel for Drugs of Abuse: Directions for Use, Biosite;* Diagnostics: San Diego, CA; 1991.

164. Wu, A. H. B.; Wong, S. S.; Johnson, K. G.; Callies, J.; Shu, D. X.; Dunn, W. E.; Wong, S. H. Y. *J. Anal. Toxicol.* **1993,** *17,* 241.

165. Koch, T. R.; Raglin, R. L.; Kirk, S.; Bruni, J. F. *J. Anal. Toxicol.* **1994,** *18,* 168.

166. Jenkins, A. J.; Mills, L. C.; Darwin, W. D.; Huestis, M. A.; Cone, E. J.; Mitchell, J. M. *J. Anal. Toxicol.* **1993,** *17,* 292.

167. Jenkins, A. J.; Darwin, W. D.; Huestis, M. A.; Cone, E. J.; Mitchell, J. M. *J. Anal. Toxicol.* **1995,** *19,* 5.

Chapter 6
Chromatographic Methods

Classification and Complementary Nature of Chromatographic Methods

Chromatographic methods of analysis involve the introduction of a multicomponent mixture onto a stationary phase, followed by separation of the mixture by the constant passage of a mobile phase. The stationary phase can be an open sheet of paper (paper chromatography), a plate (thin-layer chromatography), a confined column (column chromatography) packed with adsorbents or inert particles coated (physically or chemically) with selected chemicals, or a capillary column with the inner wall coated with stationary phases. The mobile phase can take the form of a liquid (liquid chromatography, LC) or a gas (gas chromatography, GC), with the limitation that the latter mobile phase is applicable only to column chromatography. These variations are shown schematically in Figure 6.1 [1].

Throughout this text, the "HP" in the commonly used abbreviation HPLC is used to mean "high-pressure" instead of "high-performance" for the following reasons:

1. "Performance" is a subjective appraisal and not an objective descriptive word.

2. In certain applications, LC techniques that cannot be classified as HPLC are still the methods of choice.

The underlying operational conditions of GC limit its use to volatile compounds, i.e., small and nonpolar or small but derivatized polar compounds. The use of a liquid mobile phase in LC extends the amenable analytes to larger and to small but polar compounds, whereas most of the macrobiomolecules can only be analyzed by electrophoresis techniques. Thus, GC, LC, and electrophoresis are complementary instrumental approaches providing separation for the complete spectrum of molecules.

GC performed with capillary columns can provide the highest degree of resolution of the analytes, whereas the superior compatibility of the liquid mobile phase with many analytes renders the use of LC advantageous in many applications. Thus, liquid and capillary GC techniques are widely used for the analysis of drugs of abuse.

Chromatographic methodologies by themselves can only resolve analytes into separate components; they do not provide adequate identification data to meet the requirements of a modern confirmatory test. Therefore, many analytical methodologies that are capable of providing more definitive data are adopted as the detector to provide more specific or conclusive identification or to enhance the limit of detection. Modern utilization of chromatographic methods is characterized as follows:

1. Conventional column LC methods are now widely utilized as a sample-pretreatment technique in solid-phase extraction approaches (*see* Chapter 2).

2. Both HPLC and capillary GC are used as preliminary screen or semiconclusive analytical tools.

3. Chromatographic methods are used in combination with other analytical methodologies capable of providing structural information, to provide definite "fingerprint" data for unambiguous identification or enhanced detection of the analytes.

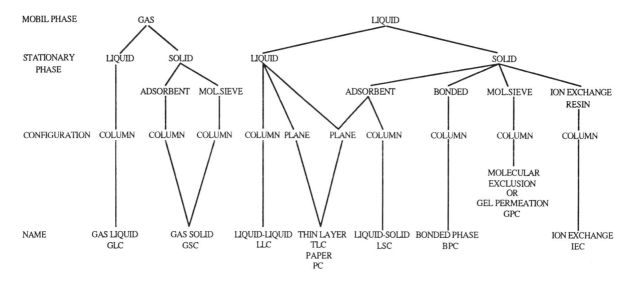

Figure 6.1. Classification of chromatographic systems. (Reproduced with permission from Ref. [1].)

The use of LC methods as solid-phase extraction approaches is discussed in Chapter 2. Materials included in this chapter emphasize the basic separation approaches and practical considerations of non-mass-spectrometric detection methods. GC–MS and LC–MS applications are discussed in Chapter 9.

Operational Characteristics of GC and HPLC

Because chromatographic techniques are mainly used for analyte resolutions that are achieved in the column, the column is considered the heart of a chromatographic system. The selection of a column with a suitable stationary phase that may effectively resolve the mixture components is essential. The advance in capillary column technology in recent years has revolutionized the chromatographic methods of analysis. Because a much longer capillary column can be used to achieve the desired degree of separation without seriously restricting the flow rate, the separation ability (number of theoretical plates) per unit column length is not as critical in capillary chromatography. A wide variety of compounds can be resolved with the use of only a limited number of columns (stationary phases) of sufficient length. On the other hand, a wide variety of liquids with different affinity to the analytes can be used as the mobile phase in HPLC. Thus, analyte separation can be achieved through careful selection of the mobile phase using a limited number of stationary phases.

With a selected column, the approaches used in optimizing analyte separation in GC and HPLC operations are different. With a gas as the mobile phase, the column temperature can be conveniently varied. The parallel HPLC operation involves varying the mobile-phase polarity by changing the relative amounts of a multiple-component mobile phase. It should be noted, however, that the increase in GC column temperature increases the viscosity of the carrier gas, which results in a decrease of flow rate, whereas the HPLC gradient elution changes the mobile-phase viscosity, requiring a higher operation pressure to deliver a constant flow rate.

The effectiveness of a chromatographic system is often reflected by the *column efficiency*. The column efficiency is commonly expressed in terms of the height equivalent to a theoretical plate (HETP) and related to the linear gas velocity (in GC), μ, by the Van Deemter equation:

$$HETP = A + B/\mu + C \cdot \mu \qquad (6.1)$$

where A, B, and C are constants. Thus, maximum column efficiency is reached at an optimal gas linear velocity, which varies with the carrier gas used (Figure 6.2) [2].

With the mass-transfer mechanism of the liquid mobile phase [1], the corresponding HETP curve (Figure 6.3) [1] does not reach an optimal value. Although a continuous decrease in flow rate will continue to improve the separation efficiency, the flow rate cannot be unlimitedly reduced because of the unacceptable resulting retention time.

Gas Chromatography

Basic Operational Principles and Parameters

A basic GC system includes a pressurized inert-gas reservoir with regulated flow control, a sample introduction port (injector), the analyte-resolving column (column), a detection device (detector), and a data presentation and processing unit. The injector, column, and detector are installed inside blocks in which the operation temperature can be controlled, while the data presentation and processing unit for a modern GC system is normally a computer system instead of the conventional strip chart recorder.

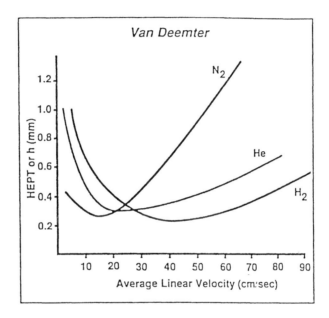

Figure 6.2. HETP vs. average linear velocity of various carrier gases. HETP (h): height equivalent to a theoretical plate. (Reproduced with permission from Ref. [2].)

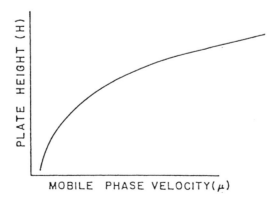

Figure 6.3. Typical plate height vs. mobile-phase velocity. (Reproduced with permission from Ref. [1].)

Stationary Phases

The most commonly used stationary phases in modern capillaries are ethylene glycol-based (Structure 6.1) and siloxane-based (Structure 6.2) polymers. Carbowax 20M is an example of the former category of stationary phases, and it is widely used for the separation of analytes with substantial polarity. Other analytes can be resolved using the polysiloxane stationary phases with various substitution groups (CH_3, phenyl, CN, and CF_3, in increasing polarity) at various percentages.

$$HO-CH_2-CH_2-[O-CH_2-CH_2]_n-OH$$

1

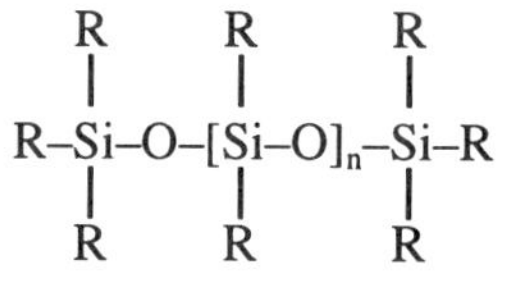

$$R-Si-O-[Si-O]_n-Si-R$$

2

Chart 6.1. Structures of ethylene glycol-based (1) and siloxane-based (2) stationary phases.

To increase the stability and lifetime of a column, modern stationary phases are covalently bonded to the inner wall of fused silica tubing, with the strands of the polymers further joined by cross-linkage. The outer wall of the column is normally finished with polyimide coatings.

Table 6.1 [3] is a list of stationary phases used by a commercial source and the equivalent products marketed by various manufacturers. This information is intended to provide a quick cross-reference for chromatographers who may find the various column names appearing in the literature confusing. These column names by themselves often do not reveal the nature of the stationary phases.

Optimization of Analysis—Selection of Column Internal Diameter, Stationary-Phase Film Thickness, Carrier Gas, and Stationary Phases

The phase ratio, β, which is the ratio of the volumes of the gas and liquid phases (equation 6.2), is controlled by the column internal diameter (ID) and the film thickness. A column with ID < 0.53 mm having a film thickness of 1 μm is considered a narrow-bore column, whereas a wide-bore column (0.53 mm ID) with a film thickness of 3–5 μm is considered a thick-film capillary [4]. The phase ratio, β, is given by

$$\beta = r/2d_f \tag{6.2}$$

where r is the column radius (μm), and d_f is the film thickness (μm).

The retention time, t_R, which determines the speed of analysis for a given analyte, is determined [5] by extrinsic parameters such as partition ratio [k = (the amount of solute in the liquid phase) ÷ (the amount of solute in the gas phase)], the column length, L, and the average linear velocity, μ, as shown in equation 6.3. It can be further related to the column radius and film thickness by using the relationship between the partition ratio and the column phase ratio shown in equation 6.4.

$$t_R = (L/\mu)(k + 1) \tag{6.3}$$

$$k = K/b = K \cdot 2d_f/r \tag{6.4}$$

where K is the partition coefficient ([concentration of solute in the stationary phase] ÷ [concentration of solute in the gas phase]), an intrinsic property of the liquid phase.

Optimization of Analysis Speed. On the basis of equations 6.2–6.4, it is apparent that the speed of analysis can be improved by using a shorter column, a higher carrier-gas average linear velocity, and a smaller amount of liquid phase that has limited retention property (smaller K). However, there are limits on what can be achieved by improving the speed of analysis without sacrificing the quality of separation.

Table 6.1. GC Phase cross reference.[a]

| | Equivalent products from other manufacturers | |
Phase	Capillary	Packed
100% Methyl	007-1 (MS), 5CB, BP-1, CB-1, CP-Sil, DB-1, HP-1, HP-101, OV-1, PE-1, RSL-150, RSL-160, Rt$_x$-1, SP-2100, SPB-1, Ultra-1	CP-Sil 5, DC-200, OV-1, OV-101, S96, SE-30, SP-2100, UC-W982
5% Phenyl–95% Methyl	007-2 (MPS-5), BP-5, CB-5, CP-Sil 8CB, DB-5, HP-2, RSL-200, OV-5, PE-5, Rt$_x$-5, SE-52, SE-54, SPB-5, Ultra-2	CP Sil 8, Dexsil 300, Florolube, OV-3, OV-17
6% Cyanopropylphenyl–14% Methyl	DB-1301	None
14% Cyanopropylphenyl–86% Methyl	007-1701, BP-10, CB-1701, CP-Sil 19CB, DB-1701, OV-1701, PE-1701, Rt$_x$-1701, SPB-7	OV-1701
50% Phenyl–50% Methyl	007-17 (MPS-50), DB-17, HP-17, PE-17, RSL-400, Rt$_x$-50, SP-2250	DC-710, HP-17, OV-11, OV-22, SP-2250
50% Trifluoropropyl–50% Methyl	DB-210, RSL-500, SP-2401	OV-202, OV-210, OV-215, QF-1, SP-2401
50% Cyanopropylphenyl–50% Methyl	007-225, BP-225, CB0225, CP-Sil 43CB, DB-225, HP-225, OV-225, PE-225, RSL-500, RTx-225, SP-2330	OV-225, Silar 5 CP
Polyethylene glycol	007-CW, BP-20, Carbowax, CB-WAX, CP-Wax 52CB, DB-WAX, HP-20M, PE-CW, Stabilwax, Supelcowax 10, SUPEROX II	Carbowax 20M, PEG, Supelcowax Supelcowax 10, Superox
Polyethylene glycol (acid modified)	007-FFAP, DB-FFAP, HP-FFAP, Nukol, OV-351, PE-FFAP, Stabilwax-DA, SUPEROX FA	Carbowax 20M–Terephthalic acid FFAP, OV-351
50% Cyanopropyl–50% Methyl	007-CPS-1, DB-23, PE-CPS-1, RTx-2330, SP-2330	CP-Sil 58, SP-2310, SP-2330
Polyethylene glycol (base modified)	CAM, Stabilwax-DB	None
(Not specified)	007-608, DB-608, NON-PAKD Pesticide, PE-608, SPB-608	None
(Not specified)	007-502, 007-624, DB-624, NON-PAKD AT-624, PE-624, Rt$_x$-502.2, Rt$_x$-VOCOL, Volatiles	None
(Not specified)	DB-5.625, PET-5	None
(Not specified)	GS-Q, PoraPLOT Q	Chromosorb 101, HayeSep Q, Porapak Q
(Not specified)	Cyclodex-B, LIPODEX C	None

[a] Data are adapted from Ref. [2].

With the use of a shorter column, the number of theoretical plates, N, available will be reduced. Fortunately, because N is inversely proportional to the column radius, as seen in equations 6.5 and 6.6, the loss of N due to the use of a shorter column can be compensated, to some extent, by using a column with a smaller column radius.

$$N = L/HETP \tag{6.5}$$

$$HETP = r(1 + 6k + 11k^2)/[3(1 + k)^2]^{1/2} \tag{6.6}$$

The use of a column with a smaller radius and a shorter length further improves the speed of analysis because the optimal μ is inversely proportional to the column radius and length. The consequential reduction of the sample loadability can be compensated by increasing the film thickness which, unfortunately, becomes counterproductive in increasing the k term in equation 6.3.

Analysis speed can also be improved by using a carrier gas with a higher optimal value of μ. Because hydrogen is cheaper and can still provide a high efficiency when the adopted μ is well above the optimum gas velocity (a common practice), it is a preferred [6] carrier gas provided that necessary safety precautions are taken. Higher efficiency means that a shorter column can be used to achieve the desired separation with sharper and higher peaks.

Limitations. The analysis time can also be improved, with limitations, through the use of a stationary phase with a lower intrinsic retention property, K, and operation at a higher temperature. The optimization limitations associated with these two factors are shown as follows:

$$n_{req} = 16R^2[(k + 1)/k]^2[\alpha/(\alpha - 1)]^2 \qquad (6.7)$$

$$R = (t_{R,2} - t_{R,1})/[0.5(W_1 + W_2)] \qquad (6.8)$$

$$\alpha = t'_{R,2}/t'_{R,1} \qquad (6.9)$$

where n_{req} is the number of plates required to achieve the desired separation, R is resolution, $t_{R,1}$ and $t_{R,2}$ are the retention times, $t'_{R,1}$ and $t'_{R,2}$ are the adjusted retention times, and W_1 and W_2 are the peak base widths. With the use of a phase of lower retention, the relative retentions of the analytes, the α term, are also usually lowered. Thus, the $[\alpha/(\alpha - 1)]^2$ term becomes large and requires more theoretical plates to achieve the same degree of resolution.

Analysis time can also be reduced by altering the extrinsic retention property, the k term, of the stationary phase by operating at a higher temperature. Again, there are counterproductive limitations: as k becomes too small, the $[(k + 1)/k]^2$ term becomes large and the number of theoretical plates required to achieve a desired resolution increases.

Chemical Derivatization

Merits of Chemical Derivatization

Ideally, an analyte should be tested in its original form so that there will not be any doubt about its identity and quantity. The conversion of an analyte to a different form (derivatization) prior to analysis involves an additional chemical step that is not only time-consuming but also introduces potential impurities, uncertainties about the completeness of the conversion, and possible interferences or interpretation difficulties associated with the adopted chemical reaction. In spite of these potential complications, many drugs are still derivatized prior to undergoing GC analysis for the following reasons [7]:

1. Conferment of volatility;
2. Improvement of stability;
3. Improvement of chromatographic properties;
4. Improvement of separations;
5. Analysis of functional groups;
6. Provision for selective detection (non-mass-spectrometric);
7. Production of mass shift in mass spectra;
8. Modification of fragmentation; and
9. Use of derivatives in conjunction with chemical ionization.

These reasons can be grouped into three basic categories: bringing the analytes to the chemical forms that are compatible with the basic *chromatographic environment,* and maximizing their chromatographic *separation* and *detection* efficiencies.

Compatibility with the Chromatographic Environment. The majority of chemical derivatizations are performed to convert the analytes to chemical forms that are more compatible with the chromatographic environment. Compatibility may be mandatory to successfully chromatograph the analytes or to simply improve performance characteristics, as by producing a convenient retention time, improving peak shape, or preventing adsorptive loss. There is, however, no clear distinction between these two categories; the use of a column with a different stationary phase may render the mandatory requirement an option.

In addition to the obvious volatility concerns, carboxylic acids and amines form strong hydrogen bonds with any of the silanol groups present in the chromatographic system or with components of sample residues left in the injector or column. These undesired interactions can result in a loss of peak response due to irreversible adsorption or peak tailing caused by reversible adsorption [8]. Thus, these hydroxyl (free or part of a carboxylic acid) or amine groups are often converted to an inactive species prior to chromatographic analysis. The chromatograms in Figure 6.4 [8], obtained using a DB-5 column (5% phenyl polysiloxane phase), show the dramatic differences in the chromatographic characteristics of the six amine and alcoholic amine drugs without (top) and with (middle and bottom) derivatization. Thus, with the DB-5 column, quantitative determinations or even qualitative identifications of these compounds cannot be achieved without prior derivatization.

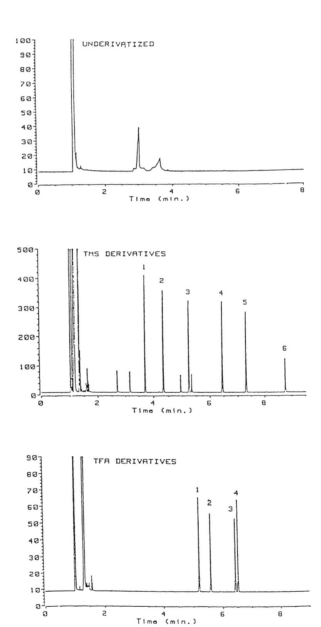

Figure 6.4. Gas chromatograms of underivatized and derivatized amphetamine drugs. (Reproduced with permission from Ref. [8].)

The derivatization of barbiturates represents an effort to improve their chromatographic characteristics. Although derivatization of barbiturates is not mandatory for their GC analysis, barbiturates in their native state tend to cause adsorption and result in material loss, column contamination, and peak tailing. Results can be improved significantly [9] with *N,N*-dimethylation prior to the chromatographic analysis. The methylation process has also been utilized [10] to add additional information for confirmatory identification purposes. In this application [10], extracts obtained from urine samples screened positive by RIA are first chromatographed without derivatization; extracts that show the presence of barbiturates are then derivatized and chromatographed again. With the conformity of two chromatographic parameters to the respective controls, the certainty in confirming the presence of these barbiturates is increased.

Another report [11] on large-scale and routine quantitative analysis of four barbiturates (butalbital, amobarbital, pentobarbital, and secobarbital) clearly demonstrated that methylation greatly improves the analysis of these compounds in the following aspects:

1. Chromatographic peak shapes of these compounds are generally better, and, more important, the interval between maintenances of the column, during which acceptable chromatograms are produced, is greatly lengthened.

2. Analyte stability is significantly improved, as reflected by the observation of more-consistent quantitative results from extraction–derivatization products that are delayed in their GC–MS analysis for different lengths of time following the reconstitution step.

3. Reproducibility in the quantification of control samples is significantly improved.

Achieving Required Separation or Improving Separation Efficiency. Enantiomeric separation by chiral stationary phases is now a common practice; however, many practical separations are still achieved through derivatization with a chiral reagent. This approach will be treated in Chapter 11. The discussion in this section is limited to improvement of separation efficiency for nonenantiomeric applications.

For the convenience of GC analysis, the compounds of interest should ideally be eluted and well resolved within an approximate 2–6 minute retention window, with the column operating isothermally at a reasonably high temperature. Analytes eluted too early may be interfered with by the solvent, whereas elution occurring too late is time-consuming. Isothermal operation is convenient, more reproducible, shows less baseline drift, and minimizes the chance of a gas leak that may develop as a result of temperature cycling, whereas operation at a high temperature helps maintain a clean chromatographic system. Derivatizations are often performed to achieve this ideal analytical condition. Thus, drugs of low molecular weight may be converted to esters or amides with acids or alcohols of higher molecular weight, whereas drugs of higher molecular weight may be converted to fluoro-containing acids or alcohols of lower molecular weight.

The derivatization of ecgonine methyl ester and benzoylecgonine with pentafluoropropionic anhydride [12] for the simultaneous analysis of these two compounds and cocaine is a good example involving the improvement of chromatographic efficiency with derivatization. Although these three compounds can be chromatographed using a DB-5 column [13], the chromatographic conditions utilized and the resulting chromatogram (Figure 6.5) are not as satisfactory as with prior derivatization with pentafluoropropionic anhydride (Figure 6.6) [12]. The chromatogram shown in Figure 6.6 was obtained using a dimethyl silicone (HP-19091-6-312) fused-silica capillary column with temperature programming from 100 to 225 °C at 50 °C/min. Judging from the observed resolution, these three derivatized analytes can be well resolved with an isothermal operation at a reasonably high temperature.

Improvement of Detection and Structure-Elucidation Efficiency. Chemical derivatizations are commonly used for enhancing analyte detection and quantification and structure elucidation. Fluorinated anhydrides are extensively used to prepare fluoroacyl derivatives from alcohols, phenols, and amines for analysis by GC and GC–MS. Although enhancing analyte volatility through the introduction of fluorine atoms may be desirable in some applications, the resulting high volatility of the derivative may prohibit the use of a higher operational temperature and may not be desirable for the analysis of low-molecular-weight analytes such as amphetamine and methamphetamine [14]. Furthermore, the negative inductive effects of the fluorine atoms in the derivatized product were found to cause the derivatized products to be more susceptible to hydrolysis in the presence of moisture [15, 16].

The introduction of these fluorine atoms, however, greatly enhances the detection effectiveness in cases where electron-capture detection [17] is used. For example, an electron-capture detector was used to achieve a 2-pg detection limit for a heptafluorobutyryl derivative of morphine in 1977 [18]. Examples of signal enhancements in applications using mass spectrometric detection are shown in Chapter 9.

In GC–MS applications, mass spectra obtained from thoughtfully designed derivatives can show distinctive characteristics that are not available from parent compounds. The advantages include the generation of ions more suitable for quantification and helpful for structural elucidation, as shown in Chapter 9.

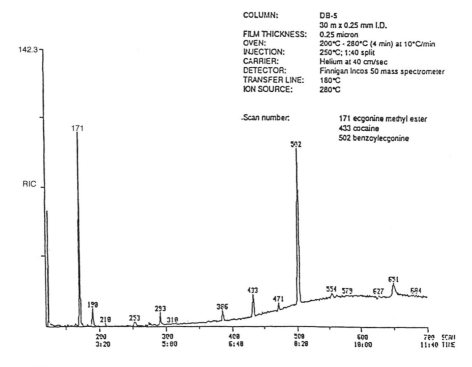

Figure 6.5. A typical chromatogram of underivatized ecgonine methyl ester, cocaine, and benzoylecgonine mixture. (Reproduced with permission from Ref. [13].)

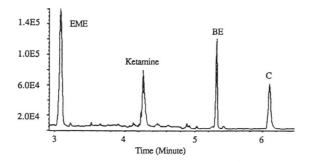

Figure 6.6. A typical chromatogram of PFP–ecgonine methyl ester, PFP-benzoylecgonine, and cocaine. Abbreviations: EME, ecgonine methyl ester; BE, benzoylecgonine; C, cocaine. (Reproduced with permission from Ref. [12].)

Chemical-Derivatization Reactions and Practical Considerations

Information on the chemical derivatization of compounds for chromatographic and related analyses is widely available in the literature [7, 19–28]. Many procedures are also available through commercial suppliers carrying derivatizing reagents [29, 30]. Because a derivatization reaction should be simple, rapid, and stoichiometric, this analytical approach is applied mainly to compounds possessing labile protons on heteroatoms with functional groups such as –COOH, –OH, –NH, and –NH$_2$, although high-yield derivatization at carbon sites has also been reported [31].

Three major categories of derivatization reactions are commonly used for drug analysis; these are silylation, acylation, and alkylation. Table 6.2 lists commonly used derivatizing reagents with brief descriptions of their main characteristics.

Table 6.2. Silylation, acylation, and alkylation derivatizing reagents and characteristics.

Reagent and reaction	Characteristics
Silylation	
N,O-Bis(trimethylsilyl)trifluoroacetamide (BSTFA)	Reacts faster and more completely than BSA
$F_3C-C(O-TMS)=N-TMS + H-Y-R \longrightarrow TMS-Y-R + F_3C-C(=O)-NH-TMS$	Combine with 1% or 10% TMCS for hindered hydroxyl and other functionalities
Trimethylchlorosilane (TMCS)	Commonly used as a catalyst
$TMS-Cl + H-R \longrightarrow TMS-R + HCl$	Reaction by-product HCl
N,O-Bis(trimethylsilyl)acetamide (BSA)	Mild reaction conditions
$H_3C-C(O-TMS)=N-TMS + H-Y-R \longrightarrow TMS-Y-R + H_3C-C(=O)-NH-TMS$	Forms stable products
	By-product TMS–acetamide may elute with analyte
N-Methyltrimethylsilyltrifluoroacetamide (MSTFA)	By-product TMS–acetamide very volatile
$F_3C-C(=O)-N(CH_3)(TMS) + H-Y-R \longrightarrow TMS-Y-R + F_3C-C(=O)-NH-CH_3$	Most suitable for volatile trace analyte
Trimethylsilylimidizole (TMSI)	Reacts with hydroxyl but not amine
$TMS-N(imidazole) + H-Y-R \longrightarrow TMS-Y-R + H-N(imidazole)$	Suitable for hindered hydroxyl group
Trimethylsilyldiethylamine (TMS-DEA)	Basic reagent for amino and carboxylic acids
$TMS-N(C_2H_5)_2 + H-Y-R \longrightarrow TMS-Y-R + H-N(C_2H_5)_2$	
Hexamethyldisilazane (HMDS)	A weak TMS donor
$TMS-NH-TMS + H-Y-R \longrightarrow TMS-Y-R + TMS-NH_2$	
N-Methyl-*N*-(*t*-butyldimethylsilyl)trifluoroacetamide (MTBSTFA)	Exceptionally strong yet mild reagent
$C(CH_3)_3-Si(CH_3)_2-N(CH_3)-C(=O)CF_3 + H-Y-R \longrightarrow$	Stable product in resisting hydrolysis
$C(CH_3)_3-Si(CH_3)_2-Y-R + F_3CC(=O)NHCH_3$	Combine with 1% *t*-butyldimethylchlorosilane catalyst for hindered alcohol and amine
Acylation	
Fluorinated anhydrides (TFA, PFPA, HFBA)[a]	Most commonly used for electron capture detector (ECD)
$O(CC_nF_{2n+1})_2 + R-CH_2OH \longrightarrow R-CH_2-OCCF_3 + F_3COH$	Most reactive and volatile product with $n = 1$, lowest analysis temperature with $n = 2$, and most sensitive to ECD with $n = 3$
	Often used with bases, such as triethylamine

Table 6.2. (Continued.)

Reagent and reaction	Characteristics
Heptafluorobutyrylimidizole (HFBI)	Often a better choice for an ECD
	Reaction fast and mild, works best for phenol, alcohol, and amine
	By-product is not acidic

$$C_3F_7\overset{O}{\overset{\|}{C}}-N\diagdown\diagup N + R-NH_2 \rightarrow C_3F_7\overset{O}{\overset{\|}{C}}NHR + HN\diagdown\diagup N$$

| *N*-Methyl-*N*-bis(trifluoroacetamide) (MBTFA) | Reacts rapidly with primary and secondary amine, slowly with alcohol, phenol, and thiol |
| | Mild reaction conditions with inert and volatile by-products |

$$(CF_3\overset{O}{\overset{\|}{C}})_2N(CH_3) + R-NH_2 \rightarrow CF_3\overset{O}{\overset{\|}{C}}NHR + CF_3\overset{O}{\overset{\|}{C}}NCH_3$$

Pentafluorobenzoyl chloride (PFBCI)	Highly reactive, forming the most sensitive ECD derivatives of amine and phenol
	Suitable for sterically hindered functionalities
	Base often used to remove HCl produced

$$+ HCl$$

Alkylation

| DMF-Dialkylacetal (*n* = 1, 2, 3, or 4) | Most commonly used for carboxyl groups, but also reacts with amine, phenol, and amino acid |
| | With *n* = 1, 2, 3, or 4 to control analyte retention time |

| Trimethylanilinium hydroxide (TMAH) | Commonly used as a flash alkylation reagent |

$$\left[\bigcirc-N(CH_3)_3\right]^+ [OH]^- + RC\overset{O}{\diagup}\diagdown_{OH} \rightarrow RC\overset{O}{\diagup}\diagdown_{OCH_3}$$

| Tetrabutylammonium hydroxide (TBH) | Especially suitable for low molecular weight amines |

$$[N(CH_3)_4]^+ [OH]^- + RC\overset{O}{\diagup}\diagdown_{OH} \rightarrow RC\overset{O}{\diagup}\diagdown_{OC_4H_9}$$

| BF$_3$/Methanol (*n*-Butanol) (*n* = 1 or 4) | Most commonly used to form methyl (butyl) ester with acid |

$$F_3B:\overset{H}{O}-C_nH_{2n+1} + RC\overset{O}{\diagup}\diagdown_{OH} \rightarrow RC\overset{O}{\diagup}\diagdown_{OC_nH_{2n+1}}$$

[a] Abbreviations: TFAA: tetrafluoroacetyl; PFPA: pentafluoropropylacetyl; HFBA: heptafluorobutylacetyl; DMF: dimethylformamide.

Several practical considerations, listed below, have to be addressed when selecting a desirable derivatization reaction and a derivatizing reagent:

1. Safe and easy formation of the derivative with a readily available and inexpensive reagent;
2. High yield of a stable product;
3. Mild reaction conditions preventing an undesirable reaction of the analyte; and
4. No undesirable by-products that may be harmful to the stationary phase.

Thus, historically important diazomethane, used to form methyl ester derivatives from carboxylic acids, is no longer popular because of its high toxicity, explosion hazard, and the formation of artifacts with unsaturated and keto acids.

Catalysts, such as HCl, BF$_3$, and BCl$_3$, are commonly used with alcohols to form ester derivatives of carboxylic acids. The HCl used or formed as a by-product when trimethylchlorosilane is used as the trimethylsilyl (TMS) derivatization reagent should be removed prior to the introduction of the derivatization product to a GC or a GC–MS system. Thus, pyridine and dimethylformamide are commonly used as the solvents because they also act as acid scavengers. Similarly, triethylamine or 5% bicarbonate is used as a neutralization agent when trifluoroacetic acid is formed in trifluoroacetyl derivatization.

Because the TMS derivatives are susceptible to hydrolysis in the presence of moisture (stability decreases in the order of TMS ethers > TMS esters > TMS amines [29]), exposure of the derivatization product to the atmosphere should be limited, especially when the derivatives are not analyzed immediately.

Representative Derivatization Approaches

Although not intended to be an exhaustive list, Table 6.3 gives a summary of references representing the commonly applied derivatization approaches. These references reflect the use of various derivatizing reagents to bring about the compatibility of the analytes with the chromatographic environment, improve chromatographic characteristics, enhance detection limits, or achieve special analytical goals. Thus, various benzodiazepines are derivatized with alkyl and trimethylsilyl groups

Table 6.3. Derivatization in drug analysis.

Compound category	Derivatization agent	Analytical procedure[a]	Analytical objective	Ref.
Amphetamine-related amine drugs	Acetyl and analogs, trichloroacetyl, trifluoroacetyl and analogs	GC–MS	Separation and SIM effectiveness	[14]
	4-Carbethoxyhexafluorobutyl	Same as above	Same as above	[32]
	Trimethylsilyl	Same as above	Same as above	[14]
	Dansyl	MS	Identification	[33]
	Fluorobenzoyl and analogs	NICI	Detection enhancement	[34, 35]
	Fluorophthaloyl	ECD	Same as above	[34]
	N-(R)-α-phenylbutyryl	GC–MS	Enantiomer separation	[36]
	1-Naphthoyl	LC–MS	Enantiomer separation	[37]
Barbiturates	Methyl	GC–MS	Improvement of chromatography	[38, 39]
	Methyl	GC, GC–MS	On-column derivatization	[40–42]
	Trimethylsilyl and d_9-analog	GC–MS	Structure elucidation	[43]
	2-Chloroethyl	GC–ECD	Detection enhancement	[44]
	Pentafluorobenzyl	GC–ECD	Same as above	[45]
Benzodiazepines	Alkyl	GC; specific detection	Improvement of detection limit	[46–48]
	Trimethylsilyl	GC–MS	Quantification in urine	[49]
Benzoylecgonine	Alkyl	GC–MS	General analysis	[59]
	Fluoroalkyl	GC; specific detection	General analysis	[60]
	Trimethylsilyl	GC-FID	General analysis	[61]
Cannabinoids	Fluorobenzyl	GC; specific detection	Improvement of detection limit	[50]
	Methyl	GC–MS (full scan)	Improvement of identification	[51]
	Trimethylsilyl	GC–MS	Effectiveness in derivatization	[52, 53]
	Trimethylsilyl and analogs, d_9-trimethylsilyl	GC–MS	Identification	[54]
	Methylboronate and trimethylsilyl	GC–MS	Comparison of derivatization	[55]
	Vinyldimethylsilyl	GC–MS	Structure elucidation	[56]
	t-Butyldimethylsilyl chloride, trimethyl-silylketene, diethylphosphate	GC; HPLC; MS	Separation and identification	[57]
	Dabsyl	HPLC, TLC; CI	Separation and detection	[58]
Ecgonine	Fluorobenzoyl	GC–MS	General analysis	[49]
Fentanyls	Heptafluorobutyryl	GC-ECD	Routine detection	[62]
Lysergic acid	Fluoroacetyl	GC/NICI	Quantification	[63]
	Trimethylsilyl	GC–MS, GC	General analysis	[64–66]
Opiates	Acetyl	GC–MS (SIM)	Comparison of derivatization method	[15, 16]
	Fluoroacetyl and analog	GC; GC–MS (SIM)	Comparison of derivatization method; detection limit	[16, 18, 67, 68]
	Trimethylsilyl	GC–FID	Routine analysis	[69, 70]
		GC–MS	Comparison of derivatization method	[15, 16]
Psilocybin	Trimethylsilyl	GC–MS	General analysis	[71]

[a] Abbreviations: NICI: negative ion chemical ionization; CI: chemical ionization; FID: flame ionization detector; SIM: selected ion monitoring.

to prevent ring construction (increase the stability) at elevated temperatures; various alkylation and trimethylsilylation approaches are adopted to improve the chromatographic characteristics and to generate mass spectra of detailed structural information; detection enhancements with electron-capture detection are studied using various fluorinated derivatizing reagents; chiral derivatizing reagents are used to differentiate enantiomers; and deuterated derivatizing reagents are used for analyte structural elucidation.

Retention Data as a Basis for Preliminary Drug Identification

Because the commonly used immunoassays are developed with specific drug categories in mind, they are not suitable for applications in which the presence of a broad spectrum of drugs needs to be identified. On the other hand, the "wide window" provided by modern high-resolution chromatography is ideal for the preliminary identification of a wide range of compounds in a totally unknown specimen. The window widens further if the sample is screened with and then without derivatization [72]. Gas–liquid chromatographic approaches for drug screening range from the observation of the relative retention time (with respect to caffeine) [73] to a sophisticated and definite identification using dual-column and mass-spectrometric methods of detection [74].

Identification Based on Relative Retention Time Data

Table 6.4 lists numerous studies reporting the use of relative retention time data for the identification of a broad range of drugs in samples of interest. Because these studies are for preliminary screen purposes, minimal sample preparation is also emphasized.

To improve identification reliability, relative retention time data are obtained from two columns. Dual-column data may be obtained sequentially by injecting a sample into two GC systems equipped with two different columns [75, 76], or obtained simultaneously by injecting a sample into a GC system equipped with a dual-column arrangement [83–87].

Identification Based on Retention Index Data

Kovats's retention indices (RIs), as defined in Equation 6.10 [88], offer several advantages over relative retention time information. With the use of references covering the full retention window, RI values are essentially calibrated every 100 units and are therefore more reproducible and less susceptible to changes in chromatographic conditions. RI is defined by

$$I = 100n + 100(N - n)[(\log RT'_{analyte} - \log RT'_n)/(\log RT'_N - \log RT'_n)] \qquad (6.10)$$

where I is the retention index; RT' is the net retention time, and n and N are the carbon numbers of the smaller and the larger n-paraffins that bracket the analyte.

Comparability of Retention Indices Obtained under Various Conditions.

To standardize chromatographic data and improve the utilization of literature and library data for drug identification, Kazyak and Permisohn [89] explored the use of Kovats's indices [88] obtained from three columns, QF-1 and OV-1 (or OV-17). This approach was further studied by other workers [90, 91] and has been proven effective in packed-column applications. Further studies have also shown that, under defined conditions, RI values derived from wide-bore capillary columns were comparable to those available in a data bank obtained on packed columns [92, 93]. Despite difficulties observed by some workers [94], it has now been concluded [95] that capillary systems can generate reproducible RI data.

Listed below are some important factors that have to be considered and established before the RI information can be systematically utilized for drug identification:

1. Factors affecting the intralaboratory reproducibility of RI information;

2. Intralaboratory comparability of RI information obtained from various types of columns; and

3. Interlaboratory reproducibility of RI information.

Table 6.4. Relative retention time data as the basis for drug identification.

Drug category	Sample type	Extraction solvent	Retention ref.	Ident. parameter	Column	Detector[a]	Ref.
Basic	Liver	Butyl acetate (pH 9)	Cyclizine, mesoridazine	Rel. ret. time	SE-52	NPD	[77]
Basic	Urine	Isoamyl alcohol/hexane (pH 9.5)	Proadifen	Rel. ret. time	DB-1701	NPD	[78]
Selected acidic	Urine	Chloroform	Aprobarbital	Rel. ret. time	SP-2250	FID	[79]
	Blood	Toluene/ether (alkaline)	Barbital, trihexyphenidyl		DB-1701	NPD	[80]
Acidic, basic	Serum, urine	Dichloromethane (pH 2 and 9)	Cyheptamide	Rel. ret. time	SE-54	FID	[81]
Basic	Postmortem blood	n-Butyl chloride (pH 9.0–9.5)	Prazepam	Rel. ret. time	3% OV-1, 3% OV-17	NPD	[75]
Basic	Blood, tissue	n-Butyl chloride (alkaline)	Cholestane, naphthalene	Rel. ret. time	OV-1, OV-17, SE-30, Apiezon L (10% KOH)	FID	[76]
Basic, neutral	Blood	Butyl chloride	Cyclizine	Rel. ret. time	BP-5	NPD	[82]
Basic	Urine	Isoamyl alcohol/n-butyl chloride	Proadifen	Dual-column; rel. ret. time	SE-54, OV-101	NPD	[83]
Basic, neutral	Blood	n-Butyl chloride (pH 9.0)	Proadifen	Dual-column; rel. ret. time	Ultra-1, HP-17	NPD	[84]
Basic, neutral	Plasma	Ether (basic pH)	RN 927	Dual-column; rel. ret. time	Ultra-1, CP Sil 19 CB	NPD	[85]
Selected	Serum	Charcoal/ether	Barbital	Dual-column; rel. ret. time	3% OV-17, 3% OV-1	NPD	[86]
Selected	Urine	Charcoal/CHCl$_3$/i-BuOH/ether	Barbital	Dual-column; rel. ret. time	3% OV-17, 3% OV-1	NPD	[87]

[a] Abbreviations: NPD: Nitrogen–phosphorus selective detector; FID: flame ionization detector.

Because high drug concentrations overloading the capillary column will affect RI values [96, 97], precautions should be taken to avoid errors that may derive from this cause. It has been reported [98], however, that the biological matrix and method of isolation may affect the detectability, but they have no significant influence on RI values.

Standard deviations of interlaboratory RI data from a packed SE-30 column were found [90] to be 15 to 20 RI units; however, RI reproducibility was found to depend on the drugs and columns used. For example, nitrazepam and strychnine showed [96] extremely large RI differences on all four columns (bonded CP-Sil 5, wide-bore CP-Sil 5, narrow-bore CP-Sil 5, and OV-1701) studied; paracetamol, butallylonal, primidone, phenytoin, MMDA (3-methoxy-4,5-methylenedioxyamphetamine), and nitrazepam were found to exhibit an RI unit shift of more than 75 between a DB-1 fused silica column and a 3% SE-30 packed column. When RI values from narrow-bore BP-1, wide-bore BP-1, and wide-bore DB-1 were compared [99] with values obtained using a packed SE-30 column, it was concluded that some compounds may differ by more than 50 RI units, and some compounds (codeine, dihydrocodeine, amitriptyline, and butriptyline) may even reverse their elution orders [99].

Perhaps the most significant factor that may affect RI reproducibility is the temperature-programming rate. When using linear temperature programmming, RIs can be calculated by replacing the logarithms of the net retention time (RT′) with the actual retention time (RT) [100]. Thus,

$$I = 100n + 100(N - n)[(RT_{analyte} - RT_n)/(RT_N - RT_n)] \qquad (6.11)$$

Despite the differences between RI data obtained from various conditions, quantitative correlations of these data are possible [99, 101]. It was found [101] that RIs from independent databases obtained under different conditions can be quantitatively related as shown in Table 6.5. Thus, it is possible to preliminarily identify a chromatographic peak using the RI data in a database that was established using different chromatographic conditions. To increase the reliability of this identification, RI data obtained from two columns [91, 95, 100] are used. For maximal efficiency, dual columns of different polarities (methyl silicone and 5% phenylmethyl silicone) are installed in the same GC system to generate data in one injection [100].

Table 6.5. Correlations of retention index data obtained under different conditions.

Database	Column type and parameters	Temperature programming	Correlation line and coefficient (with database 1)	Ref.
1	BP1: 12 m, 0.22 mm i.d., 0.25 μm thickness	120 to 270 °C (8 °C/min) 270 to 300 °C (25 °C/min)		[101]
2	DB1: 15 m, 0.22 mm i.d., 0.25 μm thickness	120 to 280 °C (8 °C/min)	$Y = -4 + 1.002X$; 0.9999	[96]
3	SE-30: 15 m, 0.25 mm i.d., 0.25 μm thickness	100 to 295 °C (5 °C/min)	$Y = 3 + 0.988X$; 0.9999	[102]
4	Ultra: 12 m, 0.2 mm i.d., 0.33 μm thickness	50 to 300 °C (20 °C/min)	$Y = 22 + 1.009X$; 0.9999	[100]

Utilization of Non-Paraffins as References. Because paraffins do not respond well toward the commonly used nitrogen–phosphorus detector and exhibit different retention characteristics in relation to the drugs of interest, other compound series have been explored as the references for RI data calculations. It is hoped that selection of other reference series may provide more-comparable detection responses and minimize the differential interaction with the stationary phase when variations in GC conditions—that are beyond the control of chromatographers—occur.

Two approaches are used in selecting reference compound series: the first approach uses homologous compound series [103–106], whereas a second approach uses selected drugs as references [107, 108]. Data listed in Table 6.6 indicate that there is a correlation between the RI data calculated using *n*-alkanes and other nitrogen-containing homologous compound series [104, 106]. It is therefore possible to use existing *n*-alkane-based RI library data and relate them to RI data based on homologous nitrogen-containing compound series. Data shown in this table also indicate that capillary columns with a thin film gave somewhat lower RI values relative to the mean, whereas long columns with a thicker film gave somewhat higher values [106].

Table 6.6. Reference compounds for the calculation of retention indices.

Reference compounds	Correlation of retention indices[a]	Ref.
n-Alkanes	RI_K	[106][b]
Diisopropylalkylamines	$RI_K = RI_D + 584$	[106][b]
Tri-n-alkylamines	$RI_K = RI_T \times 2.79 + 69$	[106][b]
Nitroalkanes	$RI_K = RI_N \times 1.02 + 431$	[106][b]
	$RI_K = RI_N \times 1.02 + 430$	[104][c]
	$RI_K = RI_N \times 1.07 + 530$	[104][d]

[a] RI_K, RI_D, RI_T, and RI_N are retention indices calculated on the basis of using n-alkanes, diisopropylalkylamines, tri-n-alkylamines, and nitroalkanes as the references.
[b] Column used: DB-1 (25 m, 3.0 μm thick).
[c] Column used: CP-Sil 5 (10 m, 5 μm thick, 0.53 mm i.d.).
[d] Column used: CP-Sil 19 (10 m, 5 μm thick, 0.53 mm i.d.).

A different approach was used in which selected drug mixtures were utilized as references for the calculation of RI. In one study [107], a mixture of d-amphetamine, phenylpropanolamine, barbital, pentobarbital, diphenhydramine, methaqualone, codeine, morphine, nalorphine, haloperidol, flurazepam, and strychnine was used as a reference standard. In this study, the slopes and intercepts of lines connecting adjacent drugs in the reference mixture were first calculated using RI values previously determined with the n-alkane reference series; the RI of analyte a, RI(a), was calculated using the following formula:

$$RI(a) = [1000 \log t_R(a) - B]/A \qquad (6.12)$$

where $t_R(a)$ is the retention time of analyte a, and B and A are the intercept and the slope of the line connecting adjacent reference drug data points. A is calculated as follows:

$$A = \{[1000 \log t_R(y)] - [1000 \log t_R(x)]\}/[RI(y) - RI(x)] \qquad (6.13)$$

where $t_R(x)$ and $t_R(y)$ are the retention times of reference drugs eluting just before and after analyte a; and $RI(x)$ and $RI(y)$ are the n-alkane-based RIs of reference drugs eluting just before and after analyte a.

In another study [108], nine chemical classes (phenylethylamine, alkaloid, glutarimide, diphenylmethane, pheniramine, dibenzazepine, pyrrolidine, benzodiazepine, and piperazine) of six distinct pharmacological categories (amphetamine, nicotine, methyprylon, pheniramine, chlorpheniramine, trimipramine, clemastine, flurazepam, and thiethylperazine) were used as the reference standard mixture. The RIs of analytes were calculated on the basis of the retentions and molecular masses (MMs) of the references as follows:

$$RI_x = 10MM(b) + [10(RT_x - RT_b)(MM_a - MM_b)]/(RT_a - RT_b) \qquad (6.14)$$

where x is the unknown analyte, and a and b are the standards eluting immediately after and before the unknown analyte, respectively.

Approaches using drug mixtures as references appear to generate RI data that are much less temperature-dependent [106]. Because molecular masses are used in equation 6.14, RI information thus calculated is more predictable and reproducible when stationary phases of low-to-medium selectivities are used (DB-1 and DB-17) [108]. More studies are needed to determine whether drug mixtures are indeed preferred reference standards for producing more-consistent interlaboratory RI data, and whether this data can be conveniently related to existing data libraries based on conventional n-alkane references. If the drug-mixture references are indeed better alternatives and the resulting data cannot be readily related to the existing library data, it would be necessary to build up databases using a well-conceived drug reference mixture. Reaching consensus on a preferred reference mixture is important because correlations and conversions among data generated from different reference mixtures would be complicated, difficult, and time-consuming, if not impossible.

Detection Systems

Among various detection systems that have been developed for GC methods of analysis, flame ionization detection (FID), electron-capture detection (ECD), thermionic nitrogen (and phophorus) selective detection (NPD), Fourier transform infrared spectrophotometry (FTIR), and mass spectrometry (MS) are effective and commonly used devices for the analysis of drugs and drug-related compounds. While the underlying principle of all detectors is based on the charged species originating from the analytes as they elute from the column, FID, FTIR, and MS are responsive to all classes of compounds, and ECD and NPD are selective.

The basic principles of operation and examples of applications of flame ionization, electron capture, and thermionic detectors are included here, whereas FTIR and GC–MS applications are included in Chapters 7 and 9, respectively.

Basic Operational Principle

A flame ionization detector measures the current, flowing through a pair of electrodes with an imposed potential difference, originating from the ionized analytes generated by the flame situated between the electrodes. In general, a flame ionization detector responds proportionally to the number of oxidizable carbon atoms, and it is therefore considered a universal detector.

Nitrogen–phophorus selective detectors utilize the same basic feature as a flame ionization detector, but their flame temperature is reduced to suppress reponses to general compound categories. Atomic alkali are introduced into the flame through an independently heated rubidium silicate bead, to produce charged nitrogen-containing particles as shown below [109]:

$$C-N= \longrightarrow \cdot C \equiv N \xrightarrow{Rb^*} Rb^+ + [C \equiv N]^-$$

This proposed mechanism is consistent with the observation that nitrogen-containing functions that cannot form the proposed cyan radical show no or only a limited response. Such molecules include $-CO-NH-CO-$ in barbiturates (but not alkylated barbiturates), $-CO-NH_2$ in acid amide, and $-O-NO_2$ in nitrate esters.

Electron-capture detectors utilize a radioactive isotope (^{63}Ni or ^3H) to generate electrons bombarding the carrier gas for the formation of thermal electrons that are collected at an electrode to produce a standing current. Electron-absorbing analytes in the carrier gas alter the current flow to generate signals. Halocarbon compounds, nitroaromatics, and some conjugated compounds containing two weak electron-capturing groups connected by specific bridges show high responses [110]. Responses of halogens decrease in the order of I > Br > Cl > F.

Applications

Nitrogen-selective detectors are commonly used for general analysis of drugs in biological specimens for the following reasons:

1. The sensitivity of NPD toward nitrogen-containing compounds is higher than that of the standard flame ionization detector.

2. The relative responses of NPD toward nitrogen-containing compounds are much higher than toward other compounds.

3. Many analytes of interest in biological specimens are present in low concentrations and contain nitrogen atoms.

Thus, NPD is incorporated in many studies aimed at establishing routine analytical procedures, as is evident by the literature cited earlier.

Although applications of NPD have mainly been emphasized for sensitivity enhancement and quantitative determinations, NPD has also been explored for compound characterization on the basis of its characteristic responses (in relation to FID) toward nitrogen-containing compounds [111]. In this study, a "response index" (R-I) is calculated using the peak heights (PHs) obtained from the dual NPD and FID chromatographic system as follows:

$$R\text{-}I = (PH_{\text{NPD, analyte}}/PH_{\text{FID, analyte}})/(PH_{\text{NPD, caffeine}}/PH_{\text{FID, caffeine}}) \qquad (6.15)$$

The effect of chemical structure on R-I is shown in Figure 6.7 [111]. Although the R-I was found to be characteristic of a given drug and had considerable exclusionary value in reducing the possible candidates for identifying a drug unknown, overlapping R-I data render specific compound identification based on R-I and retention-time information unlikely.

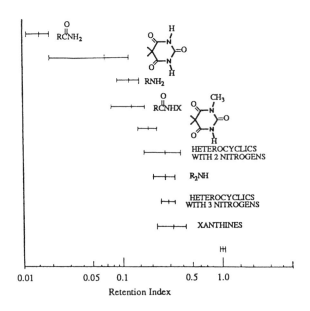

Figure 6.7. Effect of chemical structure on the response index. (Reproduced with permission from Ref. [111].)

Enhanced responses of ECD toward halocompounds encourage analysts to derivatize compounds of interest with halogen-containing moieties to form halocarbonsilyl ethers, haloacyl esters, or pentafluorophenyl derivatives [99]. The responses of the later two categories of derivatives are especially high because of the electron-stabilization capacities of the carbonyl and phenyl groups. Although responses are in the order of I > Br > Cl > F, fluorine is preferred because multiple atoms can be incorporated without unfavorably affecting the analytes' volatility.

The relationship between halocompound structure and ECD relative responses is approximated in Table 6.7 [110], while the effects of the analyte and the derivatizing-reagent structures on the ECD response are exemplified by the amphetamine-related drugs as shown in Tables 6.8 and 6.9 [112].

High-Pressure Liquid Chromatography

Among the various forms of liquid chromatographic approaches shown in Figure 6.1, the bonded-phase HPLC configuration, operating under reverse-phase mode, is most valuable for the analysis of drugs of abuse. From a practical point of view, bonded-phase columns are substantially care-free and can sustain a much longer lifetime, whereas reverse-phase operation generates more-reproducible retention information than normal-phase operation. The performance of normal-phase operation is often affected by the water content in the mobile phase.

Table 6.7. Relationship between halocompound structure and relative ECD response.[a]

Parameter	Variation	Relative response
Halogen series	I	9×10^4
	Br	3×10^2
	Cl	1
	F	«1
Substitution on carbon	Tertiary	10
	Secondary	2
	Primary	1
Frequency on carbon	Tetra-	4×10^5
	Tri-	6×10^4
	Di-	1×10^2
	Mono-	1
Positional isomer	α-	10
	β-	5
	δ-	1
Geometrical isomer	*Trans-*	4
	Cis-	1

[a] Data are taken from Ref. [110].

Table 6.8. Comparative ECD responses of various amphetamine-related drugs analyzed as pentafluorobenzamide.[a]

Compound	Structure	Rel. response
β-Phenethylamine	–CH_2–CH_2–NH_2	300
Amphetamine	–CH_2–$CH(CH_3)$–NH_2	225
α-Methylbenzylamine	–$CH(CH_3)$–NH_2	180
p-Methoxyphenethylamine	H_3CO—⬡—$CH_2CH_2NH_2$	135
Mescaline	H_3CO, H_3CO, H_3CO—CH_2–CH_2–NH_2	155
Phentermine	–CH_2–$C(CH_3)_2$–NH_2	29
Phenmetrazine	(O, CH_3, CH_3, N, H)	28
Mephentermine	–CH_2–$C(CH_3)_2$–$NH(CH_3)$	9.2
Methamphetamine	–CH_2–$CH(CH_3)$–$NH(CH_3)$	6.1
Methoxyphenamine	–CH_2–CH–NH–CH_3, OCH_3	3.2

[a] Data are taken from Ref. [112].

Table 6.9. Comparative ECD responses of amphetamine derivatized with various fluorine-containing derivatizing reagents.[a]

Compound	Structure[b]	Response[c]
N-Pentafluorobenzamide		225
N-Pentafluorobenzylidine		43
N,p-Nitrobenzamide		11
N-Pentafluorobenzylamine		3.7
N-2,4-Dinitroaniline		2.8
N-Heptafluorobutyramide	$F_3C-CF_2-CF_2-\overset{O}{\overset{\|}{C}}-NHR$	2.2
N-Heptafluorobutyrylidine	$F_3C-CF_2-CF_2-C=N-R$	2.1
N-Heptafluorobutylamine	$F_3C-CF_2-CF_2-CH_2-NH-R$	0.41

[a] Data are taken from Ref. [112] (chromatographed on OV-17).
[b] $R = C_6H_5CH_2CH(CH_3)$.
[c] In coulombs $\times 10^3$/mole.

Basic Operational Principles and Parameters of HPLC

The most important advantage of using a liquid as the mobile phase is its variable affinity to the analytes; thus, the separation can be optimized by:

1. Proper selection of *both* the stationary and the mobile phase; and
2. Separation of analytes in their solution forms.

The difficulties associated with the use of a liquid are:

1. High viscosity, which requires the application of high pressure for efficient operation; and
2. The impracticality of varying column temperature.

   With the possibility of varying the affinity of the mobile phase toward the analytes, the need for increasing the variety of the stationary phase is reduced. In reverse-phase operation, the polarity of the mobile phase is commonly increased by mixing increasing amounts of a polar liquid. This practice greatly shortens the analysis time in parallel with the temperature programming operation in GC methods of analysis. Furthermore, positive ions (e.g., tetrabutylammonium) and negative ions (e.g., alkylsulfonate) can be added into the mobile phase to achieve the desired chemistry for the separation mechanism. Just as the increase in column temperature in GC will increase the viscosity of the carrier gas, decreasing the flow rate, the gradient elution changes the mobile-phase viscosity and requires an increasingly higher operation pressure to deliver a constant flow rate.

Solvent Characteristics and Resolution Optimization

Solvent strength and solvent selectivity are the major parameters to be considered when selecting a solvent for a specific analytical system. The *strength* of a solvent depends on its polarity or the ability to preferentially dissolve more polar compounds, whereas solvent *selectivity* refers to the ability of a given solvent to selectively dissolve one compound as opposed to another, where the polarities of the two compounds are not obviously different [113].

Figure 6.8 [115] ranks some common solvents in order of increasing elution strength for normal- and reverse-phase solid-phase extraction sorbents. Common solvents are classified [113] into eight groups (Table 6.10) according to their selectivity parameters, x_e, x_d, and x_n, reflecting their abilities to function as a proton acceptor, a proton donor, and a strong dipole interactor, respectively, as shown in Figure 2.1 [114].

It is a common practice to blend miscible solvents in varying amounts to achieve the desired analyte elution patterns. For the common reverse-phase HPLC practice, the base solvent is water, and the selectivity-adjusting solvents are methanol, acetonitrile, and tetrahydrofuran. (To obtain acceptable peak shapes for many compounds with ionizable functionality, the pH and ionic strength of the aqueous phase are often adjusted [116].) In normal-phase operation, the base solvent is either nonpolar (e.g., hexane or trichlorotrifluoroethane) or very polar (e.g., methanol), depending on the polarity of the analytes. The selectivity-modifying solvents are an ether (usually *t*-butyl ether), chloroform, and dichloromethane [116]. The polarity and selectivity parameters of these and other solvents are shown in Table 6.11 [113].

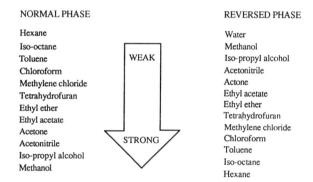

Figure 6.8. Ranking of strengths of common solvents. (Reproduced with permission from Ref. [115].)

Table 6.10. Classification of solvent selectivity.[a]

Group	Solvents
I	Aliphatic ethers, tetramethylguanidine, hexamethylphosphoric acid amide, triakylamines[b]
II	Aliphatic alcohols
III	Pyridine derivatives, tetrahydrofuran, amide (except formamide), glycol ethers, sulfoxides
IV	Glycols, benzyl alcohol, acetic acid, formamide
V	Methylene chloride, ethylene chloride
VI[c]	(a) Tricresyl phosphate, aliphatic ketones and ethers, polyethers, dioxane
	(b) Sulfones, nitriles, propylene carbonate
VII	Aromatic hydrocarbons, halo-substituted aromatic hydrocarbons, nitro compounds, aromatic ethers
VIII	Fluoroalkanols, *m*-cresol, water, chloroform[d]

[a] Data are taken from Ref. [113].

[b] Somewhat more basic than other group I solvents.

[c] This group is rather broad and can be subdivided as indicated into groups VIa and VIb; however, normally there is no point to this in practical usage of the present scheme.

[d] Somewhat less basic than other group VIII solvents.

Table 6.11. Classification of solvents, including less-volatile liquids used as stationary phases in gas chromatography.[a]

Solvent	Strength (P′)	Group	Selectivity x_e	x_d	x_n
Carbon disulfide	0.3	—[b]	—	—	—
Cyclohexane	0.2	—[b]	—	—	—
Triethylamine	1.9	I	0.56	0.12	0.32
Ethyl ether[c]	2.8	I	0.34	0.13	0.34
n-Hexane[d]	0.1	—[b]	—	—	—
i-Octane	0.1	—[b]	—	—	—
Tetrahydrofuran[e]	4.0	III	0.38	0.20	0.42
i-Propyl ether	2.4	I	0.48	0.14	0.38
Toluene	2.4	VII	0.25	0.28	0.47
Benzene	2.7	VII	0.23	0.32	0.45
p-Xylene	2.5	VII	0.27	0.28	0.45
Chloroform[c]	4.1	VIII	0.25	0.41	0.33
Carbon tetrachloride	1.6	—[b]	—	—	—
Butyl ether	2.1	I	0.44	0.18	0.38
Methylene chloride[c]	3.1	V	0.29	0.18	0.53
n-decane	0.4	—[b]	—	—	—
Chlorobenzene	2.7	VII	0.23	0.33	0.44
Bromobenzene	2.7	VII	0.24	0.33	0.43
Fluorobenzene	3.2	VII	0.24	0.32	0.45
2,6-Lutidine	4.5	III	0.45	0.20	0.36
Squalane	1.2	—[b]	—	—	—
Ethoxybenzene	3.3	VII	0.28	0.28	0.44
2-Picoline	4.9	III	0.44	0.21	0.36
Ethylene chloride[f]	3.5	V	0.30	0.21	0.49
Ethyl acetate	4.4	VIa	0.34	0.23	0.43
Iodobenzene	2.8	VII	0.24	0.35	0.41
Methylethyl ketone	4.7	VIa	0.35	0.22	0.43
Bis(2-ethoxyethyl) ether	4.6	VIa	0.37	0.21	0.43
Anisole	3.8	VII	0.27	0.29	0.43
n-Octanol	3.4	II	0.56	0.18	0.25
Cyclohexanone	4.7	VIa	0.36	0.22	0.42
t-Butanol	4.1	II	0.56	0.20	0.24
Tetramethylguanidine	6.1	I	0.47	0.17	0.35
i-Pentanol	3.7	II	0.56	0.19	0.26
Pyridine	5.3	III	0.41	0.22	0.36
Dioxane[e]	4.8	VIa	0.36	0.24	0.40
n-Butanol	3.9	II	0.59	0.19	0.25
i-Proponal	3.9	II	0.55	0.19	0.27
n-Proponal	4.0	II	0.54	0.19	0.27
Phenyl ether	3.4	VII	0.27	0.32	0.41
Acetone	5.1	VIa	0.35	0.23	0.42
Benzonitrile	4.8	VIb	0.31	0.27	0.42
Tetramethylurea	6.0	III	0.42	0.19	0.39
Benzyl ether	4.1	VII	0.30	0.28	0.42
Acetophenone	4.8	VIa	0.33	0.26	0.41
Hexamethylphosphoric acid triamide	7.4	I	0.47	0.17	0.37
Ethanol[f]	4.3	II	0.52	0.19	0.29
Quinoline	5.0	III	0.41	0.23	0.36

Table 6.11. (Continued.)

Solvent	Strength (P′)	Group	Selectivity		
			x_e	x_d	x_n
Nitrobenzene	4.4	VII	0.26	0.30	0.44
m-Cresol	7.4	VIII	0.38	0.37	0.25
N,N-Dimethylacetamide	6.5	III	0.41	0.20	0.39
Acetic acid	6.0	IV	0.39	0.31	0.30
Nitroethane	5.2	VII	0.28	0.29	0.43
Methanol[e]	5.1	II	0.48	0.22	0.31
Benzyl alcohol	5.7	IV	0.40	0.30	0.30
Dimethylformamide	6.4	III	0.39	0.21	0.40
Tricresyl phosphate	4.6	VIa	0.36	0.23	0.41
Methoxyethanol	5.5	III	0.38	0.24	0.38
Nonylphenol oxyethylate[f]	—	III	0.38	0.22	0.40
N-Methyl-2-pyrrolidone	4.7	III	0.40	0.21	0.39
Acetonitrile[e]	5.8	VIb	0.31	0.27	0.42
Aniline	6.3	VIb	0.32	0.32	0.36
Methylformamide	6.0	III	0.41	0.23	0.36
Cyanomorpholine	5.5	VIa	0.35	0.25	0.40
Butyrolactone	6.5	VIa	0.34	0.26	0.40
Nitromethane	6.0	VII	0.28	0.31	0.40
Dodecafluoroheptanol	8.8	VIII	0.33	0.40	0.27
Formylmorpholine	6.4	VIa	0.36	0.24	0.39
Propylene carbonate	6.1	VIb	0.31	0.27	0.42
Dimethyl sulfide	7.2	III	0.39	0.23	0.39
Tetrafluoropropanol	8.6	VIII	0.34	0.36	0.30
Tetrahydrothiophene-1,1- dioxide	8.6	VIb	0.33	0.28	0.39
tris-Cyanoethoxypropane	6.6	VIb	0.32	0.27	0.41
Oxydipropionitrile	6.8	VIb	0.31	0.29	0.40
Diethylene glycol	5.2	III	0.44	0.23	0.33
Triethylene glycol	5.6	III	0.42	0.24	0.34
Ethylene glycol[f]	6.9	IV	0.43	0.29	0.28
Formamide[e]	9.6	IV	0.36	0.33	0.30
Water[f,g]	10.2	VIII	0.37	0.37	0.25

[a] Data are taken from Ref. [113].
[b] Selectivity group is irrelevant because of a low P′ value.
[c] Common modifiers used in normal-phase operation.
[d] Base solvent used in normal-phase operation.
[e] Common modifiers used in reversed-phase operation.
[f] Some values are estimated or not available.
[g] Base solvent used in reversed-phase operation.

The resolution, R_s, of an LC system is controlled by selectivity, α, efficiency, N, and capacity, k', factors as shown in the following equation [117]:

$$R_s = (1/4)\,(\alpha - 1)\,(N^{1/2})[k'/(1 + k')] \tag{6.16}$$

where $\alpha = k'_2 / k'_1$

All three factors can be optimized to give adequate resolution in the shortest time. The efficiency term, N, is determined mostly by the column length, particle size, and mobile-phase velocity, which represent the physical aspects of the operation. To chemically optimize the resolution of analytes in a mixture, a solvent pair of appropriate *strength* (proportion) that will result in the desirable range ($1 \le k' \le 10$) [117] is first selected. The *selectivity* of the solvent system is then fine-tuned to provide the best resolution of the analytes. A trial-and-error procedure for selecting a solvent system of appropriate strength and selectivity is described as follows.

To select a solvent pair of appropriate *strength*, several scouting runs can be conducted by adding several different amounts of a convenient modifier (e.g., methanol for reverse-phase operation and methyl *t*-butyl ether for normal-phase operation) to a convenient base solvent, that is, water in reverse-phase operation and *n*-hexane for normal-phase operation. On the basis of the results obtained from these scouting runs, the composition of the solvent pair that will generate desirable k' values in the 1–10 range for all analytes in the test mixture is determined. The compositions of two more solvent pairs produced using two other modifiers, but with a similar strength, are calculated using the following three equations [117]:

$$k_2 / k_1 = 10^{(p_1 - p_2)/2} \text{ for normal-phase operation} \tag{6.17}$$

$$k_2 / k_1 = 10^{(p_2 - p_1)/2} \text{ for reverse-phase operation} \tag{6.18}$$

$$P' = \varphi_a P_a + \varphi_b P_b \tag{6.19}$$

where φ_a and φ_b are the volume fractions, and p_1, p_2, P_a, and P_b are the polarity values of solvents 1, 2, a, and b, respectively. Thus, in a reverse-phase operation, a solvent strength of 45% methanol in water, 52% acetonitrile in water, and 37% tetrahydrofuran in water are nearly equivalent. Similarly, in a normal-phase operation, a solvent strength of 52% ethyl ether in *n*-hexane, 35% chloroform in *n*-hexane, and 47% methylene in *n*-hexane are equivalent [117].

The next step is fine-tuning the resolution by selecting a system with an appropriate *selectivity*. A "constant-strength selectivity triangle" (Figure 6.9) is constructed using the compositions of the three solvent pairs as the apices, representing the maximal points of proton acceptor, proton donor, and dipole interaction capabilities. As shown in Figure 6.9, the respective base solvent-modifier components for these three apices are water–methanol, water–acetonitrile, and water–tetrahydrofuran for reverse-phase operation, and *n*-hexane–methanol, *n*-hexane–chloroform, and *n*-hexane–methylene chloride for normal-phase operation. The ideal solvent system is then determined by evaluating the seven chromatograms obtained using the following solvent compositions (Figure 6.9): three apex solvent-pair compositions (one from a scouting run and two by calculation), three midpoint mixtures of two of the three apex compositions, and an equal mixture of the three compositions.

Stationary Phase

Most HPLC applications utilize silica-based materials as the stationary phase [118]. These materials are characterized by a narrow size distribution of 5-μm particles for standard analytical purposes (or 3 μm for better efficiency), and a standard pore size of 100 Å (or 300 Å for biomolecular applications). To prevent the hydrolysis of silica from occurring, the operational pH should not exceed 8, and it is normally limited to 2.0–3.5 at the lower end. It is therefore advisable to use a buffered mobile phase whenever possible.

Base silica is used as the stationary phase in normal-phase HPLC. Compounds that interact with the hydrophilic silanol groups will have greater retention. In addition, base silica bonded with cyanopropyl and diol functions are also used in normal-phase chromatography.

In reverse-phase HPLC, the silanol groups in the base silica are derivatized with a selected quantity of octadecyl (or a shorter alkyl chain such as octyl, butyl, ethyl, or methyl for reduced retention), cyanopropyl, pyridyl, benzyl, or other groups. These stationary phases are characterized by the functional groups used, the carbon load (a measure of the number of functional groups bonded), and the degree of "end-capping" (subsequent reaction of the silica with trimethylchlorosilane after bonding with the selected functional groups).

Although silica-based packing materials are used in most reverse-phase HPLC separations (reportedly more than 80%), residual non-end-capped silanol groups may cause peak tailing of basic

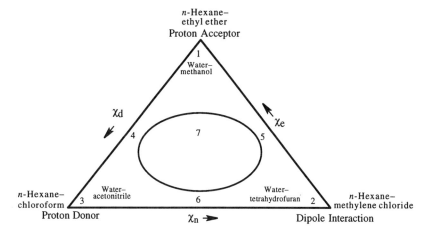

Figure 6.9. Constant-solvent-strength selectivity triangle for common solvents in reverse-phase and normal-phase chromatography. (Constructed using data taken from Refs. [116] and [117].)

compounds. To limit the solubility of a silica-based matrix under alkaline conditions, the pH of the eluents is normally kept below 8, which may not be favorable for basic compounds containing a nitrogenous moiety. Thus, alumina-based packing material coated with polybutadiene has been evaluated for systematic toxicological analysis of basic drugs under alkaline conditions. Although poor retentions were observed for compounds carrying phenolic (e.g., morphine) or carboxylic (e.g., benzoylecgonine) functions, satisfactory and consistent retention behavior was reported for other basic drugs [119].

Ion-exchange chromatography, in which the stationary phase has ionizable functional groups bonded to the silica, is also applied to the analysis of drugs of abuse. Commonly used anion and cation exchangers are quaternary amines and carboxymethyl (or benzenesulfonic acid), respectively. Chromatograms are developed on the basis of replacement of analytes bonded to the exchanger's binding sites by competing ions in the mobile phase.

The reverse-phase paired-ion mode of operation involves the use of a nonpolar stationary phase and the addition of an appropriate counterion to the mobile phase to form reversible ion-pair complexes with the analytes. Analyte separation is based on the difference in the adsorption and desorption rates of the ion-pair complexes on the surface of the stationary phase [120].

Retention Data as a Basis for Preliminary Drug Identification

HPLC requires fewer sample-pretreatment procedures than GC, and it is therefore suitable for limited and general screening, provided that the drugs of interest can be reasonably resolved within the chromatographic window. Thus, numerous studies have emphasized the ability of HPLC systems to resolve individual drugs of a particular category. HPLC systems applicable to the resolution of broader drug categories have also been developed and reported. In addition, injection of biological specimens, pharmaceutical preparations, and solutions of illicit preparations into an HPLC system with or without minimal treatment has also been emphasized.

Limited Screening
Representative reports on the applications of HPLC systems for resolving drugs within a specific category are shown in Table 6.12. These studies demonstrated that various systems can be used to analyze a particular drug category. For example, tricyclic antidepressants can be resolved using a reverse-phase system with C-18 [163], C-8 [164], CN [165, 166], and phenyl [167] functionalities, with or without ion-pairing [168]; or using a normal-phase system [169, 170], with or without in conjunction with ion-pairing [171].

Table 6.12. Representative HPLC studies for the resolution of drugs of various categories.

Drug category	Sample type	Sample pretreatment	Parameters studied	Chromatographic system	Detection	Ref.
Amines	Urine	Extraction, derivatization with 1,2-diphenylethylenediamine	Fluorometric detection of norepinephrine and epinephrine	Ultrasphere-ODS; sodium acetate buffer (pH 7)/acetonitrile/methanol	418 nm	[121]
	Liquid	Dilution with methanol	Quantification of phenyl-2-propanone	HS/C-18; methanol/phosphate buffer (pH 3) (37:63)	220 nm/254 nm	[122]
	Urine	Extraction, derivatization with phenylisothiocyanate	Resolution and detection of primary and secondary amines	µBondapak C-18; methanol/water/acetic acid	254 nm	[123]
	Liquid	Derivatization with phenyliso-thiocyanate	Resolution and detection of 3,4-methylenedioxyphenylisopropyl-amine and analogs	µBondapak C-18; phosphate buffer (pH 3)/methanol (5:1), methanol/water/acetic acid (50:49:1)	254 nm/280 nm	[124]
Barbiturates	Biological material	Extraction (and back-extraction for blood and tissues)	Resolution and quantification of 12 barbiturates	µBondapak C-18; $(NH_4)_2HPO_4$ (0.05 M)/methanol (60:40)	240 nm/290 nm	[125]
Benzodiazepines	Neat	None	Separation of 7 benzodiazepines	SUPELCOSIL LC-8; methanol/acetonitrile/0.005 M KH_2PO_4 buffer (pH 6) (26.5:16.5:57)	245 nm	[126]
	Neat	None	Effect of pH on the separation of benzodiazepines and their metabolites	DuPont Zorbax C-8; methanesulfonic acid in 0.25% H_3PO_4 (pH 3)/acetonitrile (55–58:45–42)	230, 232, 224, 250, 226 nm	[127]
	Blood	Centrifugal filtration, dilution, and solid-phase extraction	Determination of benzodiazepines in contaminated and degraded samples	Shandon ODS-Hypersil; methanol/1-propanol/aqueous phosphate (0.02 M, pH 6) (100:8:80)	Electrochemical	[128]
	Serum	Hexane/ethyl acetate extraction	Separation of 14 benzodiazepines	Novapack C-18; phosphate buffer/acetonitrile/methanol (610:292:98)	Photodiode array	[129]
	Blood and tissue (homogenized)	Hexane/ethyl acetate extraction	Separation of 8 benzodiazepines	Chromspher C8	Photodiode array	[130]
Cannabinoids	Marijuana	Extraction with chloroform	Δ^9-Tetrahydrocannabinol (Δ^9-THC) and 9-THC-COOH	Spherisorb silica; heptane/1% 2-propanol in heptane	280 nm	[131]
	Resin	Extraction with methanol/chloroform	Distribution of cannabinoids in resin	ODS-silica; 0.02 N sulfuric acid/methanol/acetonitrile (7:8:9)	220 nm	[132]
	Urine	Basic hydrolysis, then acidic extraction	Quantification of 9-THC-COOH	µBondapak C-18; acetonitrile/phosphate buffer (pH 6) (45:55)	205 nm	[133]
	Urine	Basic hydrolysis, then acidic solid-phase extraction	Quantification of 9-THC-COOH	Applied Science Excaliber adsorbosphere C-8 and Analytichem Sepralyte C-8; acetonitrile/0.05 M phosphoric acid (65:35)	214 nm	[134]
	Urine	Diluted in methanol/water, adjusted to pH 3	Profile of Δ^9-THC metabolites in marijuana user	Spherisorb-ODS; acetonitrile/water with phosphoric acid (pH 3) (36–70:64–30)	215 nm	[135]
	Plasma, urine	Made alkaline, 50 °C, then made acidic, extraction	Time course of excretion	DuPont Zorbax C-8; acetonitrile/methanol/0.02 N H_2SO_4 (35:15:50)	Electrochemical	[136]

Drug	Matrix	Sample preparation	Application	Column; mobile phase	Detection	Ref.
Cocaine	Plasma	Extraction and back-extraction	Interference and comparison with GC–MS	Spherisorb C-6; phosphate buffer (pH 3)/acetonitrile; sodium hexanesulfonate as counterion	235 nm	[137]
	Vitreous humor	Dilution and preconcentration on a precolumn	Simultaneous determination of cocaine and benzoylecgonine	CH-8 Lichrospher; acetonitrile/phosphate buffer (pH 3)	Diode array	[138]
	Biological fluids	Extraction	Simultaneous determination of cocaine, norcocaine, and benzoylecgonine	μBondapak C-18; water/acetonitrile/methanol (8:1:1) with 1% acetic acid and 0.3 M EDTA	235 nm	[139]
	Solid	None	Separation of impurities and local anaesthetic drugs	ODS-Hypersil; methanol/phosphoric solution/n-hexylamine	230 nm	[140]
	Solid	None	Separation of 4 cocaine isomers	Partisil 10-PXS; 2-propanol/heptane/diethylamine (25:75:0.1)	230 nm	[141]
Fentanyl	Synthesis product	None	Separation of fentanyl homologues and analogues	Partisil 10-ODS-3 (C-18); phosphate buffer (water/2 N NaOH/phosphoric acid, 99:3:1)/methanol/acetonitrile/tetrahydrofuran (81:4:10:5)	215 and 230 nm	[158]
	Synthesis Product	None	Effects of mobile phase composition, analyte hydrophobic substitution, and temperature on the resolution of fentanyl homologues and analogues	Partisil 10-ODS-3, PRP-1 (polystyrene-divinylbenzene); phosphate buffer (phosphoric acid/2 N NaOH/water, 1:3:16)/acetonitrile/methanol/tetrahydrofuran (various ratios)	254 nm	[159]
LSD	Biological fluids	Adjust to pH 8.5 with 0.1 M NaOH and saturate with NaCl; extract with n-heptane (containing 2% isopentanol)	Identification of LSD	Spherisorb 5-ODS, Spherisorb S5W; methanol/Na$_2$HPO$_4$ (65:35) adjusted to pH 8 with ortho-phosphoric acid, methanol/0.2 M NH$_4$NO$_3$ (60:40)	280 and 400 nm (fluorescence)	[160]
	Biological fluids	Acidify to pH 3 with HCl then extract with ether; adjust aqueous phase to pH 9 with ammonia and re-extract with ether	Identification of LSD	Partisil; methanol/0.2% NH$_4$NO$_3$ (11:9)	430 nm (fluorescence)	[161]
	Microdots, blotters	Extract with methanol/water (50:50) with ultrasonic vibration	Identification of LSD	ODS-Hypersil; methanol/phosphate buffer (pH 8.1) (60:40)	220 and 400 nm (fluorescence)	[162]
Opiates	Urine	β-Glucuronidase hydrolysis and extraction	Quantification of codeine, norcodeine, and morphine, and their excretion pattern over a period of time	Spherisorb-CN; methanol/phosphate (0.05 M, pH 6.8) (40:60)	210 nm	[142]
	Blood	Extraction	Quantification of heroin, 6-acetylmorphine, and morphine	LiChrosorb Si-60; acetonitrile/methanol/A/B (75:25:0.040:0.216); A: conc. ammonia/methanol (1:2), B: acetic acid/methanol (1:1)	218 nm	[143]

Table 6.12. (Continued.)

Drug category	Sample type	Sample pretreatment	Parameters studied	Chromatographic system	Detection	Ref.
Opiates (Continued)	Gum opium	Extraction with 2.5% acetic acid	Simultaneous determination of major opiates	μBondapak-CN; 1% sodium acetate (pH 6.78)/acetonitrile/1,4-dioxane	254 nm	[144]
	Poppy straw	Solution by sonication in acetonitrile/water (5:95)	Quantification of major opiates	Phenyl Bondapak; linear gradient of acetonitrile/water (5:95) and acetonitrile/water (20:80), both solvents with 1 mL/L acetic acid and N,N-dimethyloctylamine	275 nm	[145]
	Solid	Dissolution in water	Separation of standard opiates and their analysis in pharmaceuticals	μBondapak C-18; methanol/tetrabutyl ammonium phosphate (pH 7.5) or heptane sulfonic acid/acetic acid (pH 3)	280 nm/254 nm	[146]
	Solid	Dissolution in methanol	Separation of opiates in illicit heroin	μBondapak C-18; acetonitrile/0.85 g/100 mL ammonium acetate buffer (65:35)	280 nm	[147]
	Solid	Dissolution in acetonitrile/sulfuric acid mixture	Separation of opiates in illicit heroin	Aminopropyl bonded silica; acetonitrile/0.005 M tetrabutylammonium phosphate (85:15)	284 nm	[148]
	Solid	Dissolution in methanol	Separation of opiates in illicit heroin	μBondapak C-18; acetonitrile/water/phosphoric acid (12:87:1) with 0.02 M methanesulfonic acid (pH 2.2)	220 nm/254 nm	[149]
	Reaction mixture	Acetylation	Separation of heroin processing impurities and effect of organic modifiers on separation	Zorbax ODS, HS-5 C-18; acetonitrile/phosphate buffer with hexylamine (pH 2.2), methanol/buffer, tetrahydrofuran/buffer	210 nm/260 nm fluorescence, electrochemical	[150]
	Solid	Dissolution in methanol	Comparison of standard and fully-end-capped C-18 columns	C-18, Nova C-18; acetonitrile/0.01 potassium perchlorate with 0.005 M n-butylamine (pH 3) (10:90 to 70:30)	280 nm	[151]
	Solid	Dissolution in eluent	Analysis of low-pK_a and high-pK_a components on a coupled C-18 and alumina, respectively, column system in illicit heroin and opium samples	Alumina, C-18; methanol/acetonitrile/citrate/0.01 M tertramethylammonium hydroxide in citric acid buffer (pH 6) (28:17:55)	Diode array	[152]
Phencyclidine	Synthesis product	None	Separation of phencyclidine-related synthesis by-products	Partisil-ODS; methanol/pH 2.6 buffer (0.1 M KH$_2$PO$_4$/0.1 M K$_2$HPO$_4$, 9:1) (45:55)	215 nm	[153]
Psilocybin	Mushroom	Methanolic extraction	Determination of psilocybin and psilocin	Partisil-10 PAC; acetonitrile/water/phosphoric acid (pH 5.5–6.0) (5:94.5:0.5)	254 and 267 nm	[154]
	Mushroom	Extraction with 10% 1 N NH$_4$NO$_3$	Determination of psilocybin, psilocin and baeocystin	Partisil 5 (silica); methanol/water/1 N NH$_4$NO$_3$ (pH 9.7) (240:50:10)	254 nm	[155]
	Mushroom	Methanolic extraction	Determination of psilocybin, psilocin and baeocystin	μBondapak C-18; Water/methanol (75:25) with 0.05 M heptane-sulfonic acid (pH 3.5)	254 nm	[156]

	Sample	Extraction	Application	Chromatographic conditions	Detection	Ref.
Psilocybin (Continued)	Mushroom	Methanolic extraction	Determination of psilocybin, psilocin baeocystin	Partisil SCX-10; methanol/water (20:80) with 0.2% ammonium phosphate and 0.1% KCl (pH 4.5)	267 nm; 335 nm (fluorescence)	[157]
Tricyclic antidepressants	Standard mixture	None	Separation of tricyclic antidepressants	ODS-Hypersil; acetonitrile/phosphate (pH 3) (3:7)	230 nm	[163]
	Serum	Microprocessor-based instrumental extraction (isooctane/1-propanol) and back-extraction (50 mM/L H_2SO_4)	Assay of 7 tricyclic antidepressants and metabolites	Dimethyloctylchlorosilane; 580 mL acetonitrile with 2.55 mM H_3PO_4 and 2.55 mM KH_2PO_4 (pH 3.3)	205 nm	[164]
	Plasma	Extraction with ethyl ether/chloroform (80:20)	Assay of 8 tricyclic antidepressants	Zorbax Cyanopropylsilane; 0.5 M/L acetic acid/0.03% n-butylamine in acetonitrile (54:46)	254 nm	[165]
	Serum	Manual extraction with hexane and PREP type W cartridge	Sample preparation and assay of 9 tricyclic antidepressants	μBondapak-CN; acetonitrile/methanol/Na_2HPO_4, 5 mM/L (pH 7.0) (60:15:25)	254 nm	[166]
	Plasma	Extraction, back-extraction, re-extraction	Analysis of chlorimipramine and demethylchlorimipramine	0.01 M tetrapropylammonium hydrogen sulfate on Partisil 10; 13% 1-butanol in n-hexane	254 nm	[167]
	Serum	Clean up with precolumn, wash with phosphate buffer (pH 7.5)	Separation of tricyclic and tetracyclic antidepressants and metabolites	TSKgel ODS-80TM; acetonitrile/phosphate buffer (100 mM/L, pH 2.7) (32.5:67.5) with sodium 1-heptanesulfonate (0.2 g/L)	210 nm	[168]
	Plasma	Extraction with isoamyl alcohol/hexane	Assay of 8 tricyclic antidepressants and their metabolites	Zorbax Sil; NH_4OH/methanol (2:998)	214 and 254 nm	[169]
	Serum	Extraction with amyl alcohol/heptane (1.5:98.5)	Simultaneous determination of imipramine, amitriptyline, and their metabolites	Silica; 0.004% perchloric acid in methanol	210 nm	[170]
	Plasma	Extraction with n-hexane/isoamyl alcohol (99:1), back-extraction with dilute phosphoric acid	Analysis of 6 tricyclic antidepressants in two groups	Mini phenyl columns; acetonitrile/n-nonylamine (0.01 M) in 0.02 M NaH_2PO_4 (pH 4) (1:9)	214 nm	[171]

General Screening

The development of a system (i.e., both stationary and mobile phases) that is suitable for chromatographing all drug categories greatly simplifies the task of preliminary identification of drugs present in a totally unknown sample. A typical study [172] of this type involved a silica column used with a methanol/aqueous ammonium nitrate buffer (9:1 v/v) eluent. The results showed that 79 narcotic analgesics and related compounds, and amphetamine and its structural and pharmacological analogs, can be eluted with good peak shapes, but not dopamine, levodopa, methyldopa, methyldopate, or noradrenaline. Majors difficulties of this application included the following:

1. Although the elution order of the drugs remains the same, enormous differences in capacity factors were observed for stationary phases from different manufacturers.

2. With this system's limited discriminating power, arising from the large number of compounds eluting within a small retention range, any identification can only be considered tentative.

Reverse-phase ion-pair operation utilizing a μBondapak C-18 (Water Associates, Milford, MA) column and 1-heptanesulfonic acid sodium salt counterion (in methanol) was found effective [173] in resolving numerous drugs within the following categories: ergot alkaloids, barbiturates, phenylethyl-amines, opium alkaloids, and other drugs studied, including antipyrine, procaine, benzocaine, lidocaine, cocaine, methaqualone, mecloqualone, and tetracaine.

Recent reports have stated that the use of microbore Hyperil ODS [174] and Superspher 10 RP-18 packing [175] in conjunction with photodiode-array detection and data-searching systems was effective for general toxicological screens.

Standardization of Retention Data

The reproducibility of HPLC data is much harder to control than their GC counterpart. This is especially true in normal-phase operation. For example, in a study [176] using a mixture of dipipanone, pro-mazine, codeine, prolintane, phenylephrine, ephedrine, protriptyline, and strychnine as test drugs, the authors concluded the following:

1. Considerable care had to be taken to ensure reproducible eluent.

2. Significant differences were observed with columns containing different batches of Spherisorb S5W.

3. Large differences were found for separations on columns packed with different brands of silica.

Recent empirical data have shown the following:

1. The reproducibility of peak shape and retention time greatly depend on the preparation of silica (metal content and silanol activity) and the bonding process [177].

2. Nominally identical reverse-phase C-18 columns (with identical length, diameter, and particle size) from different manufacturers show significant differences in acidity [178] and exhibit different capacity factors, asymmetry, selectivity, and reduced plate height toward a set of test mixtures. They also behave differently when different organic modifiers (methanol, acetonitrile, and tetrahydrofuran) are used under the same polarity–selectivity eluent mixture conditions [179].

Thus, the interpretive value of HPLC retention data obtained from a single run is limited if the data cannot be related to data generated from the same laboratory at a different time and to data generated by a different laboratory. In parallel with GC applications, several approaches [180] have been attempted to convert the raw retention data into more-comparable parameters.

Capacity Factor and Relative Adjusted Capacity Factor.   The simplest attempt to generate more-comparable HPLC data involves the calculation of the capacity factor, k', as follows:

$$k' = (t_R - t_0)/t_0 \qquad (6.20)$$

where t_R and t_0 are the retention times of the analyte and an unretained compound, taking into account the column's void volume.

In theory, this adjustment should compensate for changes in the eluent flow rate and column dimensions; in practice, however, the accurate measurement of the small t_0 value is difficult, and the true column void volume may differ with different analytes. To ease this difficulty, the relative adjusted capacity factor, k'' in which the retention data of the analyte is reported relative to a reference compound with similar retention characteristics, is calculated using the following expression:

$$k'' = (t_R - t_0)/(t_Q - t_0) \qquad (6.21)$$

where t_Q is the retention time of the reference compound.

Using (*p*-methylphenyl)phenylhydantoin as the reference compound, this approach was used for the analysis of therapeutic and toxic amounts of barbiturates, anticonvulsants, diuretics, nonsteroidal anti-inflammatory drugs, sulfonylurea antidiabetic drugs, theophylline, and analgesic drugs [181]. Using a photodiode-array detector, the investigators reported satisfactory results in intralaboratory applications.

Retention Index. The use of relative capacity factors should largely take care of the factors that have a proportional effect on both the analyte and the reference compound, including flow rate and column dimension. Indeed, relative adjusted capacity factors were found to be reproducible in a barbiturate study in which a barbiturate was used as the reference compound [182]. Satisfactory results were reported, although variations of pH do dramatically affect the retention of barbiturates because they are partially ionized. However, when the reference and the analytes have different structural features, this approach cannot adequately compensate for differences in the selectivity of the system caused by differences in temperature or composition of the stationary and mobile phases. Because the use of a structurally related reference compound may not always be feasible in general screen or other applications, the use of retention indices—in parallel with GC applications—has been substantially explored.

Earlier studies of *n*-alkane and alkylbenzene series concluded that they were not suitable RI scale reference series, mainly because of their weak chromophobic characteristic with popular detection devices and their strong retention in reverse-phase chromatographic systems [180]. Using 2-oxoalkane series as the RI scale standards [183], the RI data of a seven-drug mixture (aspirin, caffeine, phenobarbital, phenacetin, methaqualone, chlordiazepoxide, and androsterone) resulting from reverse-phase C-18 columns from different batches and manufacturers were found to be more reproducible than either relative retention time or adjusted relative retention time. However, intra- and interlaboratory variations were still significant, presumably because of differences in column selectivities associated with the extent of surface coverage of each column.

Studies of alkyl aryl ketone series [184–186] reported stronger chromophobic characteristics and satisfactory results in compensating for small changes in eluent compositions [184, 186], but not for differences in packing materials [185]. Compared to capacity factors, RI data of the ten barbiturates studied [184] were found to be less susceptible to minor changes in eluent composition, pH, ionic strength, and temperature. For un-ionized compounds, RI data were virtually independent of experimental conditions.

With the retention of the lower components of the 1-nitroalkane series weaker than all drugs of interest, RI values for early-eluting drugs were reportedly more reliable when this series was used as the RI scale standards [187]. Attempts to compare RI data obtained under isocratic conditions with gradient elution failed to generate a satisfactory correlation [188]. The RI values of acidic drugs and earlier-eluting basic drugs decreased distinctly with increasing concentrations of acetonitrile, whereas opposite trends were observed for later-eluting basic drugs.

A recent comprehensive study of 469 drugs tested the effects of different HPLC systems and solvent-gradient rates on the reproducibility of RI calculated on the basis of equation 6.11, using relative analyte retention time (relative to acetophenone) and 1-nitroalkane series as references [189]. This approach was found to generate reproducible RI data under large variations in chromatographic parameters for drugs similar to the reference series, but not for compounds with different functional groups. It was suggested that laboratories wishing to use published HPLC retention databases to aid in drug identification should use HPLC systems that closely emulate the HPLC conditions under which the particular database was created.

Further Improvements.   Numerous studies have concluded that capacity factors are not precise enough for comparing data generated at different times or from different laboratories. Variations were found to derive from difficulties in duplicating exact experimental conditions and differences in column-packing materials. Differences derived from the former source can be rather successfully compensated for by reporting the retention data of the analytes relative to a structurally closely related reference compound, or calculating RI values using alkyl aryl ketone [190] or 1-nitroalkane series [191] as RI scale standards. The RI approach is preferable for universal applications.

Variations of retention data derived from the proven differences [192, 193] between packing materials are much harder to correct. Two approaches have been adopted to minimize variations from this source. The first approach involves the extension of reporting relative retention data in which several, instead of a single, reference compounds are used. Thus, retention data of barbiturates were reported as corrected capacity factors using a four-barbiturate standard mixture as the reference compounds [182]. The mixture included phenobarbital, butobarbital, amobarbital, and secobarbital, which have a wide range of capacity factors of 1.23, 3.42, 7.00, and 11.21, respectively, in the adopted HPLC system. For each laboratory, experimental retention times for these drugs were plotted against the capacity factor values from the references. A least-squares fit line was obtained and used to determine the capacity factors of other barbiturate analytes.

The second approach is the correction of RI values obtained with various RI scale standards [194]. At least three correction standards, distributed uniformly throughout the elution range, are used. The corrected RI value, RI^c, of a given analyte is calculated using the following equations [191]:

$$RI^c = aRI + b \qquad (6.22)$$

$$a = [(RI^0)_2 - (RI^0)_1]/(RI_2 - RI_1) \qquad (6.23)$$

$$b = (RI^0)_2 - aRI_2 \qquad (6.24)$$

where RI is the RI of the analyte; $(RI^0)_1$ and $(RI^0)_2$ are the literature values of RI for correction standards 1 and 2, eluted before and after the analyte; and RI_1 and RI_2 are the observed values for standards 1 and 2.

This approach has been applied to correct the RI data (1-nitroalkane series scale) of basic drugs [195] using cocaine, desipramine, and triflupromazine [195]; and morphine, chloroquine, benzoylecgonine, cocaine, diphenhydramine, haloperidol, amitriptyline, thioridazine, and meclozine [175] as correction standards. The same approach was applied to correct the RI data (alkyl aryl ketone series scale) of acidic and neutral drugs [196] using two sets of correction standards (brallobarbital, nordiazepam, and phenylbutazone; and butalbital, flunitrazepam, and prazepam); and the RI data (1-nitroalkane series scale) using paracetamol, barbital, brallobarbital, pentobarbital, secobarbital, clobazam, indomethacin, prazepam, and amiodarone as the correction standards [175]. A co-extracted biological matrix appeared to broaden the scattering of the corrected RI data of particular drugs without a noticeable systematic trend in retention characteristics [197].

The effectiveness of a system can be evaluated using a parameter called discrimination number (DN), which represents the number of retention windows, each 2 standard deviations wide, that can be fitted into a defined chromatographic range [182]. In essence, the reporting system is evaluated on the basis of the maximum number of peaks that can be resolved in an available separation space. This approach was used to evaluate the effectiveness of the corrected capacity factor of barbiturates using phenobarbital, butobarbital, amobarbital, and secobarbital as references and was found to result in a DN of 64, while that for RI data (using alkyl aryl ketones as a series scale) was only 34. This undoubtedly reflects the fact that alkyl aryl ketones and barbiturates respond differently to small variations in chromatographic conditions, and barbiturate standards can provide better compensation for the analysis of barbiturates.

This same parameter was also used to evaluate [191] the effectiveness of the corrected RI approach and was found to improve the effectiveness for the identification of basic drugs studied [195] by three times, and of the acidic and neutral drugs studied [196] by approximately two times. Although not as effective as the corrected capacity factor approach reported for the barbiturate study [194], the corrected RI approach offers the advantage of universal applicability and it does not rely on the identity of the analytes.

Detection Systems

In contrast to GC, which benefited from thermal conductivity and flame-ionization detectors almost from the beginning and the convenient interface with mass spectrometry (MS) at the later development stage, HPLC technology is still searching for a universal and practical detection system. Thus, developments in detection technology have become one of the key factors in the advances and applications of HPLC methodology [198].

Early refractive index detectors were too sensitive to changes in the environment. Ultraviolet absorbance detection was initially limited to 254 nm, which is suitable for compounds with aromatic rings, and it became applicable to other compound categories when variable-wavelength detection became available. Recently developed diode-arrays make peak identification and peak purity analysis possible. Fluorescence detection may enhance detection limits, but it is selective and may have to rely on prior derivatization with special agents. The use of electrochemical detectors is also selective. Although MS is a preferred detection device, cost considerations, interface technology, and associated mass-spectrum characteristics are currently limiting factors. Table 6.13 summarizes the features of various HPLC detectors [199]. HPLC–MS applications are discussed in Chapter 9.

Table 6.13. Comparison of detectors for liquid chromatography.[a]

Detector	Universality	Destructive	Compd ident.	Ease of use	Sensitivity	Cost
Refractive index	High	No	Low	Medium	Low	Low
UV-visible, multiple wavelength	Medium	No	Low	High	High	Low
UV-visible, diode array	Medium	No	Medium	High	High	Low
Fluorescence	Low	No	Low	High	Very high	Low
Electrochemical	Low	Yes	Low	Medium	High	Low
Mass spectrometric	High	Yes	Very high	Low	Medium	Very high

[a] Data are taken from Ref. [199].

Absorption

The UV detector was one of the earliest HPLC detectors, and it is still used in most applications (*see* Table 6.12). The concentration of the analyte is determined on the basis of Beer's law, which relates the absorption of monochromatic light to the number of absorbing molecules as follows [200]:

$$A = \log(I_0/I) = \varepsilon c l \tag{6.25}$$

where A is the absorbance, I_0 is the incident light intensity, I is the transmitted light intensity, ε is the molar absorptivity, c is the analyte concentration, and l is the path length of the flow cell.

Initially, low-pressure mercury, zinc, and cadmium lamps, which produce strong radiation at 254, 214, and 229 nm, respectively, were used as fixed-wavelength detectors. With the use of diffraction gratings and continuous-radiation-producing deuterium lamps, variable-wavelength detectors became available for the detection of various compound classes.

Recent advances in diode-array technology enable simultaneous multiwavelength detection. A typical photodiode-array detector uses a deuterium lamp to produce continuous radiation that passes through an achromatic lens system, and then through a shutter that permits calibration and dark-current measurement; the light is then focused into the flow cell. After passing through the cell, the light reflects off a holographic diffraction grating that splits it into the spectrum and disperses it across an array of 70–512 photodiodes (Figure 6.10) [201]. Because the output of the deuterium lamp is weak at wavelengths greater than 600 nm, a second tungsten lamp may be used for the 600–800-nm range.

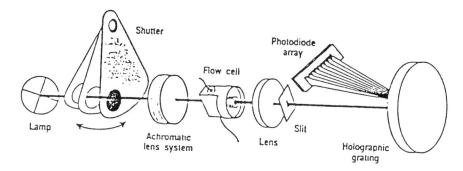

Figure 6.10. Schematic of diode-array detector optics. (Reprinted with permission from Ref. [201].)

Absorbance ratios at selected wavelengths have been explored as the basis for general screening [206]. It was reported that the 28 common drugs of abuse studied can be placed into five classes (Table 6.14) based on the following absorbance ratios: A_{230}/A_{260}, A_{280}/A_{260}, and A_{320}/A_{260}. Absorbance ratios and retention data obtained from six unknowns—representing one drug from each class and a drug not on the 28-drug list—enable the correct assignments of these unknowns.

Because a diode-array detector can be considered a series of single-wavelength detectors, complete spectra are registered and can be compared (overlaid) with either standards or previously stored spectra for compound-identification purposes. By comparing the peak areas of the standard and the sample at the two wavelengths having the highest absorbance difference, impurities can be determined by noting any deviation from the flat ratio plot [201]. The ability to compare first- and second-derivative spectra also facilitates the comparison of minima and maxima of the spectral peak.

The ability to provide certain structural information makes a diode-array a convenient detector for method development, in which the identities of chromatographic peaks in a single run can be ascertained; the elution of analyte in a solvent front and the coelution of two analytes can be diagnosed, and the confusion of peak crossover between mobile phases of different strengths can be avoided [202]. Thus, this detector has been successfully used to simultaneously identify the following: a mixture of six drug standards [203]; the presence of diphenhydramine and 8-chlorotheophylline in an illicit methaqualone tablet [202]; psilocybin and psilocin in a suspected mushroom material [202]; cocaine, basic by-products, basic impurities, and acidic by-products in an illicit cocaine sample [83]; diamorphine in illicit heroin samples [204]; compounds with low pK_a values, such as caffeine, papaverine, and noscapine; and compounds with high pK_a values, such as heroin, acetylcodeine, 6-monoacetylmorphine, procaine, codeine, morphine, and strychnine in illicit heroin and opium samples [205].

Recent advances in computer technologies have greatly facilitated the utilization of photodiode-array detectors in HPLC applications. It is now a common practice [207–210] to conduct a library search using both retention and UV spectrum information for compound identification. For example, the spectrum of an analyte peak was compared with those in the library within a "time window" (±0.5 or ±1.0 min) to generate a list of compounds with an acceptable "fit" value [207]. In another approach, retention indices and absorption maxima with the highest wavelength were used to assess the identification power of the HPLC system [208].

Because the retention characteristics of analytes often vary with slight changes in mobile-phase composition and other operational conditions, the use of a structurally sensitive diode-array detector reduces the doubts associated with compound identifications that are based on retention parameters alone. However, these absorption characteristics are more likely reflections of a general compound category than the specific structural details that can be provided by more definite identification instrumentation such as MS or NMR. Other identification techniques may still be needed for purposes of final confirmation.

Diode-array detectors are also subject to UV-absorbing interferences commonly associated with biological matrices. Although data-processing techniques may be useful for purposes of spectra correction, spectral interference can often reduce the limit of detection. For example, it has been reported [203] that the recognition of 50 mg/L of strychnine from a liver extract and 1 mg/L of amitriptyline in plasma was difficult. The latter interference is caused by a continuous background absorption.

Table 6.14. Absorbance ratios of 28 common drugs of abuse.[a]

Drug and class	Absorbance ratio			
	A_{220}/A_{260}	A_{230}/A_{260}	A_{280}/A_{260}	A_{320}/A_{260}
Class A				
1. Amphetamine	8.4	0.00	0.00	0.00
2. Ephedrine	17.0	1.66	0.00	0.00
3. Methamphetamine	10.2	0.54	0.00	0.00
4. Phentermine	16.6	0.00	0.00	0.00
5. Propoxyphene	15.2	1.05	0.00	0.00
Class B				
1. Amobarbital		6.0	0.00	0.00
2. Pentobarbital		4.8	0.00	0.00
3. Phencyclidine		5.6	0.00	0.00
4. Phenmetrazine		6.4	0.00	0.00
5. Phenobarbital		3.1	0.05	0.00
6. Secobarbital		4.2	0.00	0.00
Class C				
1. MDMA[b]		7.50	6.92	0.00
2. Pentazocine		12.0	5.0	0.00
3. STP[b]		20.7	8.72	0.00
Class D				
1. Cocaine		19.7	1.00	0.00
2. Codeine		11.3	2.53	0.00
3. Heroin		10.3	2.94	0.00
4. Hydromorphone		8.2	1.44	0.00
5. Methadone		8.0	0.55	0.00
6. Morphine		10.5	2.17	0.00
7. Oxycodone		10.1	1.48	0.00
Class E				
1. Chlordiazepoxide		0.56	0.57	0.17
2. Clorazepate		2.68	0.30	0.16
3. Diazepam		2.48	0.30	0.17
4. Flurazepam		2.92	0.27	0.18
5. Lorazepam		4.40	0.19	0.24
6. Methaqualone		3.95	0.49	0.24
7. Oxazepam		3.33	0.37	0.21
Internal standard	1.25	0.78	1.11	0.00

[a] Data are taken from Ref. [206].
[b] Abbreviations: MDMA: 3,4-methylenedioxymethamphetamine; STP: 2,5-dimethoxy-4-methylamphetamine.

Fluorescence

Fluorescence is a luminescence phenomenon that occurs when a compound absorbs radiation and then emits it at a longer wavelength. Compounds that fluoresce will emit only a fraction of the absorbed light as fluorescence, as follows:

$$F = (I_0 - I)\phi_f \tag{6.26}$$

where F is the fluorescence intensity; I_0 is the incident light intensity; I is the transmitted light intensity; and ϕ_f is the quantum efficiency of the solute. F can be related to the solute concentration, c, by substituting the exponential form of Beer's law into the above equation. Thus,

$$F = \phi_f K I_0 (1 - 10^{-\varepsilon cl}) \tag{6.27}$$

where K is the efficiency of fluorescence collection, ε is the molar absorptivity, and l is the path length of the flow cell.

Without losing too much precision, equation 6.27 can be approximated to [200]:

$$F = 2.303 \phi_f K I_0 \varepsilon cl \tag{6.28}$$

Thus, fluorescence intensity is approximately proportional to the concentration, especially at the lower concentration range.

Because many compounds of interest may not be highly conjugated and do not display natural fluorescence, fluorescence detection is often accomplished by converting the analyte to a fluorescent molecule. The most common conversion processes are through post- or precolumn derivatization, such as the derivatization of amphetamine with 2-mercaptoethanol-o-phthaldehyde [211]. Other unique applications include the irradiation of cannabinol with a high flux of UV light to produce the fluorescent photoproduct [212], and the phenolic oxidative coupling of morphine and related opiates to yield fluorescent dimers [213].

Because fluorescence detection is based on the observation of emitted radiation against a zero background level, it is inherently more sensitive than the absorption measurement, which compares the difference between the incident and transmitted light intensities. The associated derivatization process also increases the specificity of fluorescence detection. Although this approach is often limited to specific systems because of its nonuniversality and the additional step involved, it can be the method of choice, as in the detection of lysergic acid diethylamide [211].

Dual fluorescence and UV detectors have been used to provide fluorescence-to-UV response ratios, which are used along with retention information for simultaneous determination of morphine and codeine in blood and bile extracts [214].

Electrochemical Detection

The most popular form of electrochemical detector is based on anodic oxidation of analytes on a solid electrode. The first step is usually the removal of an electron from a hetero atom (commonly nitrogen) with a potential insufficient to oxidize the carbon skeleton of the molecule [211]. Although the use of a relatively high applied potential will maximize analyte response, oxidation of the eluent and the electrode will also be increased to produce a higher background current, which may actually decrease the signal-to-noise ratio.

The drug categories that can be detected through electrochemical oxidation reactions and their typical oxidation voltages are shown in Table 6.15 [215]. A common feature in the molecular structure of these compounds is an aliphatic nitrogen or oxygen atom. In the case of nitrogen, the ease of oxidation increases in the order of primary < secondary < tertiary. Because the electrochemical oxidation process involves the lone pair of electrons on the nitrogen atom, the pH must be high enough to maintain these compounds in their basic form [216].

Table 6.15. Categories of compounds detectable by electrochemical oxidation and typical oxidation potentials.[a]

Functional group	Oxidation potential (V)
Phenol, aromatic amine	0.7
Phenothiazine sulfur	0.8
Imidazolyl nitrogen, indole	0.9
Tertiary aliphatic amine	1.0
Secondary aliphatic amine	1.2
Primary aliphatic amine	1.6
Pyridyl nitrogen, quaternary ammonium compound, amide	>1.6

[a] Data are taken from Ref. [215].

Mobile-phase pH and composition have a major impact on the success of electrochemical detection. To maintain the analytes in their unprotonated forms, the adopted eluent pH is often a compromise between retention, peak shape, and detection response [216]. Increasing the eluent pH produces a higher absolute response because the unprotonated form is favored; however, the background current also increases. Because the use of an organic component enhances the basicity [217], the nitrogen atoms remain in their basic unprotonated forms at pH values lower than that present when strictly aqueous media are used. Although successful applications have been reported for the analysis of cannabinoids [218, 219], opiates [215, 220, 221], and cocaine-related compounds [215] (Table 6.16), electrochemical detections can only be used for very specific systems.

Table 6.16. Examples of electrochemical detection.

Analytes[a]	Column	Mobile phase	Electrode/potential/reference	Ref.
Morphine, codeine/thebaine, narcotine	Whatman Partisil ODS-3	Acetonitrile/0.02 M KH_2PO_4 (NaOH buffer, pH 6) (55:45)	Glassy carbon; +1.2 V; Ag/AgCl/NaCl (3 M)	[215]
Morphine, 6-MAM, 3-MAM, heroin/acetylcodeine	Same as above	Acetonitrile/0.02 M KH_2PO_4 (NaOH buffer, pH 7) (40:60)	Same as above	[215]
Ecgonine, benzoylecgonine, cocaine	Same as above	Acetonitrile/0.02 M KH_2PO_4 (NaOH buffer, pH 8) (50:50)	Same as above	[215]
Cannabinoids	μBondapak C-18	0.02 N H_2SO_4/methanol/ acetonitrile (6:7:16)	+1.2 V; Ag/AgCl	[218]

[a] MAM: monoacetylmorphine.

References

1. Miller, J. M. *An Introduction to Liquid Chromatography for the Gas Chromatographer;* Gow-Mac Instrument: Bridgewater, NJ, 1979; p 3.
2. McNair, H. M.; Bonelli, E. J. *Basic Gas Chromatography;* Varian Aerograph: Walnut Creek, CA, 1990; p 2.
3. *J&W Scientific 1990-91 GC and HPLC Chromatography Catalog;* J&W Scientific: Folsom, CA, 1990; p 22.
4. Duquet, D. *All-Chrom Newsletter* **1986,** *25,* 13.
5. *The Separation Times* **1989,** *3,* 1.
6. Houtermands, J. M. *Topics* **1984,** *11,* 6.
7. Brooks, J. W.; Edmonds, C. G.; Gaskell, S. J.; Smith, A. G. *Chem. Phys. Lipids* **1978,** *21,* 403.
8. *The Separation Times* **1990,** *4,* 5.
9. Neville, G. A. *Anal. Chem.* **1970,** *42,* 347.
10. Jain, N. C.; Budd, R. D.; Sneath, T. C.; Chinn, D. M.; Leung, W. *J. Clin. Toxicol.* **1976,** *9,* 221.
11. Liu, R. H.; McKeehan, A. M.; Edwards, C.; Foster, G. F.; Bensley, W. D.; Langner, J. G.; Walia, A. S. *J. Forensic Sci.* **1994,** *39,* 1501.
12. Mule, S. J.; Casella, G. A. *J. Anal. Toxicol.* **1988,** *12,* 153.
13. *The Separation Times* **1988,** *2,* 3.
14. Hornbeck, C. L.; Czarny, R. J. *J. Anal. Toxicol.* **1989,** *13,* 144.
15. Paul, B. D.; Mell, L. D., Jr.; Mitchell, J. M.; Irving, J.; Novak, A. J. *J. Anal. Toxicol.* **1985,** *9,* 222.
16. Edlund, P. O. *J. Chromatogr.* **1981,** *206,* 109.
17. Poole, C. F.; Zlatkis, A. *Anal. Chem.* **1980,** *52,* 1002.
18. Nicolau G.; Van Lear, G.; Kaul, B.; Davidow, B. *Clin. Chem.* **1977,** *23,* 1640.
19. Knapp, D. R. *Handbook of Analytical Derivatization Reactions;* Wiley: New York, 1979.
20. Nicholson, J. D. *Analyst (Cambridge, U.K.)* **1978,** *103,* 1.
21. Nicholson, J. D. *Analyst (Cambridge, U.K.)* **1978,** *103,* 193.
22. Drozd, J. *J. Chromatogr.* **1975,** *113,* 303.
23. Ahuja, S. *J. Chromatogr.* **1975,** *113,* 283.
24. Lochmuller, C. H.; Souter, R. W. *J. Chromatogr.* **1975,** *113,* 283.
25. Anggard, E.; Hankey, A. *Acta Chem. Scand.* **1969,** *23,* 3110.
26. Poole, C. F.; Zlatkis, A. *J. Chromatogr. Sci.* **1979,** *17,* 115.
27. Moore, J. M. *Forensic Sci. Rev.* **1990,** *2,* 79.
28. Blau, K.; King, G. *Handbook of Derivatives for Chromatography;* Heydon and Son: Philadelphia, PA, 1978.

29. *Pierce Handbook and General Catalog;* Pierce: Rockford, IL, 1989.
30. *Bring Chemistry to the Chromatographic Sciences;* Regis: Morton Grove, IL, 1984.
31. Moore, J. M.; Allen, A. C.; Cooper, D. A.; Carr, S. *Anal. Chem.* **1986,** *58,* 1656.
32. Czarny, R. J.; Hornbeck, C. L. *J. Anal. Toxicol.* **1989,** *13,* 257.
33. Danielson, T. J.; Boulton, A. A. *Biomed. Mass Spectrom.* **1974,** *1,* 159.
34. Matin, S. B.; Rowland, M. *J. Pharm. Sci.* **1972,** *61,* 1235.
35. Moffat, A. C.; Horning, E. C.; Matin, S. B.; Rowland, M. *J. Chromatogr.* **1972,** *66,* 255.
36. Gilbert, M. T.; Brooks, C. J. W. *Biomed. Mass Spectrom.* **1977,** *4,* 226.
37. Crowther, J. B.; Covey, T. R.; Dewey, E. A.; Henion, J. D. *Anal. Chem.* **1984,** *56,* 2921.
38. Harvey, D. J.; Nowlin, J.; Hickert, P.; Butler, C.; Gansow, O.; Horning, M. G. *Biomed. Mass Spectrom.* **1974,** *1,* 344.
39. Budd, R. D. *J. Chromatogr.* **1982,** *237,* 155.
40. Gambaro, V.; Mariani, R.; Marozzi, E. *J. Anal. Toxicol.* **1982,** *6,* 321.
41. Budd, R. D. *J. Chromatogr.* **1980,** *192,* 212.
42. Caddy, B.; Kidd, C. B. M.; Leung, S. C. *J. Forensic Sci. Soc.* **1982,** *22,* 3.
43. Falkner, F. C.; Watson, J. T. *Org. Mass Spectrom.* **1974,** *8,* 257.
44. Dilli, S.; Pillai, D. N. *J. Chromatogr.* **1977,** *137,* 111.
45. Walle, T. *J. Chromatogr.* **1975,** *114,* 345.
46. McCurdy, H. H.; Slightom, E. L.; Harrill, J. C. *J. Anal. Toxicol.* **1979,** *3,* 195.
47. Vessman, J.; Johansson, M.; Magnusson, P.; Stromberg, S. *Anal. Chem.* **1977,** *49,* 1545.
48. De Gier, J. J.; Hart, B. J. *J. Chromatogr.* **1979,** *163,* 304.
49. Mule, S. J.; Casella, G. A. *J. Anal. Toxicol.* **1989,** *13,* 179.
50. ElSohly, M. A.; Arafat, E. S.; Jones, A. B. *J. Anal. Toxicol.* **1984,** *8,* 7.
51. Wimbish, G. H.; Johnson, K. G. *J. Anal. Toxicol.* **1990,** *14,* 292.
52. Baker, T. S.; Harry, J. V.; Russell, J. W.; Myers, R. L. *J. Anal. Toxicol.* **1984,** *8,* 255.
53. Turner, C. E.; Hadley, K. W.; Henry, J.; Mole, M. L. *J. Pharm. Sci.* **1974,** *63,* 1873.
54. Harvey, D. J. *J. Pharm. Pharmacol.* **1976,** *28,* 280.
55. Harvey, D. J. *J. Mass Spectrom.* **1977,** *4,* 88.
56. Harvey, D. J. *J. Mass Spectrom.* **1980,** *7,* 212.
57. Knaus, E. E.; Coutts, R. T.; Kazakoff, C. W. *J. Chromatogr. Sci.* **1976,** *14,* 525.
58. Maseda, C.; Yuko, M. P.; Fukui, Y.; Kimura, K.; Matsubara, J. *Forensic Sci.* **1984,** *28,* 911.
59. Isenschmid, D. S.; Levine, B. S.; Caplan, Y. H. *J. Anal. Toxicol.* **1988,** *12,* 242.
60. Verebey, K.; DePace, A. *J. Forensic Sci.* **1989,** *34,* 46.
61. Moore, J. M. *J. Chromatogr.* **1974,** *101,* 215.
62. Moore, J. M.; Allen, A. C.; Cooper, D. A.; Carr, S. *Anal. Chem.* **1986,** *58,* 1656.
63. Papac, D. I.; Foltz, R. L. *J. Anal. Toxicol.* **1990,** *14,* 189.
64. Francom, P.; Andrenyak, D.; Lim, H. -K.; Bridges, R. R.; Foltz, R. L.; Jones, R. T. *J. Anal. Toxicol.* **1988,** *12,* 1.
65. Lerner, M. L.; Katsiaficas, M. D. *Bull. Narc.* **1969,** *21,* 47.
66. Steinhauer, F. A. *Microgram* **1980,** 46.
67. Chen, B. H.; Taylor, E. H.; Pappas, A. A. *J. Anal. Toxicol.* **1990,** *14,* 12.
68. Moore, J. M.; Allen, A. C.; Cooper, D. A. *Anal. Chem.* **1984,** *56,* 642.
69. Neumann, H. *J. Chromatogr.* **1984,** *315,* 404.
70. Sperling, A. R. *J. Pharm. Sci.* **1977,** *66,* 743.
71. Repke, D. B.; Leslie, D. T.; Mandell, D. M.; Kish, N. G. *J. Pharm. Sci.* **1977,** 66, 743.
72. Sharp, M. E. *J. Anal. Toxicol.* **1987,** *11,* 8.
73. Sine, H. E.; McKenna, M. J.; Law, M. R.; Murray, M. H. *J. Chromatogr. Sci.* **1972,** *10,* 297.
74. Finkle, B. S.; Taylor, D. M.; Bonelli, E. J. *J. Chromatogr. Sci.* **1972,** *10,* 312.
75. Pierce, W. O.; Lamoreaux, T. C.; Urry, F. M.; Kopjak, L.; Finkle, B. S. *J. Anal. Toxicol.* **1978,** *2,* 26.
76. Foerster, E. H.; Hatchett, D.; Garriott, J. C. *J. Anal. Toxicol.* **1978,** *2,* 50.
77. Edlund, A.; Jonsson, J.; Schuberth, J. *J. Anal. Toxicol.* **1983,** *7,* 24.
78. Taylor, R. W.; Greutink, C.; Jain, N. C. *J. Anal. Toxicol.* **1986,** *10,* 205.
79. Ferslew, K. E.; Manno, B. R.; Manno, J. E. *J. Anal. Toxicol.* **1979,** *3,* 30.
80. Foerster, E. H.; Dempsey, J.; Garriott, J. C. *J. Anal. Toxicol.* **1979,** *3,* 87.
81. Ehresman, D. J.; Price, S. M.; Lakatua, D. J. *J. Anal. Toxicol.* **1985,** *9,* 55.
82. Drummer, O. H.; Horomidis, S.; Kourtis, S.; Syrjanen, M. L.; Tippett, P. *J. Anal. Toxicol.* **1994,** *18,* 134.
83. Hime, G. W.; Bednarczyk, L. R. *J. Anal. Toxicol.* **1982,** *6,* 247.
84. Watts, V. W.; Simonick, T. F. *J. Anal. Toxicol.* **1986,** *10,* 198.
85. Turcant, A.; Premel-Cabic, A.; Cailleux, A.; Allain, P. *Clin. Chem.* **1988,** *34,* 1492.
86. Adams, R. F. *Clin. Chem. Newsl.* **1972,** *4,* 15.

87. Adams, R. F. *Clin. Chem. Newsl.* **1972,** *4,* 22.
88. Kovats, E. *Helv. Chim. Acta* **1958,** *41,* 1915.
89. Kazyak, L.; Permisohn, R. *J. Forensic Sci.* **1970,** *15,* 346.
90. Moffat, A. C. *J. Chromatogr.* **1975,** *113,* 69.
91. Marozzi, E.; Gambaro, V.; Saligari, E.; Mariani, R.; Lodi, F. *J. Anal. Toxicol.* **1982,** *6,* 185.
92. Franke, J. P.; de Zeeuw, R. A.; Wijsbeek, J. *J. Anal. Toxicol.* **1986,** *10,* 132.
93. Bogusz, M.; Bialka, J.; Gierz, J.; Klys, M. *J. Anal. Toxicol.* **1986,** *10,* 135.
94. Schepers, P.; Wijsbeek, J.; Franke, J. P.; de Zeeuw, R. A. *J. Forensic Sci.* **1982,** *27,* 49.
95. Lora-Tamayo, C.; Rams, M. A.; Chacon, J. M. R. *J. Chromatogr.* **1986,** *374,* 73.
96. Perrigo, B. J.; Ballantyne, D. J.; Peel, H. W. *Can. Soc. Forensic Sci. J.* **1984,** *17,* 41.
97. Bogusz, M.; Wijsbeek, J.; Franke, J. P.; de Zeeuw, R. A. *J. Anal. Toxicol.* **1983,** *7,* 188.
98. Bogusz, M.; Wijsbeek, J.; Franke, J. P.; de Zeeuw, R. A.; Gierz, J. *J. Anal. Toxicol.* **1985,** *9,* 49.
99. Japp, M.; Gill, R.; Osselton, M. D. *J. Forensic Sci.* **1987,** *32,* 1574.
100. Newton, B.; Foery, R. F. *J. Anal. Toxicol.* **1984,** *8,* 129.
101. Stowell, A.; Wilson, L. *J. Forensic Sci.* **1987,** *32,* 1214.
102. Anderson, W. H.; Stafford, D. T. *J. High Resolut. Chromatogr. Chromatogr. Commun.* **1983,** *6,* 247.
103. Watts, V. W.; Simonick, T. F. *J. Anal. Toxicol.* **1987,** *11,* 210.
104. Aderjan, R.; Bogusz, M. *J. Chromatogr.* **1988,** *454,* 345.
105. Franke, J.-P.; Wijsbeek, J.; de Zeeuw, R. A. *J. Forensic Sci.* **1990,** *35,* 813.
106. Ojanperä, I.; Rasanen, I.; Vuori, E. *J. Anal. Toxicol.* **1991,** *5,* 204.
107. Christ, D. W.; Noomano, P.; Rosas, M.; Rhone, D. *J. Anal. Toxicol.* **1988,** *12,* 84.
108. Manca, D.; Ferron, L.; Weber, J.-P. *Clin. Chem.* **1989,** *35,* 601.
109. Kolb, B.; Bischoff, J. *J. Chromatogr.* **1974,** *12,* 625.
110. Poole, C. F.; Zlatkis, A. *Anal. Chem.* **1980,** *52,* 1002.
111. Baker, J. K. *Anal. Chem.* **1977,** *49,* 906.
112. Matin, S. B.; Rowland, M. *J. Pharm. Sci.* **1972,** *61,* 1235.
113. Snyder, L. R. *J. Chromatogr. Sci.* **1978,** *16,* 223.
114. Chen, X. -H. Ph.D. Thesis, State University Groningen, Groningen, the Netherlands, 1993.
115. *J&W Scientific 1990-91 GC and HPLC Chromatography Catalog;* J&W Scientific: Folsom, CA, 1990; p 148.
116. Cox, G. B. In *Drug Determination in Therapeutic and Forensic Contexts;* Reid, E.; Wilson, I. D., Eds.; Plenum: New York, 1984; p 71.
117. Glajch, J. L.; Kirkland, J. J.; Squire, K. M.; Minor, J. M. *J. Chromatogr.* **1980,** *199,* 57.
118. *J&W Scientific 1990-91 GC and HPLC Chromatography Catalog;* J&W Scientific: Folsom, CA, 1990; p 146.
119. Lambert, W. E.; Meyer, E.; De Leenheer, A. P. *J. Anal. Toxicol.* **1995,** *19,* 73.
120. Soni, S. K.; Dugar, S. M. *J. Forensic Sci.* **1979,** *24,* 437.
121. Moleman, P.; van Dijk, J. *Clin. Chem.* **1990,** *36,* 732.
122. Sottolano, S. M. *J. Forensic Sci.* **1988,** *33,* 1415.
123. Noggle, F. T., Jr. *J. Assoc. Off. Anal. Chem.* **1980,** *63,* 702.
124. Noggle, F. T., Jr.; DeRuiter, J.; Long, M. J. *J. Assoc. Off. Anal. Chem.* **1986,** *69,* 681.
125. Mangin, P.; Lugnier, A. A.; Chaumont, A. J. *J. Anal. Toxicol.* **1987,** *11,* 27.
126. *The Supelco. Reporter* **1985,** *4,* 8.
127. Sohr, C. J.; Buechel, A. T. *J. Anal. Toxicol.* **1982,** *6,* 286.
128. Lloyd, J. B. F.; Parry, D. A. *J. Anal. Toxicol.* **1989,** *13,* 163.
129. Puopolo, P. R.; Pothier, M. E.; Volpicelli, S. A.; Flood, J. G. *Clin. Chem.* **1991,** *37,* 701.
130. Lambert, W. E.; Meyer, E.; Xue-Ping, Y.; De Leenheer, A. P. *J. Anal. Toxicol.* **1995,** *19,* 35.
131. Kanter, S. L.; Musumeci, M. R.; Hollister, L. E. *J. Chromatogr.* **1979,** *171,* 504.
132. Baker, P. B.; Gough, T. A.; Wagstaffe, P. J. *J. Anal. Toxicol.* **1983,** *7,* 7.
133. Posey, B. L.; Kimble, S. N. *J. Anal. Toxicol.* **1984,** *8,* 234.
134. ElSohly, M. A.; ElSohly, H. N.; Jones, A. B.; Dimson, P. A.; Wells, K. E. *J. Anal. Toxicol.* **1983,** *7,* 262.
135. Alburges, M. E.; Peat, M. A. *J. Forensic Sci.* **1986,** *31,* 695–706.
136. Nakahara, Y.; Sekine, H.; Cook, C. E. *J. Anal. Toxicol.* **1989,** *13,* 22.
137. Jatlow, P.; Nadim, H. *Clin. Chem.* **1990,** *36,* 1436.
138. Logan, B. K.; Stafford, D. T. *J. Forensic Sci.* **1990,** *35,* 1303.
139. Evans, M. A.; Morarity, T. *J. Anal. Toxicol.* **1980,** *4,* 19.
140. Gill, R.; Abbott, R. W.; Moffat, A. C. *J. Chromatogr.* **1984,** *301,* 155.
141. Lewin, A. H.; Parker, S. R.; Carroll, F. I. *J. Chromatogr.* **1980,** *193,* 371.
142. Posey, B. L.; Kimble, S. N. *J. Anal. Toxicol.* **1984,** *8,* 68.
143. Umans, J. G.; Chiu, T. S. K.; Lipman, R. A.; Schultz, M. F.; Shin, S.-U; Inturrisi, C. E. *J. Chromatogr.* **1982,** *233,* 213.

144. Srivastava, V. K.; Maheshwari, M. L. *J. Asoc. Off. Anal. Chem.* **1985,** *68,* 801.
145. Pettitt, B. C., Jr.; Damon, C. E. *J. Chromatogr.* **1982,** *242,* 189.
146. Soni, S. K.; Dugar, S. M. *J. Forensic Sci.* **1979,** *24,* 437.
147. Love, J. L.; Pannell, J. *Forensic Sci.* **1980,** *25,* 320.
148. Baker, P. B.; Gough, T. A. *J. Chromatogr. Sci.* **1981,** *19,* 483.
149. Lurie, I. S.; Sottolano, S. M.; Blasof, S. *J. Forensic Sci.* **1982,** *27,* 519.
150. Lurie, I. S.; Allen, A. C. *J. Chromatogr.* **1984,** *317,* 427.
151. Galewsky, G.; Nessler, C. L. *Chromatographia* **1984,** *18,* 87.
152. Billiet, H. A. H.; Wolters, R.; de Galan, L.; Huizer, H. *J. Chromatogr.* **1986,** *368,* 351.
153. Wall, G. M.; Clark, C. R. *J. Assoc. Off. Anal. Chem.* **1981,** *64,* 1431.
154. Sottolano, S. M.; Lurie, I. S. *J. Forensic Sci.* **1983,** *28,* 929.
155. White, P. C. *J. Chromatogr.* **1979,** *169,* 453.
156. Beug, M. W.; Bigwood, J. *J. Chromatogr.* **1981,** *207,* 379.
157. Perkal, M.; Blackman, G. L.; Ottrey, A. L.; Turner, L. K. *J. Chromatogr.* **1980,** *196,* 180.
158. Lurie, I. S.; Allen, A. C.; Issaq, H. J. *J. Liq. Chromatogr.* **1984,** *7,* 463.
159. Lurie, I. S.; Allen, A. C. *J. Chromatogr.* **1984,** *292,* 283.
160. Twitchett, P. J.; Fletcher, S. M.; Sullivan, A. T.; Moffat, A. C. *J. Chromatogr.* **1978,** *150,* 73.
161. Christie, J.; White, M. W.; Wiles, J. M. *J. Chromatogr.* **1976,** *120,* 496.
162. Gill, R.; Key, J. A. *J. Chromatogr.* **1985,** *246,* 423.
163. Gill, R.; Wanogho, S. O. *J. Chromatogr.* **1987,** *391,* 461.
164. Bannister, S. J.; van der Wal, S. J.; Dolan, J. W.; Snyder, L. R. *Clin. Chem.* **1981,** *27,* 849.
165. Yang, S.; Evenson, M. A. *Anal. Chem.* **1983,** *55,* 994.
166. Koteel, P.; Mullins, R. E.; Gadsden, R. H. *Clin. Chem.* **1982,** *28,* 462.
167. Wong, S. H. Y.; McHugh, S. L.; Dolan, J.; Cohen, K. A. *J. Liq. Chromatogr.* **1986,** *9,* 2511.
168. Matsumoto, K.; Kanba, S.; Kubo, H.; Yagi, G.; Iri, H.; Yuki, H. *Clin. Chem.* **1989,** *35,* 453.
169. Sutfin, T. A.; D'Ambrosio, R.; Jusko, W. J. *Clin. Chem.* **1984,** *30,* 471.
170. Blakesley, J. D.; Howse, C. G.; Spender-Peet, J.; Wood, D. C. F. *Ann. Clin. Biochem.* **1986,** *23,* 552.
171. Mellstrom, B.; Tybring, G. *J. Chromatogr.* **1977,** *143,* 597.
172. Law, B.; Gill, R.; Moffat, A. C. *J. Chromatogr.* **1984,** *301,* 165.
173. Lurie, I. *J. Assoc. Off. Anal. Chem.* **1977,** *60,* 1035.
174. Turcant, A.; Premed-Cabic, A.; Caillenx, A.; Allain, P. *Clin. Chem.* **1991,** *37,* 1210.
175. Bogusz, M.; Wu, M. *J. Anal. Toxicol.* **1991,** *15,* 188.
176. Smith, R. M.; Hurdley, T. G.; Westlake, J. P.; Gill, R.; Osselton, M. D. *J. Chromatogr.* **1988,** *455,* 77.
177. Neue, U. D.; Phillips, D. J.; Walter, T. H.; Capparella, M.; Alden, B.; Fisk, R. P. *LC-GC* **1994,** *12,* 468.
178. Leach, D.C.; Stadalius, M. A.; Berus, J. S.; Snyder, L. R. *LC-GC* **1988,** *6,* 494.
179. Claessens, H. A.; Vermeer, E. A.; Cramers, C. A. *LC-GC* **1994,** *12,* 114.
180. Smith, R. M. *Retention Indices in Reversed-Phase HPLC;* Giddings, J. C., Ed.; Marcel Dekker: New York, 1987; Vol. 26, p 277.
181. Drummer, O. H.; Kotsos, A.; McIntyre, I. M. *J. Anal. Toxicol.* **1993,** *17,* 235.
182. Gill, R.; Moffat, A. C.; Smith, R. M.; Hurdley, T. G. *J. Chromatogr. Sci.* **1986,** *24,* 153.
183. Baker, J. K.; Cates, L. A.; Corbett, M. D.; Huber, J. W.; Lattin, D. L. *J. Liq. Chromatogr.* **1982,** *5,* 829.
184. Smith, R. M.; Hurdley, T. G.; Gill, R.; Moffat, A. C. *Chromatographia* **1985,** *19,* 401.
185. Smith, R. M.; Hurdley, T. G.; Gill, R.; Moffat, A. C. *Chromatographia* **1985,** *19,* 407.
186. Smith, R. M.; Murilla, G. A.; Burr, C. M. *J. Chromatogr.* **1987,** *388,* 37.
187. Bogusz, M.; Aderjan, R. *J. Chromatogr.* **1988,** *435,* 43.
188. Bogusz, M. *J. Anal. Toxicol.* **1991,** *15,* 174.
189. Hill, D. W.; Kind, A. J. *J. Anal. Toxicol.* **1994,** *18,* 233.
190. Smith, R. M.; Hurdley, T. G.; Gill, R.; Moffat, A. C. *LC-GC* **1986,** *4,* 314.
191. Bogusz, M. *LC-GC* **1991,** *9,* 290.
192. Verzele, M.; Dewaele, C. *Chromatographia* **1984,** *18,* 84.
193. Daldrug, T.; Kardel, B. *Chromatographia* **1984,** *18,* 8.
194. Bogusz, M. *J. Chromatogr.* **1987,** *387,* 404.
195. Bogusz, M.; Neidl-Fischer, G.; Aderjan, R. *J. Anal. Toxicol.* **1988,** *12,* 325.
196. Bogusz, M.; Aderjan, R. *J. Anal. Toxicol.* **1988,** *12,* 67.
197. Bogusz, M; Erkens, M. *J. Anal. Toxicol.* **1995,** *19,* 49.
198. Stevenson, R. American Biotechnology Laboratory: 1990; p 8.
199. Newton, P. *LC-GC* **1990,** *8,* 706.
200. White, P. C. *Analyst (Cambridge, U. K.)* **1984,** *109,* 677.
201. Pickering, M. V. *LC-GC* **1990,** 8, 846.

202. Logan, B. K.; Nichols, H. S.; Fernandez, G. S.; Stafford, D. T. *Crime Laboratory Digest* **1990,** *17,* 5.
203. de Zeeuw, R. A.; Franke, J. P.; Bogusz, M. In *Analytical Methods in Forensic Chemistry;* Ho, M. H., Ed.; Ellis Horwood: New York, 1990.
204. White P. C.; Etherington A.; Catterick T. *Forensic Sci. Int.* **1988,** *37,* 55.
205. Lurie, I. S.; Moore, J. M.; Cooper, D. A.; Kram, T. C. *J. Chromatogr.* **1987,** *405,* 273.
206. Reuland, D. J.; Trinler, W. A. *Forensic Sci. Int.* **1988,** *37,* 37.
207. Tracqui, A.; Kintz, P.; Mangin, P. *J. Forensic Sci.* **1995,** *40,* 254.
208. Maier, R. D.; Bogusz, M. *J. Anal. Toxicol.* **1995,** *19,* 79.
209. Puopolo, P. R.; Volpicelli, S. A.; Johnson, D. M.; Flood, J. G. *Clin. Chem.* **1991,** *37,* 2124.
210. Koves, E. M.; Wells, J. *J. Forensic Sci.* **1992,** *37,* 42.
211. Jane, I.; McKinnon, A.; Flanagan, R. J. *J. Chromatogr.* **1985,** *323,* 191.
212. Twitchett, P. J.; Williams, P. L.; Moffat, A. C. *J. Chromatogr.* **1978,** *149,* 683.
213. Nelson, P. E.; Nolan, S. L.; Bedford, K. R. *J. Chromatogr.* **1982,** *234,* 407.
214. Crump, K. L.; McIntyre, I. M.; Drummer, O. H. *J. Anal. Toxicol.* **1994,** *18,* 208.
215. Schwartz, R. S.; David, K. O. *Anal. Chem.* **1985,** *57,* 1362.
216. Flanagan, R. J.; Jane, I. *J. Chromatogr.* **1985,** *323,* 173.
217. Popovych, O.; Tomkins, R. P. T. *Nonaqueous Solution Chemistry;* Wiley: New York, 1981; p 197.
218. Nakahara, Y.; Sekine, H. *J. Anal. Toxicol.* **1985,** *9,* 121.
219. Isenschmid, D. S.; Caplan, Y. H. *J. Anal. Toxicol.* **1986,** *10,* 170.
220. Kim, C.; Fats, T. *J. Anal. Toxicol.* **1984,** *8,* 135.
221. Bedford, K. R.; White, P. C. *J. Chromatogr.* **1985,** *347,* 398.

SECTION III

Identification Methods

Analytical methods used for drug analysis can be classified into two categories: those based on the *separation* of the analytes on a retention medium, and those based on the *identification* of the analytes' structural characteristics. The former category includes various chromatographic and electrophoretic techniques, and the latter category includes mass spectrometry, nuclear magnetic resonance (NMR) spectrometry, and classic molecular spectrometry, such as UV–visible and IR spectrophotometry.

The analytical parameters obtained using separation methods do not generally provide adequate resolution for conclusive identification. Although the analytical parameters obtained using identification methods reflect the analytes' structural characteristics, they are, in most cases, inadequate for unequivocal identification. Some identification methods, however, provide much more definite information than others. For example, UV–visible spectra can only provide structural-framework information, whereas mass spectra often offer near-conclusive matches when intensity ratios of major ions are also evaluated. Nevertheless, when unequivocal identification is needed, it is necessary to combine information obtained from more than one method; to take into account sample history, such as synthesis, extraction, and derivatization; and to include an authentic standard in parallel analysis for comparison under the identical analytical conditions.

Some characteristic features of several identification methods are summarized in the table below. The applications of these methodologies in drug analysis are topics of this chapter and the two that follow.

Characteristic features of identification methods commonly used for drug analysis.

Methodology	Underlying principle	Characteristic parameter for identification
UV–visible spectrometry	Absorption at 200–400 and 400–800 nm regions	Vibrational and rotational transitions of outer orbital and valence electrons
Infrared spectrometry	Absorption at 2,500–25,000 nm regions	Transitions between rotational and vibrational energy levels in the ground electronic state
Nuclear magnetic resonance	Magnetic field and radio frequency	Transition to a higher nuclear spinning energy state
Mass spectrometry	Production and separation of charged particles	Measurement of particle mass/charge ratio

The operational characteristics of several classic molecular spectrometric techniques and their variations for improved applications in drug analysis have been briefly and well summarized in a recent review [1]. It was pointedly stated that "while these techniques have been the cornerstones of laboratory analysis, there are certain problems that result when complex mixtures are analyzed." These problems are associated with the underlying principles on which these methodologies are based.

Because the transitions between electronic states are accompanied by transitions between rotational and vibrational energy levels that impart fine structure to the molecular electronic absorption bands, the absorption or emission bands in the UV region of the electromagnetic spectrum, which correspond to the molecular electronic-state transitions, are typically broad. The resulting overlapping spectra from compounds with similar chemical structures are difficult to differentiate. Although the IR absorption spectra, which are characteristic vibrational transitions of particular bonds or functional groups, often provide "a kind of spectral fingerprint for a particular compound, it is not always easy to differentiate between similar compounds or to distinguish all of the components in a mixture"[1]. Thus, it is not surprising that recent analytical approaches emphasize the use of other instrumentation, such as mass spectrometry, when a definite identification is required.

Reference

1. Allen, R. O. Abstracts of Papers, International Symposium on the Forensic Aspects of Controlled Substances, Quantico, VA; FBI Academy: Quantico, VA, 1988; p 55.

Chapter 7
Molecular Spectrophotometry

UV–Visible Spectrophotometry

Although UV–visible spectrometry may provide important structural information, the broad nature of the absorption bands observed in this region of the electromagnetic spectrum severely limits the resolution of the resulting spectra, thereby reducing the technique's effectiveness for conclusive identification of specific compounds.

In addition to their broad absorption characteristics, UV spectra of a drug may significantly shift as a result of the drug's interaction with the solvent or a change in the pH of the medium [1–4]. For example, the spectra of caffeine and isocaffeine are different in chloroform and water [5]. Furthermore, because a drug may exist in its free form or as a salt, depending on the pH of the medium, the spectrum may also vary with the pH of the medium.

It is possible, however, to take advantage of the spectral characteristics displayed in different solvents and medium pHs to add to the specificity of an identification. For example, differences in spectra obtained at pH 1 and pH 6 for chlorpheniramine maleate and pyrilamine maleate were used to characterize these compounds [6].

The strength of UV–visible spectrometry rests on its Beer's Law behavior:

$$A = \log(P_0/P) = \varepsilon b C \tag{7.1}$$

where A is absorbance, P_0 is the radiant power of the monochromatic radiation transmitted through a cell that contains only pure solvent, P is the radiant power of the transmitted radiation that emerges from the cell that contains the sample, ε is the molar absorptivity in units of liters per mole per centimeter, b is the cell length in centimeters, and C is the concentration of the analyte in moles per liter. Alternatively, ε can be replaced with a (absorptivity) in units of liters per mole per centimeter, and C in units of grams per liter.

The technique provides good quantitative information when the analytes are sufficiently pure. "Even in the case of a mixture with overlapping spectra, quantitative analysis is at least theoretically possible. For a simple mixture of two compounds where the absorption spectrum of each is known, simultaneous equations can be used to calculate the concentrations of each of the components" [1], as follows:

$$A_1 = a_{1,M}bC_{1,M} + a_{1,N}bC_{1,N} \tag{7.2}$$

$$A_2 = a_{2,M}bC_{2,M} + a_{2,N}bC_{2,N} \tag{7.3}$$

where the subscripts 1 and 2 represent the corresponding parameters obtained at two appropriate wavelengths, and the subscripts M and N represent the two components in the mixture.

For a more complex mixture, however, this approach becomes less effective. This is especially the case if one of the components is present at a very low or a very high concentration, which increases the relative concentration errors associated with the analysis at the low- and high-transmittance ends [7].

Correlation of Electronic Absorption Spectra with Molecular Structure

With the limitation in providing conclusive information, UV–visible spectra "should be used in conjunction with other evidence to confirm the identity of a compound—for example, the previous history of a compound, its synthesis, auxiliary chemical tests, and other spectroscopic methods." [7] UV–visible spectra are also valuable in providing information of excluding value. "If a compound is highly transparent throughout the region from 220 to 800 nm, it contains no conjugated unsaturated or benzenoid system, no aldehyde or keto group, no bromide or iodine" [7].

The electron transitions associated with the absorption of electromagnetic radiation in the range of 185 to 800 nm are of the following types: $p \rightarrow \sigma^*$, $\pi \rightarrow \pi^*$, and $p \rightarrow \pi^*$. Specifically, molecules that contain single absorbing groups, called chromophores, undergo transitions at approximately the wavelengths listed in Table 7.1 [7].

Table 7.1. Approximate electronic absorption bands for representative chromophores.[a]

Chromophore	System	λ_{max} (nm)	ε_{max}	λ_{max} (nm)	ε_{max}	λ_{max} (nm)	ε_{max}
Ether	—O—	185	1,000				
Thioether	—S—	194	4,600	215	1,600		
Amine	—NH$_2$	195	2,800				
Thiol	—SH	195	1,400				
Disulfide	—S—S—	194	5,500	255	400		
Bromide	—Br	208	300				
Iodide	—I	260	400				
Nitrile	—C≡N	160					
Acetylide	—C≡C—	175–180	6,000				
Sulfone	—SO$_2$—	180					
Oxime	—NOH	190	5,000				
Azido	—C=N—	190	5,000				
Ethylene	—C=C—	190	8,000				
Ketone	—C=O	195	1,000	270–285	18–30		
Thioketone	—C=S	205	Strong				
Ester	—COOR	205	50				
Aldehyde	—CHO	210	280–300				
Carboxyl	—COOH	200–210	50–70				
Sulfoxide	—S→O	210	1,500				
Nitro	—NO$_2$	210	Strong				
Nitrite	—ONO	220–230	300–4000				
Azo	—N=N—	285–400	3–25				
Nitroso	—N=O	302	100				
Nitrate	—ONO$_2$	270 (shoulder)	12				
	—(C=C)$_2$— (acyclic)	210–230	21,000				
	—(C=C)$_3$—	260	35,000				
	—(C=C)$_4$—	300	52,000				
	—(C=C)$_5$—	330	118,000				
	—(C=C)$_2$— (alicyclic)	230–260	3,000–8,000				
	C=C–C≡C	219	65,000				
	C=C–C=N	220	23,000				
	C=C–C=O	210–250	10,000–20,000	300–350	Weak		
	C=C–NO$_2$	229	9,500				

Table 7.1. (Continued)

Chromophore	System	λ_{max} (nm)	ε_{max}	λ_{max} (nm)	ε_{max}	λ_{max} (nm)	ε_{max}
Benzene		184	46,700	202	6,900	255	170
Diphenyl				246	20,000		
Naphthalene		220	112,000	275	5,600	312	175
Anthracene		252	199,000	375	7,900		
Pyridine		174	80,000	195	6,000	251	1,700
Quinoline		227	37,000	270	3,600	314	2,750
Isoquinoline		218	80,000	266	4,000	317	3,500

^{a} Data are taken from Ref. [7].

The fact that many compounds have similar or nearly identical UV–visible spectra was advocated for the systematic identification of an unknown. Thus, compounds are categorized in such a manner that those with the same unsaturated, conjugated molecular skeleton or parts of the molecule responsible for UV absorption are listed in the same group. Compounds in each group therefore have similar UV-absorption characteristics and show similar pH and solvent effects. In a series of three papers [2–4], more than 500 drugs and toxic compounds were spectrophotometrically characterized into 51 groups on the basis of (1) an absorption profile, which is a graph of wavelength versus absorbance from 200 to 340 nm, (2) the effect of pH change, (3) the effect of decreased solvent polarity relative to water, and (4) the intensity of the absorption bands.

Because similarities in UV spectra are indications of structural similarities, and biological properties of drugs are likewise a consequence of chemical structure, a classification system based on structural groups may also provide information on pharmacological action.

For analytical purposes, when an unknown is found to match the spectrum of a certain known, a list of other compounds that may have identical or nearly identical UV spectra also needs to be evaluated. In no case should an identification be made on the basis of UV spectra alone; UV spectra more often than not will assist in narrowing down the possibilities.

Applications Involving Chemical Derivatization

When a compound undergoes a chemical reaction, the UV spectrum is expected to show differences in addition to the previously mentioned spectral changes. Various chemical reactions have been explored for analyte transformations to provide added information that may improve the specificity of an identification. This same approach may also be used to convert an analyte that lacks absorption or adequate absorption intensity at the desired wavelength to a more suitable derivative.

Examples of applications that utilize chemical reactions to facilitate UV–visible spectrometric methods of analysis include:

1. The oxidation of the diphenylmethylidine group, $(C_6H_6)_2C$, to benzophenone products, which are highly absorbing at 247 or 254 nm [8, 9];

2. The use of carbon disulfide to form dithiocarbamates with primary and secondary aliphatic and heterocyclic amines [10];

3. The ion-pair complexation of codeine phosphate with methyl orange, which results in strong absorption at 418 nm [6];

4. The enhancement of absorption at 253 and 293 nm with UV radiation of acid liver extract containing *d*-propoxyphene [11]; and

5. The decrease of absorption at 235 nm after the NaOH hydrolysis of benzoylecgonine extracted from urine [12].

Extended Techniques and Their Applications

Perhaps the most revolutionary approach in UV–visible spectrophotometry is the use of photodiode array detectors for the continuous monitoring at all wavelengths. This approach is especially useful when UV–visible spectrometry is used as the detector in high-pressure liquid chromatographic (HPLC) applications. Because this approach forms a topic in the chapter on chromatography (Chapter 6), it is not discussed further here.

One of the simplest extensions of the basic techniques in UV–visible spectrometry is so-called *difference spectroscopy,* which involves the subtraction of a spectrum from the overlapping spectra. The spectrum can be obtained by simply placing two different solutions of the sample in a double-beam spectrophotometer. For example, the differential spectrum of aqueous solutions at pH 11 and pH 14 was found to show characteristic features of many barbiturates and was useful for the identification and quantification of barbiturates in blood [13].

With their intrinsic nature of broadness and similarity, UV spectra of analytes are often overlapping, thus making spectral differentiations and detection of spectrum changes difficult. Quantitative determination of the minor component in an overlapping spectrum is often difficult, if not impossible. The so-called *derivative spectroscopy*, in which the first or higher derivative of absorbance or transmittance with respect to wavelength is plotted against the wavelength, can often alleviate these difficulties.

Although derivative spectra can be recorded directly in real time either by wavelength modulation or by obtaining the time derivative of the spectrum when the spectrum is scanned at a constant rate, post-time computer-assisted transformations of stored digital data provide more flexibility. Furthermore, it has been reported [14] that electronic differentiation causes shifts of all peaks in second-derivative scans to shorter wavelengths because of the time delays associated with the capacitance units within the electronic unit circuits and other instrumental parameters.

Using a theoretical example, the improvement in clarity and usefulness of the first and second derivatives over the zero-order spectra is shown in Figure 7.1 [15]. Derivative spectroscopy offers a "better resolution of spectra, especially in the case of the second derivative, because each inflection on the normal spectrum corresponds to a maximum in the second derivative, and each maximum on the normal spectrum corresponds to a minimum in the second derivative." [15] Derivative spectroscopy offers significant advantages "in quantitative determination in a system of two or more components. In a normal spectrum, . . . relative narrow absorption bands normally appear overlapped by the broadband of a second component. The first or second derivative of any spectrum which is a mixture of two or more components permits [one] . . . under certain conditions, to see the different components clearly and to carry out quantitative determinations with very slight or even negligible errors." (*See* Figure 7.2 [15].) Several examples of applications of derivative spectroscopy are summarized in Table 7.2.

Infrared Spectrophotometry

With its ability to provide rich structural information, IR spectrophotometry was the most specific method of identification in most laboratories in the 1960s. Sample handling for IR spectrophotometry is also very flexible. For example, samples can be prepared as films, mulls, pellets, and surfaces for multiple reflections (attenuated total reflectance) to produce absorption spectra, or samples can be simply mixed or coated on the surface of KBr for diffuse reflectance to produce reflectance spectra. Recent advances in the combined use of the Michelson interferometer and the Fourier transform, described later, in so-called Fourier transform infrared (FT-IR) spectrophotometry greatly enhanced the sensitivity with higher resolving power and faster analytical speed.

Basic Operational Parameters

A conventional (dispersive) spectrophotometer records the intensity of the radiation transmitted through the specimen as specific wavelengths are selected by a monochromator. Thus, a spectrum is obtained

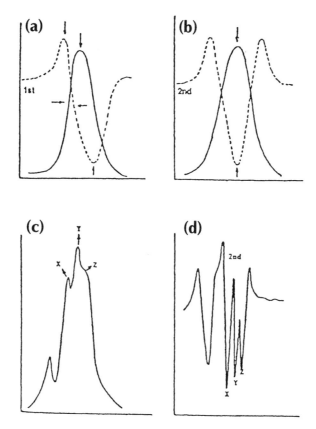

Figure 7.1. Normal spectra and their first and second derivatives (theoretical examples): (a) symmetric absorption peak and its first derivative, (b) symmetric absorption peak and its second derivative, (c) normal spectrum of a mixture, and (d) second derivative of the absorption shown in (c). (Reproduced with permission from Ref. [15].)

by simply plotting the percent transmittance as a function of the wavelengths that are sequentially scanned (passed through the monochromator). On the other hand, an FT-IR spectrophotometer repeatedly looks at all the wavelengths at all times and relies on an interferometer to construct the spectral information.

The main hardware feature of an FT-IR spectrophotometer is a device that "scans" the movable mirror in the Michelson interferometer (Figure 7.3A) [24]; the detector signal is sampled at precise intervals during the mirror scan. The interferometer includes a beam splitter that reflects half the radiation from the IR source to the fixed mirror and transmits the other half to the movable mirror. Because the movable mirror moves back and forth (in the audiofrequency range), the light beam directed to this path travels a different distance. If we hypothetically consider a beam consisting of one wavelength, the light beam directed to the movable mirror will, at a particular point in the movement of the mirror, travel half a wavelength farther than the corresponding beam that was directed to the fixed mirror. At this point, these two beams show destructive interference and the detector detects zero radiation. Different wavelengths show positive (constructive) or negative (destructive) interference at different times, which correspond to different positions of the movable mirror. "The signal recorded at the detector is then an intensity signal that is a function of mirror position. The mirror position is referenced to the point at which the two mirrors are equidistant from the beam splitter, a point known as zero path difference (ZPD). At ZPD in an ideal interferometer, all the wavelengths of radiation are in phase" [25], and the detector will record the most intense signal. As the movable mirror is translated away from ZPD, the various wavelengths of radiation are either in phase or out of phase, and the intensity of radiation recorded on the detector will change accordingly. The resulting signal as a function of distance, usually recorded in centimeters, is called the interferogram (Figure 7.3B).

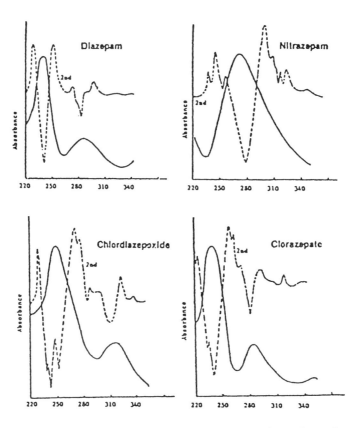

Figure 7.2. Normal spectra and second derivatives of the four benzodiazepines. (Reproduced with permission from Ref. [15].)

The information in the interferogram is then used to construct the spectrum via the mathematical process known as the Fourier transform. The Fourier transform converts the function of one independent variable to a function of another independent variable, with one variable the reciprocal of the other. Thus, the Fourier transform of the interferogram data set is a function of wavenumber (cm^{-1}).

The advantages of FT-IR over its dispersive counterpart include improved resolution, increased sensitivity, and higher certainty in wavelength measurement. The improved wavelength measurement is possible with the use of a reference laser signal for the control of the detector sampling rate and the mirror movement velocity (Figure 7.3C). With precise wavelength measurement, it is possible to reliably subtract large absorptions from two overlapping spectra and to look for small differences derived from a trace component.

Spectrum–Structure Correlation

The spectra in the mid-IR region, ranging from 2.5 μm (4000 cm^{-1}) to 25 μm (400 cm^{-1}), are the most informative part. This region is further divided into the "group frequency" region (2.5–7.69 μm or 4000–1300 cm^{-1}) and the "fingerprint" region (7.69–15.38 μm or 1300–650 cm^{-1}) [26]. The absorption bands in the group frequency region are due to vibration units consisting of only two atoms of a molecule and are informative for revealing the functional groups. For example, the frequencies for C–H, triple bonds, and double bonds fall in the regions of 3.03–3.57 μm (3300–2800 cm^{-1}), 4.00–5.00 μm (2500–2000 cm^{-1}), and 5.00–6.49 μm (2000–1540 cm^{-1}), respectively. It is also possible to distinguish among C=O, C=S, C=C, C=N, N=O, and S=O bands. For example, the IR spectrum of pentobarbital is almost identical to that of its thio counterpart (thiopental), except that one of the C=O bands near 1700 cm^{-1} is replaced by a C–S band near 1550 cm^{-1} [27].

Table 7.2. Application examples of derivative spectroscopy in drug analysis.

System studied	Order of derivative	Parameters studied	Ref.
Acetaminophen–phenacetin and degradation products	2nd	Optimal medium for simultaneous determination	[16]
Amphetamine in liver extract	2nd	Identification	[14]
Single benzodiazepines	2nd	Quantitative determination	[15]
Cocaine	2nd	Quantitative determination and study of interference compounds	[17]
Cocaine–local anesthetics mixtures	2nd	Evaluation of medium effects on resolution	[18]
Ephedrine or pseudoephedrine	2nd and 4th	Quantitative determination	[19]
Heroin	2nd	Quantitative determination and study of interference compounds	[17]
Heroin–morphine mixture	2nd	Spectral separation of mixtures with proportions	[20]
Methadone	1st and 2nd	2nd-derivative spectrum characteristics of broad and sharp absorption peaks	[14]
Phenothiazines and their sulfoxides	3rd	Optimization of derivative order and instrument conditions	[21]
Δ^9-THC–CBN mixture[a]	2nd	Simultaneous quantitative determination	[22]
Papaverine–methaqualone–caffeine	2nd	Sequential component determination by subtracting spectrum of known compound	[23]

[a] Abbreviations: Δ^9-THC: Δ^9-tetrahydrocannabinol; CBN: cannabinol.

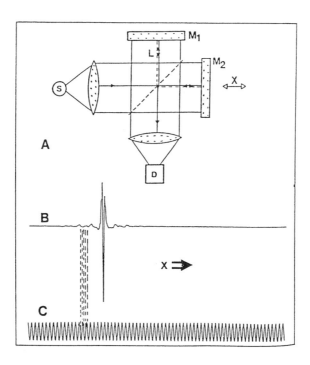

Figure 7.3. (A) Schematic of a Michelson interferometer: S, source; D, detector; M1, fixed mirror; M2, movable mirror; X, mirror displacement. (B) Signal measured by detector D—the interferogram. (C) Interference pattern of a laser source; its zero crossings define the position where the interferogram is sampled (dashed lines). (Reproduced with permission from Ref. [24].)

Although absorption bands in the group frequency region do not provide much information on the complete molecular structure, secondary structural effects do cause significant shifts of the frequencies of absorption bands caused by the functional groups. For example, the C–H stretching frequencies for C≡C–H, aromatic and unsaturated compounds, and aliphatic compounds occur around 3.03 μm (3300 cm^{-1}), 3.33–3.23 μm (3000–3100 cm^{-1}), and 3.33–3.57 μm (3000–2800 cm^{-1}), respectively.

The major absorption bands in the fingerprint region are single-bond stretching frequencies and bending vibrations (skeletal frequencies) of polyatomic systems that involve motions of bonds linking a substituting group to the remainder of the molecule. Multiplicity in this fingerprint region is too great for assured individual identification of the bands.

The interpretation of an unknown IR spectrum generally starts with determining whether the compound is aromatic, aliphatic, or both by examining the hydrogen stretching region. The group frequency region is then examined to establish the presence or absence of certain functional groups. In most cases, the examination of characteristic frequencies is not sufficient to positively identify a total unknown, but it may help in deducing the type or class of the compound. Once the category is established, the spectrum of the unknown is compared with spectra of appropriate compounds. If the exact compound is not in the file, particular structure variations within the category may assist in suggesting possible compounds or eliminating others.

With the rich structural information provided by IR spectra, the collection of a database of IR spectra is useful for making comparisons with unknown spectra. Effective spectra filing and searching mechanisms are essential if systematic matchings are to be conducted. Thus, the conventional Sadtler Standard [Infrared] Grating Spectra Spec-Finder [28] provides a mechanism for locating standard references that have absorption peaks similar to those of an unknown spectrum. An unknown spectrum is compared with the reference spectra to reach an exact match and therefore a positive identification of the unknown compound. Or, if an exact match is not obtained, the general chemical structure of the unknown compound will be revealed.

The Spec-Finder is created by first coding the wavenumber location of the strongest band within each 200 wavenumber from 3600 cm^{-1} to 2000 cm^{-1} and within each 100 wavenumber from 2000 cm^{-1} to 400 cm^{-1}. This coding procedure is performed for each spectrum in the collection of Standard Infrared Grating Spectra. The coded values are sorted in increasing numerical sequence. Within a group of spectra that have the same strongest-band value, the secondary sorting sequence is in accordance with numerical ascendancy of the band codes across the 2000 cm^{-1} to 400 cm^{-1} wavenumber regions. To compare an unknown spectrum with the spectra in the collection, the unknown spectrum is first coded with the same procedure. The strongest band in the unknown spectrum is then used to locate the section of the collection with the same strongest band, followed by matching the codings in the 2000 cm^{-1} to 400 cm^{-1} intervals of the unknown spectrum with the reference spectra in this section.

Alternatively, database development and automated search algorithms have been explored on the basis of a comparison of full FT-IR spectra. "In the simplest form, the unknown spectrum and each reference spectrum are presented as points in a multidimensional space, where each dimension corresponds to a particular wavelength location in the spectrum. The similarity of the unknown to each reference spectrum is then computed as the 'distance' between the two points in the 'hyperspace'. This is the basis of the nearest-neighbor technique, which assigns features to an unknown compound based on the features contained in the compounds whose spectra are 'closest' to it." [29]

A multivariate analysis [30] emphasizing *classification*, rather than *identification*, was conducted in which the IR spectra of six barbiturates, four amphetamines, and five other drugs were included. A spectrum is divided into five regions (4000–2000, 1999–1500, 1499–1200, 1199–700, and 699–300 cm^{-1}), taking the 3, 3, 10, 13, and 7 most intense peaks from regions 1, 2, 3, 4, and 5, respectively. Although the study has shown the feasibility of classifying representative drugs, an automatic data collection and storage mechanism enabling direct linkage to the IR spectrometer will have to be developed before a full-scale study and application can be explored.

In contrast to mid-IR spectra, which record vibrations of chemical bonds, the near-IR (NIR) region shows overtones and combination vibrations, especially the CH, NH, and OH stretching vibrations. The advantage of the NIR region is its lower absorptivity, which makes sample preparation unnecessary. Diffuse-reflectance NIR spectra can be obtained by directly measuring the sample with the fiber-optic probe in its container. Typical NIR spectra shown in Figure 7.4 [31] display strongly overlapped bands

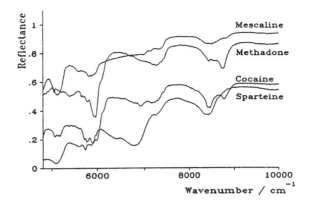

Figure 7.4. Near-infrared spectra of mescaline, methadone, cocaine, and sparteine. (Reproduced with permission from Ref. [31].)

with relatively high absorbances at lower frequencies and lower absorbances toward the visible part of the spectrum. Identification is made by comparing an unknown spectrum with either a reference spectrum or a library of reference spectra. With fewer distinct peaks, comparisons of NIR spectra rely on computerized factor analysis and may not be effective for resolving spectra resulting from multiple components. In light of no sample-preparation requirement, it can be an effective approach for excluding certain compounds.

Practical Considerations and Applications

Sample Preparation for Spectrometric Examination

The advantages of FT-IR spectrophotometry can be further extended by the use of unconventional diamond-cell and diffuse-reflectance sampling accessories [32]. A dispersive spectrophotometer is limited by the energy throughput [33] when a small sample aperture is used. Too much beam condensing will result in a thermally damaged sample. Thus, a compromise size of the smallest aperture is about 1 mm. With the improved sensitivity and the modulation of the energy from the source by an interferometer, an aperture as small as 25 μm can be used to produce satisfactory spectra in FT-IR [33]. Thus, potential contamination of the grinding and pressing process of a small amount of sample with KBr, as often used in dispersive spectrophotometry, can be avoided by simply placing the sample between two flat IIa diamonds. With the application of a great deal of pressure, the sample will be made to flow over the surface of the diamonds.

Several approaches have been explored for infrared spectrophotometric examination of multicomponent samples such as pharmaceutical products and illicit drug exhibits. One of these approaches adopts a crude extraction of the sample, followed by addition of a microcrystal reagent and examination of the micro FT-IR spectra of the resulting drug–crystal complexes. Satisfactory results were reported using mercuric iodide in HCl for opiates and chloroauric acid in HCl for cocaine, amphetamines, and PCP. Common drug adulterants that produce a crystalline reaction and absorb in the IR region exhibit sufficiently different spectra and do not cause confusion in identification [34].

The full-spectrum quantitation method was found [35] to be suitable for content-uniformity determination of pharmaceutical products, and it will presumably be useful for determining the similarity of illicit drug seizures.

The most effective approach, although it may be more time-consuming, includes the use of a gas-chromatograph component for prior component separation. Two approaches are used to adopt FT-IR as the on-line identification tool. The basic design of a "vapor phase GC–FT-IR" consists of a heated fused-silica transfer gas cell, with IR data continually collected by transmitting an IR beam through the cell as the effluent flows from the GC. The "cryogenic deposition GC–FT-IR" approach connects the transfer line to a vacuum chamber that ends with a silica restrictor. The effluent exits the transfer line and is deposited onto a Zn–Se IR transparent plate, held at liquid-nitrogen temperature by a clod block, where IR spectra continue to be obtained. The cryogenic deposit approach was reportedly capable of detecting at the picogram level, greatly improving the sensitivity of IR detection [36].

Approaches and Cautions for Comparing Unknown Spectra with Known Collections

When comparing an unknown spectrum with a known collection, care must be exercised to consider the exact status of the sample for which the spectrum was obtained. Listed below are some examples reported in the literature:

1. The salt and free acid forms of pentobarbital have exhibited different carbonyl absorptions, and the various crystalline forms and amorphous states of barbiturates might cause interpretation difficulties [32].

2. With the use of various solvents in a drug analysis process, any variations in the IR spectrum of a drug that may associate with a particular solvent deserve special attention. It has been reported [37] that codeine free-base might absorb water from the air in the KBr pellet preparation process and show the spectrum characteristic best assigned to that of the monohydrate of the free-base.

3. The IR spectrum indicated [38] the formation of a noncrystalline 6-monoacetylmorphine–ethyl acetate complex under aqueous ammonia extraction with ethyl acetate followed by evaporation of the extract to dryness under a gentle stream of nitrogen. No solvent complexation was found when saturated aqueous sodium bicarbonate (pH ≈ 8) was used instead of aqueous ammonia (pH ≈ 12). This formation of ethyl acetate complex appeared to be unique to 6-monoacetylmorphine, and it was not observed for morphine and its closely related derivatives, such as codeine, thebaine, heroin, and 3-monoacetylmorphine.

4. Using different solvents, neutralization conditions, and temperature conditions, it has been demonstrated [39] that at least two polymorphic forms of heroin base with various crystal habits exist. Crystallization from chloroform and acetone, depending on temperature conditions, yielded the polymorph (with spherulite habit), with carbonyl absorption maxima at 1728.3 and 1756.6 cm^{-1}. The plate habit was characteristic of the polymorph, with carbonyl absorption maxima at 1741.1 and 1761.7 cm^{-1}.

5. The IR transmission spectra of some hydrogen halide salts of methamphetamine displayed significant variations, which were dependent on the alkali halide matrix in which the salt was dispersed for examination [40]. A strong absorption in the hydrogen-stretching region of the IR spectrum was found to be particularly sensitive to the anion present in the salt. Chemical exchange of halide ions between the methamphetamine salt and the matrix material was thought to have caused this observed effect.

6. A detailed study [41] of the quantitative determination of cocaine and heroin based on the absorbance of the carbonyl peaks adopted a judicious selection of baseline to correct broadband absorbance interference. Corrections for spectral interferences were made on the basis of absorbance ratios with a noninterfering spectral line or, alternatively, by spectral subtraction of the interfering component. In a mixture containing the free-base and the hydrochloride salt, the preferred method for quantitative determination was by net area integration of the carbonyl peaks from 1675 to 1825 cm$^{-1.}$

Other Molecular Spectrophotometry

Fluorescence Spectrophotometry

Fluorimetry, which is based on excitation of the sample with one wavelength and observation of the emission at a different wavelength, can be inherently more sensitive than absorption spectrophotometry. The signal-to-noise ratio is greatly improved because the detector is comparing a signal over a dark background, whereas the absorption method measures the small intensity reduction of rather strong radiation. Thus, for compounds that fluoresce efficiently, improved sensitivity can be achieved for the analysis of samples in trace quantities.

In addition, fluorescence spectrophotometry can provide added information. For example, compounds absorbing at the same wavelengths may fluoresce with different intensities or lifetimes [1], thus adding specificity of an identification.

Lysergic acid diethylamide (LSD) fluoresces strongly between 400 and 530 nm, depending on the solvent and the pH employed. It is possible [42] to construct a low-cost, robust, and compact differential fluorimeter for the determination of LSD in illicit preparations. For compounds that may not have a high quantum yield, their detection can also be made very sensitive by chemical derivatization. For example, propoxyphene [43] in biological materials was reacted with 4-chloro-7-nitrobenzo-2,1, 3-oxadiazole to produce highly fluorescent derivatives, and amphetamines [44] were reacted with formaldehyde and acetyl acetone to produce fluorescing lutidine derivatives prior to their fluorometric determination.

Circular Dichroism

A circular dichroism (CD) spectrum can be viewed as a modified absorption spectrum in which ellipticity is plotted against wavelength. The ellipticity, $\Delta\varepsilon$, is directly proportional to the difference in absorption between left, ε_L, and right, ε_R, circularly polarized light. The spectra will show departures from the baseline, which is set by zero ellipticities measured outside the range of the absorption bands and crossover points between positive and negative bands, which occur whenever $\varepsilon_L = \varepsilon_R$. These spectral characteristics are the basis for discriminating among analytes. Quantitative determinations are made by measuring band heights or band areas. Most successful analytical applications have been made at wavelengths longer than 230 nm, where total absorptions are relatively weak and CD spectra are more analyte-dependent.

The significance of CD in drug analysis was well illustrated in a recent review article [45]. Out of an approximate 1850 drugs marketed in the world today, 1045 are chiral compounds, of which 570 are available as single enantiomers and the other racemates. Many of the most commonly encountered illicit drugs, such as the opiates, cannabinoids, phenethylamines, adamantane, tryptamine alkaloids, and steroids, are CD-active. Others are CD-inactive because they are either racemic, for example, barbiturates and the benzodiazepines, or inherently nonchiral, for example, phencyclidine (PCP), and some phenethylamines. CD activity for the latter category of drugs is induced only after reaction with a chiral reagent. CD determination of optical isomers for these drugs offers direct analysis, which circumvents many chromatographic procedures, and identifies the controlled substance by a physical property that, in the end, provides much more substantial evidence than retention time used in chromatographic methods.

The simplicity of the CD method has been confirmed by studies on the CD detection of cocaine, heroin, and other morphine analogs, amphetamine and methamphetamine, LSD, barbiturates, phencyclidine and analogs, cannabinoids, and diazepams. All that is necessary is to dissolve a weighed aliquot of the unseparated raw sample in dilute aqueous HCl and measure the CD spectrum for the solution from 230 to 350 nm. If it turns out that the sample contains only one CD-active substance, then the screening experiment has suddenly become quantitative.

Several advantages have been noted in the studies reported. For example, it was possible to simultaneously determine morphine, codeine, thebaine, and noscapine in unseparated extracts of opium [46]. Because structurally related CD-inactive substances, such as benzocaine, lidocaine, procaine, and dopamine, do not interfere with the determination of amphetamines, their separation was not a necessary step [47]. *l*-LSD was determined directly from spectral data obtained for a diluted aqueous HCl extract of the drug taken from the matrices of a variety of confiscated forms, such as microdots, window panes, and transfers [48]. The only interfering effect was a decrease in the S/N ratio because of the strong absorption by the achiral dyes that were added to the microdot forms. Tetrahydrocannabinol and cannabidiol have CD spectra of opposite sign and were easily determined simultaneously from data from the single spectrum of unseparated extracts of organs of the marijuana plant [49].

References

1. Allen, R. O. *Abstracts of Papers,* International Symposium on the Forensic Aspects of Controlled Substances, Quantico, VA; FBI Academy: Quantico, VA, 1988; p 55.
2. Siek, T. J. *J. Forensic Sci.* **1974,** *19,* 193.
3. Siek, T. J.; Osiewicz, R. J. *J. Forensic Sci.* **1975,** *20,* 18.
4. Siek, T. J. *J. Forensic Sci.* **1976,** *21,* 525.
5. Yanuka, Y.; Bergman, F. *Tetrahedron* **1986,** *42,* 5991.
6. El Kheir, A. A.; Belal, S. F.; El Shanwani, A. *J. Assoc. Off. Anal. Chem.* **1985,** *68,* 1048.
7. Willard, H. H.; Merritt, L. L., Jr.; Dean, J. A.; Settle, F. A., Jr. *Instrumental Methods of Analysis,* 7th ed; Wadsworth Publishing: Belmont, CA, 1988.
8. Caddy, B.; Hish, F.; Mullen, P. W.; Tranter, J. *J. Forensic Sci. Soc.* **1973,** *13,* 127.
9. Wallace, J. E. *Clin. Chem.* **1969,** *15,* 323.
10. Stevens, H. M. *J. Forensic Sci. Soc.* **1973,** *13,* 119.
11. Thompson, E.; Villaudy, J.; Plutchak, L. B.; Gupta, R. C. *J. Forensic Sci.* **1970,** *15,* 605.
12. Sweeney, W.; Goldbaum, L. R.; Lappas, N. T. *J. Anal. Toxicol.* **1983,** *7,* 235.
13. Yen, B.; Cimbura, G. *Can. J. Forensic Sci.* **1976,** *9,* 113.
14. Gill, R.; Bal, T. S.; Moffat, A. G. *J. Forensic Sci. Soc.* **1982,** *22,* 165.
15. Martínez, D.; Giménez, M. P. *J. Anal. Toxicol.* **1981,** *5,* 10.
16. Korany, M. A.; Bedair, M.; Mahgoub, H.; Elsayed, M. A. *J. Assoc. Off. Anal. Chem.* **1986,** *69,* 608.
17. Larsen, A. K.; Titus, P. R.; Browne, S. *Abstracts of Papers,* International Symposium on the Forensic Aspects of Controlled Substances, Quantico, VA; FBI Academy: Quantico, VA, 1988; p 197.
18. Arufe-Martinez, M. I.; Romero-Palanco, J. L. *J. Anal. Toxicol.* **1988,** *12,* 192.
19. Davidson, A. G.; Elsheikh, H. *Analyst (Cambridge, U. K.)* **1982,** *107,* 879.
20. Lawrence, A. H.; Kovar, J. *Anal. Chem.* **1984,** *56,* 1731.
21. Fasanmade, A. A.; Fell, A. F. *Analyst (Cambridge, U. K.)* **1985,** *110,* 1117.
22. Lawrence, A. H.; Kovar, J. *Analyst (Cambridge, U. K.)* **1985,** *110,* 827.
23. Cruz, A.; Lopez-Rivadulla, M.; Fernandez, P.; Bermejo, A. M. *J. Anal. Toxicol.* **1992,** *16,* 240.
24. Herres, W.; Gronholz, J. *Comp. Appl. Lab.* **1984,** *4,* 216.
25. De Haseth, J. A. *Spectroscopy (Eugene, Oreg.)* **1986,** *1,* 10.
26. Willard, H. H.; Merritt, L. L., Jr.; Dean, J. A.; Settle, F. A., Jr. *Instrumental Methods of Analysis,* 7th ed; Wadsworth Publishing: Belmont, CA, 1988.
27. Ulrich, W. F.; Stine, K. S.; Blecha, D. L.; Harms, D. R. *Analytical Instrumentation in the Forensic Sciences;* Beckman Instruments: Fullerton, CA, 1971; p 36.
28. *Sadtler Special Collection: Commonly Abused Drugs;* Bio-Rad: Philadelphia, PA, 1972.
29. Lowry, S. R.; Huppler, D. A.; Anderson, C. R. *J. Chem. Inf. Comput. Sci.* **1985,** *25,* 235.
30. Moss, W. W.; Posey, F. T.; Peterson, P. C. *J. Forensic Sci.* **1980,** *25,* 304.
31. Kohn, W. H.; Jeger, A. N. *J. Forensic Sci.* **1992,** *37,* 35.
32. Suzuki, E. M. *Abstracts of Papers,* The International Symposium on the Forensic Aspects of Controlled Substances, Quantico, VA; FBI Academy: Quantico, VA, 1988; p 201.
33. Shearer, J. C.; Peters, D. C.; Kubic, T. A. *Trends Anal. Chem.* **1985,** *4,* 246.
34. Wielbo, D.; Tebbett, I. R. *J. Forensic Sci. Soc.* **1993,** *33,* 25.
35. Simonian M. H.; Dinh, S.; Fay, L. A. *Spectroscopy (Eugene, Oreg.)* **1993,** *8,* 37.
36. Kalasinsky, K. S.; Levine, B.; Smith, M. L. *J. Anal. Toxicol.* **1992,** *16,* 332.
37. Kanai, F.; Inouye, V.; Goo, R. *Anal. Chim. Acta* **1985,** *173,* 373.
38. Beckstead, H. D.; Neville, G. A. *J. Forensic Sci.* **1988,** *33,* 223.
39. Ravreby, M.; Gorski, A. *J. Forensic Sci.* **1989,** *34,* 918.
40. Chappell, J. S. *Forensic Sci. Int.* **1995,** *75,* 1.
41. Ravreby, M. *J. Forensic Sci.* **1987,** *32,* 20.
42. Baudot, P.; Andre, J.-C. *J. Anal. Toxicol.* **1983,** *7,* 69.
43. Valentour, J. C.; Monforte, J. R.; Sunshine, I. *Clin. Chem.* **1974,** *20,* 275.
44. Nix, C. R.; Hume, A. S. *J. Forensic Sci.* **1970,** *15,* 595.
45. Purdie, N. *Forensic Sci. Rev.* **1991,** *3,* 1.
46. Han, S. M.; Purdie, N. *Anal. Chem.* **1986,** *58,* 113.
47. Bowen, J. M.; Crone, T. A.; Head,. V. L.; McMorrow, H. A.; Kennedy, R. K.; Purdie, N. *J. Forensic Sci.* **1981,** *53,* 2237.
48. Bowen, J. M.; McMorrow, H. A.; Purdie, N. *J. Forensic Sci.* **1982,** *27,* 822.
49. Han, S. M.; Purdie, N. *Anal. Chem.* **1985,** *57,* 1068.

Chapter 8
Nuclear Magnetic Resonance Spectrometry

Magnetic Resonance

Nuclear magnetic resonance (NMR) spectrometry is concerned with the measurement of energy differences between close-lying levels of nuclear spin states [1]. The transitions of these nuclear states occur in the presence of an external magnetic field and correspond to energy levels in the radio frequency range. In the presence of a higher magnetic field, the energy gaps, ΔE, between the spin states widen and the population difference between the lower, N_1, and the higher, N_2, nuclear spin states improves as shown in the Boltzman relation:

$$N_1/N_2 = e^{\Delta E/RT} \tag{8.1}$$

where R is the gas constant, and T is the Kelvin temperature. This phenomenon translates into the effect that the experiment can be more sensitive under a higher magnetic field.

The exact energy levels (frequencies) that are required for the occurrence of specific nuclear transitions (resonance) are related to the magnetic field strength according to the Larmor equation:

$$\nu = (\text{constant})H_0 \tag{8.2}$$

where ν is the resonance frequency, H_0 is the magnetic field, and the constant is the characteristic of the particular nucleus concerned. For example, the resonance frequency of the proton is 60 MHz at 14,092 gauss. Thus, it is possible to obtain NMR spectra through field-sweep with a constant radio frequency or through frequency-sweep with a constant magnetic field.

Chemical Shift and Spin–Spin Coupling

Chemical shift (δ) and spin–spin coupling (J) are the two most important parameters for qualitative analysis. Under the influence of H_0, electrons surrounding the nucleus circulate and produce a current which in turn generates a small magnetic field opposing the applied H_0. Thus, for a resonance to occur, a higher field is needed to overcome this "shielding" effect. The shielding effect varies with, among other factors, the hybridization and electronegativity of the groups attached to the atom containing the nucleus being studied.

Chemical shifts are normally presented in relation to a reference material, commonly tetramethyl silane (TMS), for proton spectra in nonaqueous media, so that they will be independent of the different field strengths used for different spectrometers. Thus, the chemical shifts are commonly expressed in parts per million, ppm, as:

$$\delta = [(H_{sample} - H_{TMS})/\nu] \times 10^6 \tag{8.3}$$

where H_{sample} and H_{TMS} are the positions of the absorption lines for the sample and the reference.

Because magnetic moments of nuclei interact with the strongly magnetic electrons in the intervening bonds, the otherwise sharp resonance lines can be split into multiplets given by $2nI + 1$, where n is the number of nuclei producing the splitting, and I is the spin quantum number of the concerned nucleus. The strength of the coupling, denoted by J, is given by the spacing of the multiplets and is expressed in hertz.

Proton Nuclear Magnetic Resonance (^{1}H NMR) Spectrometry

In general, NMR spectrometry cannot be used to substitute for IR or mass spectrometry as the primary tool for drug identification. It can, however, be valuable in providing complementary information when IR spectrophotoscopic and mass spectrometric data alone are not conclusive. Statements made by experienced drug analysts strongly suggest that NMR spectra are valuable in providing additional information for differentiating some members of the phenethylamine, barbiturate, and benzodiazepine drug categories:

> Obtaining an infrared (IR) spectrum of substituted phenethylamines (that is, methamphetamine, amphetamine) is often complicated by their hygroscopic and polymorphic properties. This can result in several different IR spectra for a single compound. The mass spectra of these phenethylamines exhibit weak parent peaks and very similar fragmentation patterns and are not as definitive as NMR. Similarly, barbiturates form clathrates (that is, chloroform adducts) and polymorphs, which makes their analysis by IR difficult. Mass spectral differences between most of the barbiturates are usually minimal and not easily distinguished. Benzodiazepines, tranquilizers such as diazepam (Valium), flurazepam (Dalmane), and oxazepam (Serax), break down in such a manner that they also display similar mass spectra. [2]

Even with the use of Fourier transform methods and higher magnetic fields, the detection limit of NMR is not as desirable as those of other methods, such as mass spectrometry. Typically, analytes in "0.1 mg/mL can provide valuable structural information, and sample concentrations of 1 mg/mL provide a complete NMR proton spectrum" [3]. Thus, with its ability to provide valuable structural information, but with its somewhat higher detection limits, the NMR method can be useful for the analysis of components of many pharmaceutical preparations [2, 4].

^{1}H NMR Spectrometry of Commonly Abused Drugs

The ^{1}H NMR characteristics of commonly abused substances, such as opiate-related compounds [5, 6] and nicotine [7, 8] have been examined in some detail. The application of ^{1}H NMR techniques to the analysis of illicit drug exhibits was outlined by Kram [9] in 1978, and used for the identification of amphetamines [9], look-alike drugs [10], barbiturates [9], phencyclidine [9], and cocaine [11].

It is possible to identify amphetamine, methamphetamine, 3,4-methylenedioxyamphetamine (MDA), and 3,4-methylenedioxymethamphetamine (MDMA) in drug exhibits on the basis of their chemical shift and coupling characteristics (Chart 8.1 and Table 8.1). Because illicit drug preparations are typically composed of several components, the possibility of qualitative and quantitative determinations of multiple components is often emphasized. In one approach [9], a ^{1}H NMR spectrum was first taken from the $CDCl_3$ solution of an exhibit to reveal what categories of protons were present. Spectra of both the aqueous and the organic phases were subsequently taken after the solution was extracted with deuterated water. Through this approach, it was possible to obtain clear spectra of piperidine and phencyclidine (both as salts) in the D_2O and the $CDCl_3$ phases. The concentration of phencyclidine was

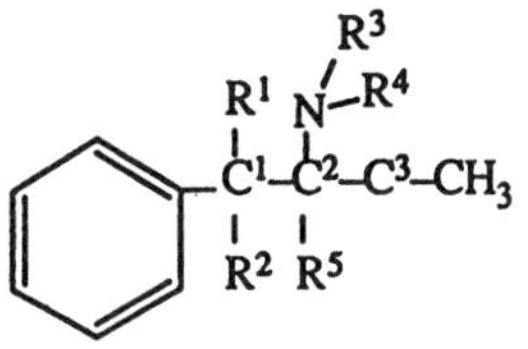

Chart 8.1. Structural framework of amphetamine-related compounds (a), and caffeine (b).

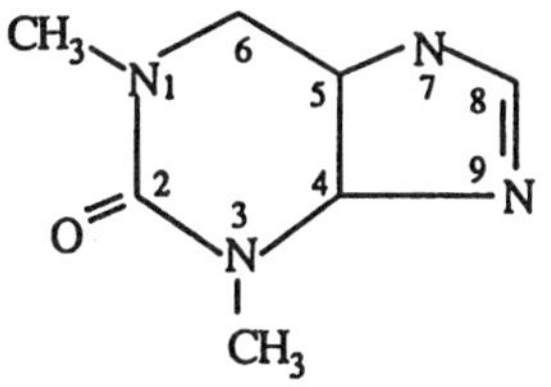

Table 8.1. Chemical shift (at various carbon positions) of amphetamine and look-alike drugs.

Sample	Solvent	Protons at various carbon positions (Chart 8.1)[a]				Methylene		Ref.
						dioxy proton	N-CH$_3$ (caffeine)	
		C^1	C^2	C^3	R^4			
Amphetamine HCl	D$_2$O	2.9 (d)	3.6 (m)	1.3 (d)	–	N/A	N/A	[9]
Methamphetamine[b]	D$_2$O				2.7 (s)	N/A	N/A	[9]
MDA[b]	D$_2$O					6.0 (s)	N/A	[9]
MDMA[b]	D$_2$O				2.7 (s)	6.0 (s)	N/A	[9]
Ephedrine	CDCl$_3$	4.76 (d)	2.78 (m)	0.85 (d)				[10]
Ephedrine salt	D$_2$O	5.14 (d)	3.56 (m)	1.6 (d)				[10]
Phenylpropanolamine	CDCl$_3$	4.51 (d)	3.56 (m)	1.16 (d)				[10]
Phenylpropanolamine salt	D$_2$O	4.98 (d)	3.70 (m)	1.20 (d)				[10]
Caffeine	CDCl$_3$						4.00, 3.58, 3.41	[10]
Caffeine salt	D$_2$O						3.89, 3.41, 3.24	[10]
Ephedrine–caffeine–	CDCl$_3$	4.74 (d)/4.52 (d)	2.77 (m)/3.13 (m)	0.85 (d)/0.97 (d)	2.46 (s)			[10]
phenylpropanolamine	D$_2$O	5.16 (d)/5.00 (d)	c	1.16 (d)/1.22 (d)	2.84 (s)			[10]
Pink tablet	CDCl$_3$	4.76 (d)/4.51 (d)	2.79 (m)/3.15 (m)	0.86 (d)/0.98 (d)	2.47 (s)			[10]
	D$_2$O	5.16 (d)/5.08 (d)	c	1.16 (d)/1.21 (d)	2.84 (s)			[10]

[a] Abbreviations: (d): doublet; (m): multiplet; (s): singlet.

[b] Abbreviations: MDA: 3,4-methylenedioxyamphetamine; MDMA: 3,4-methylenedioxymethamphetamine. Only signals in addition to that reported for amphetamine are listed.

[c] Overlap of the phenylalkylamine C^2-H's and the caffeine N-CH$_3$ resonance peaks.

further determined using maleic acid as the internal standard. The phenyl absorption of phencyclidine, representing five protons, was compared with the olefinic absorption of maleic acid, at 6.4 ppm representing two protons, for the quantification of phencyclidine. This quantification approach, along with signal subtraction, was used to quantify methamphetamine and the methylamine adulterant and mannitol diluent of an exhibit.

A similar approach was used for the identification of look-alike drug components [10]. Standard spectra of ephedrine, phenylpropanolamine, caffeine, and the three-component mixture were obtained and compared with those obtained from an exhibit (Table 8.1). It was stated that unequivocal identification of the three components was readily accomplished, although the C^2–H resonance peaks of phenylalkylamine overlap with the N–CH$_3$ resonance peaks of caffeine. ^{1}H NMR has proven to be extremely useful as a rapid screening method to distinguish licit and look-alike preparations.

Differentiation of Enantiomers

The determination of the enantiomeric identity of a drug is a significant aspect in forensic drug analysis for two reasons. Enantiomeric differentiation is essential in cases in which one enantiomer is under regulation while the other is not; on the other hand, determination of enantiomeric composition can provide valuable information as to the common origin or common trade route of different samples.

Because it was realized that the ^{1}H NMR technique could be used to differentiate diastereomeric pairs, derivatizing agents [12], chiral solvents (alone [13] or with an achiral lanthanide shift reagent [14]), and chiral lanthanide shift reagents [15–17] have been utilized to generate the diastereomers prior to obtaining ^{1}H NMR spectra. Thus, Gelsomino and Raney [18] used (–)-α-phenethyl alcohol to react with d,l-cocaine and observed distinguishable N–CH$_3$ (at ≈2.1 ppm) and O–CH$_3$ (at ≈3.4 ppm) peaks of the resulting "solution diastereomers" (Figure 8.1). The standardized procedure involved the recrystallization of the chloroform extract from a sodium bicarbonate solution. (–)-α-Phenethyl alcohol was then added to the CS$_2$ solution of the crystals. If singlets were observed for the N–CH$_3$ and O–CH$_3$ peaks, standard l-cocaine was added. The sample was determined to be l-cocaine if the resulting N–CH$_3$ and O–CH$_3$ peaks remained singlets; otherwise, it was determined to be d-cocaine. It was concluded that common adulterants, such as lidocaine and benzocaine, do not interfere because their spectra are well separated from the area of interest.

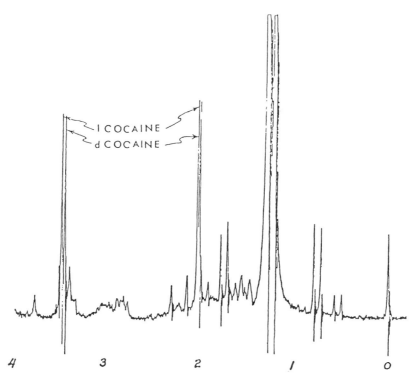

Figure 8.1. NMR spectrum of d,l-*cocaine and (–)-phenethyl alcohol dissolved in* CS$_2$. *(Reproduced with permission from Ref. [18].)*

Differences in the chemical shift of the O–CH$_3$ protons of d- and l-cocaine were also observed [19] by complexing cocaine with the chiral-shift reagent tris(3-trifluoroacetyl-d-camphorato)europium(III) (Eu(tfac)$_3$ or Eu(facam)$_3$), also called tris(3-trifluoromethylhydroxymethylene-d-camphorato)europium(III) (Eu(tfc)$_3$) (Chart 8.2). The separation of the d-cocaine (3.40 ppm) and l-cocaine (3.48 ppm) peaks allows a 5% isomeric determination limit. This chiral-shift reagent has also been used for the study of amphetamine and several methoxylated derivatives [20]; it was concluded that the enantiomeric pairs studied were not adequately resolved.

Six chiral-shift reagents have been extensively compared in the author's (R. H. Liu's) laboratory for their effectiveness in differentiating amphetamine [21] and methamphetamine [22] enantiomers. The six reagents (Chart 8.2) include tris(*d,d*-dicampholylmethanato)europium(III) (Eu(dcm)$_3$); Eu(tfac)$_3$; tris(3-heptafluorobutyryl-*d*-camphorato)europium(III) (Eu(hfbc)$_3$), also called tris(3-heptafluoro-propylhydroxymethylene-*d*-camphorato)europium(III) (Eu(hfc)$_3$); tris(3-trifluoroacetyl-*d*-camphorato)-praseodymium(III) (Pr(tfac)$_3$); tris(3-heptafluorobutyryl-*d*-camphorato)praseodymium(III) (Pr(hfbc)$_3$); and tris(3-trifluoroacetyl-*d*-camphorato)ytterbium(III) (Yb(tfac)$_3$).

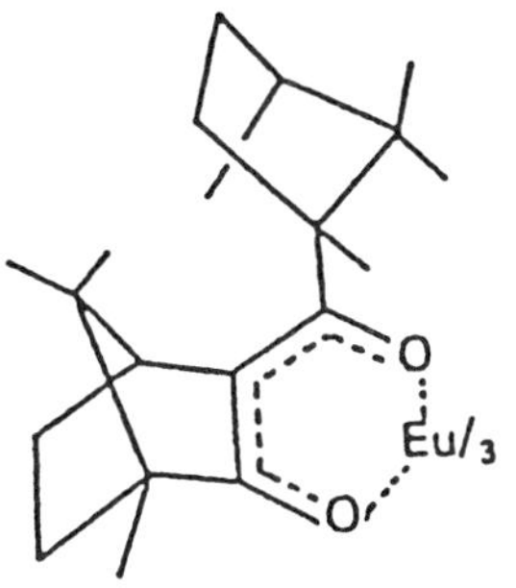

Chart 8.2. Structures of Eu(dcm)$_3$ (a); Eu(tfac)$_3$ (b), where R = CF$_3$ and Ln = Eu; Eu(hfbc)$_3$ (b), where R = C$_3$F$_7$ and Ln = Eu; Pr(tfac)$_3$ (b), where R = CF$_3$ and Ln = Pr; Pr(hfbc)$_3$ (b), where R = C$_3$F$_7$ and Ln = Pr; and Yb(tfac)$_3$ (b), where R = CF$_3$ and Ln = Yb.

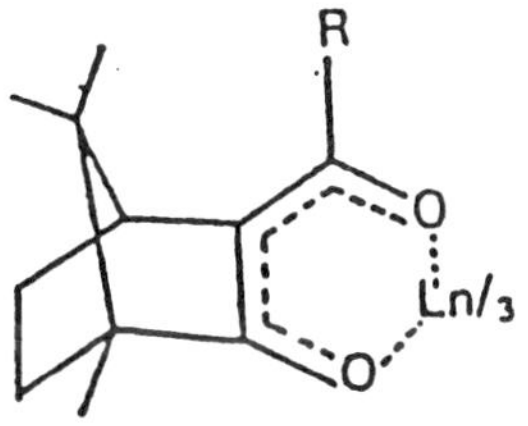

Among these six chiral-shift reagents studied, it was concluded that only Eu(dcm)$_3$ is effective for realistic applications.

Figure 8.2 shows the C–CH$_3$ ^{1}H NMR spectra of a *d*- and *l*-amphetamine mixture with various amounts of Eu(dcm)$_3$ added. The *d*- and *l*-amphetamine are completely resolved when the shift reagent to substrate molar ratio (MR) is equal to or greater than approximately 0.15. Although the coupling constants of both *d*- and *l*-amphetamine remain at 6.0 Hz, the chemical shifts, $\Delta\delta$, are changed as a function of MR. Peak broadening was observed as MR increased. Considering the magnitude of the difference in the enantiomeric pairs' changes in their chemical shifts, $\Delta\delta$, and the peak broadening, the optimum MR that should be used for the determination of these two enantiomers falls in the range of 0.15–0.40.

The other five shift reagents were also used with a similar procedure. Spectra obtained with optimum MRs of all six shift reagents are compared in Figure 8.3. As expected [23], Eu(III) and Yb(III) induced downfield shifts, whereas Pr(III) induced an upfield shift. The excessive line-broadening caused by Pr(III) and Yb(III) chelates renders the use of these shift reagents impractical for amphet-amine enantiomer determination.

The results presented in Figure 8.3 also indicate that the order of the ligands' effectiveness in inducing $\Delta\delta$ is dcm > hfbc > tfac. Figure 8.3 also indicates that the $\Delta\delta$ of *l*-amphetamine induced by Eu(tfac)$_3$ is larger than that of *d*-amphetamine; the reverse is true for Eu(dcm)$_3$ and Eu(hfbc)$_3$. Although the Yb(III) chelate induces a larger $\Delta\delta$ than Eu(III), no $\Delta\delta$ difference for *d*- and *l*-amphetamine is observed.

Figure 8.4 shows the variation of N–CH$_3$ and C–CH$_3$ ^{1}H NMR spectra of *d*- and *l*-methamphetamine with various amounts of Eu(dcm)$_3$ added. Considering the magnitude of $\Delta\delta$ and peak broadening, the optimum MR for resolving C–CH$_3$ proton spectra falls in the range of 0.02–0.15. Results obtained by using the other shift reagents are much less satisfactory. Eu(dcm)$_3$ is the only shift reagent that is effective in resolving these corresponding enantiomeric protons. Whatever $\Delta\delta$ may have been induced by other shift reagents is unresolved as a result of peak broadening.

These studies [21, 22] concluded that quantitative enantiomeric composition determinations of amphetamine and methamphetamine can be effectively performed using the integrated areas of the resolved C–CH$_3$ (for amphetamine) and N–CH$_3$ proton spectra (for methamphetamine). More recent studies have further improved the resolutions of methamphetamine enantiomers using 2,3,4,6-tetra-*O*-acetyl-β-D-glucopyranosyl isothiocyanate chiral derivatizing reagent [24] and (*R*)-(+)-1,1'-bis-2-naphthol solvating agent [25].

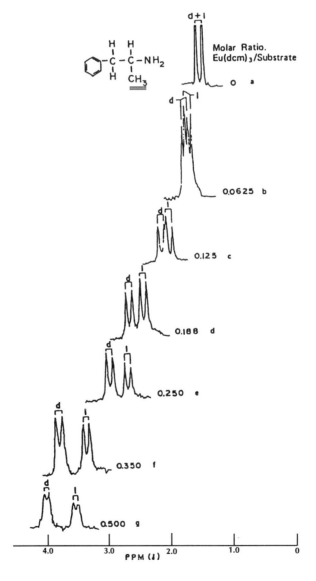

Figure 8.2. C CH$_3$ proton NMR spectra of a d- *and* l-*amphetamine mixture in the absence (a) and presence (b–g) of various amounts of Eu(dcm)$_3$. (Reproduced with permission from Ref. [21].)*

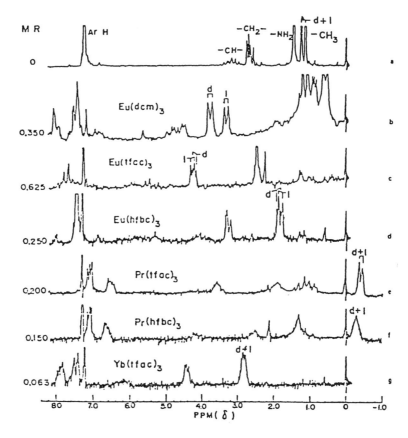

Figure 8.3. Comparison of the resolving power (with an optimum shift reagent to amphetamine ratio) of the six chiral shift reagents on the C–CH₃ protons of d- and l-amphetamine: no shift reagent (a), Eu(dcm)₂ (b), Eu(tfac)₃ (c), Eu(hfbc)₃ (d), Pr(tfac)₃ (e), Pr(hfbc)₃ (f), and Yb(tfac)₃ (g). (Reproduced with permission from Ref. [21].)

Carbon-13 Nuclear Magnetic Resonance (^{13}C NMR) Spectrometry

With the utilization of Fourier transform methods, the use of ^{13}C NMR, which is relatively insensitive because of the low abundance of the natural ^{13}C isotope, has been somewhat obviated. With the superior resolution of ^{13}C NMR spectra and the application of proton-decoupling techniques that greatly simplify the spectra, ^{13}C NMR has been found to be a useful tool in forensic analysis. Thus, ^{13}C NMR spectra, along with functional group–chemical shift correlations, have been documented in the literature for cannabinoids [26], barbiturates [27], morphine alkaloids [28], cocaine [29], methaqualone [30], phencyclidine and related compounds [31], and amphetamines [32, 33].

The superior resolution, spectral simplicity, and ability to differentiate between structurally closely related compounds are exemplified by the spectra shown in Figure 8.5 [34]. The potential in the analysis of multicomponent drug exhibits is exemplified by the analysis of a look-alike drug preparation [10] and an actual street seizure [34]. All components, except procaine, shown in Figure 8.6, were quantified.

Undoubtedly, ^{13}C NMR can be an effective screening method. Despite its relatively low sensitivity, extensive accumulation time is normally not necessary when compounds of interest are the major components in the sample examined.

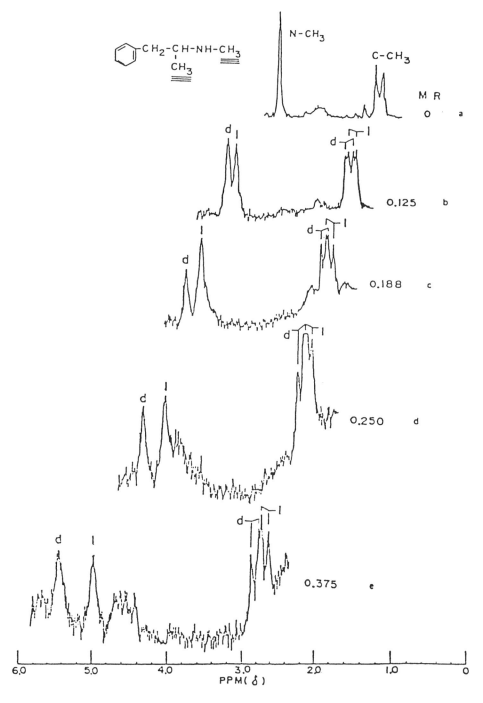

Figure 8.4. N–CH₃ and C–CH₃ proton NMR spectra of a d- and l-methamphetamine mixture in the absence (a) and presence (b–e) of various amounts of Eu(dcm)₃. (Reproduced with permission from Ref. [22].)

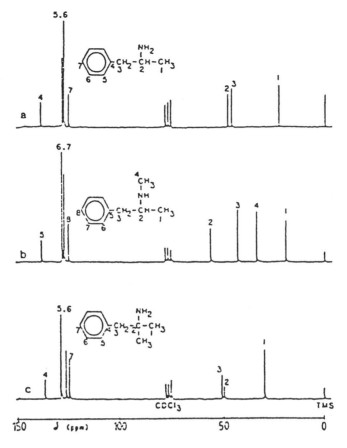

Figure 8.5. ^{13}C NMR spectra of amphetamine (a), methamphetamine (b), and phentermine (c). (Reproduced with permission from Ref. [34].)

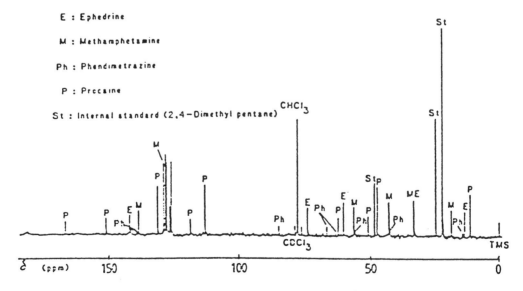

Figure 8.6. ^{13}C NMR spectra of an illicit drug mixture. All components except procaine are quantitated. (Reproduced with permission from Ref. [34].)

References

1. Willard, H. H.; Merritt, L. L., Jr.; Dean, J. A.; Settle, F. A., Jr. *Instrumental Methods of Analysis,* 7th ed; Wadsworth Publishing Company: Belmont, CA, 1988.
2. Angelos, S. A.; Janovsky, T. J.; Raney, J. K. *J. Forensic Sci.* **1991,** *36,* 358.
3. Mills, T. *Abstracts of Papers,* The International Symposium on the Forensic Aspects of Controlled Substances, Quantico, VA; FBI Academy: Quantico, VA, 1988; p 171.
4. Turczan, J. W.; Goldwitz, B. A.; Nelson, J. J. *Talanta* **1972,** *19,* 1549.
5. Jacobson, A. E.; Yeh, H. J. C.; Sargent, L. J. *Org. Magn. Reson.* **1972,** *4,* 875.
6. Allen, A. C.; Cooper, D. A.; Moore, J. M. *J. Forensic Sci.* **1985,** *30,* 908.
7. Pitner, T. P.; Edwards, W. B.; Bassfield, R. L.; Whidby, J. F. *J. Am. Chem. Soc.* **1978,** *100,* 246.
8. Whidby, J. F.; Edwards, W. B., Jr.; Pitner, T. P. *J. Org. Chem.* **1979,** *44,* 794.
9. Kram, T. C. *J. Forensic Sci.* **1978,** *23,* 456.
10. Avdovicj, H. W.; Jin, S. H.; Wilson, W. L. *Can. Soc. Forensic Sci. J.* **1985,** *18,* 24.
11. Clark, C. C. *Microgram* **1983,** *16,* 25.
12. Dale, J. A.; Dull, D. L.; Mosher, H. S. *J. Org. Chem.* **1969,** *34,* 2543.
13. Pirkle, W. H.; Beare, S. D. *J. Am. Chem. Soc.* **1969,** *91,* 5150.
14. Jennison, C. P. R.; Mackay, D. *Can. J. Chem.* **1974,** *51,* 3726.
15. McCreary, M. D.; Lewis, D. W.; Wernick, D. L.; Whitesides, G. M. *J. Am. Chem. Soc.* **974,** *96,* 1038.
16. Goering, H. L.; Eikenbeffy, J. N.; Koermer, G. S.; Lattimer, C. J. *J. Am. Chem. Soc.* **1974,** *96,* 1493.
17. Sullivan, G. R. In *Topics in Stereochemistry;* Eliel, E. L.; Allinger, N. L., Eds.; Interscience: New York, 1978; Vol. 10, p 287.
18. Gelsomino, R.; Raney, J. K. *Microgram* **1979,** *12,* 1.
19. Kroll, J. A. *J. Forensic Sci.* **1979,** *24,* 303.
20. Smith, R. V.; Erhardt, P. W.; Rusterholz, D. B.; Barfknecht, C. F. *J. Pharm. Sci.* **1976,** *65,* 412.
21. Liu, J. R. H.; Tsay, J. T. *Analyst (Cambridge, U. K.)* **1982,** *107,* 544.
22. Liu, J. R. H.; Ramesh, S.; Tsay, J. T.; Ku, W. W.; Fitzgerald, M. P.; Angelos, S. A.; Lins, C. L. K. *J. Forensic Sci.* **1981,** *26,* 656.
23. Kime, K. A.; Sievers, R. E. *Aldrichimica Acta* **1977,** *10,* 54.
24. Kram, T. C.; Lurie, I. S. *Forensic Sci. Int.* **1992,** *55,* 131.
25. LeBelle, M. J.; Savard, C.; Dawson, B. A.; Black, D. B.; Katyal, L. K.; Zrcek, F.; By, A. W. *Forensic Sci. Int.* **1995,** *71,* 215.
26. Archer, R. A.; Johnson, D. W.; Hagaman, E. W.; Moreno, L. N.; Wenkert, E. *J. Org. Chem.* **1977,** *42,* 490.
27. Asada, S.; Nishijo, J. *Bull. Chem. Soc. Jpn.* **1978,** *51,* 3379.
28. Carroll, F. I.; Moreland, C. G.; Brine, G. A.; Kepler, J. A. *J. Org. Chem.* **1976,** *41,* 996.
29. Baker, J. K.; Borne, R. F. *J. Heterocycl. Chem.* **1978,** *15,* 165.
30. Singh, S. P.; Parmer, S. S.; Stenberg, V. I.; Akers, T. K. *J. Heterocycl. Chem.* **1978,** *15,* 53.
31. Bailey, K.; Legault, D. *Anal. Chim. Acta* **1980,** *113,* 375.
32. Bailey, K.; Legault, D. *J. Forensic Sci.* **1981,** *26,* 27.
33. Bailey, K.; Legault, D. *J. Forensic Sci.* **1981,** *26,* 368.
34. Alm, S.; Bomgren, B.; Boren, H. B.; Karlsson, H.; Maehly, A. C. *Forensic Sci. Int.* **1982,** *19,* 271.

Chapter 9
Mass Spectrometry

Mass spectrometric (MS) methods of analysis are highly valued and widely used because of their intrinsic capabilities of providing analytical results with high specificities. The addition of a mixture-handling capacity through the use of appropriate separation techniques, and the recent development of multiple-stage MS, undoubtedly renders MS-based methods the most effective approaches for the analysis of organic compounds. As a result of the recent availability of this instrumentation in most crime, toxicological, and clinical laboratories, MS-based techniques have also become the methods of choice in drug analysis. These applications and potentials have long been recognized and discussed in general papers and monographs authored by leading scientists with an interest in solving problems related to drugs of abuse [1–7].

Comprehensive MS literature with an emphasis on basic understanding [8–21], applications [22–29], and data compilation [30–40] is readily available. An impressive number of mass spectra of abused drugs have been collected by Mills et al. [39] and Pfleger et al. [34]. The database of Mills et al. [39] includes various derivatives of commonly abused drugs. Artifact formations of certain drugs during the extraction, hydrolysis, or analysis processes (e.g., N-oxidation, decarboxylation, methylation, and cleavage and rearrangement) are noted in the spectra in the database of Pfleger et al. [34]. The latter collection is also available in a computer data bank permitting automatic searches of spectra. Interested readers are referred to these sources for further information. This chapter addresses the applications of MS to drug analysis.

The chapter provides (1) a brief overview of the instrumentation and the trend and development of applications, (2) summaries of MS characteristics of common drugs of abuse, (3) unique analytical approaches, and (4) applications of chromatography–MS and specific gas chromatography–MS (GC–MS) analytical protocols tailored to the analysis of specific drugs.

General Trend and Development

The first step of a typical MS method of analysis is the introduction of the sample of interest into the ion source of an MS. If the sample is not already in the gaseous form, it is converted to this state and then ionized by (1) electron impact (EI), or (2) reagent ions produced by EI from the selected reagent gas, such as methane, isobutane, and ammonia. The latter process is conventionally called chemical ionization (CI). Charged particles thus generated are then successively selected by an analyzer and detected by a detector.

In the handling of a complex mixture, the sample is introduced through a gas or liquid chromatographic device. For the analysis of compounds of poor volatility or thermal instability, "soft" ionization techniques are used, such as laser ionization, field desorption, thermal-spray ionization, fast-atom bombardment, and secondary-ion MS, in which ions can be generated from condensed phases [16, 41, 42]. The recently developed multiple-stage MS technologies have found successful applications in structure elucidation and the identification of compounds of interest in complex matrices. The newly commercialized ion-trap technology has also been explored for the analysis of commonly abused drugs.

Modified with permission from *Forensic Mass Spectrometry;* Jehuda Yinon, Ed.; 1987, Chapter 1. Copyright 1987 CRC Press, Boca Raton, Florida.

To facilitate the enormous amount of data generated by MS experiments, computer technologies are essential for data acquisition, processing, interpretation, storage, and retrieval.

Sample Introduction

The selection of an appropriate sample-introduction method depends on the nature and purity of the sample under examination. The use of a direct-insertion probe involves minimal instrumentation and analysis time [43]. It can also be used to introduce relatively nonvolatile compounds. With rapid heating [44], proper positioning of the probe [45], and the use of a suitable sample holder [46, 47], this technique is extremely valuable in the analysis of labile compounds. However, this approach often yields spectra of multicomponent origin when samples have not been sufficiently purified. This difficulty may be somewhat minimized by sample fractionation facilitated by gradient heating [48] of the probe or by the use of a tandem MS [49–57].

Although tandem MS has been successfully used for the analysis of complex drug mixtures [52–57], a chromatograph is usually used as the sample inlet device [58]. Because of their common use and comparability, gas chromatographs equipped with a packed [59] or a capillary [60] column are interfaced to mass spectrometers and widely used in forensic laboratories. The applications of GC–MS to the analysis of drug samples in forensic laboratories are well documented [61–63]. CI approaches are commonly used to facilitate the determination of the molecular weights of compounds under examination and the analysis of relatively impure samples without using a separation device [64]. The same gas, such as methane, can be used as the carrier for the gas chromatograph and, at the same time, as the reagent gas in CI-MS applications [65–68].

High-pressure liquid chromatographic (HPLC) technologies, which are complementary to GC, are gradually becoming practical inlet devices in MS applications. Compounds not suitable for GC analysis may be introduced through a liquid chromatograph. Because the mobile phase used in liquid chromatography is not compatible with the ion-source condition of an MS, various interface techniques have been explored [69–81]. The use of a liquid chromatograph as the inlet device for a tandem MS system has also been applied to the analysis of samples of interest to forensic science [78, 79]. Because the HPLC–MS interface approach is far from reaching the maturity achieved by the GC–MS counterpart, various approaches are still being explored.

Pyrolysis [82, 83], pyrolysis GC [84, 85], and thin-layer chromatographic plates [86–88] have also been used as sample-introduction devices for MS analysis, but these applications are rare and hold far less potential.

Ionization Methods

Following the general application trend of MS in other fields, the most established EI technique is widely used for drug analysis. The stable fragmentation patterns provided by 70-eV EI are conveniently used for the comparison of spectra generated by different laboratories. On the other hand, when molecular weight information is the primary concern, or when thermally labile or high-boiling compounds are encountered, "soft" ionization techniques [16, 41, 42, 89] are needed. Despite the availability of several soft ionization methods, only CI has been used to a significant extent in forensic sciences [90]. Rare applications of other ionization techniques include the exploratory use of secondary-ion MS for the analysis of drugs in salt forms [91], and a ^{63}Ni foil at atmospheric pressure [92]. The latter approach is used to obtain both positive- and negative-ion mass spectra for cocaine and barbiturates.

Spectra obtained by EI, positive CI (PCI), negative CI (NCI), and charge exchange (CE) ionization methods have been compared in several studies [93–98]. Figure 9.1 is an example of this comparison. Because these spectra were generated by different laboratories under different conditions, a quantitative comparison may not be valid. However, their comparative value in revealing molecular weight or structural information is apparent. CI techniques often result in a certain degree of fragmentation that can supplement information [99] available from EI MS. The highest mass peak of significant intensity resulting from NCI is often, but not always, derived from the loss of one (or sometimes two or more) hydrogen from the molecular anion [97]. It appears that spectra produced by this ionization technique are less predictable (or less well understood) than their positive-ion counterparts. However, enhanced sensitivities are often achievable when compounds with elements having high electron affinities are

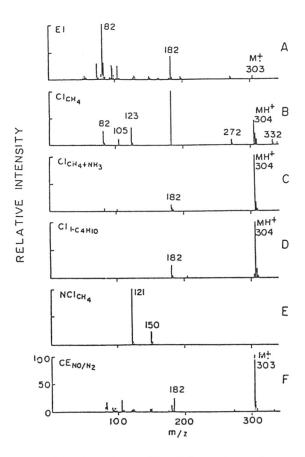

Figure 9.1. Mass spectra of cocaine obtained by different ionization methods. A: EI, B: CI (methane), C: CI (methane + ammonia), D: CI (isobutane), E: NCI (methane), F: CE (10% nitric oxide in nitrogen). (A, B, C, and E are from Ref. [93]. D is constructed from a table with permission from Ref. [99]; F is from Ref. [94].)

analyzed [100]. It is interesting to note that ammonium carbonate and sodium bicarbonate have been used for the in situ generation of reagent gases in positive [101] and negative [102, 103] CI methods.

Recent Developments in MS

In addition to new ionization methodologies, recent developments in MS include tandem MS and ion-trap technologies. These new approaches have also been applied to the analysis of drugs. Examples of these applications include Fourier transform MS (FT-MS) [104], mass-analyzed ion kinetic energy spectrometry (MIKES) [49–57], and triple-quadrupole MS [105–108].

Tandem MS has been shown to provide valuable information for the identification of a compound of interest in a complex matrix [109], elucidation of an unknown structure, and differentiation of compounds. MIKES has been used to map the concentration ratio of cocaine and cinnamoylcocaine in entire coca plant tissues without sample preparation or prefractionation [55, 56]. The experiments employ direct probe vaporization, CI, and collision-induced dissociation. As the sample is evaporated, both compounds are characterized by their molecular weights and the dissociation involving the loss of benzoic acid and cinnamic acid from the protonated molecules. Similar approaches have also been used to identify papaverine in raw opium [54], and mescaline in crude plant extracts [52, 53].

Molecular structure determinations of China White and related fentanyl derivatives illustrate the potential of tandem MS in structure elucidation [110]. The structure is pieced together through the successive pairing of EI "complementary ions", whose sum of masses or elemental compositions equals that of the molecular ion, and the determination of the substructures of these ion pairs by their collisionally activated dissociation spectra.

Spectra of the protonated or ethylated molecules generated by methane CI also provide complementary information not available from other ionization methods. This latter application is nicely illustrated in a study concerning the analysis of barbiturates [57].

Ion-trapping technologies are being applied in the form of the Ion Trap Detector (trademark of Finnigan MAT) [111]. This approach differs from quadrupole MS in that charged particles are trapped and then successively destabilized so that their trajectories impinge on a detector. The use of this form of MS has been reported for the analysis of China White [112], tetrahydrocannabinol metabolite, and other abused drugs [113].

The use of an Ion Trap system for the routine analysis of drugs of abuse has also been promoted [114]. In contrast to the selected ion monitoring (SIM) approach in the drug urinalysis industry, this system emphasizes the merits of providing full-scan MS data.

Computer Applications: Data Acquisition and Conversion, Data Retrieval, Spectrum Interpretation, Mathematical and Statistical Treatment of MS Data, and Targeted-Compound Analysis

Computer technologies are now used in every stage of MS-related operations. The use of a computer in place of photographic plates or oscillographic displays has greatly facilitated instrument calibration and data acquisition. The development and description of these technologies are well documented in a monograph [115] and in commercial-instrument operation manuals.

Although experienced mass spectrometrists often propose the structure of the compound being examined on the basis of the interpretation of the observed mass spectrum, routine analyses rely heavily on the assistance of a computer. The most common practice is based on a *retrieval system*, the matching of the observed mass spectra with those collected in a database. With the development of databases specific to forensic applications [40, 116, 117] and comprehensive databases consisting of more than 68,000 compounds [118], and the availability of sophisticated computer algorithms [119], this approach is particularly useful.

Both "forward" and "reverse" searching algorithms [17] are used to match an unknown spectrum with those in a computer database. The forward-searching approach matches peaks in the unknown spectrum with those in the reference spectrum. The effectiveness of this approach is improved by giving more "weights" to the matching of higher-mass ions. The reverse-searching approach "ascertains whether the peaks of the reference spectrum are present in the unknown spectrum, not whether the unknown's peaks are in the reference" [17]. This approach ignores peaks present in the unknown spectrum because of the presence of other components; thus, it is effective for interpreting the spectra of mixtures. The forward-searching approach can also handle spectra containing more than one component by a two-step process: the initially retrieved best-matching reference spectrum is subtracted from the unknown spectrum, and the residual spectrum is in turn matched to another reference spectrum [120].

Users are cautioned that a computerized search of an unknown spectrum against a comprehensive database will generally yield an answer with sufficient matching quality. However, the quality of the result depends on (1) the quality of the spectra in the database, (2) whether the spectra in the database are searched as complete or condensed spectra, (3) the search algorithms, (4) the quality of the unknown spectrum, and (5) whether the spectrum of the unknown compound is in the database [121].

When a sample is introduced through a chromatographic device, the observed retention behavior of the compound offers additional information for a positive identification. However, one should always be on guard against placing too much confidence in an identification based on this approach alone, and one should observe recommended guidelines [122]. When the mass spectrum of the compound under examination is not available in a database, *spectra-interpretation systems* may be helpful. Pattern recognition, artificial intelligence, and the search for structural analogs are the basis of the interpretative mechanism [119, 123, 124]. The "self-training interpretive and retrieval system" [17] uses an interpretive algorithm to find at least partial structural information, relying on a knowledge of 26 classes of mass-spectral fragmentation rules and an empirical search of reference spectra. The algorithm matches the unknown spectral data in each of the 26 classes against the corresponding data in all reference spectra to determine whether one or more of 600 common substructures are present in a significant portion of these compounds. This approach is a useful aid to a trained interpreter for identifying an unknown with no reference spectrum in the database.

A commercial software package [125] is also available for automatic identification of limited numbers of targeted drugs. A common approach includes the monitoring of a few characteristic ions for each drug of interest; the identification is based on GC retention time, the presence of these characteristic ions, and their intensity ratios. Computer technologies are further utilized for mathematical and statistical treatment of mass-spectral data. Qualitative [126–128] and quantitative [129, 130] identification of components in a mixture can be achieved through the aid of these treatments. Other computer techniques include the possible correlation between mass spectra and biological activity [131]. Although only limited information is available in the literature, this approach seems to present an interesting research topic.

A recently adopted workplace drug-testing policy emphasizes large-scale analysis of a limited number of drugs (and/or metabolites) with well-established chromatographic and MS characteristics. Targeted-compound analysis approaches have become common in the drug-testing industry. These approaches have prompted the development of software packages that incorporate automated protocols for (1) SIM GC–MS data collection from specimens that have been pretreated with a specific procedure tailored to the drugs of interest, (2) conclusive drug identification based on the appearance of the monitored ions at a correct retention time with acceptable intensity ratios among these ions, and (3) quantification by an internal standard (IS) procedure.

For compound identification, the retention time and ion-intensity ratios observed in the test sample are compared with those established by an authentic standard incorporated in the same analytical batch. They are also evaluated on the basis of "intrasample" information, that is, the retention time of the selected ions of the analyte relative to those of the same parameters of the IS within the test sample should be comparable to those observed in the authentic standard. Considering the fact that the identified analyte has also survived the chemical processes (pretreatment, extraction, and derivatization) designed for the targeted drug and proven effective for the IS (often a deuterated analog with chemical properties practically identical to those of the analyte), these identification criteria should not lead to an incorrect conclusion if they are utilized by an able analyst.

The quantification procedure typically adopts a deuterated analog of the analyte as the IS. Quantification is achieved by comparing a selected analyte-to-isotopic analog ion-intensity ratio observed in the test sample with the same ratio observed in the calibration standard. The calibration standard contains the same amount of the IS and a known amount of the analyte and is processed in parallel with the test sample. The analyte concentration in the test sample can be calculated using a one-point calibration approach, as shown in Figure 9.2.

This procedure can be used directly to (1) determine the analyte concentration in a test sample (one-point calibration), or (2) determine the responses of a set of standards, from which a calibration curve can be established and used for analyte concentration determination in a test sample (multiple-point calibration). Multiple-point calibration approaches can be implemented with different regression models, including linear or nonlinear, weighted or unweighted, and with or without forcing the regression through the origin point.

MS Characteristics of Commonly Abused Drugs

When they have a common structural feature, drugs of the same category often undergo a similar fragmentation pattern. Thus, the basic EI fragmentation pattern for each drug category is emphasized here along with the structural characteristics of common drugs in the category. Common and unique analytical approaches, including CI and GC–MS applications, are also briefly addressed.

$$\frac{\left[\frac{(\text{Selected Ion Intensity})_{\text{Analyte}}}{(\text{Selected Ion Intensity})_{\text{Internal Std.}}}\right]_{\substack{\text{Test} \\ \text{Sample}}}}{\left[\frac{(\text{Selected Ion Intensity})_{\text{Analyte}}}{(\text{Selected Ion Intensity})_{\text{Internal Std.}}}\right]_{\substack{\text{Calibration} \\ \text{Standard}}}} \times \left[\begin{array}{c}\text{Standard} \\ \text{Concn}\end{array}\right]_{\substack{\text{Calibration} \\ \text{Standard}}} = \left[\begin{array}{c}\text{Analyte} \\ \text{Concn}\end{array}\right]_{\substack{\text{Test} \\ \text{Sample}}}$$

Figure 9.2. Equation for quantification with one-point calibration.

Amphetamines and Related Amine Drugs

During the 45-month period ending in September, 1981, the Drug Enforcement Administration of the U.S. Department of Justice seized a total of 751 clandestine drug laboratories in the United States. More than half of these clandestine activities involved the manufacture of amine drugs—mainly methamphetamine [132].

Owing to their possible psychotomimetic properties [133], compounds with methoxy-group substitution on the benzene ring of amphetamine and related compounds are encountered in forensic laboratories [134], and their identifications and differentiation have been reported [135–138]. Tryptamines represent another category of amine drugs of interest to forensic chemists. Other amine drugs or related compounds often encountered in forensic laboratories include phenyl-2-propanone [139] (a primary precursor for amphetamine and methamphetamine manufacture), side-chain positional isomers of amphetamine [140], psilocybin [141–143], mescaline [144], and ephedrines [144–146]. The structures of these compounds are shown in **1** through **11** (Chart 9.1).

MS procedures are used in the analysis of the drugs cited above. Fragmentation of these compounds under EI is emphasized in several studies [145–153] and is found to follow a common pattern, as summarized in Figure 9.3 [148, 150]. The general fragmentation pattern of these compounds is the "β-fission" process (β to the amino function and to the aromatic system), with charge retention predominantly on the side chain, as observed for amphetamine and methamphetamine. However, the charge tends to be retained mainly in the aromatic fragment if the benzene is substituted by a methoxy

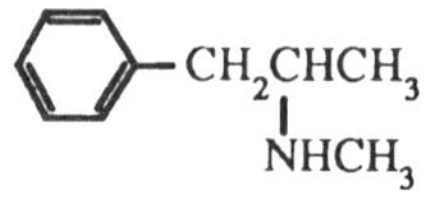

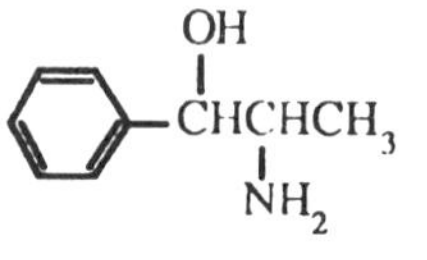

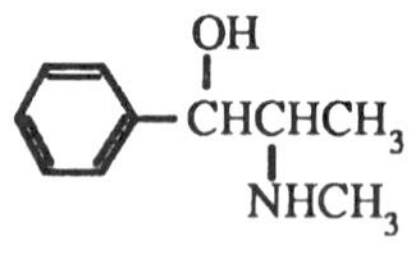

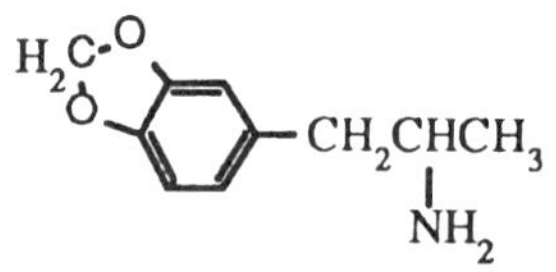

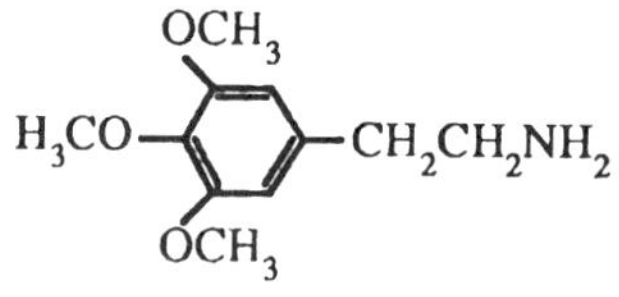

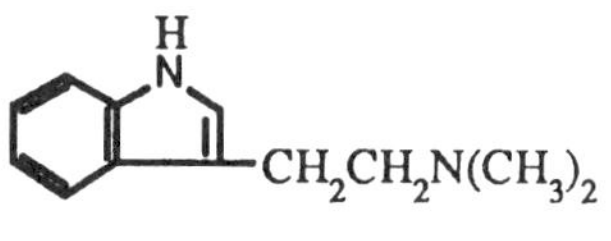

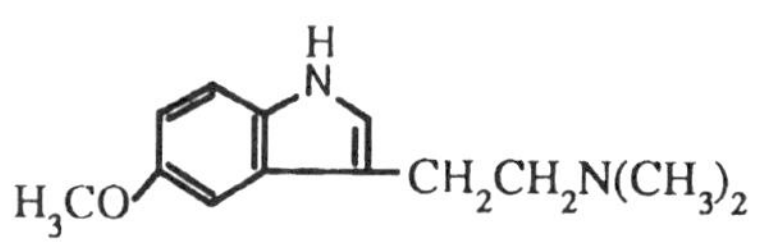

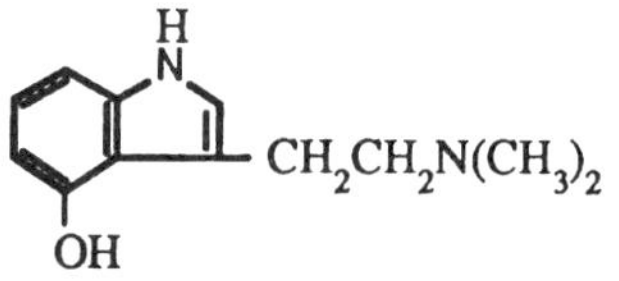

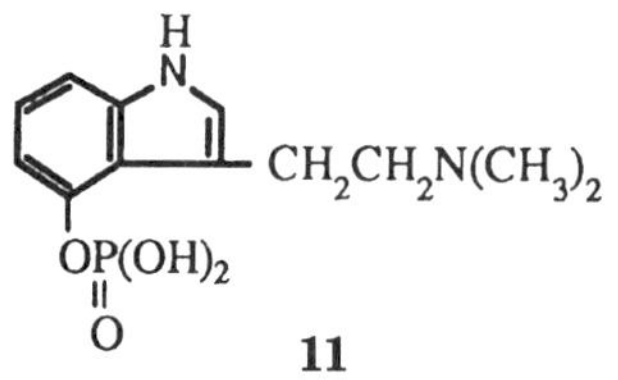

*Chart 9.1. Structures of eleven amine drugs: amphetamine (**1**), methamphetamine (**2**), norephedrine (**3**), ephedrine (**4**), 3,4 methylenedioxyamphetamine (**5**), mescaline (**6**), N,N-dimethyltryptamine (**7**), N,N-diethyltryptamine (**8**), N,N-dimethyl-5-methoxytryptamine (**9**), psilocin (**10**), and psilocybin (**11**).*

group [144, 147, 148]. Although the positions of substitution may alter the spectra to some extent, differentiation of these isomers based on mass spectra alone is difficult [136, 137, 149]. The presence of a hydroxyl group, as in ephedrine, appears to enhance the retention of the charge in the aromatic fragment [144, 149–151] and may also include an arrangement process, as shown in Figure 9.4.

CI has also been applied to the analysis of ephedrine and norephedrine [142, 150, 154]. As desired, [M + 1]$^+$ ions are the base peaks in most cases in which isobutane is used as the reagent gas. Although the [M + 1]$^+$ ions of ephedrine and norephedrine isomers are not always the base peaks, their intensities are substantially higher than the corresponding M$^+$ ions obtained under EI conditions. The CI mode of operation has also been used to improve the detection limit of amphetamine and methamphetamine in a single hair analyzed by a CI GC–MS procedure [155].

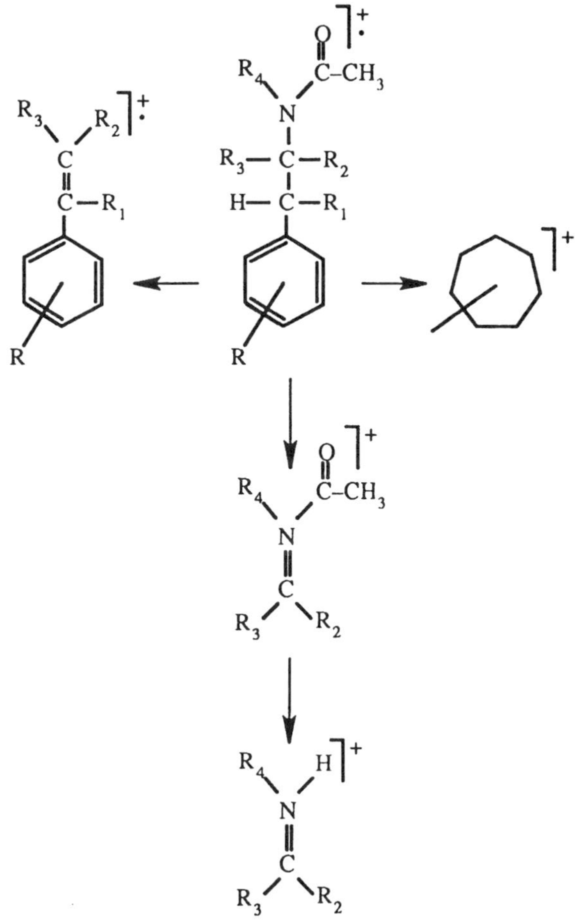

Figure 9.3. Major EI fragmentation pathways of amphetamine-related compounds [147].

NCI has been applied to the analysis of ephedrine (reagent gas, CO$_2$) [103], amphetamine (reagent gas, methane) [100], and amphetamine congeners (reagent gas, ammonia) [154]. The CO$_2$ NCI spectrum showed M$^-$ as the base peak. Derivatized with pentafluorobenzoyl chloride or tetrafluorophthaloyl anhydride, amphetamine at the attomole level (10^{-18} mole) can be detected with NCI GC–MS SIM methodology. The M$^-$ ion was found to be the base peak in this case. The ammonia NCI studies showed [M − 1]$^-$ as the base peaks, with small peaks corresponding to the loss of two hydrogens from the side chains [154].

Many of the MS studies mentioned above were conducted in conjunction with GC. The incorporation of KOH (2%) to Apiezon L [156, 157] and Carbowax 20M [148] in packed columns is

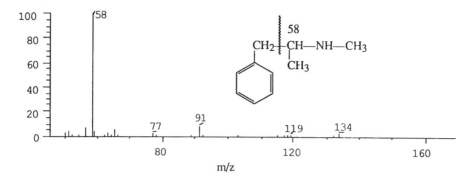

Figure 9.4. EI mass spectrum of methamphetamine, showing the "β-fission" process.

successfully used for the analysis of underivatized and derivatized amine drugs. When properly derivatized, these compounds are well separated by Carbowax 20M [148], OV-101 [148], OV-1 [146], and OV-17 [146] in packed columns, and SP-2100 and Chirasil-Val (a chiral phase) in capillary columns [158, 159].

The combination of GC–MS and derivatization techniques is used for the separation and identification of enantiomeric ephedrine [144–146] and other related drug enantiomers [144, 145]. A mixture of nine amine drugs, including possible enantiomers, and caffeine was separated by an SE-54 fused-silica-glass capillary [144]. Most of these compounds were analyzed as *N*-trifluoroacetyl-*l*-prolyl derivatives (see Structure **12**). The structures of these compounds are shown in **1** through **9,** and the resulting total-ion and single-ion current chromatograms are shown in Figure 9.5 [144]. It is unlikely that all these compounds and enantiomers will be encountered in one sample, and the separation of these compounds in a single chromatographic run is apparently beyond practical needs. However, with the establishment of the GC retention and MS characteristics of all these compounds in a single GC–MS condition, the identification of one or several amine drugs in a sample can be routinely performed.

Understandably, MS procedures are developed for the identification of these compounds and the associated synthesis by-products and impurities [156–165]. Earlier studies [160] used conventional extraction-based procedures for the purification and isolation of these compounds prior to the use of MS (EI) for identification. Packed [156, 157] and capillary [158, 159] column GC–MS procedures are now preferred.

12

N-Trifluoroacetyl-1-prolyl derivative of amphetamine.

Major impurities found in association with the Leuckart synthesis of methamphetamine [156–161] are summarized in **13** through **25** (Chart 9.2). The identification of major impurities in illicit tryptamine synthesis [138] has also been reported.

Morphine and Related Alkaloids

The chemical structures of ten common morphine-related compounds are shown in **26** through **35** (Chart 9.3). With their pharmaceutical significance, morphine alkaloids have been widely studied using

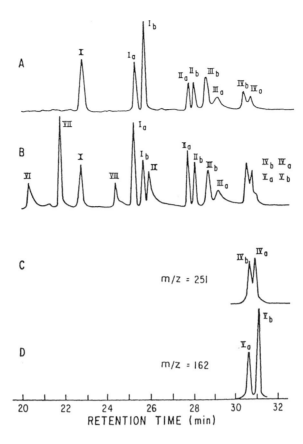

Figure 9.5. Total ion chromatograms of (A) a mixture containing caffeine (I), N-trifluoroacetyl (TFA)-l-prolyl-l-amphetamine (Ia), N-TFA-l-prolyl-d-amphetamine (Ib), N-TFA-l-prolyl-d-methamphetamine (IIa), N-TFA-l-prolyl-d-methamphetamine (IIb), N-TFA-l-prolyl-d-norephedrine (IIIb), N-TFA-l-prolyl-l-norephedrine (IIIa), N-TFA-l-prolyl-d-ephedrine (IVb), and N-TFA-l-prolyl-l-ephedrine (IVa); and (B) a mixture containing the compounds in (A) plus mescaline (VI), N,N-dimethyltryptamine (VII), N,N-diethyltryptamine (VIII), N,N-dimethyl-5-methoxytryptamine (IX), N-TFA-l-prolyl-l-3,4-methylenedioxyamphetamine (Va), and N-TFA-l-prolyl-d-3,4-methylenedioxyamphetamine (Vb). Single-ion current chromatograms of (C) compounds IVb and IVa (m/z = 251); and (D) compounds Va and Vb (m/z = 162). (Reprinted with permission from Ref. [144].)

MS-related techniques [166–169]. The fragmentation mechanisms under EI conditions have long been established mainly on the basis of the observed mass shifts (with respect to the parent morphine) of morphine-related compounds. Major fragments derive from the cleavage of the two bonds in the β-position to the nitrogen atom, as summarized in Figure 9.6 [169, 170]. It should be noted that compounds without the ether ring undergo much more rapid fragmentation [166]. It has also been reported that the relative abundance of the molecular ion and the $m/z = 59$ ion (or its analogous counterparts in other compounds) in morphine depends on the stereochemistry (*cis*- or *trans*-) of the B and C rings in the compound under examination [168].

CI MS studies [47, 171–176] of morphine alkaloids emphasize the simplicity of the spectra that meet the analytical need. This advantage allows for the identification of heroin in the presence of its diluents [174] and the analysis of raw opium [175]. It is further used to study an extended list of morphine-related compounds [172]. Because spectra obtained from conventional CI MS often contain base peaks other than the $[M + H]^+$ ion, which may complicate spectrum interpretation, CI MS in the form of CE, using 10% NO in N_2, has also been investigated [94, 95]. Relative intensities of the molecular ions (or "pseudomolecular ions") observed under EI, isobutane CI, and 10% NO (in N_2) CE ionization methods are extracted from the original studies [95] and summarized in Table 9.1. The data clearly show that, under CE conditions, the molecular ions are the base peaks in all 12 compounds studied, whereas this is true only for 5 compounds under EI and isobutane CI conditions.

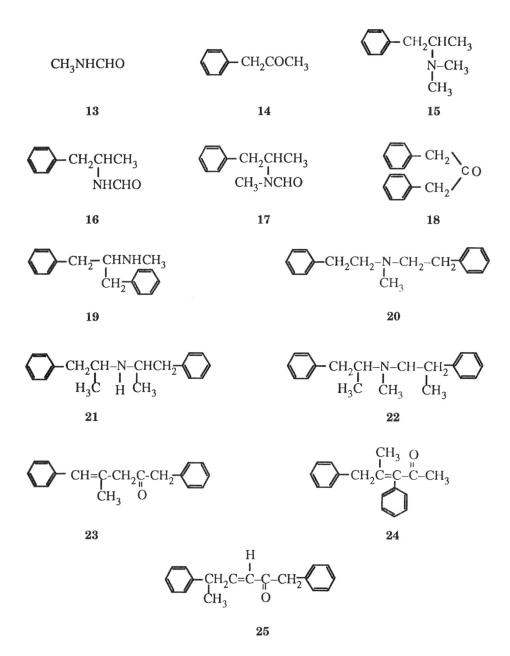

Chart 9.2. Major impurities found in Leuckart synthesis of methamphetamine: N-methylformamide (13), methyl benzyl ketone (14), N,N-dimethylamphetamine (15), N-formylamphetamine (16), N-formylmethamphetamine (17), dibenzyl ketone (18), α-benzyl-N-methylphenethylamine (19), N-methyldiphenethylamine (20), α,α'-dimethyldiphenethylamine (21) (diastereoisomer pair), N,α,α'-trimethyldiphenethylamine (22), 1,5-diphenyl-4-methyl-4-penten-2-one (23), 3,5-diphenyl-4-methyl-3-penten-2-one (24), 1,5-diphenyl-4-methyl-3-penten-2-one (25). Some of these compounds may be artifacts produced during the analysis process.

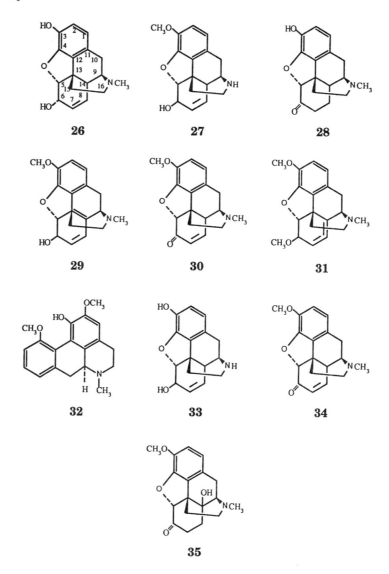

Chart 9.3. *Structures of ten morphine alkaloids: morphine (**26**), norcodeine (**27**), hydromorphone (**28**), codeine (**29**), hydrocodone (**30**), thebaine (**31**), isothebaine (**32**), normorphine (**33**), codeine (**34**), oxycodone (**35**).*

It should be noted that certain unsaturated compounds undergo reduction reactions [177, 178] under CI conditions, thus producing protonated molecular ions of the reaction products. This phenomenon is observed for codeinone [178] and should be recognized when qualitative and quantitative analysis of unsaturated compounds is addressed.

CI MS has been used with various deuterated reagent gases to probe the number and nature of active hydrogens attached to heteroatoms in the compounds being examined [177, 178]. This approach may reveal more structural information and increase the analytical value of CI MS. On the basis of the mass shifts of the "protonated" ion observed when switching from ammonia to d_3-ammonia, the number of active hydrogens in morphine and several other drugs is nicely demonstrated [179]. The combined information provided by (1) mass shifts, (2) the tendencies of the protonated and adduct ions to lose molecules of water, and (3) the adduct ion to protonated ion intensity ratios, further reveals the nature of the active hydrogens involved [178]. To demonstrate the use of this information, the CI spectra of the ten morphine alkaloids shown in **26** through **35** are summarized in Table 9.2 [178].

The total number of active hydrogens in a compound is derived from the observed mass shifts (n) of the "protonated" ion when ammonia is replaced with d_3-ammonia as the reagent gas:

$$n = M_{ND_3} - M_{NH_3} - 1 \qquad (9.1)$$

where M_{ND_3} and M_{NH_3} are the mass-to-charge ratios of the "protonated" ions observed with ND_3 and NH_3 as the reagent gases.

Thus, compounds **26** and **27** can be differentiated from **28** on the basis of the mass shift shown in Equation 9.1. Under the methane CI conditions used throughout this series of alkaloids, an allyl hydroxyl group generates an $[M + H - 18]^+$ ion intensity 3 times that of the $[M + H]^+$ ion, whereas a phenolic OH group generates an $[M + H]^+$ ion intensity 30 times that of the $[M + H - 18]^+$ ion. These $[M + H]^+/[M + H - 18]^+$ ratios suggest that the hydrogen in compound **28** is phenolic. Furthermore, as

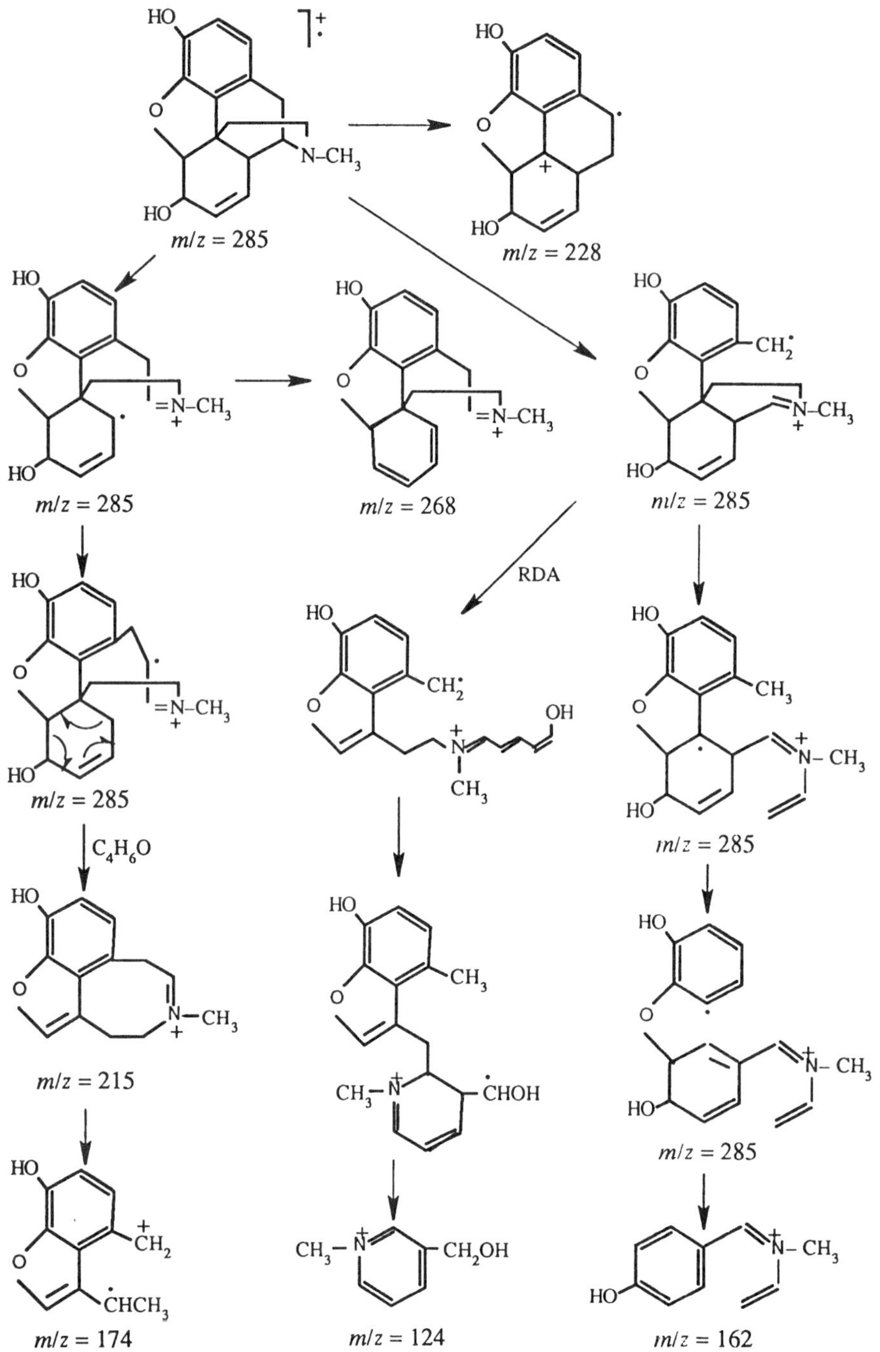

Figure 9.6. Major EI fragmentation patterns of morphine alkaloids [170].

Table 9.1. Relative intensities of molecular ions observed under EI, isobutane CI, and CE (10% NO in N$_2$) mass spectrometry.[a]

Compound	Relative molecular ion intensity			
	MW	EI	CI [M+1]$^+$	CE
Heroin	369	25	15	100
3-Monoacetylmorphine	327	96	18	100
6-Monoacetylmorphine	327	98	20	100
Morphine	285	100	15	100
Nalorphine	311	100	20	100
Codeine	299	100	13	100
Hydrocodone	299	100	100	100
Levorphanol	257	64	100	100
Levallorphan	283	100	100	100
Meperidine	247	22	100	100
Cocaine	303	26	100	100
Atropine	289	15	16	100

[a] Data are taken from Ref. [95].

Table 9.2. Number of active hydrogen atoms and relative intensities of protonated and adduct ions.[a]

Compound designation[b]		26	27	28	29	30	31	32	33	34	35
Molecular weight		285	285	285	299	311	311	311	271	297	315
Total no. of active H		2	2	1	1	0	0	1	3	0	1
No. of allyl OH		1	1	0	1	0	0	0	1	0	1
No. of phenolic OH		1	0	1	0	0	0	1	1	0	0
Observed *m/z* of "protonated" ion											
NH$_3$		286	286	286	300	300	312	312	272	298	316
ND$_3$		289	289	288	302	301	313	314	276	299	318
Reagent gas[c]	Ion										
CH$_4$	[M + C$_2$H$_5$]$^+$, D	7.2	8.3	14	8.1	13	8.9	16	5.3	13	13
	[M + C$_2$H$_5$ – 18]$^+$, C	4.3	—[d]	—	—	—	—	—	2.7	—	1.0
	[M + H]$^+$, B	31	35	100	28	100	100	100	28	100	100
	[M + H – 18]$^+$, A	100	100	3.3	100	—	—	2.6	100	—	4.9
	[C + D]$^{+e}$	8.8	6.1	14	6.3	13	8.9	16	6.3	13	13
	[A + B]$^+$	100	100	100	100	100	100	100	100	100	100
NH$_3$	[M + NH$_4$]$^+$	81	77	100	98	100	—	—	32	98	—
	[M + H]$^+$	100	100	34	100	32	100	100	100	100	100

[a] Data are taken from Ref. [178].
[b] *See* Chart 9.3.
[c] Data are obtained using isobutane as the reagent gas are parallel to those obtained under methane conditions and are not presented or discussed.
[d] Less than 1.0% of the base peak.
[e] This represents a better estimate of the adduct ion to protonated ion intensity ratio.

a result of differences in proton affinity, compound **28** differs from **26** and **27** by showing a much more intense adduct ion intensity under either the methane or ammonia CI conditions. Compound **26** differs from compound **27** by showing a substantial intensity (4.3%) of the [M + C$_2$H$_5$ – 18]$^+$ ion. The two nonalcoholic hydrogens in compounds **26** and **27** cause a small (8.8% vs. 6.1%) but proven difference in their adduct ion to protonated ion intensity ratios. This difference (81% vs. 77%) is again observed when the more basic ammonia is used as the reagent gas. This subtle difference shown by the two different nonalcoholic hydrogens in compounds **26** and **27** undoubtedly will not alert casual observers during routine analysis. On the other hand, when structural elucidations are emphasized and studied under precisely controlled conditions, this difference is recognizable.

Anions in NCI are generally formed through electron capture and reactant-ion CI mechanisms. This technique has been applied to the analysis of morphine and several related compounds [96, 102]. In the study of morphine performed using CO_2 as the reagent gas, the most intense peak represents the loss of one hydrogen from the molecular anion. When methane is used, the most intense peak is $[M - 2]^-$. Under the methane conditions, the most intense peaks for codeine and dihydrocodeinone are $[M - 1]^-$ and M^-.

GC–MS has been applied to the analysis of morphine, codeine, thebaine, and papaverine in opium samples [180]. It is also useful in the identification of impurities and by-products in illicit heroin preparations [181, 182]. Using heptafluorobutyric anhydride as the derivatizing reagent, this technique is used to identify the preparation of 3-monoacetylmorphine in an illicit heroin preparation. Columns packed with OV-17 [62, 181, 182] or OV-101 [180] are used in these applications.

Marijuana and Other Cannabinoid-Containing Materials

Cannabinoid-containing samples in the forms of marijuana (*Cannabis sativa* L.), hashish, and hashish oil are commonly encountered in forensic science laboratories [183]. Although the use of microscopic examination [184], a modified Duquenois–Levine color test [185], and chromatographic examination of extracts [186] is usually sufficient to determine whether the sample under examination is indeed marijuana, GC–MS, preferably capillary column [187, 188], is probably needed to satisfy a critical chemist. GC–MS procedures have been applied to the analysis of trace marijuana components in hand swabs [189] and to the identification of cannabinoids in pyrolytic products [190, 191] and smoke condensates [187, 188, 192].

Major compounds of general interest are shown in **36** through **43** (Chart 9.4). Basically, cannabinoids contain a pyran ring with four consecutive carbons as parts of a toluene and an alkyl phenol system. The toluene system is replaced by a monocyclic terpene moiety in cannabidiols and tetrahydrocannabinols. In cannabidiols, the ether linkage in the pyran ring opens to form a hydroxyl group. Many of these three classes of cannabinoids also exist with different carbon numbers, mainly odd numbers, in the side chain [196–199], and/or with one or more carboxyl groups. The position of the double bond in the terpene moiety may also vary. The phenolic group may also be replaced by an alkoxy group [200]. Many cannabinoids derived from other variations of these basic structural features have also been identified [201, 202].

Fragmentation of major cannabinoids has long been studied [203–206]. Fragmentation occurs preferentially in the alicyclic portion of the molecules, whereas the benzene ring functions as a charge-stabilizing element [203]. The most important fragmentation processes start with a retro Diels–Alder reaction of the cyclohexene ring [204]. On the basis of earlier studies and the effects of the ionization energy on the intensities of major ions (and their analogs in derivatives), the mechanisms for the formation of commonly observed higher-mass ions (m/z = 314, 299, 271, 258, 246, and 231) are described in great detail [207]. Using Δ^1-tetrahydrocannabinol (Δ-1-THC) as an example, the major fragmentation pathways are summarized in Figure 9.7.

MS is used extensively in the structural characterization of compounds derived from *Cannabis sativa* L. *Cannabis* has been shown to contain over 400 compounds, of which 61 are known to be cannabinoids [207]. Because many of these cannabinoids differ only in the position of the double bond or the number of carbons in the side chain, and many have the same molecular weights, MS characterization of these compounds is examined under several ionization energy levels, ranging from 5.5 to 21 eV [193–198, 200, 207–210]. "It is assumed that variation of electron ionization energy will reflect certain characteristic aspects of the stability of the parent compound. The rate at which a certain fragment may originate from different molecular ions will depend upon the activation energy required for fragmentation and upon the pathway available for the fragmentation" [210]. Thus, the technique of plotting relative intensities of specific ions against ionization electron voltage often reveals the characteristic feature of each compound and serves as a useful identification tool.

Δ-1-THC, the principal active component of the plant material, has been analyzed by the methane and methane–ammonia CI methods [211]. Under a mixture of 2×10^{-4} mm ammonia and 8×10^{-4} mm methane, trimethylsilylated Δ-1-THC shows a single $[M + 1]^+$ ion. The pentafluorobenzoyl derivative of Δ-1-THC can be detected at a 20-femtogram level (10^{-15} g) by NCI using methane as the reagent gas [100]. The intensity of the most intense ion in NCI is reported to be more than 300 times higher than that observed under positive CI (PCI) conditions.

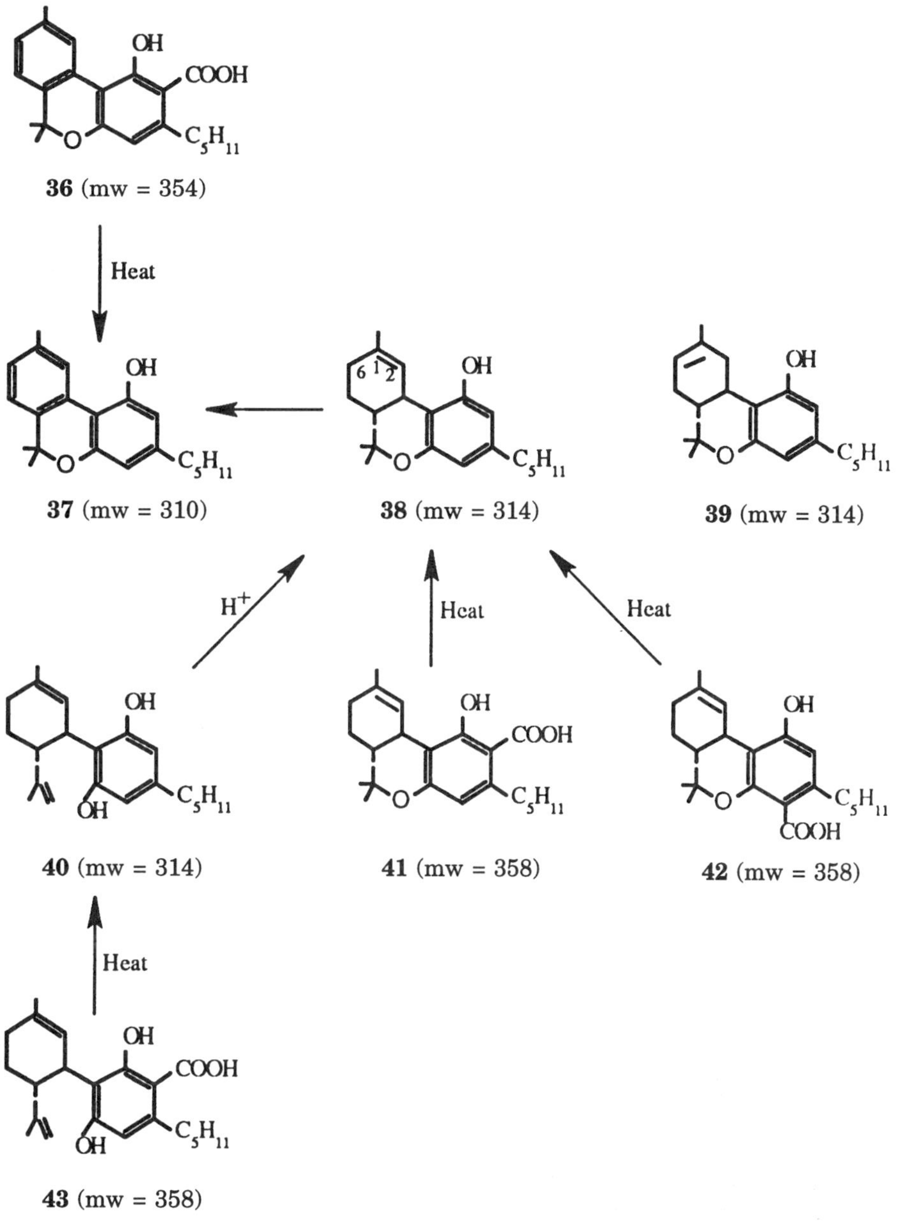

Chart 9.4. *Chemical structure, molecular weight, and possible interconversions of cannabinoic acid (36), cannabinol (37), Δ-1-THC (38), Δ-6-THC (39), cannabidiol (40), Δ-1-THC acid A (41), Δ-6-THC acid (42), and cannabidiolic acid (43).*

An ideal method for the analysis of cannabinoid-containing samples in forensic laboratories should require minimum sample preparation, yet be specific enough to produce a definite conclusion. MS procedures [193, 194] were developed to meet these goals. For this approach, a minute amount (around 0.2 mg) of the pulverized raw sample is introduced to the MS through a direct inlet probe. Intensities of selected ions (m/z = 314, 310, 299, 295, 271, 258, 246, 143, 138, 231, and 193), obtained from the sample under examination, are related to the intensities of these ions obtained from standard compounds as follows:

$$Y_1 = X_{11}r_1C_1 + X_{12}r_2C_2 + \ldots + X_{1m}r_mC_m$$

$$Y_2 = X_{21}r_1C_1 + X_{22}r_2C_2 + \ldots + X_{2m}r_mC_m$$

$$Y_n = X_{n1}r_1C_1 + X_{n2}r_2C_2 + \ldots + X_{nm}r_mC_m \tag{9.2}$$

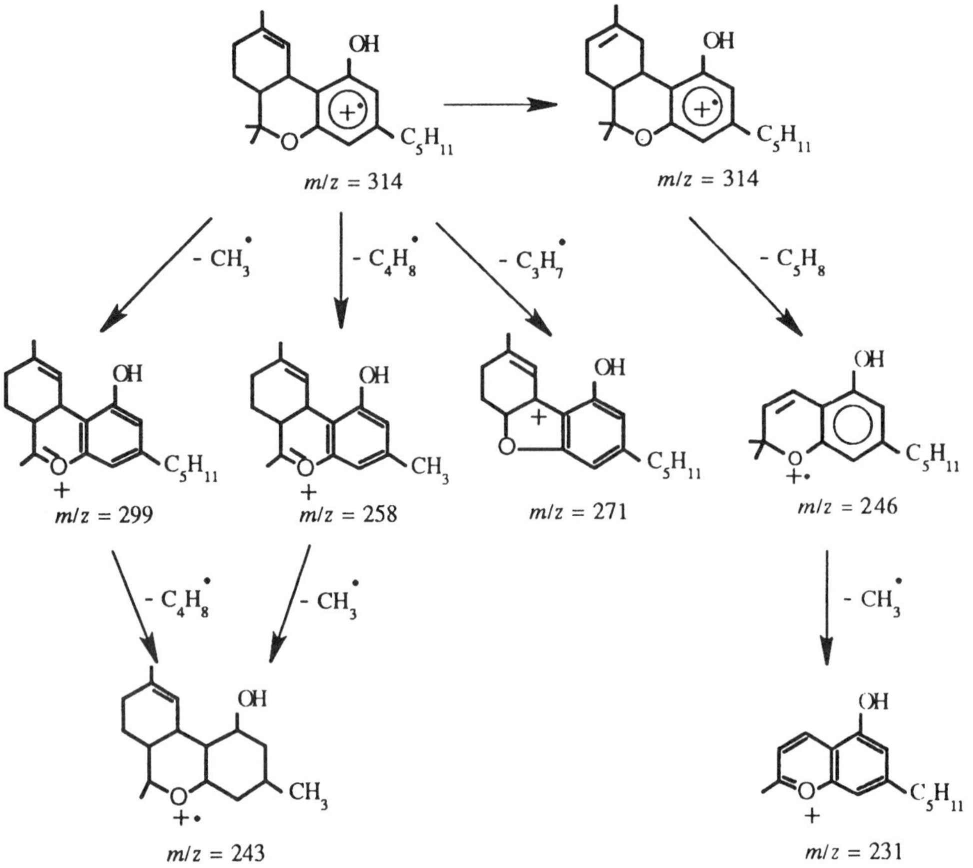

Figure 9.7. Major EI fragmentation pathways of Δ-1-THC [196].

where Y_n is the observed relative intensity of the ion, n, in the sample; X_{nm} is the relative intensity of the ion, n, obtained from the standard cannabinoid, m; r_m is the relative sensitivity factor of the cannabinoid, m, under the experimental conditions; and C_m is the concentration of the cannabinoid, m, in the sample under examination. An established statistical package, SPSS [195], is used to perform the regression analysis with X's and Y's obtained from MS experiments. In theory, data obtained from an unlimited number of compounds, m, can be fed into Equation 9.2 for regression analysis. In these cited studies, m is limited to the four major cannabinoids, Δ-1-THC (**38**), Δ⁶-tetrahydrocannabinol (Δ-6-THC) (**39**), cannabinol (**37**), and cannabidiol (**40**).

The presence of ions common to those found in standard cannabinoids indicates, qualitatively, that the sample under examination may indeed contain cannabinoids. Because samples normally contain different proportions of several cannabinoids that often generate common ions, the conventional approach in comparing the relative intensities of these ions cannot offer definite conclusions. Using the approach described, a high coefficient of determination, normally higher than 97%, can be considered as a "quantitative" indication of the certainty of the qualitative analysis. The conclusiveness of the qualitative analysis is further affirmed by the observation of high coefficients of determination with MS data obtained under several different levels of ionization energy, ranging from 14 to 20 eV. This represents a quantitative application of the semiquantitative "electron voltage mass fragmentation" approach [196–198, 200, 207–210].

The regression coefficient (r_m terms in equation 9.2) obtained for each compound by the regression analysis indicates the relative concentration of the compound in the sample. It is further possible to use this parameter for differentiation [194]. Because standardization is not done in these studies [193, 194], the concentration term, C_m, cannot be extracted from the regression coefficient term, $r_m C_m$. (This extraction was done in a later study applied to the analysis of polychlorinated biphenyl formulations, Aroclors [130].)

Barbiturates

The ready availability of barbiturates, through both prescription and illicit sources, necessitates the development of reliable procedures for the analysis of these compounds. The molecular weights of these compounds and their substituents in the ring system (Structure **44**) of common barbiturates are listed in Table 9.3. EI spectra of barbiturates [212–225] and their *N*-derivatized counterparts [217–225] have been shown to follow similar patterns.

44

Framework of barbiturates. (See Table 9.3 for the names and substitution groups of individual compounds.)

Table 9.3. Structures and major NCI ions of barbiturates.

Name	MW	Y_1	Y_2	R_3	R_1; mass	R_2; mass	Relative intensities of major NCI ions[a]			
							[M – H]	[M – R_1]	[M – R_2]	[M – R_1 – R_2 + H]
Phenylmethyl-barbituric acid	219	O	O	H	Methyl; 15	Phenyl; 77	9	22	100	2
Barbital	184	O	O	H	Ethyl; 29	Ethyl; 29	82	—	100	12
Probarbital	198	O	O	H	Ethyl; 29	Isopropyl; 43	33	9	100	1
Butethal	212	O	O	H	Ethyl; 29	Butyl; 57	65	45	100	8
Butabarbital	212	O	O	H	Ethyl; 29	1-Methylpropyl; 57	14	5	100	2
Pentobarbital	226	O	O	H	Ethyl; 29	1-Methylbutyl; 71	9	13	100	3
Amobarbital	226	O	O	H	Ethyl; 29	3-Methylbutyl; 71	65	49	100	8
Vinbarbital	224	O	O	H	Ethyl; 29	1-Methyl-1-butenyl; 69	16	8	100	11
Cyclobarbital	236	O	O	H	Ethyl; 29	1-Cyclohexen-1-yl; 81	11	11	100	4
Heptabarbital	250	O	O	H	Ethyl; 29	1-Cyclohepten-1-yl; 95	17	12	100	14
Phenobarbital	232	O	O	H	Ethyl; 29	Phenyl; 77	6	43	100	7
Aprobarbital	210	O	O	H	Ellyl; 41	2-Methylethyl; 43	1	100	6	5
Idobutal	224	O	O	H	Allyl; 41	Butyl; 57	0.8	100	2	3
Butalbital	224	O	O	H	Allyl; 41	2-Methylpropyl; 57	0.5	100	3	3
Talbutal	224	O	O	H	Allyl; 41	1-Methylpropyl; 57	1	100	13	6
Nealbarbital	238	O	O	H	Allyl; 41	2,2-Dimethylpropyl; 71	1	100	2	2
Secobarbital	238	O	O	H	Allyl; 41	1-Methylbutyl; 71	1	100	13	4
Allobarbital	208	O	O	H	Allyl; 41	Allyl; 41	0.2	100	—	39
Cyclopal	234	O	O	H	Allyl; 41	2-Cyclopenten-1-yl; 67	4	100	27	19
Ibomal	288	O	O	H	–CH_2–CBr=CH_2; 107	Isopropyl; 43	2.5	100	3	1
Sigmodal	316	O	O	H	–CH_2–CBr=CH_2; 107	1-Methylbutyl; 71	—[b]	100	3	17
Brallobarbital	286	O	O	H	–CH_2–CBr=CH_2; 107	Allyl; 41	—[b]	100	13	30
Hexobarbital	236	O	O	CH_3	Methyl; 15	1-Cyclohexen-1-yl; 81	35	18	100	22
Metharbital	198	O	O	CH_3	Ethyl; 29	Ethyl; 29	100	—	28	3
Mephobarbital	246	O	O	CH_3	Ethyl; 29	Phenyl; 77	11	41	100	11
Enallylpropymal	224	O	O	CH_3	Allyl; 41	Isopropyl; 43	17	100	6	1
Methohexital	262	O	O	CH_3	Allyl; 41	1-Methyl-2-pentynyl; 81	6	15	100	15
Thiopental	242	O	S	H	Ethyl; 41	1-Methylbutyl; 71	13	4	100	18
Thialbarbital	264	O	S	H	Allyl; 41	2-Cyclohexen-1-yl; 81	10	13	100	32

[a] Data are taken from Ref. [229].

[b] 0.0–0.2.

Major EI fragmentation pathways of barbiturates, as depicted for *N,N*-dimethyl derivatives, are summarized in Figure 9.8 [218]. EI spectra of these compounds show the following characteristics [215]:

1. With the exception of sulfur analogs, molecular ions are absent or show low intensities [219].

2. The ring system is stable. Only when R_1 or R_2 is aromatic does cleavage of the ring system occur.

3. The loss of the longer, saturated alkyl side chain from the molecular ion is preferred. Thus, there are intragroup similarities between spectra obtained from butabarbital, butethal, pentobarbital, and amobarbital; and secobarbital, itobarbital, and butalbital.

4. When R_1 is an ethyl group and R_2 is not α-branched, route A becomes more significant; thus, the differentiation between amobarbital and pentobarbital, and butethal and butabarbital, is possible.

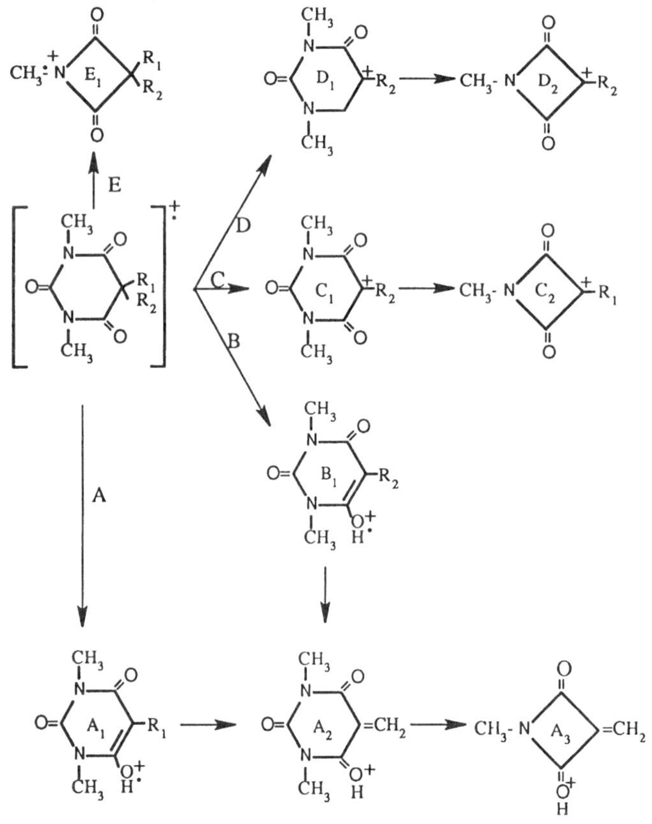

Figure 9.8. Major EI fragmentation pathways of N,N-*dimethylbarbiturates [218].*

With these basic fragmentation patterns, spectra obtained from aprobarbital, talbutal, and secobarbital are hard to differentiate. This difficulty has indeed been reported in the case of aprobarbital and secobarbital [226]. Methane CI studies [226, 227] were thus conducted to meet this need. Partly because of the large cross section for proton capture possessed by the barbituric acid nucleus, protonated ions derived from these compounds are generally intense and useful for revealing molecular weight information. The CI approach, however, fails to differentiate between isomeric barbiturates such as amobarbital and pentobarbital [226]. The combined use of EI and CI is therefore needed for successful identification of all barbiturates. With this in mind, argon–water CI studies on amobarbital and pentobarbital have been conducted [227], and they were found to produce abundant protonated and fragment ions. The intensities of the latter ions follow the characteristic EI patterns. With the molecular weight information and the EI characteristics, the differentiation of these two compounds becomes possible [227]. Because this approach has not been applied to the analysis of any extended list of barbiturates, it is not clear whether it will produce a unique spectrum for each compound.

Reactant-ion [97] and electron capture [228] NCI has also been applied to the analysis of barbiturates. Anions derived from the loss of side-chain radicals and a hydrogen atom provide valuable information concerning the molecular weight and side-chain substituents. Perceiving this advantage, isobutane NCI studies of 30 underivatized barbiturates have been conducted [229]. By observing the intensities of $[M - H]^-$, $[M - R_2]^-$, and $[M - R_1 - R_2 + H]^-$ ions, all but two barbiturates are identified and differentiated. The relative intensities of these ions observed in the 30 compounds studied are listed in Table 9.3.

The following observations are generalized from this isobutane NCI study [229]:

1. NCI spectra from 5-allyl-substituted barbiturates are relatively more intense than the corresponding 5-ethyl-substituted ones.

2. *N,N*-Dimethylated barbiturates generate anion spectra of lower abundances compared to their parent compounds.

3. Spectra from bromoallyl barbiturates are comparable to those of their nonbrominated analogs.

4. Thiobarbiturates are characterized by base peaks corresponding to $[M - R_2]^-$ ions.

5. $[M - R_2]^-$ ions are the base peaks for 5-ethyl barbiturates, while $[M - R_1]^-$ ions are the most intense ones [with the exception of cyclopentobarbital (Cyclopal)].

6. For 5-allyl compounds, branching at the carbon adjacent to the ring appears to enhance the $[M - R_2]^-$ ion. However, branching at the B position does not appear to make any difference. Spectra of idobutal and butalbital are virtually identical.

The structural similarities of this group of compounds make them ideal examples for comparing EI, CI, and NCI spectral characteristics. The studies just cited strongly suggest the complementary nature of these techniques; only under rare and unusual occasions will one single ionization technique provide all the information obtainable by every ionization method. Despite the claim that anion MS combines the advantages of EI and CI in providing structurally informative fragmentation patterns and molecular weight indications, complete identification of all barbiturates still cannot be done!

The merits of a chromatographic component and a chemical derivatization process cannot be overemphasized. Although barbiturates themselves are amenable to chromatography, *N,N*-dimethyl derivatives provide improved results and are commonly used in GC–MS analysis [217–222, 230]. SE–52 [217], OV–1 [218, 219], XE–60 [220], SE–30 [221], and OV–101 [222] have all been successfully used for this application.

Cocaine

At the present time, cocaine is probably one of the most abused controlled substances. The unique nature of cocaine use in our society necessitates sophisticated analysis skills in handling this drug in forensic laboratories [231]. One of the major characteristics of forensic analysis is to provide information leading to the determination of samples from a common origin [232]. The analysis of cocaine illustrates this point well [233–235]. Defense attorneys, perhaps with the assistance of expert witnesses, have been reported to raise the issue of natural versus synthetic origin [233].

Cocaine of natural origin may come from either the extraction of the coca leaf followed by successive recrystallizations, or the hydrolysis of all ecgonine-based alkaloids of the leaf with subsequent esterification. The presence of cinnamoylcocaine is often used as an indication of the former procedure, while the latter procedure, as a result of incomplete esterification of the ecgonine base, often produces ecgonine, methylecgonine, and benzoylecgonine. However, one should be cautioned that decomposition of cocaine may also produce these latter impurities. On the other hand, cocaine derived from total synthesis, unless scrupulously purified, contains diastereoisomers and enantiomers [234–236]. With these considerations, various techniques have been developed for the separation and identification of these compounds [233–242]. The chemical structures of cocaine enantiomers, diastereoisomers, and common impurities are shown in **45** through **55** (Chart 9.5).

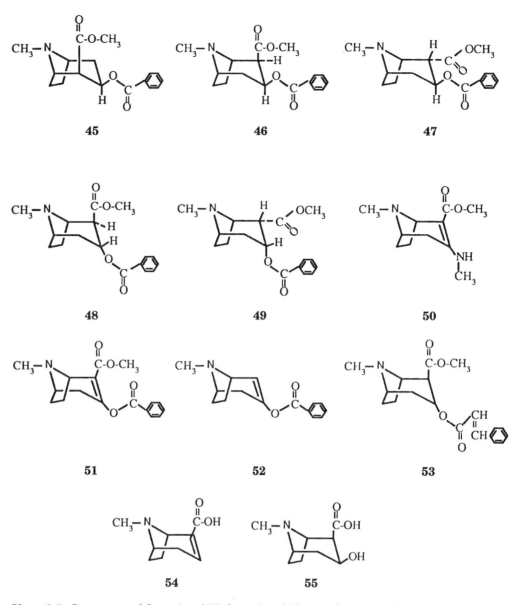

Chart 9.5. Structures of d-*cocaine (45),* l-*cocaine (46), pseudococaine (47), allococaine (48), pseudoallococaine (49), 3-aminomethyl-2-methoxycarbonyl-8-methyl-8-azabicyclo(3.2.1)oct-2-ene (50), 3-benzoyloxy-2-methoxycarbonyl-8-methyl-8-azabicyclo(3.2.1)oct-2-ene (51), 3-benzoyloxy-8-methyl-8-azabicyclo(3.2.1)oct-2-ene (52), cinnamoylcocaine (53), ecgonidine (54), and ecgonine (55).*

EI spectra of cocaine-related compounds have been presented in several articles for sample-identification purposes [233–238]. Fragmentation pathways have also been explored [233, 242, 243]. Fragmentation patterns of cocaine are summarized in Figure 9.9. The establishment of these pathways is based on the following considerations:

1. The charge of the molecular ion, which is initially localized on the N atom and, to a lesser extent, on the two carbonyl O atoms [242]; and

2. Mass shifts of ions derived from deuterated analogs [242, 243] and related compounds, such as cinnamoylcocaine, benzoylecgonine, methylecgonine, and ethylecgonine [233].

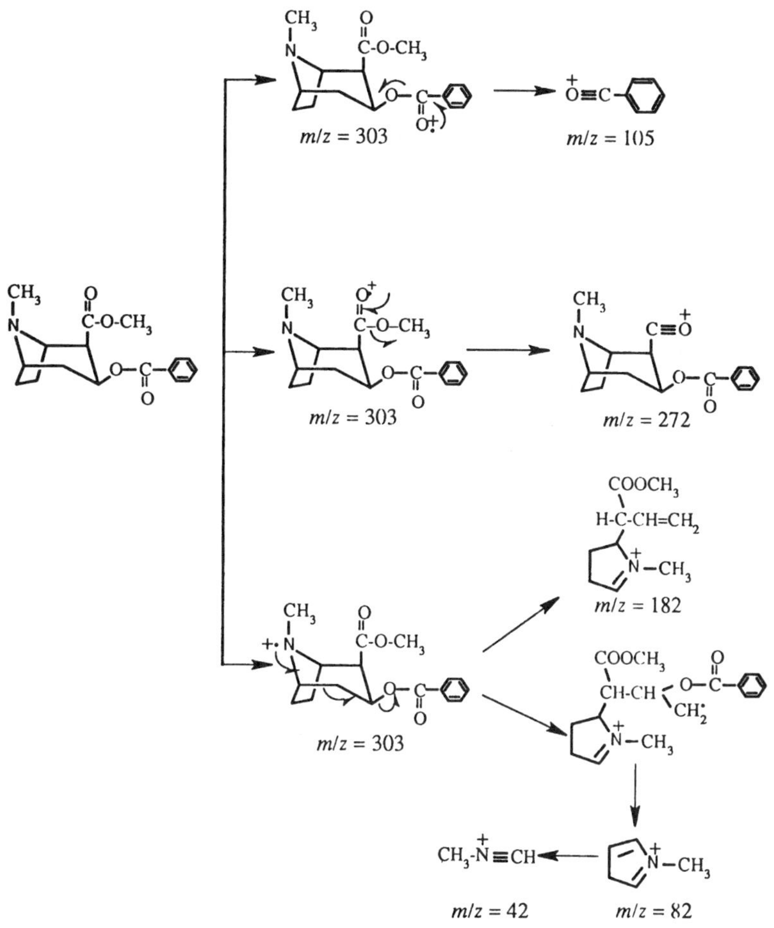

Figure 9.9. Major EI fragmentation pathways of cocaine [242].

High-resolution MS is also used to confirm chemical formulas proposed for specific ions [243]. Because of the olefinic bond at C_2, the mass spectrum of methylecgonidine differs somewhat [233] from the spectra of substituted ecgonines. The base peak at $m/z = 152$ represents the formation of the substituted pyridine structure. Aromatization provides the driving force for this pathway. The spectrum does contain the common characteristic ions at $m/z = 82$ and 42.

Isobutane [241], methane [244, 245], ammonia [244], and methane–ammonia [245] CI spectra of these compounds have also been reported. The energetic protonated molecular ion derived from the methane undergoes more fragmentation and produces ions of lower intensities [245].

Because the identification of impurities in a cocaine sample provides valuable information, GC–MS analysis is often applied. Stationary phases used include OV-1 [233, 235, 245, 246] and OV-17 [236, 243].

Phencyclidine and Analogs

When a particular drug is regulated by government agencies, substances with a similar structure are often introduced. The appearance of phencyclidine (PCP) analogs in the drug abuse scene reflects this characteristic. Many of these analogs [247–253] (Table 9.4 and Chart 9.6) have actually been found in street samples. Most of them are reported to produce a similar but weaker effect in users.

EI spectra of these compounds have been systematically studied [234, 252, 254]. These spectra are informative in revealing the structural characteristics of these compounds. All compounds give molecular ions, with intensities ranging from 10% to 60%: $[M - 43]^+$ ions (through the loss of $-C_3H_7$

from the cyclohexyl ring), and, to a lesser degree, $[M - 29]^+$ and $[M - 57]^+$ ions. Nitrogen heterocyclic compounds are characterized by the presence of the ions corresponding to the nitrogen heterocyclic fragment and the molecular ion minus the corresponding nitrogen heterocyclic radical. This phenomenon is even more pronounced in the thiophene-containing series. Thiophene and benzene compounds are characterized by the presence of $[C_5H_5S]^+$ and $[M - 83]^+$, and $[C_7H_7]^+$ and $[M - 77]^+$ ions.

Methane [248, 254–255], isobutane [248], and methane–ammonia [256] CI have been used in combination with GC in the analysis of these compounds. Among the several phases used, (OV-7 [251], OV-207 [248, 250, 251, 254], OV-25 [250], OV225 [248, 250], SP-2250 [252], and SE-30 [248, 250, 255]) appears to give the best overall separation of this series of compounds [248].

Table 9.4. Structures and major EI ions of PCP-related compounds.[a]

Designation; MW		Name	R_1[b]; mass	R_2[b]; mass	Characteristic EI ions[c,d]
PCP;	243	1-(1-Phenylcyclohexyl)piperidine	C_6H_6; 77	Pi; 84	$M - R_1$; C_7H_7; R_1; $M - R_2$
PCMeP;	257	1-(1-Phenylcyclohexyl)-4-methylpiperidine	C_6H_6; 77	4-CH_3-Pi; 98	$M - R_1$; C_7H_7; R_2; $M - R_2$
PCHP;	259	1-(1-Phenylcyclohexyl)-4-hydroxypiperidine	C_6H_6; 77	4-HO-Pi; 100	$M - R_1$; C_7H_7; R_2; $M - R_2$
PCPY;	229	1-(1-Phenylcyclohexyl)pyrrolidine	C_6H_6; 77	Py; 70	$M - R_1$; C_7H_7; R_2; $M - R_2$
PCM;	245	1-(1-Phenylcyclohexyl)morpholine	C_6H_6; 77	M; 86	$M - R_1$; C_7H_7; R_2; $M - R_2$
PCE;	203	N-Ethyl-1-phenylcyclohexylamine	C_6H_6; 77	NHC_2H_5; 64	$M - R_1$; C_7H_7
PCDE;	231	N,N-Diethyl-1-phenylcyclohexylamine	C_6H_6; 77	$N(C_2H_5)_2$; 112	$M - R_1$; C_7H_7
PCMe;	189	N-Methyl-1-phenylcyclohexylamine	C_6H_6; 77	$NHCH_3$; 30	$M - R_1$; C_7H_7
PCDMe;	203	N,N-Dimethyl-1-phenylcyclohexylamine	C_6H_6; 77	$N(CH_3)_2$; 44	$M - R_1$; C_7H_7
PCPr;	217	N-Propyl-1-phenylcyclohexylamine	C_6H_6; 77	NHC_3H_7; 58	$M - R_1$; C_7H_7
PCiPr;	217	N-Isopropylphenylcyclohexylamine	C_6H_6; 77	NH-i-C_3H_7; 58	$M - R_1$; C_7H_7
PCBu;	231	N-Butylphenylcyclohexylamine	C_6H_6; 77	NHC_4H_9; 72	$M - R_1$; C_7H_7
PCC;	192	1-Piperidinocyclohexanecarbonitrile	CN; 26	Pi; 84	$M - R_2$; R_2
MCC;	194	1-Morpholinocyclohexanecarbonitrile	CN; 26	M; 86	$M - R_2$; R_2
PYCC;	178	1-Pyrrolidinocyclohexanecarbonitrile	CN; 26	Py; 70	$M - R_2$; R_2
DEACC;	152	1-Diethylaminocyclohexanecarbonitrile	CN; 26	$N(C_2H_5)_2$; 112	
TCP;	249	1-[1-(2-Thienyl)cyclohexyl]piperidine	2-TH; 83	Pi; 84	$M - R_1$; C_5H_5S; $M - R_2$; R_2
TCPy;	235	1-[1-(2-Thienyl)cyclohexyl]pyrrolidine	2-TH; 83	Py; 70	$M - R_1$; C_5H_5S; $M - R_2$; R_2
TCM;	251	1-[1-(2-Thienyl)cyclohexyl]morpholine	2-TH; 83	M; 86	$M - R_1$; C_5H_5S; $M - R_2$; R_2

[a] See Structure **56** for the general structure of these compounds. Data are taken from Refs. [247], [252], [254], and [255].

[b] See Structures **57–60** for the structure of the following designations: Pi, Py, M, 2-TH.

[c] Ions are not listed in any intensity order.

[d] m/z M^+, $[M - 43]^+$, $[M - 29]^+$, and $[M - 57]^+$ are common to all compounds.

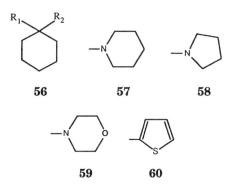

56   **57**   **58**

59   **60**

*Chart 9.6. Framework of PCP (**56**) and structural designations for Pi (**57**), Py (**58**), M (**59**), and 2-TH (**60**) in Table 9.4. (See Table 9.4 for the names and substitution groups of individual compounds.)*

Lysergic Acid and Related Compounds

The exact structures of various lysergic acid related compounds can be derived from the structural framework shown as Structure **61** and the substitution groups included in Table 9.5 [152, 257–259]. The basic structure shown in Structure **61** is depicted for the *d*-enantiomer. Only one *iso*-compound is shown in Table 9.5. The three-dimensional structures of *l*-enantiomers and *iso*-compounds can be derived from the basic frameworks provided in Structure **61** and Table 9.5. Among these compounds, *d*-LSD (first synthesized in 1943 [260]) was reported to be the most effective hallucinogen.

61

Framework of lysergic acid related compounds. (See Table 9.5 for the names and substitution groups of individual compounds.)

Table 9.5. Structures and EI MS characteristics of lysergic acid related compounds.

Name	MW	Double bond	R_1	R_2	R_3	R_4	R_5	a	b	c	d	e	f	g	h	Ref.
d-Lysergic acid	268	9,10	H	H	CH_3	COOH	H	21	20	—	43	68	20	21	41	[223–225, 227, 228]
d-iso-Lysergic acid	268	9,10	H	H	CH_3	H	COOH									[224, 228]
d-Lysergic acid amide	267	9,10	H	H	CH_3	$CONH_2$	H									[223]
d-Lysergic acid dimethylamide	295	9,10	H	H	CH_3	$CON(CH_3)_2$	H									[224]
d-Lysergic acid ethylamide	295	9,10	H	H	CH_3	$CONH(C_2H_5)$	H	—	8	36	32	—	—	21	51	[178, 223, 224]
d-Lysergic acid diethylamide (LSD)	323	9,10	H	H	CH_3	$CON(C_2H_5)_2$	H	—	5	17	32	—	—	36	64	[178]
								—	7	61	9	—	—	11	15	[223]
d-Lysergic acid methylpropylamide	323	9,10	H	H	CH_3	$CON(CH_3)C_3H_7$	H									[224]
d-Lysergic acid ethylpropylamide	337	9,10	H	H	CH_3	$CON(C_2H_5)C_3H_7$	H									[224]
d-Lysergic acid dipropylamide	351	9,10	H	H	CH_3	$CON(C_3H_7)_2$	H									[224]
2-Bromo-*d*-LSD	401	9,10	H	Br	CH_3	$CON(C_2H_5)_2$	H	—	1	19	74	—	—	40	53	[178]
1-Methyl-*d*-LSD	415	9,10	CH_3	H	CH_3	$CON(C_2H_5)_2$	H	—	3	21	38	—	—	32	52	[178]
1-Acetyl-*d*-LSD	365	9,10	$COCH_3$	H	CH_3	$CON(C_2H_5)_2$	H	—	11	9	14	—	—	28	41	[178]
N^6-Demethyl-*d*-LSD	309	9,10	H	H	H	$CON(C_2H_5)_2$	H	—	30	56	44	—	—	43	100	[223, 229]
N^6-Cyano-N^6-demethyl-*d*-LSD	334	9,10	H	H	CN	$CON(C_2H_5)_2$	H	—	3	55	55	—	—	44	15	[223, 229]
N^6-Carbamoyl-N^6-demethyl-*d*-LSD	352	9,10	H	H	$CONH_2$	$CON(C_2H_5)_2$	H	—	93	4	5	—	—	3	1	[223, 229]
Lysergol	254	9,10	H	H	CH_3	CH_2OH	H									[223]
1-Methylmethylergonovine	253	9,10	CH_3	H	CH_3	CONHCH–$(C_2H_2OH)C_2H_5$	H	—	1	88	26	—	—	61	100	[178]
Argroclavine	238	8,9	H	H	CH_3	CH_3	—									[223]
Elymoclavine	254	8,9	H	H	CH_3	CH_2OH	—									[223, 227]
9,10-Dihydro-LSD	325	None	H	H	CH_3	COOH	H	—	0.7	—	5	2	1	3	9	[225]

EI spectra of these compounds have been well characterized [152, 257–259, 261–265]. The basic fragmentation patterns are shown in Figure 9.10 [259], as illustrated for lysergic acid. All compounds display intense molecular ions, which nicely provide the needed molecular weight information. Characteristic pathways include:

1. The loss of a proton, yielding a highly conjugated immonium ion in the D ring (route a);

2. The loss of a –C$_2$H$_5$N group (the 5–6 and 7–8 cleavages) derived from the retro Diels–Alder reaction (route b);

3. The loss of a –C$_3$H$_4$O$_2$ group (the 6–7 and 8–9 cleavage) (route c); and

4. Various ways of eliminating the whole or part of the side chain at the 8 position (routes d, e, f, and g).

Fragments produced by routes a, d, e, and f are subjected to further fragmentation, as shown in Figure 9.10.

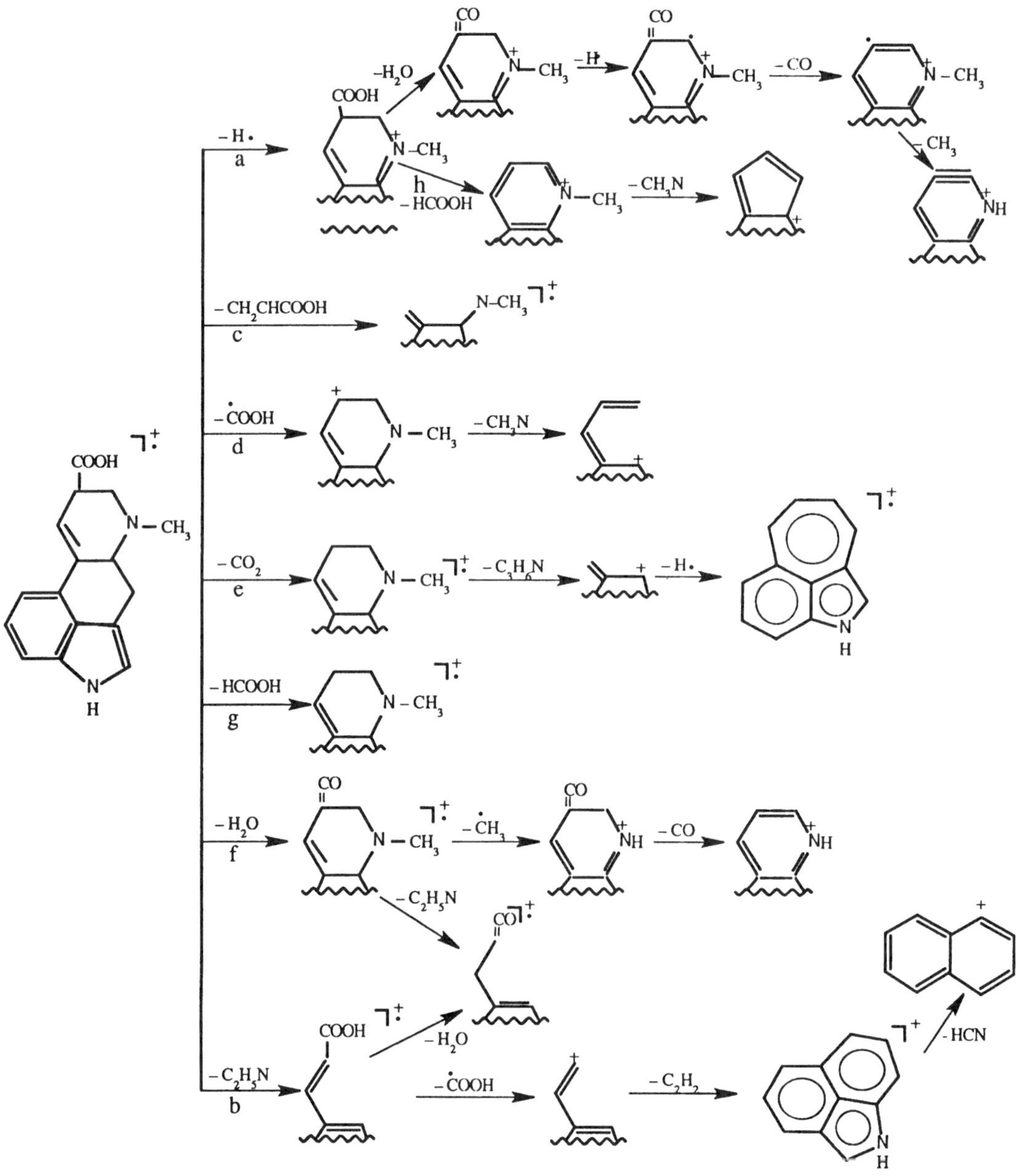

Figure 9.10. Major EI fragmentation pathways of lysergic acid [259].

The relative significance of these routes is drastically influenced by the side chain at the 8 position, the double-bond position, and the substituent at the N6 position. Intensities of ions derived from various fragmentation routes as reported in the literature are summarized in the right-hand portion of Table 9.5. These data indicate the following fragmentation characteristics:

1. The relative significance of the retro Diels–Alder reaction is affected by the substituent at the N6 position in the following order: $CONH_2$ > H > CH_3 > CN, and it becomes inoperative when the double bond is shifted to the 8,9 position.

2. Routes b and c become insignificant when the substituent at the 8 position is an alkyl or an alkyl derivative group.

3. Route c is less favorable when the carboxylic group in the 8 position is not derivatized.

4. Saturation at the D-ring eliminates (or at least lessens) several fragmentation routes and produces a much more intense molecular ion, which is easily recognized by the mass shifts of the corresponding ions.

$[M - COOH]^-$ and the $[M - H]^-$ ions are the most significant peaks in the NCI spectra [259] of lysergic acid and its saturated counterpart.

α-Methylfentanyl

α-Methylfentanyl was first found in the illicit narcotic trade of southern California streets in 1981 under the name of China White, a street term for very pure Southeast Asian heroin [264]. The identification of this compound [266] by the Drug Enforcement Administration's scientific team represents a modern version of a Sherlock Holmes episode.

α-Methylfentanyl, shown in **62**, has been reported [266] to be a potent analgesic agent. The structure of this compound resembles that of fentanyl (Structure **63**), which has been well studied [267–270] and illicitly used for "doping" racehorses [268].

62: $R_1 = CH_3$; $R_2 = H$
63: $R_1 = R_2 = H$
64: $R_1 = H$; $R_2 = CH_3$

α-Methylfentanyl (62), fentanyl (63), and N-[1-(2-phenethyl-4-piperidinyl)] acetanilide (64).

The EI spectrum of this compound is parallel to that reported for *N*-[1-(2-phenethyl-4-piperidinyl)]acetanilide, **64** [270], and has been thoroughly studied in a tandem MS investigation [271], which is discussed in this chapter in the section "Tandem MS". The secondary methyl group at the α-carbon greatly enhances the fragmentation pathway involving the loss of the benzyl group. Consequently, the molecular weight information is obtained through CI techniques [267].

Methaqualone and Mecloqualone

The spectrometric and chromatographic characteristics of 16 substituted quinazolinones (Structure 9.5 and Table 9.6) have recently been reported [272]. The fragmentation patterns of this group of compounds are similar to those reported for silylated methaqualone metabolites [273]. All compounds display significant intensities of M$^+$ and [M − 15]$^+$ ions under 70-eV EI conditions. The relative intensities of characteristic ions, such as m/z = 143 (Structure **65**) and 235 [M − R]$^+$, R = H, CH$_3$, F, Cl, Br, or I, are used to differentiate positional isomers [272].

Benzodiazepines

The analysis of oxazepam, **66** (Chart 9.7), has been widely reported mainly because monitoring of this compound can be used as a convenient mechanism for detecting the use of oxazepam or many other benzodiazepines, such as those shown in **66** through **78** [274]. (However, the use of many other benzodiazepines, such as those shown in Table 9.7, cannot be revealed through the monitoring of oxazepam.) Mainly excreted as the glucuronide, the analysis of this compound is normally preceded by acid hydrolysis [275] or enzymatic digestion [276]. Because acid hydrolysis results in the generation of benzophenones, enzymatic digestion, which preserves the intact structure, is often preferred [274].

Oxazepam may undergo ring contraction at elevated temperatures. Therefore, derivatization of the 3-hydroxy group with a trimethylsilyl (TMS) [277] or alkyl [278] group is helpful for GC and MS analysis. It has been reported [274] that methylation generated more stable products than the trimethylsilylation procedure.

65

Framework of substituted quinazolinones. (See Table 9.6 for the names and substitution groups of individual compounds.)

Table 9.6. Structure and EI MS characteristics of methaqualone congeners.^{a}

Name	MW	R$_1$	R$_2$	R$_3$	Int.^{b} of [M − R]$^+$	Int. ratio^b [M − R]$^+$:M$^+$:143
2-Methyl-3-phenyl-4-quinazolinone	236	H	H	H	72	1:0.86:0.33
2-Methyl-3-o-tolyl-4(3H)-quinazolinone	250	CH$_3$	H	H	100	1:0.70:0.03
2-Methyl-3-m-tolyl-4(3H)-quinazolinone	250	H	CH$_3$	H	93	1:0.96:0.97
2-Methyl-3-p-tolyl-4(3H)-quinazolinone	250	H	H	CH$_3$	81	1:0.90:0.54
2-Methyl-3-o-fluorophenyl-4-quinazolinone	254	F	H	H	100	1:0.90:0.54
2-Methyl-3-m-fluorophenyl-4-quinazolinone	254	H	F	H	21	1:4.50:2.50
2-Methyl-3-p-fluorophenyl-4-quinazolinone	254	H	H	F	8.3	1:11.0:7.20
2-Methyl-3-o-chlorophenyl-4-quinazolinone	270	Cl	H	H	100	1:0.19:0.33
2-Methyl-3-m-chlorophenyl-4-quinazolinone	270	H	Cl	H	57	1:1.50:1.10
2-Methyl-3-p-chlorophenyl-4-quinazolinone	270	H	H	Cl	38	1:2.60:2.30
2-Methyl-3-o-bromophenyl-4-quinazolinone	315	Br	H	H	100	1:0.06:0.08
2-Methyl-3-m-bromophenyl-4-quinazolinone	315	H	Br	H	86	1:0.29:0.53
2-Methyl-3-p-bromophenyl-4-quinazolinone	315	H	H	Br	25	1:1.80:3.00
2-Methyl-3-o-iodophenyl-4-quinazolinone	362	I	H	H	100	1:0.11:0.13
2-Methyl-3-m-iodophenyl-4-quinazolinone	362	H	I	H	43	1:1.30:0.77
2-Methyl 3-p-iodophenyl-4-quinazolinone	362	H	H	I	19	1:3.10:4.00

^{a} The positions of the R$_1$, R$_2$, and R$_3$ groups are designated in Structure **65**.

^{b} Intensity and intensity ratios are measured and calculated from the spectra in Ref. [272]. R represents CH$_3$, F, Cl, or I.

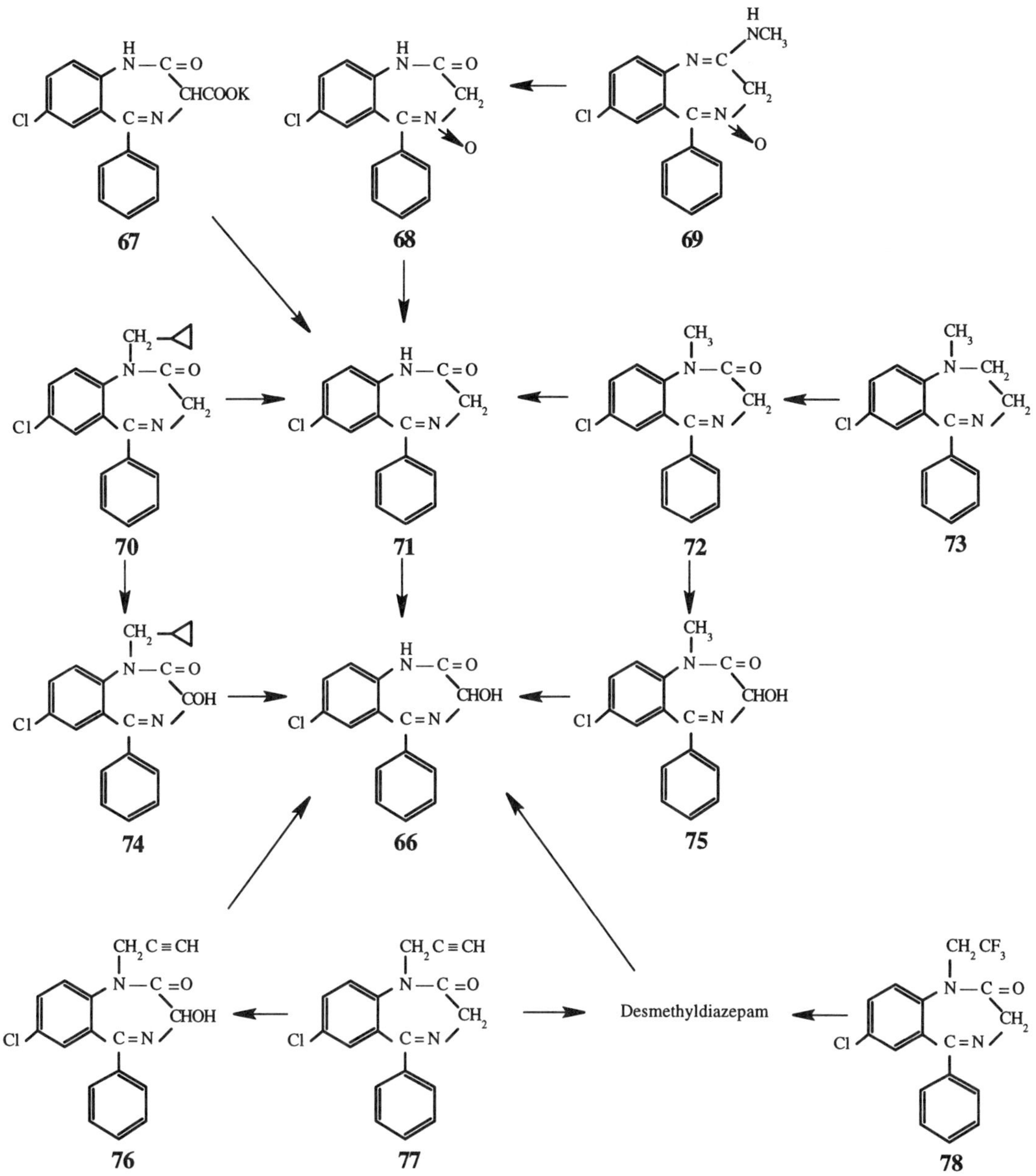

Chart 9.7. Oxazepam (**66**), clorazepate (**67**), demoxepam (**68**), chlordiazepoxide (**69**), prazepam (**70**), desmethyldiazepam (**71**), diazepam (**72**), medazepam (**73**), 3-hydroxyprazepam (**74**), temazepam (**75**), 3-hydroxypinazepam (**76**), pinazepam (**77**), and halazepam (**78**).

Using a solid-phase extraction–methylation-selected ion monitoring GC–MS procedure [274], it was possible to simultaneously monitor oxazepam, diazepam, and lorazepam in urine samples. Ions monitored for the methylated products of these three compounds and the IS (d_5-oxezapam) were $m/z = 271, 255, 314; 221, 256, 305, 283; 289, 348;$ and $276, 319$. With the first ion listed as the quantification ion, excellent linearity can be established for oxazepam, as shown in Table 9.7.

Methadone

Methadone is metabolized by mono- and di-*N*-demethylation, with subsequent cyclization to form 2-ethyl-1,5-dimethyl-3,3-diphenylpyrrolidine (EDDP) and 2-ethyl-5-methyl-3,3-diphenylpyrrolidine (EMDP) (Chart 9.8). Because about 33% of ingested methadone is excreted in the urine unchanged and

Table 9.7. Linearity study of oxazepam.[a]

Theoretical oxazepam concn (ng/mL)	Quantification ion intensity ratio	Observed[b]	
		Oxazepam concn (ng/mL)	Deviation from theoretical value
0	—	0	—
25	—	0	—
50	0.1198	54	+ 8.0%
75	0.1760	79	+ 5.3%
150	0.3184	143	− 4.7%
300	0.6668	300	0.0%
600	1.2769	574	− 4.3%
1,000	2.2661	1019	+ 1.9%
2,000	4.4041	1981	− 1.0%
4,000	8.9007	4004	+ 0.1%
8,000	17.647	7939	− 0.8%

[a] Data are taken from Ref. [274].
[b] Linear regression analysis result: $Y = -0.0026 + 0.0022X$; $R^2 = 1.000$. (Y is the intensity ratio of analyte to internal standard quantification ion and X is the analyte concentration.)

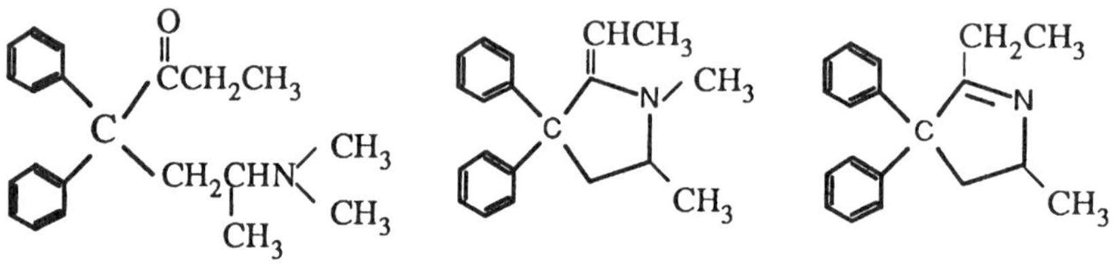

Chart 9.8. Methadone (79), 2-ethyl-1,5-dimethyl-3,3-diphenylpyrrolidine (80), and 2-ethyl-5-methyl-3,3-diphenylpyrrolidine (81).

about 43% as EDDP [278], the simultaneous determination of both methadone and EDDP provides evidence of methadone use and allows pharmacokinetic studies.

As shown in Figure 9.11, the ion intensities of EI mass spectra for methadone and its deuterated analog with $m/z > 200$ are low. Because the $m/z = 72$ ion is present in both methadone and d_3-methadone (the intended IS), and these two compounds are poorly resolved by the gas chromatograph, the $m/z = 72$ ion could not be used in the GC–MS analysis. A closer examination of the spectra indicated that the following ions (m/z) might be suitable: methadone 294, 223, and 295; and d_3-methadone 297 and 226. (Because EDDP was sufficiently separated, and deuterated EDDP was not used as the IS, the selection of ions to be monitored for EDDP was less critical; the following ions were selected: $m/z = 277, 262,$ and 276.) The first ion listed for each compound is used as the quantification ion. The ion-intensity ratios of the analyte and IS obtained from urine controls of various concentration levels are plotted in Figure 9.12. Excellent linearity was observed [279].

Analytical Approaches

Tandem MS

Examples of utilizing the various forms of the recently developed tandem MS for drug analysis include Fourier transform MS (FT-MS) [280], mass-analyzed ion-kinetic-energy spectrometry (MIKES) [51-57], and triple-quadrupole MS [281-287]. The liquid chromatograph [79, 285], gas chromatograph [286], and solid probe (for hair) [287] have also been coupled to MS–MS systems as sample-introduction devices.

Tandem MS has been shown to provide valuable information for the identification of a compound of interest in a complex matrix, and for the elucidation of an unknown structure and the differentiation of compounds. MIKES has been used to map the concentration ratio of cocaine and cinnamoylcocaine in whole coca plant tissues without sample preparation or prefractionation [55, 56]. The experiments employ direct probe vaporization, CI, and collision-induced dissociation. As the sample is evaporated, both compounds are characterized by their molecular weights and the dissociation involving the loss of benzoic acid and cinnamic acid from the protonated molecules (*see* **45** through **55** for the structural characteristics of these compounds). Similar approaches are used to identify papaverine in raw opium [54] and mescaline in crude plant extracts [52, 53].

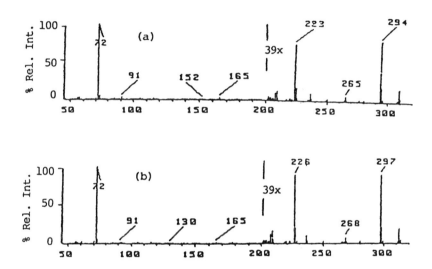

Figure 9.11. Mass spectra of methadone (top), and d_5-methadone (bottom). The ion intensities of all ions greater than m/z = 200 were enhanced 39 times. (Reprinted with permission from Ref. [279].)

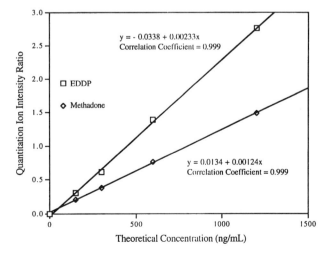

Figure 9.12. Correlation of quantitation ion intensity ratio for EDDP (m/z = 277/297) and methadone (m/z = 294/297) with theoretical concentrations [279].

Molecular structure determinations of China White and related fentanyl derivatives best illustrate the potential of tandem MS in structure elucidation [271]. The structure is pieced together through successive pairing of EI "complementary ions", whose sum of masses or elemental compositions equals that of the molecular ion, and determination of the substructures of these pair ions by their collisionally activated dissociation spectra. Spectra of the protonated or ethylated molecules generated by methane CI also provide complementary information not available from other ionization methods. This application is nicely illustrated in a study concerning the analysis of barbiturates [57].

Procedures for positive CI (PCI) and subsequent MS–MS analysis of the protonated molecular ions of TMS derivatives of LSD, iso-LSD, and N-demethyl-LSD were developed to provide a high degree of specificity for identification of these compounds in urine and blood at low-pg/mL concentrations. NCI and subsequent MS–MS analysis of the molecular anion of the trifluoroacetyl derivative provide higher sensitivity for the analysis of N-demethyl-LSD [286].

Chemical Derivatization for Enhancing Detection and Structure Elucidation

Generation of Favorable Derivatives to Improve the Limit of Detection

The formation of fluoroacyl derivatives from alcohols, phenols, and amines, an approach described in Chapter 6 and used to improve the limit of detection in GC applications, has also been applied to negative CI (NCI) in GC–MS applications [288, 289]. For example, the NCI method generated a signal that was 200-fold greater than the positive CI (PCI) counterpart when Δ^9-tetrahydrocannabinol-11-oic acid was analyzed as its pentafluoropropyl–pentafluoropropionyl derivative [289]. Similarly, when NCI was used instead of PCI [288], a 328-fold increase in signals was observed for the pentafluorobenzoyl derivative of Δ^9-tetrahydrocannabinol, and 100-fold and 678-fold increases for the pentafluorobenzoyl and tetrafluorophthaloyl derivatives, respectively, of amphetamine.

Generation of Favorable Mass Spectra through Derivatization

Generation of Ions More Suitable for Quantification. Mass spectra obtained from thoughtfully designed derivatives can show distinctive characteristics not available from parent compounds. Several advantages can result from the alteration of the characteristics of the mass spectra, as illustrated below. For the example shown in Figure 9.13 [290], improved detection of amphetamine can be obtained through the measurement of ions obtainable only through derivatization. The spectrum of the parent compound exhibits low intensities of ions at a higher mass range. Considering the probability of contributions from interfering compounds, the low-mass $m/z = 44$ ion is not suitable for quantification.

Generation of Ions Helpful for Structure Elucidation. Chemical derivatization can be used to preserve the structural characteristics that are necessary for producing interpretable mass spectra. For example, to prevent ring contraction that may occur at elevated temperatures, the 3-hydroxy group in oxazepam, **66,** is derivatized with a TMS [277] or alkyl [278] group in GC–MS analysis.

Mass shifts in the spectra produced by different derivatizing agents can provide extremely useful information for identifying an unknown compound. For example [291], the number of TMS groups attached to the parent compound is deduced on the basis of the mass shifts resulting from replacing N,O-bis(trimethylsilyl)acetamide (BSA) with d_9-BSA as the derivatizing agent. This information facilitates the identification of desoxymorphine A, monoacetyldesoxymorphine A, and diacetyldesoxymorphine A as the impurities in an illicit heroin sample. The same approach is used to characterize 6- and 3-monoacetylmorphine [32].

Similarly, compared to the mass spectrum (Figure 9.14, top) of the parent compound, the 24-amu mass shift observed in the mass spectrum of the derivatized pentobarbital (Figure 9.14, bottom) indicates the replacement of two hydrogens by two methyl groups.

As a third example, compared to parent compounds, TMS derivatives of N-substituted barbiturates are found to generate less olefin-radical elimination ($[M - 41]^+$ and $[M - 55]^-$). Instead, the formation of the $[M - 15]^+$ ion is favored, making it easier to recognize the molecular weight of the compound under examination [292].

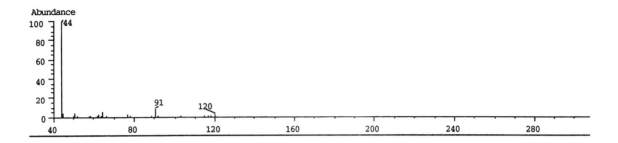

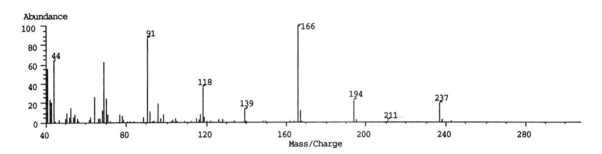

Figure 9.13. EI mass spectra of amphetamine (top), and N-trifluoroacetyl-l-prolyl derivative of amphetamine (bottom). (Reprinted with permission from Ref. [290].)

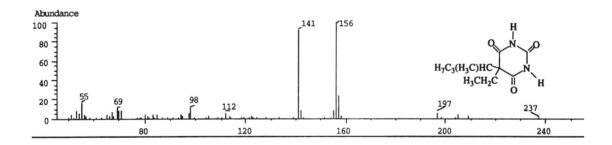

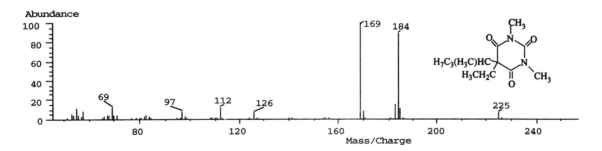

Figure 9.14. EI mass spectra of pentobarbital (top), and methyl derivative of pentobarbital (bottom). (Reprinted with permission from Ref. [230].)

Quantitative Analysis

The use of MS and allied methods for quantitative analysis is now a routine analytical procedure. This topic is treated in numerous publications, including several monographs [18, 293, 294] and review articles [295, 296].

The IS method is the most commonly adopted approach for quantitative analysis. When an MS device is used as the detection mechanism, as in GC–MS and HPLC–MS applications, an appropriate *isotopic analog* of the analyte is often a preferred IS [18]. With its practically identical chemical properties and MS fragmentation mechanisms, an isotopic analog (of the analyte) offers the following advantages when used as the IS:

1. Errors derived from (1) incomplete recovery of the analyte in the sample-preparation process, or (2) varying GC and MS conditions are compensated for.

2. The presence of interfering materials (or mechanisms) causes the absence of the IS in the final chromatogram [297] or altered response and ion-intensity ratios [298], thus alerting the analyst to the need for further investigation.

Not all deuterium-labeled isotopic analogs (of the analyte) can be used as the IS. Because the analyte and the IS are rarely separated adequately, the proposed isotopic analog must generate at least one (preferably two or three) ions relatively free of cross-contribution by the analyte. There must also be at least three ions from the analyte that are relatively free from a cross-contribution by the proposed isotopic IS. If these two requirements are not met, then the quantitative result and the ion-intensity ratio data (which are commonly used as important parameters for analyte identification) may become unreliable.

Using an Isotope-Labeled Analog as an IS

Criteria for Selecting an Isotopic Analog as the IS. A specific isotopic analog can be adopted as an IS in a GC–MS application if the following conditions are met:

1. The isotopic analog is labeled with a *sufficient number of atoms of a selected isotope* (typically deuterium) so that the corresponding ions from the IS and from the analyte will have a significant difference in their masses. If the difference is not sufficient, the [M + n] ion (designated for the analyte) may, because of the naturally occurring isotopic abundance, make a significant contribution to the intensity of the ion (designated for the isotopic analog) that corresponds to the [M] ion of the analyte. (M is the mass of the ion derived from the analyte and selected for monitoring; n is the nominal mass difference between the ions designated for the analyte and the isotopic analog.) If deuterium, as in most realistic applications, is used as the labeling isotope for the IS, a difference in 3 mass units between the analyte and the isotopic analog is sufficient under normal circumstances. It should be noted, however, that if the concentration of the analyte is disproportionally higher than the concentration of the IS included in the assay process, the intensity of the [M + 3] ion originating from the analyte may become significant enough to require an additional analysis of a diluted aliquot.

2. The analyte and the isotopic analog undergo an appropriate fragmentation process to generate several high-intensity ions, which include the labeling isotopes with an insignificant intensity of [M − nH] ions. To meet this requirement, the labeling isotopes must be positioned at appropriate locations in the molecular structure so that, after the fragmentation process, a sufficient number of high-mass ions (with significant intensities) that retain the labeling isotopes are present and will not interfere with the intensity measurement of the corresponding ions derived from the analyte. These ions (from the isotopic analog) and their counterparts in the analyte may then be monitored for ion-ratio evaluation and further used for qualitative compound identification and quantitative determination.

3. The isotopic analog is manufactured with sufficient *isotopic purity*. Otherwise, the addition of the IS may result in the observation of a significant amount of the analyte in a true negative sample and may also introduce errors in quantitation. This will become a problem, especially when a high concentration of the IS is used [299].

Another important factor that needs to be considered is the EI fragmentation process of compounds containing hydrogens. The unavoidable "[M − xH]" phenomena [300, 301] and the naturally abundant isotopes in the analyte make one wonder how accurate the measured ion intensity truly reflects the molecular population. Quantifications based on the direct comparison of the ion intensity derived from the analyte and the corresponding ion derived from the IS may not be valid. This approach makes the assumption that no isotopic effect [302, 303] takes place in the fragmentation process. Thus, normal quantification approaches compare the analyte/IS ion-intensity ratio observed in one or more calibration standards to that observed in the test sample.

One-Point Calibration. A typical quantitative GC–MS protocol usually involves monitoring several selected ions (by selected ion monitoring, SIM) from the analyte and from the isotopic analog. Quantification is achieved by comparing a selected analyte-to-isotopic analog ion-intensity ratio observed from the test sample and the same ratio observed from the calibration standard. The calibration standard contains the same amount of the IS and a known amount of the analyte, and it is processed in parallel with the test sample. The analyte concentration in the test sample can be calculated using a one-point calibration approach, as shown in Figure 9.2.

This procedure can be used directly to determine the analyte concentration in a test sample (one-point calibration) or to determine the responses of a set of standards, from which a calibration curve can be established and used for analyte-concentration determination in a test sample (multiple-point calibration). Multiple-point calibration approaches can be implemented with different regression models, including linear or nonlinear, weighted or unweighted, and with or without forcing the regression through the origin point.

Calibration Curve. Quantification based on an isotopically labeled IS follows the same principle of the isotope-dilution method. The basic principle has been discussed in several publications [304–307]. Readers are referred to these and other more general literature sources for the derivation of the following relationship:

$$y = R + Px - Qxy \qquad (9.3)$$

where y is the response ratio of the ion representing the analyte and its isotopically labeled counterpart derived from the IS, R, P, and Q are constants, and x is the mass of the analyte. Thus, the general form of a calibration curve, relating ratio to mass of analyte, is a hyperbola. This is mainly because the analyte itself also contains a small but significant fraction of the isotope-labeled analog; and the isotope-labeled IS is not, in all practicality, isotopically pure. Only if the contribution of the analyte to the signal accounting for the IS is neglected, will equation 9.3 be reduced to a linear form.

In practical application, a decision has to be made on the use of one-point or multiple-point calibration. In the latter case, the use of linear or polynomial fitting of calibration data also shows differences. A study [308] on the quantification of amphetamine and methamphetamine was conducted using concentrations ranging from 250 to 4000 ng/mL. With the same amount of the selected IS (d_5-amphetamine for amphetamine, d_9-methamphetamine for methamphetamine) used in all samples, the response–concentration relationship is examined by plotting the analyte-to-IS quantification ion-intensity ratio (m/z = 190/194 for amphetamine and 204/211 for methamphetamine) obtained from these samples. These data were fitted with first- and second-order polynomial equations, as shown in Table 9.8. The resulting standard-deviation and mean-error data indicate the nonlinear nature of the response–concentration relationship, and the suitability of the second-order polynomial model. As an example, the second-order polynomial fit for methamphetamine is shown in Figure 9.15.

Table 9.9 compares the theoretical value with the observed values obtained using the 500-ng/mL control as the one-point calibrator and all five controls fit into a second-order polynomial model. With one-point calibration, the observed concentrations for the controls at higher concentration are lower than the expected values, indicating a "curving" phenomenon. This is also evident from the second-order polynomial fit shown in Figure 9.15.

Table 9.8. First order (linear regression) and second-order polynomial fit of the data observed from five controls of known concentrations.[a]

Drug	Regression equation[b]	Std dev	Mean error
Amphetamine	$Y = 0.1179 + 7.6522 \times 10^{-4}X$	0.0567	0.0776
Methamphetamine	$Y = 0.1575 + 7.6290 \times 10^{-4}X$	0.0812	0.110
Amphetamine	$Y = 0.0675 + 7.7865 \times 10^{-4}X - 1.2751 \times 10^{-8}X^2$	5.28×10^{-4}	9.82×10^{-4}
Methamphetamine	$Y = 0.0131 + 1.0200 \times 10^{-3}X - 5.9608 \times 10^{-8}X^2$	4.17×10^{-3}	5.97×10^{-3}

[a] Data are taken from Ref. [308].

[b] Y: analyte to internal standard quantification ion ratio; X: concentration of analyte.

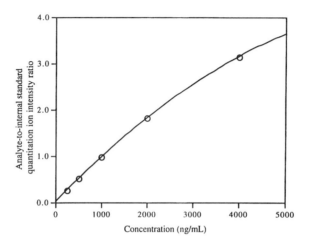

Figure 9.15. Second-order polynomial fit of methamphetamine calibration data [308].

Table 9.9. Theoretical versus observed values of controls using one-point calibration and five-point calibration curves fit into a second-order polynomial equation.[a]

Theoretical concentration (ng/mL)	Amphetamine		Methamphetamine	
	1-point calibration, observed (% dev)	2nd order polynomial, observed (% dev)	1-point calibration, observed (% dev)	2nd order polynomial, observed (% dev)
500	500 (calibrator)	499 (−0.2%)	500 (calibrator)	497 (−0.6%)
250	256 (+2.4%)	254 (+1.6%)	258 (+3.2%)	256 (+2.4%)
1,000	966 (−3.4%)	993 (−0.7%)	970 (−3.0%)	993 (−7%)
2,000	1,814 (−9.3%)	2,006 (+0.3%)	1,792 (−10.4%)	2,007 (+0.4%)
4,000	3,266 (−18.4%)	3,999 (−0.03%)	3,111 (−22.2%)	3,999 (−0.03%)

[a] Data are taken from Ref. [308].

As a second example, using d_5-pentobarbital as the IS, quantification of pentobarbital resulting from one-point calibration, linear regression, and hyperbolic curve procedures are compared in Table 9.10 [230]. These data indicate that the effectiveness of the calibration procedure is in the order of hyperbolic curve > linear regression > one-point calibration.

Further Considerations. Data shown in Tables 9.9 and 9.10 clearly indicate that multiple-point calibration provides more reliable results over a wider concentration range, especially at the higher-concentration end. However, one-point calibration may provide results that are not necessarily inferior to other more-sophisticated approaches for samples with concentrations in the immediate vicinity of the concentration of the single calibrator. In light of the heavy emphasis on workplace drug-testing programs, in which a sample is reported as positive only if the analyte's concentration exceeds a "cutoff" level, the merit of using a one-point calibrator at the cutoff concentration should not be overlooked.

The barbiturate study [230] cited above also compared the effectiveness of using a single IS (d_5-pentobarbital) for the quantitation of four barbiturates (butalbital, amobarbital, secobarbital, and pentobarbital). Results summarized in Table 9.11 illustrate an important point: the use of a deuterated analog of the analyte does not guarantee the generation of the best possible quantitation result. These data indicate that one-point calibration quantification results for pentobarbital were actually inferior to those obtained for the other three barbiturates—even though d_5-pentobarbital was used as the single IS for all analytes. Factors that should be carefully evaluated include (1) cross-contribution of quantitation ions selected for the analyte and the IS, and (2) selection of an appropriate calibration model to take into account the cross-contribution interference—a common phenomenon. Thus, more sophisticated approaches (see Table 9.9) are needed to improve the quantification of pentobarbital using d_5-pentobarbital as the IS.

Table 9.10. Comparison of pentobarbital quantitation results using different calibration methodologies.[a]

Theoretical concn (ng/mL)	Observed quantification ion int. ratio	One-point calibration		Linear regression		Hyperbolic curve	
		Observed concn	% Dev	Observed concn	% Dev	Observed concn	% Dev
0	—	—	—	—	—	—	—
13	0.0112	22.5	80	10.5	−16	11.5	−8.3
25	0.0176	35.4	42	27.5	10	27.6	10
50	0.0259	52.1	4.2	49.6	−0.79	48.5	−3.1
100	0.0452	90.9	−9.1	100.9	0.90	97.2	−2.9
200	0.0888	179	−11	216.8	8.39	207.7	3.9
400	0.1660	334	−17	422.0	5.50	405.4	1.3
800	0.3090	621	−22	802.1	0.26	778.0	−2.8
1,600	0.6055	1,217	−24	1,590	−0.61	1,578	−1.3
3,200	1.162	2,336	−27	3,069	−4.1	3,192	−0.26
6,400	2.144	4,310	−33	5,680	−11	6,451	0.80

[a] Data are taken from Ref. [230].

Table 9.11. Quantification using one-point calibration methodology.[a,b]

	Butalbital				Amobarbital				Secobarbital				Pentobarbital			
Theor concn	Ion int. ratio	Observed concn	Dev (%)	Theor concn	Ion int. ratio	Observed concn	Dev (%)	Theor concn	Ion int. ratio	Observed concn	Dev (%)	Theor concn	Ion int. ratio	Observed concn	Dev (%)
0	—	—	—	0	—	—	—	0	—	—	—	0	—	—	—
13	0.029	13.8	6.2	13	0.048	12.1	−6.9	13	0.039	13.0	0	13	—	—	—
25	0.059	27.6	10	26	0.097	24.6	−5.4	27	0.081	27.0	0	25	0.0168	32.6	30
54	0.107	50.3	−6.9	51	0.199	50.4	−1.2	53	0.162	54.0	1.9	50	0.0288	56.2	12
108	0.189	89.4	−17	102	0.386	98.1	−3.8	106	0.340	114	7.5	100	0.0534	104	4.0
208	0.454	215	3.4	204	0.801	204	0	216	0.633	212	−1.9	199	0.104	202	1.5
430	0.878	416	−3.3	408	1.56	395	−3.2	424	1.28	427	0.7	398	0.203	394	−1.0
860	1.78	879	2.2	816	3.07	777	−4.8	848	2.58	864	1.9	796	0.382	741	−6.9
1,720	3.82	1,805	4.9	1,632	6.40	1626	−0.4	1,697	5.01	1 679	−1.1	1,592	0.765	1,485	−6.5
3,440	7.62	3,608	4.9	3,264	10.9	3075	−5.8	3,393	9.48	3,173	−6.5	3,184	1.47	2,853	−10

[a] Data are taken from Ref. [230].
[b] All concentrations are in ng/mL.

Instrumental Parameters

For the best accuracy and precision of SIM measurements, instrumental parameters such as instrument resolution, ion-monitoring position, measurement of threshold setting, and dwell time must also be carefully considered. Instrument drift from the peak center and the fluctuations in intensities of the selected ion beams will affect measurement precision, mainly because of the greater variation in the intensities of the weaker signals detected [295]. Improper threshold settings will also affect the weaker signals to a greater extent [309, 310]. To achieve the best results, the signal level of the IS should be comparable to that of the analyte [310–312]. For GC–MS applications, problems associated with differences in ion residence time in the ion source must also be addressed [246, 313–315].

Chromatography–MS Applications

Serving as a chromatograph detector, an MS provides distinct advantages with regard to both the qualitative and quantitative aspects of an analytical procedure. The MS detector can resolve and display the charged particles (originating from the analytes eluting from a chromatograph) that are characteristic of individual analytes, thus providing an additional dimension of "fingerprint" information. Together with the chromatographic retention data, MS detection provides test results that are normally considered conclusive for the identification of the analyte in question.

With regard to the quantitative aspect, the use of an MS detector permits the use of isotopic analogs of the analytes as the ISs in routine practice. Although MS by itself is not necessarily the most quantitative detector for a chromatographic system, the use of an isotopic analog as the IS alleviates the effects of many variations in the analytical process that may affect the accuracy of quantitative results. Thus, with the use of an isotopic analog as the IS, an MS detector provides the best overall quantitative results for chromatographic methods of analysis.

GC–MS

A direct capillary interface has been proven to be an effective approach for interfacing GC and MS systems for contemporary capillary-column GC analysis of drugs. With this hardware issue resolved, major parameters that require careful consideration are mainly associated with the operation of the MS either in the scan or SIM modes.

For fully credible general qualitative analysis (as opposed to target-compound analysis), the MS has to be operated using the scan mode. The resulting spectra are then compared with those in the database or with an in-house standard; together with the GC retention data, a conclusive identification of the analyte may be made.

Data collection with SIM can provide enhanced sensitivity and is also commonly used as an integral part of a well-designed target-compound analysis (qualitative and quantitative) scheme, which includes well-studied extraction–derivatization steps.

Both full-scan and SIM data-collection modes have been extensively applied to the analysis of drugs in dosage form [316], body fluids [317, 318], and hair [319]. Examplar applications are summarized in Table 9.12.

Operational Parameters

Precautions and Minimal Requirements for Using SIM Data as Qualitative Analysis Criteria. Without the availability of complete spectra, the use of SIM data for qualitative-determination purposes requires the exercise of extreme caution. As many ions that are characteristic of the analyte should be selected. Ideally, the molecular ion should be used if it exists with a reasonable intensity. If the analyte is derivatized with a derivatizing reagent prior to the GC–MS analysis, at least one of the ions selected should include the complete or a characteristic moiety of the analyte. The exclusive use of only ions that are derived from the derivatizing reagent is not acceptable.

Table 9.12. Examplar GC–MS analysis of drugs of abuse.

Drug category	Derivatization	Major emphasis	Column	Scan mode	Ref.
Amphetamines	Trichloroacetyl	Methodology improvement	DB-5	SIM[a]	[306]
	t-Butyldimethylsilyl	Separation with other amine drugs	HP-1	Full scan	[320]
	Trifluoroacetyl-l-prolyl	Enantiomeric separation	SP-2100	Full scan	[156]
	(–)-Menthyl chloroformate	Enantiomeric separation	DB-5	SIM	[321]
	None	Differentiation of side chain isomers	Ultra-1	Full scan	[322]
	Trifluoroacetyl and analog	Microwave-induced derivatization	Ultra-2	Full scan	[323]
	4-Carbethoxyhexafluorobutyryl	Artifact peak as methamphetamine	Ultra-1	Full scan	[324]
Barbiturates	Methyl	Methodology improvement	DB-5	SIM	[228]
	None	Solid-phase extraction	DB-5	Full scan	[325]
Benzodiazepines	Methyl	Methodology improvement	DB-5	SIM	[272]
	t-Butyldimethylsilyl	Wide range of metabolites	DB-5	SIM	[326]
	Trimethylsilyl	Ionization methods	DB-1	SIM	[327]
	Propyl	Wide range of metabolites	DB-1	Full scan	[328]
Cannabinoids	Methyl	Method development	HP-1	SIM	[329]
	t-Butyldimethylsilyl	Derivatization	DB-1	SIM	[330]
	Trimethylsilyl	New solid-phase extraction	DB-5	SIM	[331]
	Methyl	Internal standard	DB-1	SIM	[332]
Cocaine	Pentafluoropropionyl	Wide range of cocaine metabolites	Ultra-2	SIM	[333]
	Trimethylsilyl	Metabolites; pyrolysis product	HP-1	SIM	[334]
Lysergic acid diethylamide	Trimethylsilyl	Method development	HP-1	SIM	[335]
	Trimethylsilyl	Methodology development	DB-5	SIM	[336]
	Trifluoroacetyl	Negative-ion chemical ionization		SIM	[337]
	Trifluoroacetyl; Trimethylsilyl	Ionization methods and MS–MS	Ultra-2	Full scan; SIM	[284]
Opiates	Trifluoroacetyl	Assay of heroin and metabolites	Rtx-5	SIM	[338]
	Trifluoroacetyl	Stability and resolution of derivatives	HP-1	SIM	[339]
	Trimethylsilyl	Interference (by hydromorphone, etc.)	Ultra-2	SIM	[340]
	Trimethylsilyl	Hydrolysis methods	DB-5	Full scan	[341]
Phencyclidine	None	Analogs	DB-5	Full scan	[342]

[a] SIM: selected ion monitoring.

The intensity ratios of the selected ions should be closely monitored and compared with the corresponding ratios obtained from a standard (or control) that is analyzed under the identical operational conditions.

The use of SIM data for qualitative-analysis purposes should only be applied to well-studied systems in which isotopic analogs of the analytes are normally available and incorporated in the analytical process. The relationship between the ion-intensity ratios of the IS and the corresponding ratio in the analyte should also be closely monitored.

Other than the obvious requirement of coincidental retention times for all monitored ions of an analyte, the relationship between the retention time of the isotopic analog IS and the analyte should be closely monitored. Furthermore, the GC operation condition should be optimized so that compounds closely related by structure are chromatographed with distinguishable retention data.

Precautions and Minimal Requirements for Using SIM Data for Quantitative Determination.
To assure adequate accuracy, the number of data points that define the peak must be sufficient, preferably about 20. Because of the sequential nature of the ion-monitoring process of the MS system widely used as the GC detector, the number of data points for a GC peak depends not only on the chromatographic peak width, but also on the ion-monitoring cycle time. The cycle time depends on the number of ions monitored, the time spent on monitoring each ion (dwell time), and the "overhead time" needed by the system for effective switches.

An empirical study [310] on the parameters affecting quantitative determination indicated that the peak width, peak-monitoring position, window size, and dwell time may have a secondary effect on the accuracy of a quantitative determination.

A recent report [298] provided empirical data suggesting that coeluting compounds at high concentration may reduce the response of the analyte (or IS), rendering quantitation results inaccurate. Thus, prior removal or chromatographic resolution of the coeluting compound may be necessary even when the coeluting compound does not cause difficulties in qualitative identification of the analyte.

Application of GC–MS Protocols in Drug Urinalysis: Targeted-Compound Identification

The recently adopted policy of conducting drug urinalysis as a mechanism for monitoring drug abuse in the workplace has prompted the development of many robust extraction–derivatization GC–MS procedures for routine and large-scale analysis of targeted drugs in urine specimens [343]. Examples of these protocols are summarized in this section. Detailed procedures can be readily found in a book by R. H. Liu [344].

Urine specimens are pretreated according to the schemes shown in Figure 9.16. Evaporated residues are reconstituted and analyzed by GC–MS protocols (Table 9.13), with the MS operated under SIM mode.

The resulting SIM ion chromatograms and MS data are used for qualitative and quantitative analysis of the targeted drugs. Criteria adopted for conclusive drug identification include the appearance of the monitored ions at a correct *retention time* with acceptable *intensity ratios* among these ions. The retention time and ion-intensity ratios observed in the test sample are compared with those established by an authentic standard incorporated in the same analytical batch. They are also evaluated on the basis of "intrasample" information, that is, the retention time of the analyte relative to that of the IS within the test sample should be comparable to those observed in the authentic standard. Considering the fact that the identified analyte has also survived the chemical processes (pretreatment, extraction, and derivatization) designed for the targeted drug and proven effective for the IS (often a deuterated analog with chemical properties practically identical to those of the analyte), these identification criteria should not lead to an incorrect conclusion if they are utilized by an able analyst.

For most reliable quantitative results, precautions and criteria related to the selection of deuterated ISs, quantification ions, and the calibration procedure (one-point vs. multiple-point, and linear versus nonlinear) discussed in an earlier section have to be observed.

Liquid Chromatography–MS

The merits of an MS detector are well demonstrated by the wide applications of GC–MS in drug analysis. Unfortunately, the on-line coupling of MS to an HPLC system is not an easy task, not only because the operational pressures of these two technologies are not comparable, but also because of the liquid state of the mobile phase used in HPLC. Several approaches have been explored for the introduction of the analytes in eluent and the generation of ions. These approaches either deliver analyte-enriched particles to the ion source for ionization or extract analyte ions into the MS from the charged droplets. The characteristics of these interfaces are summarized in Table 9.14 [345–347]. The original reference should be referred to for further details.

Because of the late development of this technology, its application in forensic science is still rare. Examples of limited applications include (1) direct liquid introduction of split effluent using the liquid chromatographic eluent hexane–isopropyl alcohol as the chemical ionization (CI) reagent gas for the analysis of optical isomers of ibuprofen derivatives from equine urine extract [348], (2) the direct liquid introduction–API (atmospheric pressure ionization interface) approach for the analysis of sulfa drugs (including sulfadimethoxine from racehorse urine) [349], (3) moving-belt interface analysis of bromazepam, clopenthixol, and reserpine overdose [350], (4) enantiomeric composition analysis of amphetamine and methamphetamine derivatives [351], and (5) thermospray analysis of enantiomeric methamphetamine derivatives [352].

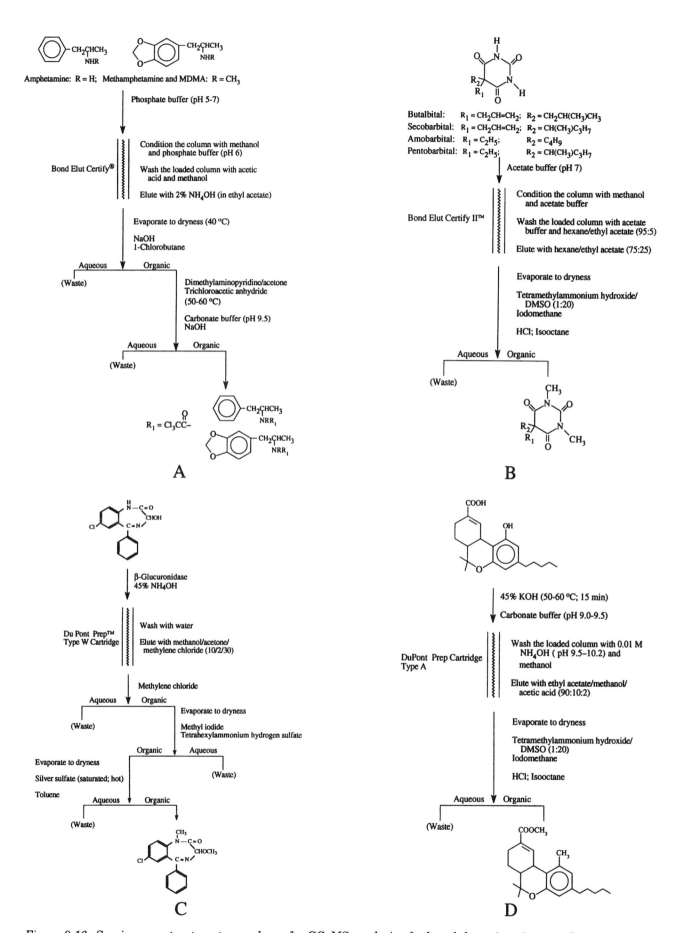

Figure 9.16. Specimen pretreatment procedures for GC–MS analysis of selected drugs in urine: amphetamines (A), barbiturates (B), benzodiazepines (C), and cannabis metabolite (D). (Reprinted with permission from Ref. [344].) Continued on next page.

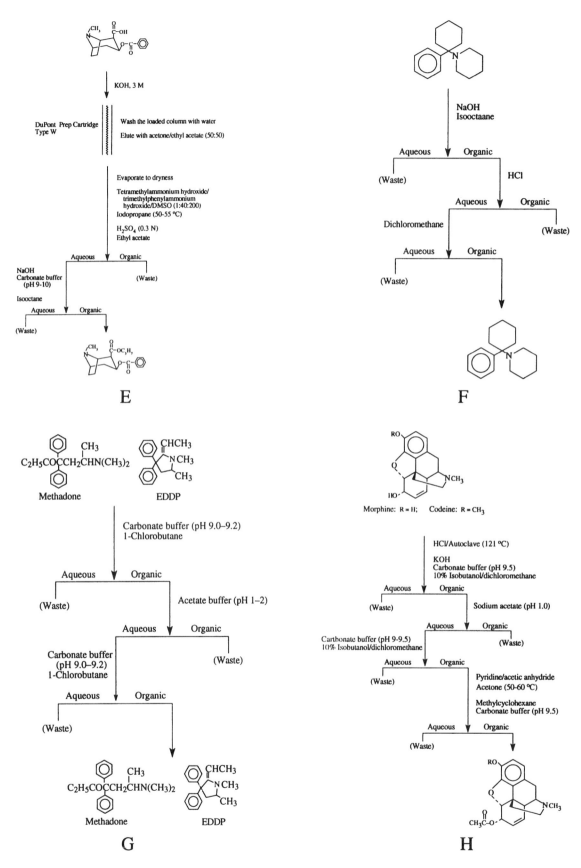

Figure 9.16. Continued. *Specimen pretreatment procedures for GC–MS analysis of cocaine metabolite (E), phencyclidine (F), methadone (G), and opiates (H). (Reprinted with permission from Ref. [344].)*

Table 9.13. Important parameters for SIM GC–MS analysis of targeted drugs.[a]

Targeted drug category and drug	Recon. solvent (approx vol)	GC retention time range (min)	Oven temp range (°C)	Ions monitored (m/e)[b] Analyte; int std[c]	Ion ratios monitored	Cutoff (ng/mL)
Amphetamines	Ethyl acetate					
Amphetamine	(40–100 µL)	2.5 ± 0.2	150–200	118,188,190; 123,194	118/190, 188/190; 123/194	500
Methamphetamine		3.1 ± 0.4		91,202,204; 209,211	91/204, 202/204; 211/209	500
MDMA[d]		5.2 ± 0.4		204,162,202; 164,208	204/162, 202/162; 208/164	500
l-Amphetamine	Ethyl acetate	8.6 ± 0.2	100–250	237; 241		500
d-Amphetamine	(10 µL)	8.7 ± 0.2	100–250	237; 241		
l-Methamphetamine		9.6 ± 0.2	100–250	251; 258		
d-Methamphetamine		9.7 ± 0.2	100–250	251; 258		
Barbiturates	Ethyl acetate					
Butalbital	(20 µL)	2.0 ± 0.2	140–190	181, 195, 196; 171, 189	181/196, 195/196; 171/189	200
Amobarbital		2.4 ± 0.2		169, 184, 185; 171, 189	184/169, 185/169; 171/189	200
Pentobarbital		2.7 ± 0.2		169, 184, 185; 171, 189	184/169, 185/169; 171/189	200
Secobarbital		3.1 ± 0.2		181, 195, 196; 171, 189	153/168, 195/168; 171/189	200
Cannabis	Cyclohexane					
9-THC-COOH	(20 µL)	4.1 ± 0.2	220–270	313, 357, 372; 360, 375	313/372, 357/372; 360/375	15
Cocaine	Cyclohexane					
Benzoylecgonine	(20 µL)	4.1 ± 0.2	190–240	210, 226, 331; 213, 334	210/331, 226/331; 213/334	100
Opiates	Ethyl acetate					
Codeine	(20 µL)	3.1 ± 0.2	200–270	229, 282, 341; 232, 344	229/341, 282/341; 232/344	300
Morphine		4.0 ± 0.3		310, 327, 369; 330, 372	310/369, 327/369; 372/330	300
Phencyclidine	Methanol (20 µL)	3.7 ± 0.2	170–210	186, 200, 243; 205, 248	186/243, 200/243; 205/248	10
Oxazepam	Cyclohexane (20 µL)	3.5 ± 0.2		255, 271, 314; 276, 319	255/271, 314/271; 276/319	300
Methadone	Ethyl Acetate (20 µL)	2.8 ± 0.2		223, 294, 295; 226, 297	223/294, 295/294; 226/297	300

[a] Data are taken from Ref. [344].

[b] Underlined ions are used for quantification.

[c] 1-Phenyl-2-aminopropane-1,2,3,3,3-d_5, 1-phenyl-2-(methyl-d_3-amino)propane-1,1,2,3,3,3-d_6, 1-(3,4-methylenedioxy)phenyl-2-(methylamino)propane-1,2,3,3,3-d_5, pentobarbital-5-ethyl-d_5, 11-nor-Δ^9-tetrahydrocannabinol-9-carboxylic acid-5′-d_3, benzoylecgonine-N-methyl-d_3, codeine-N-methyl-d_3, morphine-N-methyl-d_3, phencyclidine-phenyl-d_5, oxazepam-phenyl-d_5, and methadone-1,1,1-d_3 are used as the internal standards for respective analytes.

[d] Abbreviation: MDMA: methylenedioxymethamphetamine.

To fully utilize the potentials that cannot be offered by GC, HPLC is used to analyze drug metabolites in conjugate forms, especially as glucuronides, as they are excreted, thus alleviating the need for enzymatic or chemical hydrolysis and often the subsequent derivatization needed for GC analysis. Direct HPLC analysis of intact glucuronides [353] and polar metabolites [354] has been studied and considered a promising approach [355]. It is thus very important for the MS detector to identify these analytes. The ability to preserve the intact molecules through the process of nebulization, transfer, and other processes occurring in the interface will greatly enhance the merits of the MS detector. Thus, the identification [356] of intact glutathione (the most abundant intracellular nucleophile) adducts of chlorambucil, a cancer chemotherapeutic agent, using a thermospray interface was an example of the successful application of this technique. Unfortunately, metabolite adducts cannot always survive the interface device. For example, in the case of testosterone glucuronide, the glucuronide moiety was reported [347] to be lost during the nebulization, transfer, and ionization processes occurring in the Particle Beam Separator (Hewlett-Packard: Palo Alto, CA).

Before the HPLC–MS technique can be used on a routine basis, some sort of standardization of the ionization–fragmentation has to be established, so that standard spectral data systems can be universally used. (Undoubtedly, the 70-eV EI spectra contributed greatly to the wide application of GC–MS technology.) Although particle-beam separator and filament-on thermospray interfaces provide significant fragmentations that may be useful for spectrum-comparison purposes, much more study still needs to be conducted. Because the cost of an MS detector is substantial, average laboratories will not make a major investment of this sort before its use in routine applications is assured.

It is probably fair to conclude that HPLC still does not have an effective, practical, and universal detector equivalent to flame ionization and interfaced MS in GC technology. In the meantime, the combined use of various detectors has been explored [357].

Table 9.14. Characteristics of various HPLC–MS interfaces.[a]

Interface	Principle of operation	Volume reduction	Analyte polarity requirement	Heat decomposition	Ionization[b]
Direct liquid introduction	Liquid jet expansion through 0.5-μm orifice	Evaporation in ion source; flow rate 5–50 μL/min	Moderate	Minimal	CI
Moving belt	Deposition of analyte on belt	Evaporation of liquid from belt in oven; flow rate 0.2–0.5 mL/min for reversed phase	Low	Yes	EI or CI
Particle beam	Pneumatic or thermally assisted nebulization of the eluent into a desolvation chamber to produce small analyte-enriched particles	Differentially pumped beam separator; flow rate 0.5–1.0 mL/min	Moderate	Minimal	EI or CI
Thermospray	Expansion of eluent in heated capillary into vacuum resulting in charge-retaining droplets	Small region of solvent plume is sampled and evaporated continuously until ions are ejected from droplets; flow rate 0.5–1.5 mL/min	High	Minimal	CI
Ion spray	Ion production in high-voltage-charged capillary in combination with coaxial N_2 flow for pneumatic nebulization	API technique[c]; flow rate 1–100 μL/min (up to 0.8 mL/min with thermal assist)	High	Minimal	"FI"

[a] Data are taken from Refs. [345–348].

[b] Abbreviations: EI: electron impact; CI: chemical ionization; FI: field desorption; whether the ionization mechanism is indeed FI is not certain.

[c] Abbreviation: API: atmospheric pressure ionization. The key feature of the API technique is based on the curtain of N_2 gas, which prevents un-ionized interferents (such as solvent vapors) from entering the mass spectrometer and also disrupts clusters of solvent and analyte. Ionic species produced in the area in front of the orifice are attracted into the mass spectrometer by potential difference.

References

1. *Forensic Applications of Mass Spectrometry;* Yinon, J., Ed.; CRC Press: Boca Raton, FL, 1995.
2. *Forensic Mass Spectrometry;* Yinon, J., Ed.; CRC Press: Boca Raton, FL, 1987.
3. Papa, V. M.; Ou, D. W; Bederka, J. P. In *Recent Developments in Mass Spectrometry in Biochemistry, Medicine and Environmental Research;* Frigerio A. Ed.; Elsevier: Amsterdam, Netherlands, 1983; p 87.
4. Saferstein, R. In *Handbook of Forensic Sciences;* Saferstein, R., Ed.; Prentice Hall: Englewood Cliffs, NJ, 1982; p 91.
5. Milne, G. W. A.; Fales, H. M.; Law, N. C. In *Instrumental Applications in Forensic Drug Chemistry;* Klein, M.; Kruegel, A. V.; Sobol, S. P., Eds.; U. S. Government Printing Office: Washington, DC, 1987; p 91.
6. Klein, M. In *Instrumental Applications in Forensic Drug Chemistry;* Klein, M.; Kruegel, A. V.; Sobol, S. P., Eds.; U. S. Government Printing Office: Washington, DC, 1978; p 14.
7. Law, N. C.; Aandahl, V.; Fales, H. M.; Milne, G. W. A. *Clin. Chim. Acta* **1971**, *32,* 221.
8. Yost, R. A.; Fetterolf, D. D. In *Analytical Methods in Forensic Chemistry;* Ho, M. H., Ed.; Ellis Horwood: New York, 1990; p 1.
9. Messa, G. *Practical Aspects of Gas Chromatography/ Mass Spectrometry;* Wiley: New York, 1984.
10. Lai, S. T. F. *Gas Chromatography/Mass Spectrometry Operation;* Realistic Systems: East Longmeadow, MA, 1989.
11. Busch, L.; Glish, G. L.; McLuckey, S. A. *Mass Spectrometry/Mass Spectrometry;* VCH: New York, 1989.
12. Harrison, A. G. *Chemical Ionization Mass Spectrometry;* CRC Press: Boca Raton, FL; 1983.
13. *Tandem Mass Spectrometry;* McLafferty, F. W., Ed.; Wiley: New York, 1983.
14. Rose, M. E.; Johnsto, R. A. W. *Mass Spectrometry for Chemists and Biochemists*; Cambridge University: Cambridge, U.K., 1982.

15. Beynon, J. H.; Brenton, A. G. *An Introduction to Mass Spectrometry;* University of Wales: Cardiff, U.K., 1982.
16. Morris, H. R. *Soft Ionization Biological Mass Spectrometry;* Wiley: New York, 1981.
17. McLafferty, F. W.; Turecek, F. *Interpretation of Mass Spectra,* 4th ed.; University Science Book: Mill Valley, CA, 1993.
18. Millard, B. J. *Quantitative Mass Spectrometry;* Heyden: London, 1978.
19. Gudzinowi, B. J.; Gudzinow, M. J.; Martin, H. F. *Fundamentals of Integrated GC-MS;* Marcel Dekker: New York, 1977.
20. Gudzinowicz, B. J.; Gudzinowicz, M. J.; Martin, H. F. *Fundamentals of Integrated GC-MS;* Marcel Dekker: New York, 1976.
21. Merritt, C. J.; McEwen, C. N., Eds.; *Mass Spectrometry;* Marcel Dekker: New York, 1979.
22. Watson, J. T. *Introduction to Mass Spectrometry: Biomedical, Environmental, and Forensic Applications;* Raven: New York, 1976.
23. Fetterolf, D. D. In *Proceedings of the International Symposium on the Forensic Aspects of Controlled Substances;* FBI Academy: Quantico, VA, 1988; p 67.
24. Yost, R. A.; Fetterolf, D. D. In *Analytical Methods in Forensic Chemistry;* Ho, M. H., Ed.; Ellis Horwood: New York, 1990; p 1.
25. Gloger, G. In *Analytical Methods in Forensic Chemistry;* Ho, M. H., Ed.; Ellis Horwood: New York, 1990; p 29.
26. Liu, R. J. H. In *Analytical Methods in Forensic Chemistry;* Ho, M. H., Ed.; Ellis Horwood: New York, 1990; p 40.
27. Elledge, B. W.; Charpentier, B. A. In *Analytical Methods in Forensic Chemistry;* Ho, M. H., Ed.; Ellis Horwood: New York, NY, 1990; p 52.
28. Wolfe, R. R. *Tracers in Metobalic Research: Radioactive and Stable Isotope/Mass Spectrometry Methods;* Alan R. Liss: New York, 1983.
29. *Applications of Mass Spectrometry to Trace Analysis;* Facchetti, S., Ed.; Elsevier: Amsterdam, Netherlands, 1982.
30. Waller, G. R.; Dermer, O. C. *Biochemical Applications of Mass Spectrometry;* Wiley: New York, 1980.
31. *The Eight Peak Index of Mass Spectra,* 3rd ed.; Royal Society of Chemistry: London, 1983.
32. Ardrey, R. E.; Brown, C.; Allen, A. R.; Bal, T. S.; Moffat, A. C., Eds.; *An Eight Peak Index of Mass Spectra of Compounds of Forensic Interest;* Scottish Academic: London, U.K., 1983.
33. *EPA/NIH Mass Spectra Data Base;* Heller, S. R.; Milne, G. W. A., Eds.; U. S. Government Printing Office: Washington, DC, 1978.
34. Pfleger, K.; Maurer, H.; Weber, A. *Mass Spectral and GC Data of Drugs, Poisons, and Their Metabolites,* 2nd ed.; VCH Publishers: New York, 1991.
35. *Mass Spectra Correlations,* 2nd ed.; McLafferty, F. W.; Venkataraghvan, R., Eds.; American Chemical Society: Washington, DC, 1982.
36. Mills, T.; Price, W. M.; Price, P. T.; Roberson, J. C., Eds.; *Instrumental Data for Drug Analysis,* 2nd ed.; Elsevier: New York, Vol. 1, 1987.
37. *Instrumental Data for Drug Analysis,* 2nd ed.; Mills, T.; Price, W. M.; Roberson, J. C., Eds.; Elsevier: New York, Vol. 2, 1982.
38. Mills, T., III; Roberson, J. C. *Instrumental Data for Drug Analysis,* Elsevier: New York, Vols. 1–4, 1987.
39. Mills, T., III; Roberson, J. C.; McCurdy, H. H.; Wall, W. H. *Instrumental Data for Drug Analysis;* Elsevier: New York, Vol. 5, 1991.
40. *CRC Handbook of Mass Spectra of Drugs;* Sunshine, I., Ed.; CRC Press: Boca Raton, FL, 1981.
41. Cotter, R. J.; Tabet, J.-C. *Am. Lab. (Shelton, Conn.)* **1984,** *16,* 10.
42. Hercules, D. M. *Analytical Pyrolysis: Techniques and Applications;* Voorhees, K. J., Ed.; Butterworths: London, 1984; p 1.
43. Nilsson, C.-A.; Norstrom, A.; Andersson, K. *J. Chromatogr.* **1972,** *73,* 270.
44. Hansen, G.; Munson, B. *Anal. Chem.* **1980,** *52,* 245.
45. Cotter, R. J. *Anal. Chem.* **1980,** *52,* 1589.
46. Henion, J. D. *Hewlett-Packard GC/MS Application Note;* Hewlett-Packard: Palo Alto, CA, 1979; AN 176-31.
47. Reinhold, V. N.; Carr, S. A. *Anal. Chem.* **1982,** *54,* 499.
48. Davis, D. V.; Cook, R. G. *J. Agric. Food Chem.* **1982,** *30,* 495.
49. Cooks, R. G. *Am. Lab. (Shelton, Conn.)* **1978,** *10,* 111.
50. Kondrat, R. W.; Cooks, R. G. *Anal. Chem.* **1978,** *50,* 81.
51. Pardanani, J. H.; McLaughlin, J. L.; Kondrat, R. W.; Cooks, R. G. *Lloydia* **1977,** *40,* 585.
52. Kruger, T. L.; Cooks, R. G.; McLaughlin, J. L.; Ranieri, R. L. *J. Org. Chem.* **1977,** *42,* 4161.
53. McClusky, G. A.; Cooks, R. G.; Knevel, A. M. *Tetrahedron Lett.* **1978,** *46,* 4471.
54. Kondrat, R. W.; Cooks, R. G. *Science (Washington, D.C.)* **1978,** *199,* 978.
55. Youssefi, M.; Cooks, R. G.; McLaughlin, J. L. *J. Am. Chem. Soc.* **1979,** *101,* 3400.

56. Kondrat, R. W.; McClusky, G. A.; Cooks, R. G. *Anal. Chem.* **1978,** *50,* 2017.
57. Soltero-Rigau, E.; Kruger, T. L.; Cooks, R. G. *Anal. Chem.* **1977,** *49,* 435.
58. Fenselau, C. *Anal. Chem.* **1977,** *49,* 563.
59. Gudzinowicz, B. J.; Gudzinowicz, M. J.; Martin, H. F. *Fundamentals of Integrated GC-MS;* Marcel Dekker: New York, 1976; p 55.
60. Jensen, T. E.; Kaminsky, R.; McVeety, B. D.; Wozniak, T. J.; Hites, R. A. *Anal. Chem.* **1982,** *54,* 2388.
61. Smith, R. M. *Am. Lab. (Shelton, Conn.)* **1978,** *10,* 53.
62. Kirchgessner, W. G.; DiPasqua, A. C.; Anderson, W. A.; Delaney, G. V. *J. Forensic Sci.* **1974,** *19,* 313.
63. Sullivan, R. C.; Dugan, S.; Fava, J.; McDonnel, E.; Yarchak, M. *J. Police Sci. Adm.* **1974,** *2,* 185.
64. Callieux, A.; Allain, P. *J. Anal. Toxicol.* **1979,** *3,* 39.
65. Michnowicz, J. A. *Methane: Its Usefulness as Both Carrier and Reagent Gs in Chemical Ionization GC/ MS;* Hewlett-Packard Publication Note, Hewlett-Packard: Palo Alto, CA, 1978, AN 176-10.
66. Hatch, F.; Munson, B. *Anal. Chem.* **1977,** *49,* 169.
67. Hatch, F.; Munson, B. *Anal. Chem.* **1977,** *49,* 731.
68. Munson, B. *Anal. Chem.* **1977,** *49,* 773.
69. Games, D. E.; McDowall, M. A.; Levsen, K.; Schafer, K. H.; Dobberstein, P.; Gower, J. L. *Biomed. Mass Spectrom.* **1984,** *11,* 87.
70. Covye, T. R.; Crowther, J. B.; Dewey, E. A.; Henion, C. D. *Anal. Chem.* **1985,** *57,* 474.
71. Stout, S. J.; daCunha, A. R. *Anal. Chem.* **1985,** *57,* 1783.
72. Hayes, M. J.; Lankmayer, E. P.; Vouros, P.; Karger, B. L.; McGuire, J. M. *Anal. Chem.* **1983,** *55,* 1745.
73. Arpino, P. J. In *Instrumental Applications in Forensic Drug Chemistry;* Klein, M.; Kruegel, A. V.; Sobol, S. P., Eds.; U.S. Government Printing Office: Washington, DC, 1978; p 151.
74. Guiochon, G.; Arpino, P. C. *Anal. Chem.* **1979,** *51,* 682.
75. McFadden, W. H. *J. Chromatogr. Sci.* **1980,** *18,* 97.
76. *LC/MS Interface Methods;* Games, D. E., Ed.; Spectra: San Jose, CA, 1983; 8.
77. Games, D. E. *Biomed. Mass Spectrom.* **1981,** *8,* 454.
78. Kenyon, C. N.; Melera, A.; Erni, F. *J. Anal. Toxicol.* **1981,** *5,* 216.
79. Henion, J. D.; Thomson, B. A.; Dawson, P. H. *Anal. Chem.* **1982,** *54,* 451.
80. Games, D. E.; Alcock, N. J.; Cobelli, L.; Eckers, C.; Games, M. P. L.; Jones, A.; Lant, M. S.; mcDowall, M. A.; Rossiter, M.; Smith, R. A.; Westwood, S. A.; Wong, H. Y. *Int. J. Mass Spectrom. Ion Phys.* **1983,** *46,* 181.
81. Dobberstein, P.; Korte, E.; Meyerhoff, G.; Pesch, R. *Int. J. Mass Spectrom. Ion Phys.* **1983,** *46,* 185.
82. Cotter, R. J. *Analytical Pyrolysis: Techniques and Applications;* Voorhees, K. J., Ed.; Butterworths: London, 1984; p 42.
83. Gutteridge, C. S.; Sweatman, A. J.; Norris, J. R. In *Analytical Pyrolysis: Techniques and Applications;* Voorhees, K. J., Ed.; Butterworths: London, 1984; p 324.
84. Slack, J. A.; Irwin, W. J. *Proc. Anal. Div. Chem. Soc.* **1977,** *14,* 215.
85. Hughes, J. C.; Wheals, B. B.; Whitehouse, M. J. *Forensic Sci.* **1977,** *10,* 217.
86. Down, G. J.; Gwyn, S. A. *J. Chromatogr.* **1975,** *103,* 208.
87. Ramaley, L.; Nearing, M. E.; Vaughan, N. A.; Ackman, R. G.; Jamieson, W. D. *Anal. Chem.* **1983,** *55,* 2285.
88. Henion, J.; Maylin, G. A.; Thomson, B. A. *J. Chromatogr.* **1983,** *271,* 107.
89. *Ionization Methods;* Sphon, J. A.; Harvey, T. M., Eds.; Spectra: San Jose, CA, 1982; 8.
90. Beggs, D. P.; Day, A. G., III *J. Forensic Sci.* **1974,** *19,* 891.
91. Unger, S. E.; A. Vincze, A.; Cooks, R. G.; Chrisman, R.; Rothman, L. D. *Anal. Chem.* **1981,** *53,* 976.
92. Horning, E. C.; Horning, M. G.; Carroll, D. I.; Dzidic, I.; Stillwell, R. N. *Anal. Chem.* **1973,** *45,* 936.
93. Foltz, R. L.; Fentiman, A. F., Jr.; Foltz, R. B. *GC/MS Assays for Abused Drugs;* U.S. Government Printing Office: Washington, DC, 1980.
94. Jardine, I.; Fenselau, C. *J. Forensic Sci.* **1975,** *20,* 373.
95. Jardine, I.; Fenselau, C. *Anal. Chem.* **1975,** *47,* 730.
96. Foltz, R. L. In *Quantitative Mass Spectrometry in Life Sciences II;* De Leenheer, A. P.; Foncucci, F. F.; van Peteghem, C., Eds.; Elsevier: Amsterdam, Netherlands, 1978; p 39.
97. Brandenberger, H. In *Instrumental Applications in Forensic Drug Chemistry;* Klein, M.; Kruegel, A. V.; Sobol, S. P., Eds.; U. S. Government Printing Office: Washington, DC, 1978; p 48.
98. Fales, H. M.; Milne, G. W. A.; Winkler, H. U.; Beckey, H. D.; Damico, U. N.; Barron, R. *Anal. Chem.* **1975,** *47,* 207–218.
99. Milne, G. W. A.; Fales, H. M.; Axenrod, T. *Anal. Chem.* **1971,** *43,* 1815.
100. Hunt, D. F.; Crow, F. W. *Anal. Chem.* **1978,** *50,* 1781.
101. Bose, A. K.; Fujiwara, H.; Pramanik, B. N., Lazaro, E.; Spillte, C. R. *Anal. Biochem.* **1978,** *89,* 284.
102. Caldwell, G.; Bartmess, J. E. *Org. Mass Spectrom.* **1982,** *17,* 456.
103. Madhusdanan, K. P. *Org. Mass Spectrom.* **1984,** *19,* 517.

104. McLafferty, F. W. In *Frontiers of Analytical Techniques and Their Application;* Lunsford, C. G., Ed.; Philip Morris: New York, 1982; p 113.
105. Ghaderl, S.; Kulkarnl, P. S.; Ledford, E. B., Jr.; Wilkins, C. L.; Gross, M. L. *Anal. Chem.* **1981,** *53,* 428.
106. Yost, R. A.; Perchalski, R. J.; Brotherton, H. O.; Johnson, J. V.; Budd, M. B. *Talanta* **1984,** *31,* 929.
107. Brotherton, H. O.; Yost, R. A. *Am. J. Vet. Res.* **1984,** *45,* 2436.
108. Perchalski, R. J.; Yost, R. A.; Wilder, B. J. *Anal. Chem.* **1982,** *54,* 1466.
109. Brotherton, H. O.; Yost, R. A. *Anal. Chem.* **1983,** *55,* 549.
110. Cheng, M. T.; Kruppa, G. H.; McLafferty, F. W.; Cooper, D. A. *Anal. Chem.* **1982,** *54,* 2204.
111. Stafford, G. C., Jr.; Kelley, P. E.; Syka, J. E. P.; Reynolds, W. E.; Todd, J. F. J. *Int. J. Mass Spectrom. Ion Proc.* **1984,** *60,* 85.
112. *The Ion Trap Newsletter* **1985,** *1,* 1.
113. *Ion Trap Detector Application Data Sheet;* Finnigan MAT: San Jose, CA, 1985.
114. *ITS40™ Application Data Sheet;* Finnigan MAT: San Jose, CA, 1989.
115. Chapman, J. R. *Computers in Mass Spectrometry;* Academic: London, 1978.
116. Finkle, B. S.; Taylor, D. M.; Bonelli, E. J. *J. Chromatogr. Sci.* **1972,** *10,* 312.
117. Finkle, B. S.; Foltz, R. L.; Taylor, D. M. *J. Chromatogr. Sci.* **1974,** *12,* 304.
118. *Wiley/NBS Mass Spectral Database;* Wiley: New York, 1984.
119. McLafferty, F. W.; Stauffer, D. B. *J. Chem. Inf. Comput. Sci.* **1985,** *25,* 245.
120. Stauffer, D. B.; McLafferty, F. W.; Ellis, R. D.; Peterson, D. W. *Anal. Chem.* **1985,** *57,* 1056.
121. Sparkman, O. D. *J. Am. Soc. Mass Spectrom.* **1996,** *7,* 313.
122. Christman, R. F. *Environ. Sci. Technol.* **1982,** *16,* 143.
123. Lebedev, K. S.; Tormyshev, V. M.; Derendy, B. G.; Koptyug, V. A. *Anal. Chim. Acta* **1981,** *133,* 517.
124. Siegel, M. M.; Bauman, N. *Abstracts of Papers,* 33rd Annual Conference on Mass Spectrometry and Allied Topics, San Diego, CA; American Society of Mass Spectrometry: East Lansing, MI, 1985; p 433.
125. *Target Compound Analysis Software,* Thru-Put Systems: Orlando, FL, 1994
126. Liu, J. R. H.; Fitzgerald, M. P.; Smith, G. V. *Anal. Chem.* **1979,** *51,* 1875.
127. Flory, D. A.; Lichtenstein, H. A.; Biemann, K.; Biller, J. E.; Barker, C. *Technology (Sindri, India)* **1983,** *3,* 91.
128. Harper, A. M.; Meuzalaar, H. L. C.; Metcalf, G. S.; Pope, D. L. In *Analytical Pyrolysis – Techniques and Applications;* Voorhees, K. J., Ed.; Butterworths: London, 1984; p 157.
129. Liu, J. R. H.; Fitzgerald, M. P. *J. Forensic Sci.* **1980,** *25,* 815.
130. Liu, J. R. H.; Ramesh, S.; Liu, J. Y.; Kim, S. *Anal. Chem.* **1984,** *56,* 1808.
131. Abe, H.; Kumazawa, S.; Taji, T.; Sasaki, S.-I. *Biomed. Mass Spectrom.* **1976,** *3,* 151.
132. Frank, R. S. *J. Forensic Sci.* **1983,** *28,* 18.
133. Shulgin, A. T.; Sargent, C.; Naranjo, C. *Nature (London)* **1969,** *221,* 537.
134. Poklis, A.; Mackell, M. A.; Drake, W. K. *J. Forensic Sci.* **1979,** *24,* 70.
135. Clark, C. C. *J. Forensic Sci.* **1984,** *29,* 1056.
136. Soine, W. H.; Shark, R. E.; Agee, D. T. *J. Forensic Sci.* **1983,** *28,* 386.
137. Bailey, K.; By, A. W.; Legault, D.; Verner, D. *J. Assoc. Off. Anal. Chem.* **1975,** *58,* 62.
138. Cowie, J. S.; Holtham, A. L.; Jones, L. V. *J. Forensic Sci.* **1982,** *27,* 527.
139. Dal Cason, T. A.; Angelos, S. A.; Raney, J. K. *J. Forensic Sci.* **1984,** *29,* 1187.
140. Soine, W. H.; Thomas, M. N.; Shark, R. E.; Jane, S.; Agee, D. T. *J. Forensic Sci.* **1984,** *29,* 177.
141. Repke, D. B.; Leslie, T.; Mandell, D. M.; Kish, N. G. *J. Pharm. Sci.* **1977,** *66,* 743.
142. White, P. C. *J. Chromatogr.* **1979,** *169,* 453.
143. Casale, J. F. *J. Forensic Sci.* **1985,** *30,* 247.
144. Liu, R. H.; Ku, W. W.; Fitzgerald, M. P. *J. Assoc. Off. Anal. Chem.* **1983,** *66,* 1443.
145. Frank, H.; Nicholson, G. J.; Bayer, E. *J. Chromatogr.* **1978,** *146,* 197.
146. Gilbert, M. T.; Brooks, C. J. W. *Biomed. Mass Spectrom.* **1974,** *4,* 226.
147. Bailey, K. *Anal. Chim. Acta* **1972,** *60,* 287.
148. Coutts, R. T.; Jones, G. R.; Benderly, A.; Mak, A. L. C. *J. Chromatogr. Sci.* **1979,** *17,* 350.
149. Coutts, R. T.; Dawe, R.; Jones, G. R.; Liu, D. F.; Midha, K. K. *J. Chromatogr.* **1980,** *190,* 53.
150. Brettell, T. A. *J. Chromatogr.* **1983,** *257,* 45.
151. Reisch, J.; Pagnucco, R.; Alfes, H.; Jantos, N.; Hollmann, H. *J. Pharm. Pharmacol.* **1968,** *20,* 81.
152. Bellman, S. W. *J. Assoc. Off. Anal. Chem.* **1968,** *51,* 164.
153. Narasimhachari, N.; Spaide, J.; Heller, B. *J. Chromatogr. Sci.* **1971,** *9,* 502.
154. Marde, Y.; Ryhage, R. *Clin. Chem.* **1978,** *24,* 1720.
155. Suzuki, O.; Hattori, H.; Asano, M. *J. Forensic Sci.* **1984,** *29,* 611.
156. Kram, T. C.; Kruegel, A. V. *J. Forensic Sci.* **1977,** *22,* 40.
157. Low, I. A.; Piotrowski, E.; Furner, R. L.; Liu, R. J. H. *Biomed. Mass Spectrom.* **1986,** *13,* 531.
158. Liu, J. R. H.; Ku, W. W. *Anal. Chem.* **1981,** *53,* 2180.

159. Liu, J. R. H.; Ku, W. W.; Tsay, J. T.; Fitzgerald, M. P.; Kim, S. *J. Forensic Sci.* **1982,** *27,* 39.
160. Barron, R. P.; Kruegel, A. V.; Moore, J. M.; Kram, T. C. *J. Assoc. Off. Anal. Chem.* **1974,** *57,* 1147.
161. Verweij, A. M. A. *Forensic Sci. Rev.* **1989,** *1,* 1.
162. Lomonte, J. N.; Lowry, W. T.; Stone, I. C. *J. Forensic Sci.* **1976,** *21,* 575.
163. van der Ark, A. M.; Verweij, A. M. A.; Sinnema, A. *J. Forensic Sci.* **1978,** *23,* 693.
164. Kram, T. C. *J. Forensic Sci.* **1979,** *24,* 596.
165. Theeuwen, A. B. E.; Verweij, A. M. A. *Forensic Sci. Int.* **1980,** *15,* 237.
166. Audier, H.; Fetizon, M.; Ginsburg, D.; Mandelbaum, A.; Rull, T. *Tetrahedron Lett.* **1965,** 13.
167. Nakata, H.; Hirata, Y.; Tatematsu, A.; Toda, H.; Sawa, Y. K. *Tetrahedron Lett.* **1965,** 829.
168. Mandelbaum, A.; Ginsburg, D. *Tetrahedron Lett.* **1965,** *1965,* 2479.
169. Wheeler, D. M. S.; Kinstle, T. H.; Rinehart, K. L., Jr. *J. Am. Chem. Soc.* **1967,** *89,* 4494.
170. Sastry, S. D. In *Biochemical Applications of Mass Spectrometry;* Waller, G. R., Ed.; Wiley: New York, 1972.
171. Fales, H. M.; Lloyd, H. A.; Milne, G. W. A. *J. Am. Chem. Soc.* **1970,** *92,* 1590.
172. Saferstein, R.; Chao, J.-M. *J. Assoc. Off. Anal. Chem.* **1973,** *56,* 1234.
173. Saferstein, R.; Chao, J.-M.; Manura, J. *J. Forensic Sci.* **1974,** *19,* 463.
174. Chao, J.-M.; Saferstein, R.; Manura, J. *Anal. Chem.* **1974,** *46,* 296.
175. Zitrin, S.; Yinon, J. *Anal. Lett.* **1977,** *10,* 235.
176. Saferstein, R.; Manura, J.; Brettell, T. A. *J. Forensic Sci.* **1979,** *24,* 312.
177. Issachar, D.; Yinon, J. *Anal. Chem.* **1980,** *52,* 49.
178. Liu, R. J. H.; Low, I. A.; Smith, F. P.; Piotrowski, E. G.; Hsu, A.-F. *Org. Mass Spectrom.* **1985,** *20,* 511.
179. Lin, Y. Y.; Smith, L. L. *Biomed. Mass Spectrom.* **1979,** *6,* 15.
180. Smith, R. M. *J. Forensic Sci.* **1973,** *18,* 327.
181. Nakamura, G. R.; Noguchi, T. T.; Jackson, D.; Banks, D. *Anal. Chem.* **1972,** *44,* 408.
182. Moore, J. M.; Klein, M. *J. Chromatogr.* **1978,** *154,* 76.
183. Coutts, R. T.; Jones, G. R. *J. Forensic Sci.* **1979.** *24,* 291.
184. Nakamura, G. R. *J. Assoc. Off. Anal. Chem.* **1969,** *52,* 5.
185. Thornton, J. I.; Nakamura, G. R. *J. Forensic Sci. Soc.* **1972,** *12,* 461.
186. Wheals, B. B.; Smith, R. N. *J. Chromatogr.* **1975,** *105,* 396.
187. Lee, M. L.; Novotny, M.; Bartle, K. D. *Anal. Chem.* **1976,** *48,* 405.
188. Merli, F.; Wiesler, D.; Maskarinec, M. P.; Novotny, M.; Vassilaros, D. L.; Lee, M. L. *Anal. Chem.* **1981,** *53,* 1929.
189. Thibault, R.; Stall, W. J.; Master, R. G.; Gravier, R. R. *J. Forensic Sci.* **1983,** *28,* 15.
190. Heerma, W.; Terlouw, J. K.; Laven, A.; Dijkstra, G.; Kuppers, F. J. E. M.; Lousberg, J. J. C.; Salemink, C. In *Mass Spectrometry in Biochemistry and Medicine;* Frigerio, A.; Castagnoli, N., Jr., Eds.; Raven: New York, 1974; p 219.
191. Salemink, C. In *Marihuana: Chemistry, Biochemistry, Cellular Effects;* Nahas, G. G.; Paton, W. D. M.; Idanpaan-Heikkila, J. E., Eds.; Springer-Verlag: New York, 1976; p 31.
192. Kephalas, T. A.; Kiburis, J.; Michael, C. M.; Miras, C. J.; Papadakis, D. P. In *Marihuana: Chemistry, Biochemistry, Cellular Effects;* Nahas, G. G.; Paton, W. D. M.; Idanpaan-Heikkila, J. E., Eds.; Springer-Verlag: New York, 1976; p 39.
193. Liu, J. R. H.; Fitzgerald, M. P.; Smith, G. V. *Anal. Chem.* **1979,** *51,* 1875.
194. Liu, J. R. H.; Fitzgerald, M. P. *J. Forensic Sci.* **1980,** *25,* 815.
195. Nie, N. H.; Hull, C. H.; Jenkins, J. A.; Steinbrenner, K.; Bent, D. H. *Statistical Package for the Social Sciences,* 2nd ed; McGraw-Hill: New York, 1975; p 320.
196. Vree, T. B.; Breimer, D. D.; van Ginneken, A. M.; Rossum, J. M.; de Witte, R. *Clin. Chim. Acta* **1971,** *34,* 365.
197. Vree, T. B.; Breimer, D. D.; van Ginneken, C. A. M.; van Rossum, J. M. *J. Pharm. Pharmacol.* **1972,** *24,* 7.
198. Vree, T. B.; Breimer, D. D.; van Ginneken, C. A. M.; van Rossum, J. M. *J. Chromatogr.* **1972,** *74,* 124.
199. Harvey, D. J. *J. Pharm. Pharmacol.* **1976,** *28,* 280.
200. Brecht, C. A. L.; Lousberg, R. J. J.; Kuppers, E. J. E. M.; Salemink, C. A.; Vree, T. B.; van Rossum, J. M. *J. Chromatogr.* **1973,** *81,* 163.
201. Friedrich-Fiechtl, J.; Spiteller, G. *Tetrahedron* **1975,** *31,* 479.
202. Elsohly, M. A.; El-Feraly, F. S.; Turner, C. E. *Lloydia* **1977,** *40,* 275.
203. Budzikiewicz, H.; Alpin, R. T.; Lightner, D. A.; Djerassi, C.; Mechoulam, R.; Gaoni, Y. *Tetrahedron* **1965,** *21,* 1881.
204. Claussen, U.; Fehlhaber, H.-W.; Korte, F. Haschisch *Tetrahedron* **1966,** *22,* 3535.
205. Terlouw, J. K.; Heerma, W.; Burgers, P. C.; Dijkstra, G.; Boon, A.; Kramer, H. F.; Salemink, C. A. *Tetrahedron* **1974,** *30,* 4243.
206. Vree, T. B.; Nibbering, N. M. M. *Tetrahedron* **1973,** *29,* 3852.
207. Vree, T. B. *J. Pharm. Sci.* **1977,** *66,* 1444.

208. Turner, C. E.; Hadley, K. W.; Holley, J. H.; Billets, S.; Mole, M. L. *J. Pharm. Sci.* **1975,** *64,* 810.
209. Waller, C. W.; Hadley, K. W.; Turner, C. E. In *Marihuana: Chemistry, Biochemistry, Cellular Effects;* Nahas, G. G., Ed.; Springer-Verlag: New York, 1976.
210. Turner, C. E.; Bouwsma, O. J.; Billets, S.; Elsohly, M. A. *Biomed. Mass Spectrom.* **1980,** *7,* 247.
211. Foltz, R. L.; Fentiman, A. F., Jr.; Foltz, R. B. *GC/MS Assays for Abused Drugs in Body Fluids;* U. S. Government Printing Office: Washington, DC, 1980.
212. Liu, R. H. *Elements and Practice in Forensic Drug Urinalysis;* Central Police University Press: Taipei, Taiwan, 1994; p 107.
213. Costopanagiotis, A.; Budzikiewicz, H. *Monatsh. Chem.* **1965,** *96,* 1800.
214. Grutzmacher, H-F.; Arnold, W. *Tetrahedron Lett.* **1966,** *13,* 1365.
215. Coutts, R. T.; Lolock, R. A. *J. Pharm. Sci.* **1968,** *57,* 2096.
216. Coutts, R. T.; Locock, R. A. *J. Pharm. Sci.* **1969,** *58,* 775.
217. Gilbert, J. N. T.; Millard, B. J.; Powell, J. W. *J. Pharm. Pharmacol.* **1970,** *22,* 897.
218. Skinner, R. F.; Gallaher, E. G.; Predmore, D. B. *Anal. Chem.* **1973,** *45,* 574.
219. Thompson, R. M.; Desiderio, D. M. *Org. Mass Spectrom.* **1973,** *7,* 989.
220. Falkner, F. C.; Watson, J. T. *Org. Mass Spectrom.* **1974,** *8,* 257.
221. Harvey, D. J.; Nowlin, J.; Hickert, P.; Butler, C.; Cansow, O.; Horning, M. G. *Biomed. Mass Spectrom.* **1974,** *1,* 340.
222. van Langenhove, A.; Biller, J. E.; Biemann, K.; Browne, T. R. *Biomed. Mass Spectrom.* **1982,** *9,* 201.
223. Rautio, M.; Lounasmaa, M. *Acta Chem. Scand.* **1977,** *31,* 528.
224. Watson, J. T.; Falkner, F. C. *Org. Mass Spectrom.* **1973,** *7,* 1227.
225. Klein, M. In *Recent Developments in Mass Spectrometry in Biochemistry and Medicine;* Frigerio, A., Ed.; Plenum: New York, 1978; Vol. 1, p 471.
226. Fales, H. M.; Milne, G. W. A.; Axenrod, T. *Anal. Chem.* **1970,** *42,* 1432.
227. Hunt, D. F.; Ryan, J. F. III *Anal. Chem.* **1972,** *44,* 1306.
228. Hunt, D. F.; Stafford, G. C., Jr.; Crow, F. W.; Russell, J. W. *Anal. Chem.* **1978,** *48,* 2098.
229. Jones, L. V.; Whitehouse, M. J. *Biomed. Mass Spectrom.* **1981,** *8,* 231.
230. Liu, R. H.; McKeehan, A. M.; Edwards, C.; Foster, G.; Bensley, W. D.; Langner, J. G.; Walia, A. S. *J. Forensic Sci.* **1994,** *39,* 1504.
231. *State v. McNeal,* 288 N.W. 2d 874 (Wis. App. 1980).
232. Liu, J. H. *J. Forensic Sci.* **1981,** *26,* 651.
233. Lukaszewski, T.; Jeffery, W. K. *J. Forensic Sci.* **1980,** *25,* 499.
234. Moore, J. M. *J. Assoc. Off. Anal. Chem.* **1973,** *56,* 1199.
235. Cooper, D. A.; Allen, A. C. *J. Forensic Sci.* **1984,** *29,* 1045.
236. Siegel, J. A.; Cormier, R. A. *J. Forensic Sci.* **1980,** *25,* 357.
237. Allen, A. C.; Cooper, D. A.; Kiser, W. O.; Cottrell, R. C. *J. Forensic Sci.* **1981,** *26,* 12.
238. Findlay, S. P. *J. Am. Chem. Soc.* **1954,** *76,* 2855.
239. Moore, J. M. *J. Chromatogr.* **1975,** *101,* 215.
240. Jindal, S. P.; Lutz, T.; Vestergaard, P. *J. Chromatogr.* **1979,** *179,* 357.
241. Lewin, A. H.; Parker, S. R.; Carroll, F. I. *J. Chromatogr.* **1980,** *193,* 371.
242. Jindal, S. P.; Vestergard, P. *J. Pharm. Sci.* **1978,** *67,* 811.
243. Jindal, S. P.; Lutz, T.; Vestergaard, P. *Biomed. Mass Spectrom.* **1978,** *5,* 658.
244. Chinn, D. M.; Crouch, D. J.; Peat, M. A.; Finkle, B. S.; Jennison, T. A. *J. Anal. Toxicol.* **1980,** *4,* 37.
245. Foltz, R. L.; Fentiman, A. F., Jr.; Foltz, R. B. *GC/MS Assays for Abused Drugs in Body Fluids;* U.S. Governmental Printing Office: Washington, DC, 1980.
246. Clark, C. C. *J. Assoc. Off. Anal. Chem.* **1981,** *64,* 884.
247. Bailey, K.; Gagne, D. R.; Pike, R. K. *J. Assoc. Off. Anal. Chem.* **1976,** *59,* 81.
248. Cone, E. J.; Darwin, W. D.; Yousefnejad, D.; Buchwald, W. F. *J. Chromatogr.* **1979,** *117,* 149.
249. Smialek, J. E.; Monforte, J. R.; Gault, R.; Spitz, U. *J. Anal. Toxicol.* **1979,** *3,* 209.
250. Legault, D. *J. Chromatogr.* **1980,** *202,* 309.
251. Soine, W. H.; Balster, R. L.; Berglund, K. E.; Martin, C. D.; Agee, D. T. *J. Anal. Toxicol.* **1982,** *6,* 41.
252. Kelly, R. C.; Christmore, D. S. *J. Forensic Sci.* **1982,** *27,* 827.
253. Wong, L. K.; Biemann, K. *Clin. Toxicol.* **1976,** *9,* 583.
254. Lin, D. C. K.; Fentiman, A. F., Jr.; Foltz, R. L.; Forney, R. D., Jr.; Sunshine, I. *Biomed. Mass Spectrom.* **1975,** *2,* 206.
255. Cone, E. J.; Buchwald, W.; Yousefnejad, D. *J. Chromatogr.* **1981,** *223,* 331.
256. Foltz, R. L.; Fentiman, A. F.; Foltz, R. B. *GC/MS Assays for Abused Drugs in Body Fluids;* U.S. Government Printing Office: Washington, DC, 1980.
257. Inoue, A.; Nakahara, Y.; Niwaguchi, T. *Chem. Pharm. Bull.* **1972,** *20,* 409.

258. Bailey, K.; Verner, D.; Legault, D. *J. Assoc. Off. Anal. Chem.* **1973,** *56,* 88.

259. Schmidt, J.; Kraft, R.; Voigt, D. *Biomed. Mass Spectrom.* **1978,** *5,* 674.

260. Stoll, A.; Hofmann, A. *Helv. Chim. Acta* **1943,** *26,* 944.

261. Barber, M.; Weisbach, J. A.; Douglas, B.; Dudek, G. O. *Chem. Ind. (London)* **1965,** 1072.

262. Crawford, K. W. *J. Forensic Sci.* **1970,** *15,* 588.

263. Nakahara, Y.; Niwaguchi, T. *Chem. Pharm. Bull.* **1971,** *19,* 2337.

264. Vokoun, J.; Rehacek, Z. *Collect. Czech. Chem. Commun.* **1975,** *40,* 1731.

265. Urich, R. W.; Bowerman, D. L.; Wittenberg, P. H.; McGaha, B. L.; Schisler, D. K.; Anderson, J. A.; Levisky, J. A.; Pflug, J. L. *Anal. Chem.* **1975,** *47,* 581.

266. Kram, T. C.; Cooper, D. A.; Allen, A. C. *Anal. Chem.* **1981,** *53,* 1379.

267. Stinson, S. *Chem. Eng. News* **1981,** *59,* 71.

268. Van Bever, Q. F. M.; Niemegeers, C. J. E.; Janssen, P. A. J. *J. Med. Chem.* **1974,** *17,* 1047.

269. Gardocki, J. F.; Yelnosky, J. *Toxicol. Appl. Pharmacol.* **1964,** *6,* 48.

270. Frincke, J. M.; Henderson, G. L. *Drug Metab. Dispos.* **1980,** *8,* 425.

271. Cheng, M. T.; Kruppa, G. H.; McLafferty, F. W.; Cooper, D. A. *Anal. Chem.* **1982,** *54,* 2204.

272. Dal Cason, T. A.; Angelos, S. A.; Washington, O. *J. Forensic Sci.* **1981,** *26,* 793.

273. Bonnichsen, R.; Fri, C. G.; Negoita, C.; Ryhage, R. *Clin. Chim. Acta.* **1972,** *40,* 309.

274. Langner, J. G.; Gan, B. K.; Liu, R. H.; Baugh, L. D.; Chand, P.; Weng, J.-L.; Edwards, C.; Walia, A. S. *Clin. Chem.* **1991,** *37,* 1595.

275. Seno, H.; Suzuki, O.; Kumazawa, T.; Hattori, H. *J. Anal. Toxicol.* **1991,** *15,* 21.

276. Vessman, J.; Johansson, M.; Magnusson, P.; Strömberg, S. *Anal. Chem.* **1977,** *49,* 1545.

277. Mulé, S. J.; Casella, G. A. *J. Anal. Toxicol.* **1989,** *13,* 179.

278. Moffat, A. C., Ed.; *Clarke's Isolation and Identification of Drugs,* 2nd ed.; Pharmaceutical: London, 1986; p 742.

279. Baugh, L. D.; Liu, R. H.; Walia, A. S. *J. Forensic Sci.* **1991,** *36,* 549.

280. Ghaderl, S.; Kulkarnl, P. S.; Ledford, E. B., Jr.; Wilkins, C. L.; Gross, M. L. *Anal. Chem.* **1981,** *53,* 428.

281. Yost, R. A.; Perchalski, R. J.; Brotherton, H. O.; Johnson, J. V.; Budd, M. B. *Talanta* **1984,** *31,* 929.

282. Brotherton, H. O.; Yost, R. A. *Am. J. Vet. Res.* **1984,** *45,* 2436.

283. Perchalaski, R. J.; Yost, R. A.; Wilder, B. J. *Anal. Chem.* **1982,** *54,* 1466.

284. Brotherton, H. O.; Yost, R. A. *Anal. Chem.* **1983,** *55,* 549.

285. Verweij, A. M. A.; Lipman, P. J. L.; Zweipfenning, G. M. *Forensic Sci. Int.* **1992,** *54,* 67.

286. Nelson, C. C.; Foltz, R. L. *Anal. Chem.* **1992,** *64,* 1578.

287. Kidwell, D. A. *J. Forensic Sci.* **1993,** *38,* 272.

288. Hunt, D. F.; Crow, F. W. *Anal. Chem.* **1978,** *13,* 1781.

289. Karlsson, L.; Jonsson, J.; Aberg, K.; Roos, C. *J. Anal. Toxicol.* **1983,** *7,*198.

290. Liu, R. J. H. In *Advances in Forensic Sciences;* Lee, H. C.; Gaensslen, R. E., Eds.; Year Book Medical: Chicago, IL, 1989; p 75.

291. Moore, J. M. In *Instrumental Applications in Forensic Drug Chemistry;* Klein, M.; Kruegel, A. V.; Sobol, S. P., Eds.; U. S. Governemnt Printing Office: Washington, DC, 1978; p 180.

292. Watson, J. T.; Falker, F. C. *Org. Mass Spectrom.* **1973,** *7,* 1227.

293. *Quantitative Mass Spectrometry in Life Science;* De Ieenheer, A. P.; Roncucci, R. R., Eds.; Elsevier: New York, 1977.

294. *Quantitative Mass Spectrometry in Life Science II;* De Ieenheer, A. P.; Roncucci, R. R., Eds.; Elsevier: New York, 1978.

295. Lehmann, W. D.; Schulten, H. R. *Angew. Chem. Int. Ed. Engl.* **1978,** *17,* 221.

296. Garland, W. A.; Powell, M. L. *J. Chromatogr. Sci.* **1981,** *19,* 392.

297. Brunk, S. D. *J. Anal. Toxicol.* **1988,** *12,* 290.

298. Wu, A. H.; Ostheimer, D.; Cremese, M.; Forte, E.; Hill, D. *Clin. Chem.* **1994,** *40,* 216.

299. Liu, R. H.; Baugh, L. D.; Allen, E. E.; Salud, S. C.; Fentress, J. G.; Ghadha, H.; Walia, A. S. *J. Forensic Sci.* **1989,** *34,* 986.

300. Low, I. A.; Liu, R. H.; Barker, S. A.; Fish, F.; Settine, R. L.; Piotrowski, E. G.; Damert, W. C.; Liu, J. Y. *Biomed. Mass Spectrom.* **1985,** *11,* 638.

301. Benz, W. *Anal. Chem.* **1980,** *52,* 248.

302. Derrick, P. J. *Mass Spectrom. Rev.* **1983,** *2,* 285.

303. Derrick, P. J. In *Current Topics in Mass Spectrometry and Chemical Kinetics;* Beynon, J. H.; McGlashan, M. L., Eds.; Heyden: London, 1982; p 61.

304. Pickup, J. F.; McPherson, K. *Anal. Chem.* **1976,** *48,* 1885.

305. Jonckheere, J. A.; DeLeenheer, A. P.; Steyaert, H. L. *Anal. Chem.* **1983,** *55,* 154.

306. Moler, G. C.; Delongchamp, R. R.; Korfmacher, W. A.; Pearce, B. A.; Mitchum, R. K. *Anal. Chem.* **1983,** *55,* 835.

307. Thorne, G. C.; Gaskell, S. J.; Payne, P. A. *Biomed. Mass Spectrom.* **1984,** *11,* 415.
308. Gan, B. K.; Baugh, L. D.; Liu, R. H.; Walia, A. S. *J. Forensic Sci.* **1991,** *36,* 1331.
309. Millard, P. J. *Quantitative Mass Spectrometry;* Heyden: London, 1978; p 51.
310. Liu, R. H.; Smith, F. P.; Low, I. A.; Piotrowski, E. G.; Damert, W. C.; Phillips, J. G.; Liu, J. Y. *Biomed. Mass Spectrom.* **1985,** *11,* 638-641.
311. Farjo, K. *Isotopenpraxis* **1980,** *61,* 235.
312. Farjo, K.; Haase, G. *Isotopenpraxis* **1980,** *16,* 137.
313. Jerpe, J. H.; Bena, F. E.; Morris, W. *J. Forensic Sci.* **1975,** *20,* 557.
314. Low, I. A.; Liu, R. H.; Furner, R.; Piotrowski, E. G. *Biomed. Mass Spectrom.* **1986,** *13,* 531–534.
315. Mathews, D. W.; Hayes, J. M. *Anal. Chem.* **1976,** *48,* 1375.
316. Liu, R. H. In *Forensic Mass Spectrometry;* Yinon, J. Ed.; CRC Press: Boca Raton, FL, 1987; p 1.
317. Klein, M. In *Forensic Mass Spectrometry;* Yinon, J. Ed.; CRC Press: Boca Raton, FL, 1987; p 51.
318. Cody, J. T.; Foltz, R. L. In *Forensic Applications of Mass Spectrometry;* Yinon, J., Ed.; CRC Press: Boca Raton, FL, 1995; p 1.
319. Baumgartner, W. A.; Cheng, C.-C.; Donahue, T. D.; Hayes, G. F.; Hill, V. A.; Scholtz, H. In *Forensic Applications of Mass Spectrometry;* Yinon, J., Ed.; CRC Press: Boca Raton, FL, 1995; p 61.
320. Melgar, R.; Kelly, R. C. *J. Anal. Toxicol.* **1993,** *17,* 399.
321. Hughes, R. O.; Bronner, W. E.; Smith, M. L. *J. Anal. Toxicol.* **1991,** *15,* 256.
322. Madden, J. E.; Pearson, J. R.; Rowe, J. E. *Forensic Sci. Int.* **1993,** *61,* 169.
323. Thompson, W. C.; Dasgupta, A. *Clin. Chem.* **1994,** *40,* 1703.
324. Hornbeck, C. L.; Carrig, J. E.; Czarny, R. J. *J. Anal. Toxicol.* **1993,** *17,* 257.
325. Pocci, R.; Dixit, V.; Dixit, V. M. *J. Anal. Toxicol.* **1992,** *16,* 45.
326. Dickson, P. H.; Markus, W.; McKernan, J.; Nipper, H. C. *J. Anal. Toxicol.* **1992,** *16,* 67.
327. Fitzgerald, R. L.; Rexin, D. A.; Herold, D. A. *J. Anal. Toxicol.* **1993,** *17,* 342.
328. Meatherall, R. *J. Anal. Toxicol.* **1994,** *18,* 369.
329. Moeller, M. R.; Doerr, G.; Warth, S. *J. Forensic Sci.* **1992,** *37,* 969.
330. Clouette, R.; Jacob, M.; Koteel, P.; Spain, M. *J. Anal. Toxicol.* **1993,** *17,* 1.
331. Wu, A. H. B.; Liu, N.; Cho, Y.-J.; Johnson, K. G.; Wong, S. S. *J. Anal. Toxicol.* **1993,** *17,* 215.
332. ElSohly, M. A.; Little, T. L., Jr.; Stanford, D. F. *J. Anal. Toxicol.* **1992,** *16,* 188.
333. Abusada, G. M.; Abukhalaf, I. K.; Alford, D. D.; Vinzon-Bautista, I.; Pramanik, A. K.; Ansari, N. A.; Manno, J. E.; Manno, B. R. *J. Anal. Toxicol.* **1993,** *17,* 353.
334. Cone, E. J.; Hillsgrove, M.; Darwin, W. D. *Clin. Chem.* **1994,** *40,* 1299.
335. Francom, P.; Lim, H. K.; Andrenyak, D.; Jones, R. T.; Foltz, R. L. *J. Anal. Toxicol.* **1988,** *12,* 1.
336. Paul, B. D. Mitchell, J. M.; Burbage, R.; Moy, M.; Sroka, R. *J. Chromatogr.* **1990,** *529,* 103.
337. Lim, H. K.; Andrenyak, D.; Francom, P.; Jones, R. T.; Foltz, R. L. *Anal. Chem.* **1988,** *60,* 1420.
338. Goldberger, B. A.; Darwin, W. D.; Grant, T. M.; Allen, A. C.; Caplan, Y. H.; Cone, E. J. *Clin. Chem.* **1993,** *39,* 670.
339. Grinstead, G. F. *J. Anal. Toxicol.* **1991,** *15,* 293.
340. Fenton, J.; Mummert, J.; Childers, M. *J. Anal. Toxicol.* **1994,** *18,* 159.
341. Lin, Z.; Lafolie, P.; Beck, O. *J. Anal. Toxicol.* **1994,** *18,* 129.
342. Lodge, B. A.; Zamecnik, J.; MacMurray, P.; Brousseau, R. *Forensic Sci. Int.* **1992,** *55,* 13.
343. Goldberger, B. A.; Cone, E. J. *J. Chromatogr.* **1994,** *674,* 73.
344. Liu, R. H. *Elements and Practice in Forensic Drug Urinalysis;* Central Police University Press: Taipei, Taiwan, 1994.
345. Newton, P. *LC-GC* **1990,** *8,* 706.
346. Covey, T. R.; Lee, E. D.; Bruins, A. P.; Henion, J. D. *Anal Chem.* **1985,** *58,* 1451.
347. Bowers, L. D. *Clin. Chem.* **1989,** *35,* 1282.
348. Crowther, J. B.; Covey, T. R.; Dewey, E. A.; Denion, J. D. *Anal. Chem.* **1984,** *56,* 2921.
349. Henion, J. D.; Thomson, B. A.; Dawson, P. H. *Anal. Chem.* **1982,** *54,* 451.
350. Tas, A. C.; van der Greef, J.; de Brauw, M. C. T. N.; Plomp, T. A.; Maes, A. R. R.; Hohn, M.; Rapp, U. *J. Anal. Toxicol.* **1986,** *10,* 46.
351. Hayes, S. M.; Liu, R. H.; Tsang, W.-S.; Legendre, M. G.; Berni, R. J.; Pillion, D. J.; Barnes, S.; Ho, M. H. *J. Chromatogr.* **1987,** *398,* 239.
352. Lee, E. D.; Henion, J. D.; Brunner, C. A.; Wainer, I. W.; Doyle, T. D.; Gal, J. *Anal. Chem.* **1988,** *58,* 1349.
353. Baker, J. K. *J. Liq. Chromatogr.* **1981,** *4,* 271.
354. Hufford, C. D.; Capiton, G. A.; Clark, A. M.; Baker, J. K. *J. Pharm. Sci.* **1981,** *70,* 151.
355. Baker, J. K. In *Analytical Methods in Forensic Chemistry;* Ho, M. H., Ed.; Ellis Horwood: New York, 1990.
356. Dulik, D. M.; Colvin, O. M.; Fenselau, C. *Biomed. Enviorn. Mass Spectrom.* **1990,** *19,* 248.
357. Lurie, I. S. *LC-GC* **1990,** *8,* 454.

SECTION IV

Developing Technologies and Analytical Issues

Chapter 10
Developing Analytical Technologies

Recent developments in analytical technologies include applications in the analysis of drugs of abuse. Some of these new developments are closely related to technologies discussed in other chapters and described there, whereas others are treated as separate entities and discussed in this chapter. Examples of the former category include the application of MS–MS for structural elucidation of drugs that are newly discovered on the street (Chapter 9), and tunable diode laser spectrometry for isotopic analysis (Chapter 11). This chapter includes capillary electrophoresis and supercritical fluid extraction and chromatography.

Capillary Electrophoresis

Electrophoresis (the transport of electrically charged particles in liquid media under the influence of a direct current electrical field) has traditionally been used for the separation of biopolymers in solid support media such as gels and cellulose acetate. The most recent development of this technique is capillary electrophoresis (CE), in which electrophoretic and electrokinetic separations are performed in tiny capillaries under high electric fields. The highly limited sample diffusion achieved by the capillary has allowed the extension of electrophoresis to small molecules, such as drugs, which are not suitable for analysis in slab gels. Furthermore, the introduction into CE of chromatography-like separation mechanisms in addition to electrophoresis allows the analysis of neutral molecules, which otherwise would not be susceptible to any electrophoretic method. A schematic of CE instrumentation is shown in Figure 10.1 [1]. Typically, a high-voltage power supply is used to deliver 15–30 kV and 10–150 μA across an outer-coated polyimide, fused-silica capillary (20–100 μm i.d., 15–100 cm). The capillary is filled with an appropriate separation buffer and immersed in two reservoirs, both of which are also filled with separation buffer.

Because the highly acidic silica imparts a negative charge to the inner surface of the capillary, cations in proximity to the wall are attracted to it from the bulk buffer solution. These cations, under an applied field, are propelled through the capillary and cause a net movement of the bulk buffer toward the cathode (electro-osmotic flow), which transports all the analytes in this direction despite their individual electrophoretic mobilities. With a high surface-to-volume ratio of the capillary, which facilitates heat dissipation and allows the use of higher current, capillary electrophoresis can perform better over conventional gel electrophoresis, in which the heating in the conducting medium caused by the current flow poses a significant limitation on achievable separation speed and resolution. The open tubular format of CE and the underlying mass-to-charge ratio separation principle of electrophoresis can overcome the inconveniently long retention time of high-pressure liquid chromatography (HPLC) and, being based upon separation mechanisms substantially different from those of LC, lead to complementary results.

Among various forms of CE, micellar electrokinetic capillary chromatography (MECC) and capillary zone electrophoresis (CZE) have found viable applications in the analysis of small molecules, such as drugs and their metabolites. CZE is conducted in a continuous buffer, and the sample zones migrate without exhibiting any steady-state behavior. In this technique, separation is based on differences in net electrophoretic mobilities, and, because of the electro-osmotic flow directed toward the detector, cations and anions can often be simultaneously analyzed. MECC is a further development of CZE in which charged surfactants, such as sodium dodecyl sulfate (SDS) or sodium *N*-lauroyl-*N*-methyltaurate, are added to the buffer at or above their critical micelle concentration, with the micelles in equivalent concentration to the monomer. The movement of the electro-osmotic flow is in the opposite direction to the migration of the charged micelles. Un-ionic solutes partition between the micellar phase and the surrounding aqueous phase moving in opposite directions, and the separation is achieved largely on the basis of differences in hydrophobicity, that is, affinity for the hydrophobic core of micelles. Ionic solutes are separated (1) on the basis of electrophoresis, and (2) with ionic interactions that they can establish with the charged surface of the micelles.

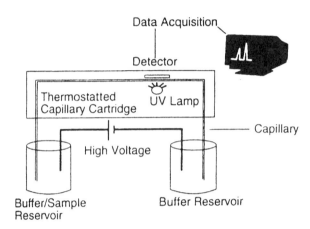

Figure 10.1. Schematic of a typical CE instrument. (Reprinted with permission from Ref. [1].)

Applications of CZE and MECC

With separation mechanisms based on differences in net electrophoretic mobilities, CZE has been applied [2] to the analysis of diuretics at two pHs (10.6 and 4.5). Compounds **1** through **13** (Chart 10.1), containing a carboxylic and/or sulphonamide group (Group A), were separated as anions under basic conditions. Compounds **14** through **16,** containing primary, secondary, or tertiary amine groups (Group B), were separated at acidic conditions as cations. A representative electropherogram is shown in Figure 10.2.

With the flexibility in separation improvement through proper selection of a micelle-forming agent and the buffer pH, and the use of additives including organic modifiers and salts, MECC has been applied to the analysis of many therapeutic and abused drugs, including benzodiazepines [3], cocaine-related compounds [4], barbiturates [5], and mixtures of several drug categories [4, 6, 7]. Potential applications of this developing technology for analyzing a wide variety of seized drugs in crime laboratories have been nicely documented [8, 9].

Because chemical derivatization is generally not required, this technique has great potential in the simultaneous analysis of the entire spectrum of parent drugs and their metabolites found in unconventional specimens, as shown in a preliminary study [10] of the analysis of hair for compounds derived from cocaine or heroin exposure. The merits, limitations, and characteristics of MECC were nicely outlined in a recent article [6], which gives an overview of the analysis of drugs in serum, saliva, and urine (with and without sample-pretreatment approaches such as ultrafiltration and extraction), as follows:

1. Untreated samples can be analyzed, and drugs eluting between creatinine and uric acid are free of general interference. However, the bulk protein in serum (and to a lesser extent in saliva) may interfere with the detection of many drugs, including cefpiramide, aspoxicillin, phenobarbital, caffeine, and theophylline.

2. Simple ultrafiltration is effective in removing interference caused by the presence of bulk protein. However, drugs such as phenytoin and valproate, which have high protein-binding capacities, may be removed along with protein and not detected.

3. At this stage of technology development, analyte retention time cannot be used as a reliable parameter for zone assignment. Analyte retention time may change dramatically with a slight variation in buffer pH and with variations in sample matrices; thus, multiwavelength or other more specific detection methods are required.

4. Derivatization is generally not required, and morphine-3-glucuronide is readily detectable. However, hydrolysis is needed for the analysis of 9-THC-COOH.

5. Although direct analysis of urine for benzodiazepines was found to be more sensitive than EMIT [11], this technique is generally not as sensitive as immunoassay. An extraction step resulting in analyte concentration is generally beneficial.

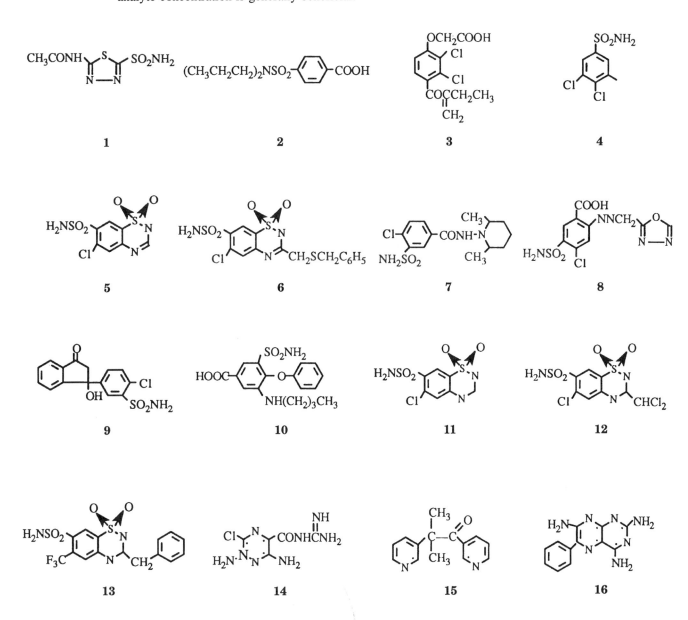

Chart 10.1. Structures of diuretics. Group A: acetazolamide (1), probenecid (2), ethacrynic acid (3), dichlorphenamide (4), chlorothiazide (5), benzthiazide (6), clopamide (7), furosemide (8), chlorthalidone (9), bumetanide (10), hydrochlorothiazide (11), trichlormethiazide (12), and bendroflumethiazide (13). Group B: amiloride (14), metyrapone (15), and triamterene (16).

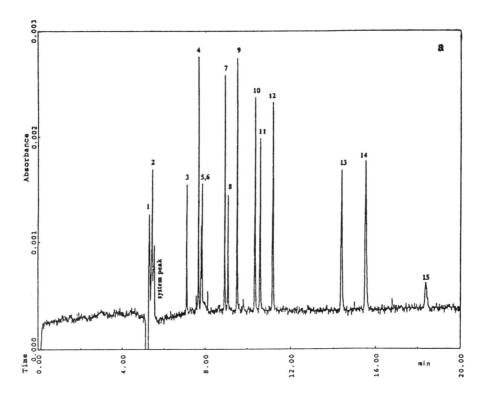

Figure 10.2. A representative CZE electropherogram of serum spiked with 10 ppm of diuretics. Conditions: uncoated silica capillary, 67 cm (60 cm effective length), 50 μm i.d., 360 μm o.d.; UV detection, 220 nm; 3-(cyclohexylamino)-1-propanesulphonic acid buffer, pH 10.6; temperature, 20 °C; applied voltage, 25 kV; hydrostatic injection by pressure, 5 s. Elution order: 1, metyrapone and caffeine; 2, triamterene and amiloride; 3, clopamide; 4, chlorthalidone; 5, ethacrynic acid; 6, probenecid; 7, bumetanide; 8, bendroflumethiazide; 9, furosemide; 10, trichlormethiazide; 11, benzthiazide; 12, hydrochlorothiazide; 13, dichlorphenamide; 14, chlorothiazide; and 15, acetazolamide. (Reprinted with permission from Ref. [2].)

With its ability to separate both neutral and ionic species, MECC has also been applied to the analysis of β-blockers (**17–27,** Chart 10.2) [12], some of which are hydrophilic while others are lipophilic. At physiological pH (pH 7.4), β-blockers exist as single cations, which permit their separation by methods based on ion-pair and micellar formations. A representative electropherogram is shown in Figure 10.3.

CE methods may prove to be the most effective approaches for the separation of enantiomers, which is achieved by adding chiral selectors to the buffer, trapping them in a gel matrix, or immobilizing them in the capillary. Although chiral surfactants and crown ethers have been explored, most applications use native and derivatized cyclodextrins. β-Cyclodextrin (Chart 10.3) [13], consisting of seven D-(+)-glucopyranose units (all in $^4C_1$ chair conformation) connected via α-1,4 linkages, is most commonly used. Cyclodextrins are relatively hydrophilic on the surface and have a hydrophobic cavity; they are therefore soluble in aqueous systems. In one application [14], β-cyclodextrin was added to the background electrolyte along with the micelle-forming surfactant SDS to achieve baseline separation of mephenytoin, 4-hydroxymephenytoin, and 4-hydroxyphenytoin enantiomers. It was believed that cyclodextrin did not interact with SDS and migrate at the velocity of the electro-osmotic flow; the uncharged enantiomeric forms of the analyte separated on the basis of the achirally selective partitioning into the SDS-micelles and of the stereoselective inclusion into the cyclodextrins. In another application [15], enantiomeric resolution efficiencies observed using mixtures of neutral and anionic cyclodextrins (heptakis(2,6-di-O-methyl)-β-cyclodextrin and sulfobutyl ether β-cyclodextrin, respectively), were evaluated and compared with that resulting from using neutral cyclodextrin alone. The authors postulated that the anionic cyclodextrin acted as a counter-migrating complexing agent. Neutral and anionic cyclodextrin mixtures with appropriate concentration ratios provided better efficiency in terms of resolution, retention time, and peak sharpness than can be achieved using neutral or anionic cyclodextrin alone, for the enantiomeric pairs methamphetamine and propoxyphene.

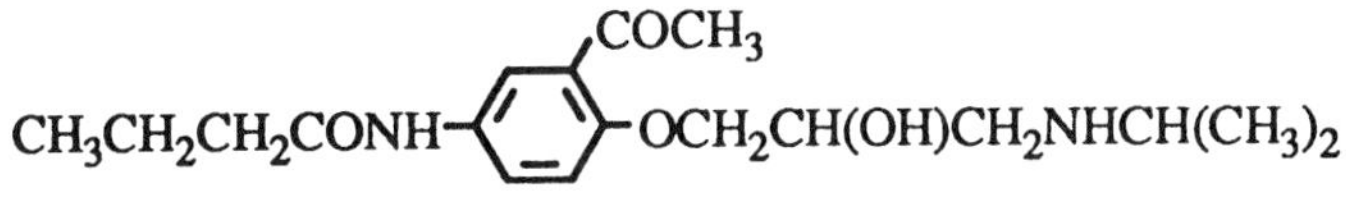

CH$_3$CH$_2$CH$_2$CONH—⟨ring with COCH$_3$⟩—OCH$_2$CH(OH)CH$_2$NHCH(CH$_3$)$_2$

17

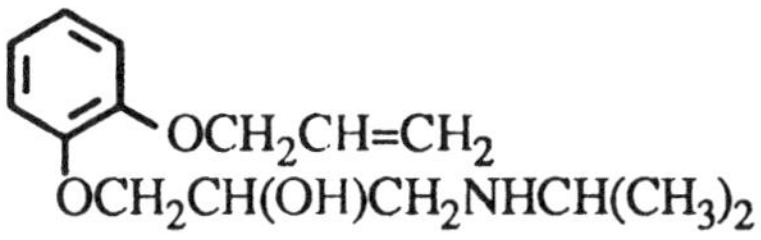

⟨ring⟩ OCH$_2$CH=CH$_2$ / OCH$_2$CH(OH)CH$_2$NHCH(CH$_3$)$_2$

23

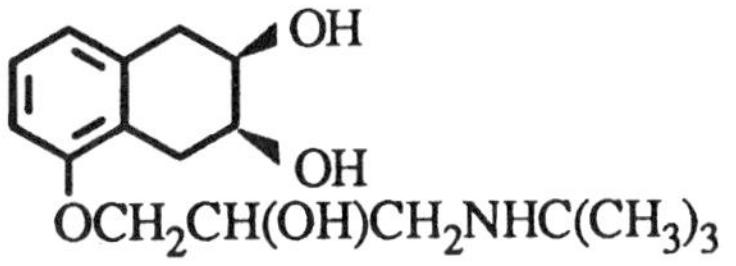

⟨tetralin ring⟩ OH, OH / OCH$_2$CH(OH)CH$_2$NHC(CH$_3$)$_3$

18

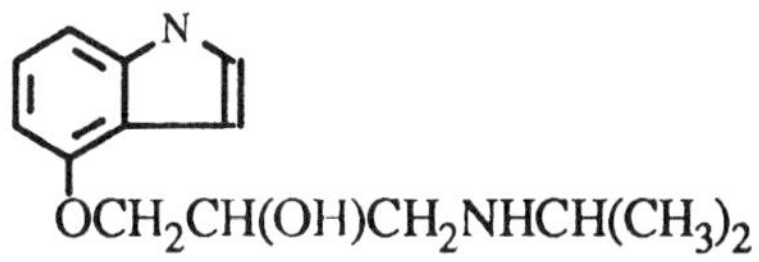

⟨indole ring⟩ OCH$_2$CH(OH)CH$_2$NHCH(CH$_3$)$_2$

24

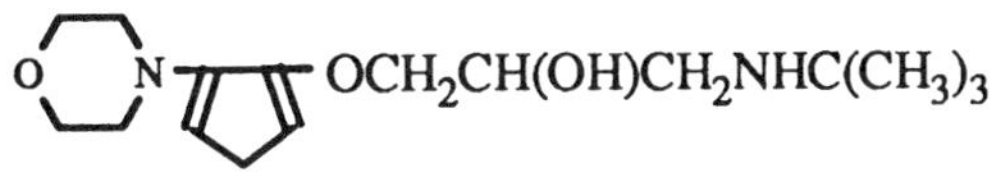

O⟨morpholine⟩N—⟨ring⟩OCH$_2$CH(OH)CH$_2$NHC(CH$_3$)$_3$

19

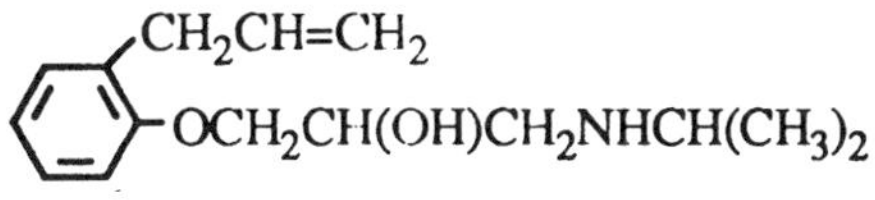

⟨ring⟩ CH$_2$CH=CH$_2$ / OCH$_2$CH(OH)CH$_2$NHCH(CH$_3$)$_2$

25

H$_2$NCOCH$_2$—⟨ring⟩—OCH$_2$CH(OH)CH$_2$NHCH(CH$_3$)$_2$

20

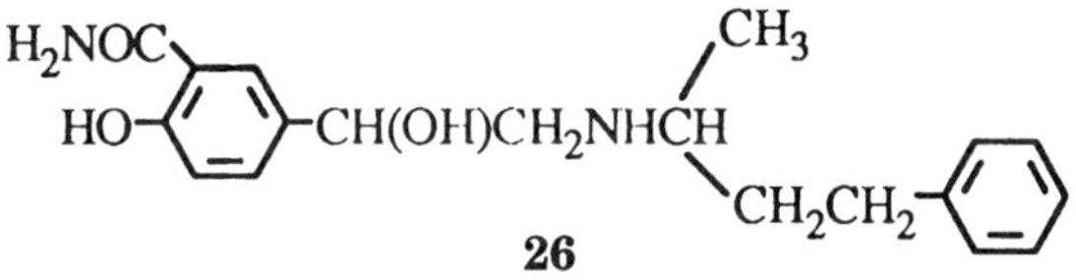

H$_2$NOC⟨ring⟩HO—CH(OH)CH$_2$NHCH CH$_3$ / CH$_2$CH$_2$—⟨ring⟩

26

CH$_3$SO$_2$NH—⟨ring⟩—CH(OH)CH$_2$NHCH(CH$_3$)$_2$

21

⟨naphthalene ring⟩ OCH$_2$CH(OH)CH$_2$NHCH(CH$_3$)$_2$

27

CH$_3$OCH$_2$CH$_2$—⟨ring⟩—OCH$_2$CH(OH)CH$_2$NHCH(CH$_3$)$_2$

22

*Chart 10.2. Structures of β-blockers: acebutolol (**17**), nadolol (**18**), timolol (**19**), atenolol (**20**), soyalol (**21**), metoprolol (**22**), oxprenolol (**23**), pindolol (**24**), alprenolol (**25**), labetalol (**26**), and propranolol (**27**).*

Important Operational Parameters: pH, Buffer Additives, and Organic Modifiers

The flexibility of adjusting many operational parameters in CE allows for effective separation of a wide variety of compounds. Thus, by operating CZE under pH 10.6 and 4.5, it was possible to analyze two groups of diuretics as anions and cations, as shown previously [2].

With the addition of N-cetyl-N,N,N-trimethylammonium bromide (CTAB) to the buffer, the negative charge of the wall is changed to positive, thus reversing the electro-osmotic flow toward the anodic end of the silica capillary. This electro-osmotic flow may be regulated by the buffer pH: it increases as the pH decreases. Optimal pHs for the resolution of 11 β-blockers were reported in the ranges of 6.6–7.0 and 7.4–7.8 [12]. The migration window was wider at the basic end of the pH range, but the total analysis time was much shorter at lower pH values. It was noted, however, that small variations in pH can cause unexpected changes in migration times of solutes and, especially in the analysis of complex mixtures, changes in migration order relative to other analytes.

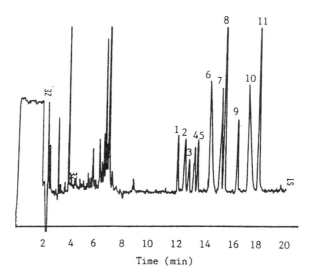

Figure 10.3. A representative micellar electrokinetic capillary chromatography (MECC) electropherogram of 75 μg/mL (150 μg/mL for 3 and 6) β-blockers-spiked urine. Conditions: fused-silica capillary, 68 cm (60 cm effective length), 50 μm i.d.; 0.08 M sodium phosphate buffer (pH 7.0) with 10 mM CTAB (N-cetyl-N,N,N-trimethylammonium bromide); UV detection, 214 nm; applied voltage, −26 kV; temperature, ambient; hydrodynamic injection mode, 30 s at 10 cm height. Elution order: 1, acebutolol; 2, nadolol; 3, timolol; 4, atenolol; 5, metoprolol; 6, 2,6-dimethylphenol (internal standard); 7, oxprenolol; 8, pindolol; 9, alprenolol; 10, labetalol; and 11, propranolol. (Reprinted with permission from Ref. [12].)

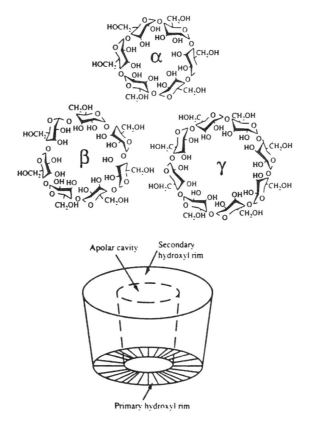

Chart 10.3. Native cyclodextrins α, β, and γ (top), and basic features of the three native cyclodextrins (bottom). The primary hydroxyl rim is made up of hydroxyls attached to the C_6 carbons, and the secondary hydroxyl rim is made up of hydroxyl groups attached to the C_2 and C_3 carbon atoms. (Reprinted with permission from Ref. [13].)

The resolution of analytes can also be greatly improved through the selection of organic modifiers. For example, the separation of a model drug mixture was greatly improved through the use of 5% 2-propanol (Figure 10.4) [6], and contaminants in illicit heroin seizures were nicely separated with the use of acetonitrile [11, 16] (Figure 10.5) [16].

Development in Detection Technologies

UV-based detectors are currently the most widely used detection devices. Single-wavelength detection at 254 nm [17] was found to be effective for the analysis of several benzodiazepams and related compounds. Photodiode array detection is informative in providing absorption characteristics that can be helpful for structural elucidations of unknown compounds. When comparing single- (195-nm) and multiple-wavelength (195–320-nm) detections for three groups of drug mixtures (benzoylecgonine, morphine, heroin, and methamphetamine; codeine, amphetamine, cocaine, and methadone; and meth-aqualone, flunitrazepam, oxazepam, diazepam, and benzoic acid, with an impurity of benzoylecgonine [17]), three-dimensional multiple-wavelength electropherograms (A, B, and C in Figure 10.6) appear to provide additional but not essential information over their single-wavelength counterparts (A, B, C in Figure 10.4) [17].

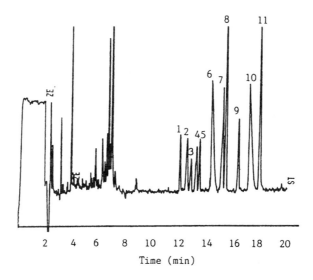

Figure 10.4. Single-wavelength (195-nm) electropherograms of a model mixture (50 μg/mL of each compound) analyzed in borate–phosphate buffer (pH 9) with 75 mM SDS (A), and 71.25 mM SDS–5% 2-propanol (B). Peaks: 1, benzoylecgonine; 4, methamphetamine; 6, amphetamine; 8, methadone; 12, diazepam; 15, morphine-3-glucuronide; M, methanol. (Reprinted with permission from Ref. [6].)

Because retention-time data of the current CE technology can change significantly with slight variations in operation conditions [6], on-line detection devices that provide definite identification information are especially valuable. Thus, mass spectrometry (MS)-based detection is most desirable and has been explored. In this regard, CZE is more suitable than MECC because the latter requires the presence of surfactants in the separation buffer that may not be compatible with the MS operation conditions. MS detection also limits the use of buffers with sufficient volatility (such as ammonium acetate). CZE with simple buffer systems is thus most compatible with this type of detection approach. In order to accommodate the hydrophobic nature of drug analytes, addition of organic solvents (such as methanol and acetic acid) and the use of nonaqueous solvents in CZE applications were found useful and compatible with atmospheric pressure ionization MS detection [18, 19]. Examples of selected ion monitoring and full-scan electropherograms, obtained from four benzodiazepines and flurazepam (and its metabolite, *N*-1-hydroxyethylflurazepam), are shown in Figure 10.7 [17]. A recently improved

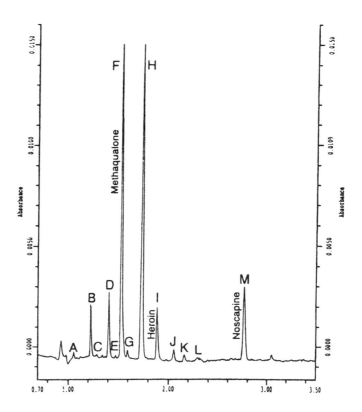

Figure 10.5. Separation of an illicit heroin sample. Conditions: uncoated silica capillary, 27 cm (20 cm effective length), 50 μm i.d.; UV detection, 214 nm; 8.5 mM sodium borate–8.5 mM sodium phosphate buffer (85:15 water:acetonitrile v/v, pH 10.6) containing 40 mM SDS; temperature, 30 °C; applied voltage, 20 kV; injection by pressure, 1 s. (Reprinted with permission from Ref. [16].)

on-line CZE–electrospray ionization–MS system is shown schematically in Figure 10.8 [19]. Further development in MS-based detection approaches is needed, and the adoption of CE as a routine analytical tool will depend greatly on the availability of a device that can be conveniently used for full-scan detection of analytes at a level expected from capillary elution.

Supercritical Fluid Extraction and Chromatography

Environmental concerns and rising solvent acquisition and disposal costs have already made solid-phase extraction (SPE) the preferred sample-pretreatment approach in most large-scale drug-testing laboratories for the analysis of abused drugs in urine. Supercritical fluid extraction (SFE) can further reduce the use of organic solvents, and it may become a valuable sample-pretreatment method for drug analysis in various matrices. Additional advantages of SFE include the speed of extraction and the ease of supercritical fluid removal. The solvent strength can also be conveniently controlled by adjusting extraction pressure and temperature.

A substance becomes a "supercritical fluid" when both temperature and pressure equal or exceed those of the substance's critical point. The phase diagram (Figure 10.9 [20]) for the most commonly used substance, CO_2, shows its critical point (critical temperature = 31 °C; critical pressure = 73 atm or 1050 psi) and the supercritical region. A typical SFE system, shown in Figure 10.10, includes a carbon dioxide source, a pumping system (liquid carbon dioxide), an extraction thimble, a restriction device, an analyte-collection device, temperature-control systems for several zones, and an overall system controller [21].

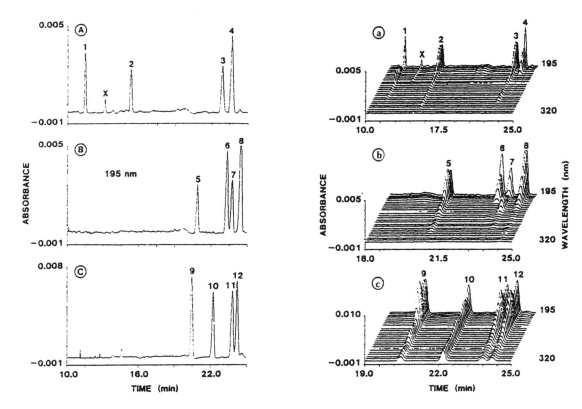

Figure 10.6. Single-wavelength (195-nm) electropherograms (A–C), and three-dimensional (195–320-nm) electropherograms (a–c) of three model mixtures of different drugs: benzoylecgonine (1), benzoic acid (x), morphine (2), heroin (3), and methamphetamine (4) in electropherograms A and a; codeine (5), amphetamine (6), cocaine (7), and methadone (8) in electropherograms B and b; and methaqualone (9), flunitrazepam (10), oxazepam (11), and diazepam (12) in electropherograms C and c. (Reprinted with permission from Ref. [17].)

Operational Parameters

Solid samples, such as pharmaceutical tablets ground to an optimum particle diameter, are the most suitable matrices for ease of solvent diffusion and mass transfer. Water in semisolid samples, such as human tissues, can be removed by applying freeze-drying techniques to the finely blended product. An adsorbent such as anhydrous sodium sulfate has also been used to remove water. Liquid samples and samples with a high content (>15%) of water or another liquid can be absorbed or dispersed on packing materials such as Celite [22] and octadecylsilane [23], as shown in Figure 10.11 [21]. The analyte can then be eluted in a similar fashion to elution by SPE using a supercritical fluid as the eluant.

A novel assembly connecting a standard SPE column to an SFE vessel that forms a leak-tight seal with the vessel's end-cap up to pressure of 680 bar was constructed for in-line trapping of anabolic steroids (nortestosterone, testosterone, and methyltestosterone) [24]. Homogenized chicken-liver tissue was mixed with Celite 566 and extracted with supercritical CO_2 at 40 °C and 272 bar. Neutral alumina (150 mesh) was used and found to retain the targeted steroids with excellent selectivity and recoveries.

Supercritical fluid CO_2 is known to dissolve analytes that are soluble in solvents such as hexane, benzene, and methylene chloride. The modulation of pressure and temperature is an effective operational parameter to adjust the *diffusivity* and *density* of the supercritical solvent for controlling extraction efficiency. For example, a recovery study [22] of active components (sulfamethoxazole and trimethoprim) in a drug formulation revealed that both analytes can be effectively extracted at a CO_2 density of 0.85 g/mL. Trimethoprim and sulfamethoxazole could not be extracted with a CO_2 density at or below 0.40 and 0.85 g/mL, respectively.

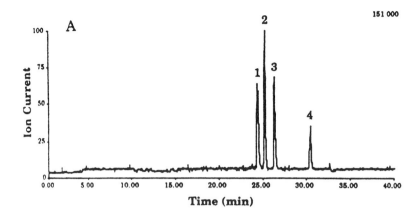

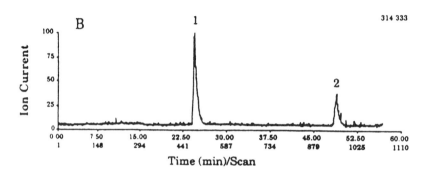

Figure 10.7. Selected-ion-monitoring and full-scan (atmospheric pressure ionization) electro-pherograms of (A) chlordiazepoxide (1), flurazepam (2), diazepam (3), and prazepam (4); and (B) flurazepam (1) and its metabolite, N-1-hydroxyethylflurazepam (2). (Reprinted with permission from Ref. [18].)

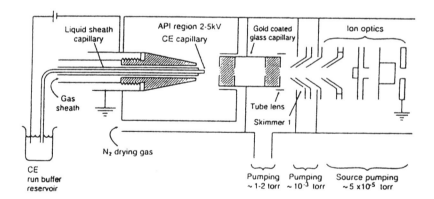

Figure 10.8. Schematic of the CE–ESI–MS interface used on the MAT 900 sector mass spectrometer. The electrospray assembly is shown enlarged for clarity. API is the atmospheric ionization interface. (Reprinted with permission from Ref. [19].)

Another important operational parameter involves the incorporation of various types and concentrations of modifiers into the superfluid CO_2 to increase the solvating power and the polarity. For example, pure supercritical fluid CO_2 was used to clean up a fly-ash sample by removing weakly adsorbed organic material followed by the extraction of dioxins using 2% methanol-modified supercritical fluid CO_2 [25]. In another example, when comparing the conventional SPE and SFE–SPE approaches for the recoveries and selectivities of a drug metabolite (mebeverine alcohol) that was spiked onto dog plasma [26], the investigators observed higher selectivities in the SFE–SPE approach using CO_2/MeOH (95:5) as the supercritical fluid at 350 atm, 40 °C, and a flow rate of 0.3 mL/min.

Liquid modifier can also be added directly to the sample when performing SFE. For example, the recovery of pesticides (atrazine) was increased from 0 to 100% by adding 1 mL of water to 5 g of spiked soil [27]. This study reported that polar modifiers increased recoveries much more than nonpolar modifiers, and the authors hypothesized that the solute–matrix interactions, rather than the solute–solvent interactions, were being modified.

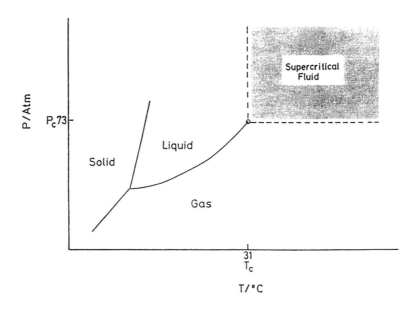

Figure 10.9. Phase diagram for carbon dioxide. Carbon dioxide becomes a supercritical fluid at a critical pressure, P_c, of 73 atm (1050 psi) and a critical temperature, T_c, of 31 °C. (Reprinted with permission from Ref. [20].)

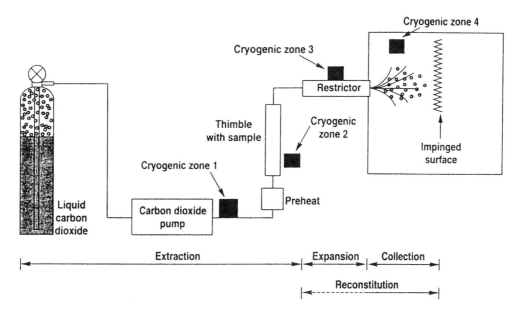

Figure 10.10. Schematic of a typical supercritical fluid extraction system. (Reprinted with permission from Ref. [21].)

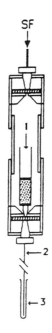

Figure 10.11. Schematic of a typical extraction thimble for supercritical fluid extraction. (Reprinted with permission from Ref. [23].)

Supercritical Fluid Chromatography

The adoption of supercritical fluid CO_2 as the mobile phase in chromatography overcomes many difficulties encountered in conventional HPLC and GC approaches. With respect to HPLC, the low viscosity of supercritical fluid CO_2 allows efficient operation at a higher flow velocity, resulting in an improvement in separation speed. Universal detectors such as flame ionization detectors (FIDs) are also compatible with supercritical fluid chromatography (SFC) and will detect analytes at a lower level. With respect to GC, the solvating strength of supercritical fluid CO_2 offers higher solute diffusity (compared with inert carrier gas), which allows for the separation of polar and higher-molecular-weight compounds at a low temperature.

Among the still limited literature reports on this developing SFC technology in drug analysis are applications of derivatized β-cyclodextrins, bonded to polysiloxane or copolymerized with siloxanes as the stationary phases, for enantiomeric resolutions of hexobarbital [28] and ibuprofen [28, 29]. The SFC approaches were applied under a low temperature (typically ~60 °C), which can enhance enantioselectivity and prevent thermal decomposition and racemization.

References

1. Tomlinson, A. J.; Benson, L. M.; Naylor, S. *Am. Lab.* **1994**, *26*, 29.
2. Jumppanen, J.; Sirén, H.; Riekkola, M.-L. *J. Chromatogr. A* **1993**, *652*, 441.
3. Schafroth, M.; Thormann, W.; Allenmann, D. *Electrophoresis* **1994**, *15*, 72.
4. Trenerry, V. C.; Robertson, J.; Wells, R. J. *Electrophoresis* **1994**, *15*, 103.
5. Thormann, W.; Meier, P.; Marcolli, C.; Binder, F. *J. Chromatogr.* **1991**, *545*, 445.
6. Thormann, W.; Lienhard, S.; Wernly, P. *J. Chromatogr.* **1993**, *636*, 137.
7. Wernly, P.; Thormann, W. *Anal. Chem.* **1992**, *64*, 2155.
8. Lurie I. S. *Am. Lab.* **1996**, *28*, 26.
9. Lurie, I. S. In *Analysis of Addictive and Misused Drugs;* Adamovics, J. A., Ed.; Marcel Dekker: New York, 1995.
10. Tagliaro, F.; Poiesi, C.; Aiello, R.; Dorizzi, R.; Ghielmi, S.; Marigo, M. *J. Chromatogr.* **1993**, *638*, 303.
11. Trenerry, V. C.; Wells, R. J.; Robertson, J. *J. Chromatogr. Sci.* **1994**, *32*, 1.

12. Lukkari, P.; Vuorela, H.; Riekkola, M.-L. *J. Chromatogr.* **1993,** *632,* 143.

13. Rogan, M. M.; Altria, K. D. *Introduction to the Theory and Applications of Chiral Capillary Electrophoresis;* Beckman Instruments: Fullerton, CA, 1993; p. 5.

14. Desiderio, C.; Fanali, S.; Küpfer, A.; Thormann, W. *Electrophoresis* **1994,** *15,* 87.

15. Lurie, I. S.; Klein, R. F. X.; Cason, T. A. D.; LeBelle, M. J.; Brebbeisen, R.; Weinberger, R. E. *Anal. Chem.* **1994,** *66,* 4019.

16. Walker, J. A.; Krueger, S. T.; Lurie, I. S.; Marché, H. L.; Newby, N. *J. Forensic Sci.* **1995,** *40,* 6.

17. Wernly, P.; Thormann, W. *Anal. Chem.* **1991,** *63,* 2878.

18. Johansson, I. M.; Pavelka, R.; Henion, J. D. *J. Chromatogr.* **1991,** *559,* 515.

19. Tomlinson, A. J.; Benson, L. M.; Johnson, K. L.; Naylor, S. *Electrophoresis* **1994,** *15,* 62.

20. Myer, L.; Tehrani, J.; Thrall, C.; Gurkin, M. *International Labmate Magazine* **1990,** *15,* 15.

21. Gere, D. R.; Derrico, E. M. *LC-GC* **1994,** *12,* 352.

22. Mulcahey, L. J.; Taylor, L. T. *Anal. Chem.* **1992,** *64,* 981.

23. Liu, H.; Wehmeyer, K. R. *J. Chromatogr.* **1992,** *577,* 61.

24. Maxwell, R. J.; Lightfield, A. R.; Stolker, A. A. M. *J. High Resolut. Chromatogr.* **1995,** *18,* 231.

25. Alexandrous, N.; Pawliszyn, J. *Anal. Chem.* **1989,** *61,* 2770.

26. Liu, H.; Cooper, L. M.; Raynie, D. E.; Pinkston, J. D.; Wehmeyer, K. R. *Anal. Chem.* **1992,** *64,* 806.

27. Knipe, C. R.; Gere, D. R.; McNally, M. E. In *Supercritical Fluid Technology;* Brig, F.; McNally, M. E. Eds.; ACS Symposium Series 488; American Chemical Society: Washington, DC, 1992; p 215.

28. Jung, M.; Mayor, S.; Schurig, V. *LC-GC* **1994,** *12,* 458.

29. Bradshaw, J. S.; Yi, G.; Rossiter, B. E.; Reese, S. L.; Petersson, P.; Markides, K. E.; Lee, M. L. *Tetrahedron Lett.* **1993,** *34,* 79.

Chapter 11
Sample Differentiation

The need to differentiate similar samples in forensic sciences is best described by the following statement characterizing the nature of criminalistics science:

> A major and overriding characteristic that sets criminalistics apart from other scientific disciplines is its unique concern with the process of individualization. Other sciences are satisfied when an object is classified into a unit place in the discipline's taxonomy. Criminalistics strives to relate the object to a particularized source. [1]

Perhaps the oldest and still the most successful and familiar example of sample differentiation is the characterization and comparison of fingerprints. Recent advances in all fields of forensic science have also achieved limited success in the differentiation of various samples. For example, by grouping various genetic markers—especially the applications of DNA-profiling technologies—forensic serologists are now in a much better position to discriminate the donor of various body fluids and secretions, such as human semen and bloodstains.

Forensic scientists working on various aspects of drug analysis are also concerned with drug-sample differentiation, correlation, and individualization. They characterize these analytical objectives using terminologies such as "ballistics" [2], "physical test" [3, 4], "fingerprint pattern" [5], "chemical signatures" [6], "drug-sample differentiation" [7–9], and "drug signature profiling" [10, 11]. Information obtained is used to establish the degree of commonality of origins among drug exhibits. Comparative analyses are performed on both the macro- and microlevel to derive "strategic" and "tactical" intelligence, allowing the establishment of national- and global-scale trafficking patterns and conspiracies on a local level [12].

Four categories of analytical approaches are utilized for comparative analysis. Drug tablets, capsules, and lysergic acid diethylamide (LSD) blotter paper are often identified and differentiated by the "ballistics" method, which relies mainly on macro- or microlevel morphological comparisons. For other types of samples, the traditional and still the most common method involves the qualitative identification and quantitative determination of major and minor components and impurities. A third method of differentiation involves the measurement of isotopic ratios of naturally occurring isotopes. The fourth approach is the determination of diastereoisomeric and enantiomeric composition of certain drug samples such as the widely abused cocaine and amphetamines.

Physical Characterization of Tablets and Capsules

Macro- and microlevel morphological comparison of tablets, capsules, and LSD blotter paper were performed [3, 4, 12–14] to provide "ballistic" characteristics that are independent of and complementary to those forms of information derived from chemical-ingredient analysis. Unique "toolmarks" are attributed to the nature of the formulations and the condition of the punches used in clandestine laboratories. Examinations mainly involve microscopic analysis of parameters such as size, shape, score marks or symbols, score depth, width and angle, unusual surface imperfections, offset punch marks, and color.

With the collection of reference material and with experience, examination of perforation patterns was found to be extremely valuable in establishing the commonality of LSD blotter paper exhibits [12].

Impurity and Component Identification and Quantification

The comparison of qualitative and quantitative composition is the most common approach used for sample differentiation. Factors associated with both drug "manufacturing" and "handling" can contribute to variations in sample compositions. Differences due to additions of adulterants and diluents and environmental degradation are caused by postproduction handling processes. Variations originating at the manufacturing step depend largely on whether the drug is of "natural" or "synthetic" origin. Natural drugs may be further subdivided—on the basis of the amount of processing adopted to arrive at the final form—into plant material (e.g., marihuana or khat), isolative (e.g., cocaine or hash oil), or semisynthetic (e.g., heroin or LSD) [12]. Chemical-reaction routes and the exact chemicals used for producing synthetic, semisynthetic, and isolative drugs can leave "signature" information in qualitative and quantitative compositions. Subspecies or environmental variations may cause botanical differences in all three categories of natural drugs.

Numerous studies comparing the qualitative and quantitative compositions of diluents, adulterants, impurities, and minor and major components have been reported for the following drug categories: amphetamine [10, 15–18], cocaine [9–11, 17–25], marijuana [18, 26], and morphine alkaloids [10, 17, 18, 26–33]. The effectiveness of this approach increases if the comparison includes a large number of components whose concentrations can be accurately determined. Parameters selected for comparison should have minimal intrabatch variations but significant interbatch variations. Criteria recommended [18] for the selection of components for comparison include the following:

1. Components should have low volatility and high stability.
2. There should be little or no correlation between the quantities of the components selected.
3. Components should be produced along with the main drug and not from its decomposition.

It is further recommended [18] that the ratios of the components to the main drug, rather than their absolute concentrations, be used.

Amphetamines

Frequent manufacturing of amphetamine and methamphetamine in clandestine laboratories [34, 35] has prompted numerous studies concerning the determination of impurities observed in these types of samples. Common routes of synthesis and the occurrences of route-characteristic impurities have been summarized in a recent review article [36].

Methamphetamine serves as a good example to illustrate the differentiation of synthesized drugs on the basis of the detection and quantification of impurities present in the samples. Differentiations are possible because impurities and their quantities in a particular illicit preparation reflect the adopted synthetic paths, precursor chemicals, and reaction conditions. Using methamphetamine samples prepared by the Leuckart reaction (Figure 11.1) as an example, the following compounds have been reported [15–18]: methylamine, formic acid, N-methylformamide, N-formylmethamphetamine, phenyl-2-propanone, N,α,α'-trimethyldiphenethylamine, dibenzyl ketone, N-formyl-α-benzyl-N-methylphenethylamine, and α-benzyl-N-methylphenethylamine. Listed below is information that can be derived on the basis of the analytical data on these impurities [18]:

1. N-Methylformamide, phenyl-2-propanone, and N-formylmethamphetamine are starting materials and the reaction intermediate.

2. *N*,α,α′-Trimethyldiphenethylamine is produced by the reaction of *N*-formylmethamphetamine with excess phenyl-2-propanone.

3. The presence of methylamine and formic acid indicates the use of these compounds to produce *N*-methylformamide.

4. The presence of dibenzyl ketone, *N*-formyl-α-benzyl-*N*-methylphenethylamine, and α-benzyl-*N*-methylphenethylamine indicates the use of phenylacetic acid and acetic anhydride to produce phenyl-2-propanone; dibenzyl ketone is the by-product of phenyl-2-propanone synthesis, whereas *N*-formyl-α-benzyl-*N*-methylphenethylamine is formed by the reaction of dibenzyl ketone with *N*-methylformamide, and α-benzyl-*N*-methylphenethylamine is the acid hydrolysis product of *N*-formyl-α-benzyl-*N*-methylphenethylamine.

Thus, the detection and quantification of these compounds may provide information related to precursor chemicals used, the purity of these precursor chemicals, and reaction conditions. Not only can samples prepared from different precursor chemicals be differentiated, those prepared from different batches of operation can also be revealed. Batch-to-batch variations are common in most clandestine laboratory operations because the reaction conditions are not precisely controlled and the purifications of products and precursor chemicals are not consistent.

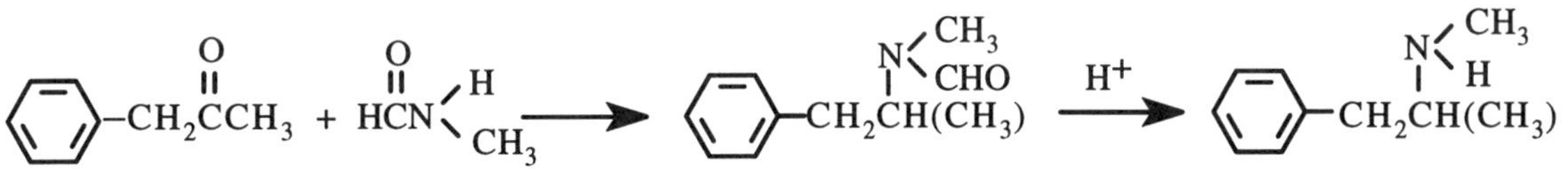

Figure 11.1. Leuckart reaction for the synthesis of methamphetamine.

Cocaine

The analysis of cocaine samples serves as a good example of the differentiation of sample origins [20–22]. Defense attorneys, with the assistance of their scientific experts, have been reported to raise the natural vs. synthetic origin issue [20]. Cocaine of natural origin may come from either the extraction of coca leaves followed by successive recrystallizations, or the hydrolysis of all ecgonine-based alkaloids of the leaf with subsequent esterification. The presence of cinnamoylcocaine is often used as an indication of the former procedure, whereas the latter procedure, as a result of incomplete esterification of the ecgonine base, often produces ecgonine, methylecgonine, and benzoylecgonine. Thus, the presence of minor alkaloids of *E. coca*, *cis*- and *trans*-cinnamoylcocaine [10] and truxillines [19], provides indirect proof of natural origin. (Interpretation based on the presence of ecgonine, methylecgonine, benzoylecgonine, methylanhydroecgonine, and ethylbenzoylecgonine should be exercised with caution, as these compounds may be decomposition products or artifacts generated during the analytical process [20].) On the other hand, cocaine derived from total synthesis, unless scrupulously purified, contains diastereoisomers and enantiomers [23–25]. The observation of impurities, such as 3-aminomethyl-2-methoxycarbonyl-8-methyl-8-azabicyclo[3.2.1]oct-2-ene, 3-benzoyloxy-2-methoxycarbonyl-8-methyl-8-azabicyclo[3.2.1]oct-2-ene, and 3-benzoyloxy-8-methyl-8-azabicyclo[3.2.1]oct-2-ene, further confirms the synthetic origin.

The analysis of cocaine samples to determine whether they are derived from the same "batch" has been a topic of interest for some time. As reviewed in a recent symposium presentation [9], samples can be profiled on the basis of the composition differences of the following components in the samples examined:

1. Cocaine and its decomposition products;

2. Absolute and relative quantities of naturally occurring impurities, such as cinnamoylcocaine, truxillines, truxillic acids, truxinic acids, *N*-norcocaine, tropacocaine, hydrococaine, hygrine, and cuscohygrine; and

3. Residual treatment chemicals.

Most sample-differentiation attempts that are based on comparisons of component compositions were conducted using manual calculations. This approach is convenient and feasible if only one or a few components are selected. Different approaches are needed if the overall sample composition is to be evaluated. For example, thanks to modern software capabilities, it was possible to show [11] differential chromatograms derived from the same and different batches of samples. Further attempts have been based on the use of elementary statistical information and pattern-recognition techniques using data obtained from capillary gas chromatographic (GC) analysis [11]. With further software development and further study on the selection of characteristic parameters, pattern-recognition approaches may provide effective sample differentiation. Because many components are derived from others (Figure 11.2) under certain specific conditions, sophisticated examination of these compositions may even be able to reveal certain information related to sample history.

Because it is much easier and more economical to produce cocaine from the natural source, there is no meaningful number of illicit samples that are produced through the synthetic route. Therefore, a database that is needed for sample differentiation based on components derived from the synthetic process is lacking. This is in contrast to the abundant information and data that are available for the synthesis of amphetamine drugs [36].

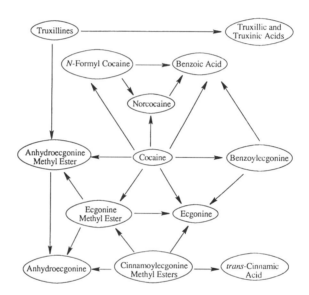

Figure 11.2. Cocaine-related impurities and pathways. (Reprinted with permission from Ref. [11].)

Cannabis

The investigation conducted by Novotny and his co-workers [26] on the use of combined high-resolution gas chromatography–mass spectrometry (GC–MS) highlighted the analysis of marijuana samples from different origins. Thirty-eight profile constituents were identified using this approach. With extended sample-pretreatment and analysis procedures, the high resolution of this method offered higher specificity.

Composition determinations of major components may also constitute a basis for sample differentiation. A simple, direct inlet probe MS procedure [37] was used to identify cannabinoids, and it was further used to differentiate various samples by the measurement of the relative concentrations of the major cannabinoids present [38].

In order to differentiate hashish oil, hashish, and marijuana-leaf samples, the samples are introduced into the ion source of an MS and spectra are continually collected until the sample is exhausted. The intensities of selected ions in these spectra from standard compounds and the actual samples are summed, respectively, and the relative intensities are calculated as shown in Table 11.1. Multiple-regression analysis is performed as shown in the section "Marijuana and Other Cannabinoid-Containing Materials" in Chapter 9.

Regression coefficients, which reflect the relative concentrations of the respective cannabinoids obtained from several samples, are listed in Table 11.2. These results indicate that among the four hashish samples investigated, the relative concentrations of cannabinol, cannabidiol, and Δ^1-THC in samples 2 and 3 are distinctly different from each other and from samples 1 and 4.

However, samples 1 and 4 are similar. The MS aspect of this approach required minimal effort. With the data system available to most MS devices and the widespread availability of personal computers, the combined use of regression analysis and MS procedures adopted here appears to have potential in routine analytical work.

Table 11.1. Relative intensities of selected ions (m/z) of CBN, CBD, Δ^1-THC, Δ^6-THC, and hashish oil.[a]

Sample	314	310	299	295	271	258	246	243	238	231	193
CBN	0	13.0	7	100	0	0	0	0	30.9	0	0
CBD	8.2	0	1.3	0	0	1.1	22.8	0	0	100	13.7
Δ^1-THC	43.7	0	49.7	0	37.5	33.0	15.3	37.9	0	100	31.4
Δ^6-THC	41.5	0	5.6	0	27.1	33.9	17.9	4.9	0	100	25.3
Hashish oil	14.2	0.6	11.0	4.2	7.1	4.9	17.4	6.8	1.3	100	13.2

[a] Data are taken from Ref. [38].

Table 11.2. Comparison of multiple regression coefficients obtained from various cannabinoid-containing samples.[a]

Sample	CBN	CBD	Δ^1-THC	Δ^6-THC
Hashish oil	0.76	0.79	0.23	0
Hashish-1	0.68	0.91	0.10	0.032
Hashish-2	1.0	0.50	0.071	0
Hashish-3	1.0	0.20	0.35	0.28
Hashish-4	0.63	0.90	0.092	0.050
Leaf	0.50	0.17	0.99	0

[a] Data are taken from Ref. [38].

Opiates

Heroin is produced by the reaction of acetic anhydride and morphine, which may or may not be isolated from opium prior to the acetylation process. It is thus apparent that the presence of significant amounts of opium components such as codeine, papaverine, noscapine, thebaine, narcotine, and norlaudanosine is indicative of the direct opium acetylation procedure.

Impurities derived from the acetylation process include synthetic by-products of the opium alkaloids thebaine, norcotine, morphine, codeine, and norlaudanosine generated during acetylation of crude morphine–opium [29]. Typical examples of these impurities are 6-monoacetylmorphine (6-MAM), acetylcodeine, thebaol, acetylthebaol, and 3,6-dimethoxyphenanthrene-4,5-epoxide [26]. Capillary GC profiles are now widely used for comparing samples derived from different origins, production processes, and reaction conditions [12, 26–31] (Figure 11.3). Although impurity profiles have been grouped into four classes (Figure 11.4) [39], it was cautioned that totally different reaction conditions may result in similar impurity profiles [40].

Quantitative determination of selected components in illicit heroin samples has been used to draw conclusions concerning the possible common origin or common trade route of different samples [32]. The compositions of heroin, 6-MAM, and acetylcodeine as shown in Table 1.3 (Chapter 1) were used [33] for sample-differentiation purposes.

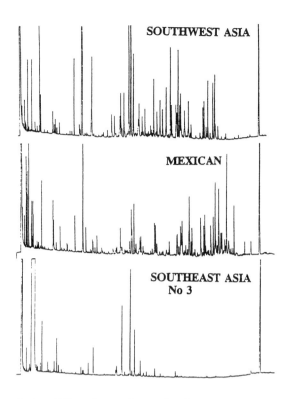

Figure 11.3. High-resolution GC analyses of trace-level, neutral processing impurities present in illicit heroin. (Reprinted with permission from Ref. [12].)

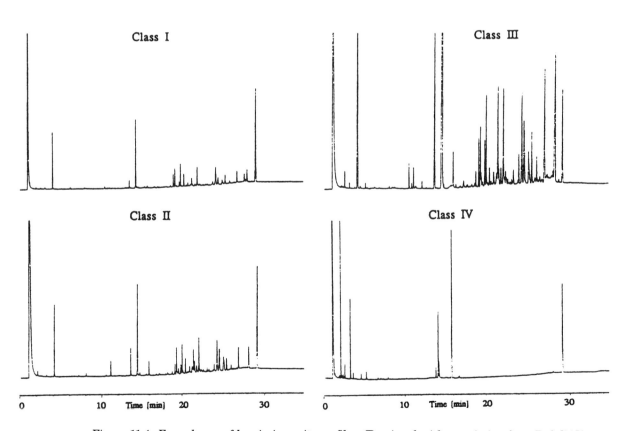

Figure 11.4. Four classes of heroin impurity profiles. (Reprinted with permission from Ref. [39].)

Analysis of Naturally Occurring Isotopes

Sample decomposition, successive dilution, and contamination during the storage and distribution process may hinder the application of sample-differentiation methods based solely on component identification and quantification. A complementary approach is based on isotopic-composition measurements. Stable-isotope ratio measurements are frequently used in studying geochemical samples [41], carbon dioxide fixation pathways in plants [42], biomedical metabolism of drugs [43, 44], environmental research [45], food adulteration [46, 47], and the differentiation of explosives [48] and hairs [49]. This approach has also been applied by this author (R. H. Liu) and others to the differentiation of marijuana [50], caffeine [51], heroin [52], diazepam [53], and confiscated methylenedioxy methamphetamine [54]. The potential application of this approach to the differentiation of various samples of forensic science interest is the topic of two book chapters [55, 56].

Although stable-isotope compositions of oxygen, nitrogen, and hydrogen have been determined, carbon-isotope measurement is most practical for its simplicity of sample preparation and potential for on-line applications.

Basic Methodology

The essential features of this approach are the unfractionated conversion of the sample into a gas that includes the element to be determined, and subsequent isotope-ratio determination of the element of interest by an MS.

Sample Preparation

Procedures used to prepare the gaseous samples for isotope-ratio determinations vary with the elements of interest. A typical procedure used for the preparation of carbon dioxide from an organic sample is as follows. Milligram quantities of the test sample are pulverized and mixed with cupric oxide. The mixture is then combusted inside a quartz tube that composes part of a vacuum line. The combustion is carried up to 900 °C and maintained at this temperature for 10 min. To ensure complete conversion of carbon into carbon dioxide, platinum gauze is used and the gas generated is recycled through the quartz tube with a Toepler pump during the entire combustion process. At the end, the carbon dioxide generated is trapped with liquid nitrogen after passing through an acetone–dry ice trap. The gas can be further purified by passing it through a series of appropriate traps. A typical sample combustion and purification system is shown in Figure 11.5.

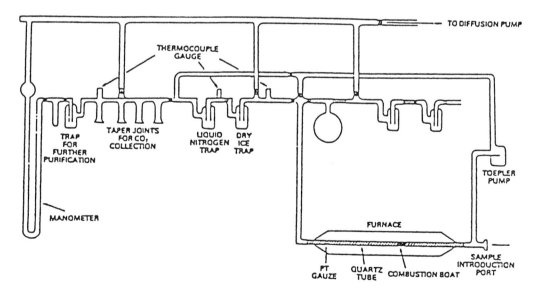

Figure 11.5. Schematic of a typical vacuum line for the combustion of organic samples and purification and collection of combustion product for the analysis of stable-carbon-isotope composition. (Reprinted with permission from Ref. [7].)

Hydrogen gas is commonly used for hydrogen-isotope ratio measurements. The water produced in the preceding combustion process can be reduced to hydrogen gas by passing it over uranium turnings at approximately 600 °C [57]. The oxygen-isotope ratio is conveniently measured in the form of carbon dioxide. Samples are sealed in a glass tube with $HgCl_2$ and heated to 500 °C for 3 h. Under these conditions, the oxygen in organic matter is converted to CO_2 and CO. The HCl also formed is removed by reacting it with quinoline. The CO is converted to CO_2 via an electric discharge and then combined with the CO_2 portion. For nitrogen-isotope ratio measurement, the sample is converted to nitrogen gas by exposing the sample to cupric oxide at 650 °C in the presence of copper wire. The small amount of nitrogen oxide resulting from this procedure is reduced by passing it through a copper furnace at 400 °C [58].

Increased interest in isotope-ratio MS has resulted in the recent development of automated or semiautomated sample-preparation procedures for carbon- and nitrogen-isotope composition measurements [59–61]. An on-line combustion, purification, and isotope-ratio MS system will undoubtedly facilitate isotope-ratio measurements and applications.

Isotope-Ratio Determination

Tunable diode IR laser spectrometry has been developed for $^{13}C/^{12}C$ measurements with an accuracy of 0.4% at this time [62]. Much better accuracy can be achieved by MS methods. Two MS approaches are commonly used to determine the stable isotopic abundance of organic compounds. The "differential" approach uses an isotope-ratio MS to measure the abundance of two selected, isotopically analogous ions from a standard gas and the same two ions from the test gas, which is derived from the sample of interest. The intensity ratios of these two ions obtained from the standard gas and the test gas are compared to determine the isotopic abundance of the test sample. Gas samples are introduced with steady flow for minimal population fluctuation in the ion source. A typical sample-introduction system is shown in Figure 11.6. The system is divided into two symmetrical halves for pressure and leak adjustment of the

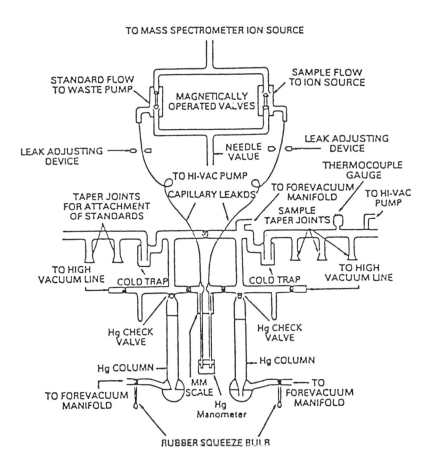

Figure 11.6. Gas-handling and dual-inlet system for isotope-ratio MS. (Reprinted with permission from Ref. [112].)

standard and the test sample. These two gases are alternately introduced into the ion source through the proper switching of the magnetically operated valves. Peak scanning is not used during MS measurements; rather, the ion beam is held stable and two (or three) ion signals are measured simultaneously with separate detectors. On the other hand, the "direct" approach for isotope-ratio measurement uses a conventional MS to measure the abundances of molecular or fragment ions and the isotopic variants of these ions. The intensity ratio of isotopically analogous ions is used directly for the calculation of the isotopic abundance.

The double inlet–double collector instrumentation generally used in the first approach provides an extremely high degree of precision. However, the process of converting the sample into a suitable gas for "differential" isotope-ratio measurement often involves laborious steps. Although some elements in selected positions of various classes of compounds can be chemically isolated [63, 64] and converted into a suitable gas for MS measurement, in most cases the whole sample is combusted and only the total isotopic abundance of the compound is determined. Consequently, chemical impurities in the test sample may cause poor accuracy of the determination. Because the "direct" approach actually detects the molecular or fragment ions of the test compound, the position identity can be preserved, and interferences originating from chemical impurities are more likely to be eliminated. Furthermore, the required instrumentation is widely available. The major disadvantage of this method is its poor precision, generally ±1–2% or more. Therefore, this method cannot be used for the comparison of isotopic abundances at the natural occurrence level; however, it is widely used for the analysis of isotopically enriched compounds. The increasing availability of isotopically enriched compounds and their widespread use in biomedical studies and in MS analysis render the "direct" approach desirable for many applications, and prompt methodology studies [65, 66] that were designed to improve the precision and accuracy of this approach.

The precision of the "differential" method allows the detection of natural variations of the stable isotopic ratios, and sometimes the determination of the origin of a given substance. Because the natural variations of stable isotopic ratios are generally small, the results of isotope-ratio MS measurements are conventionally expressed in terms of relative difference with respect to a selected standard, using the delta notation [67]:

$$\delta\ {}^{13}C\ {}^{0}/_{00}\ =\ \left(\frac{R_{sample}}{R_{standard}}\ -\ 1\right)\ \times\ 1000 \tag{11.1}$$

where

$$R\ =\ \frac{\text{atoms of minor isotope}}{\text{atoms of major isotope}} \tag{11.2}$$

Several standards have been adopted for the comparison of delta values of various isotopes [68, 69]. PDB (calcium carbonate—a fossil of Belemnitella americana from the Peedee formation in South Carolina—$^{13}C/^{12}C$ ratio, 0.011237) [70], SMOW (Standard Mean Ocean Water) [71], and air [72] are commonly used as the primary standards for $^{13}C/^{12}C$, D/H, and $^{15}N/^{14}N$ ratios. PDB and SMOW are also used as the primary standards for $^{18}O/^{16}O$ and $^{17}O/^{16}O$ ratios. PDB is calcium carbonate, a fossil (*Belemnitella americana*) from the Peedee formation in South Carolina. PDB originates from a marine carbonate shell and as such contains one of the heavier varieties of carbon in the terrestrial environment. Compared with carbon from PDB ($^{13}C/^{12}C$ = 0.011237), most natural carbon gives a negative delta value. The range of carbon-isotope variations in selected carbon-cycle reservoirs is summarized in Figure 11.7 [73].

Applications

Intrinsic Sample Isotope-Ratio Variation as a Basis for Differentiation. The

natural stable-isotope variations have been used as the basis for the differentiation of marijuana [50], heroin [52], diazepam [53], methylenedioxy methamphetamine [54], and caffeine [51]. The ratios of $^{13}C/^{12}C$ in several marijuana samples studied are listed in Table 11.3 [50]. These results indicate that ambient air seems to control the isotopic composition of the plant. The ^{13}C content in the samples

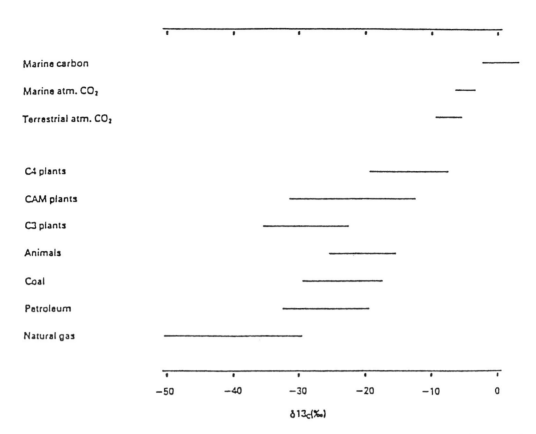

Figure 11.7. Range of carbon-isotope variation in selected carbon-cycle reservoirs. (Reprinted with permission from Ref. [73].)

Table 11.3. Ratio of $^{13}C/^{12}C$ in samples of *Cannabis sativa* L.[a]

Sample	Sample code	Run	^{13}C	Average
A	ME(3)-MI-M-L	1	−28.11	−27.99
		2	−27.82	
		3	−28.05	
B	RU(2)-MI-M-L	1	−28.85	−28.80
		2	−28.76	
		3	−28.79	
C	NO-OA-NO-L	1	−29.85	−30.25
		2	−30.64	
D	NO-NA-M-L	1	−31.64	−31.72
		2	−31.79	
E	NO-NA-II-F	1	−34.15	−34.13
		2	−34.26	
		3	−33.97	
F	NO-NA-F-L	1	−32.82	−32.80
		2	−32.79	
		3	−32.78	

[a] Data are taken from Ref. [50].

analyzed decreases in the order of A, B (from a rural area), C (from a metropolitan area), D, F (an indoor plant). This order coincides with the order of the ^{13}C content in the ambient air under which these plants are grown. Because a heater using natural gas containing less-than-usual amounts of ^{13}C is used in the greenhouse, the carbon dioxide inside the greenhouse would be the lightest. The reabsorption of carbon dioxide derived from plants confined in the greenhouse might have led to further depletion in ^{13}C content [74]. Because of the heavy use of petroleum products, it is reasonable to assume that the ^{13}C content in metropolitan areas is less than that of rural areas. Whether the difference in samples D (a male plant) and F (a female plant) is due to a sex difference is not certain, as the origin of the seeds is not well established. This preliminary study certainly provides information of interest, but a more extended survey and more specific analysis of pure chemical components are needed for a better understanding.

Variations of carbon-isotope ratio at the molecular level have been used as a basis for the differentiation of heroin samples from various geographic regions [52]. Reported data have been converted into a graphical presentation by this author (R. H. Liu) and are shown in Figure 11.8 [75]. One further step that may be useful for sample differentiation is to look into the differences of the ^{13}C-enrichment level in the acetyl groups, which are introduced during the morphine-to-heroin conversion step.

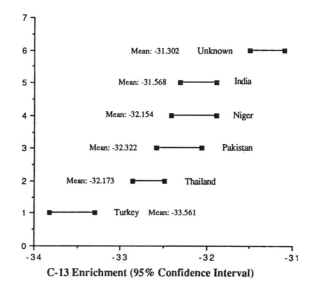

Figure 11.8. Means and ranges of δ ^{13}C (%) of heroin samples from different geographic regions, constructed on the basis of data reported in Ref. [52].

Natural-isotope abundances of carbon and hydrogen have been studied in several production batches of the commercial drug diazepam (7-chloro-1,3-dihydro-1-methyl-5-phenyl-2H-1,4-benzodiazepin-2-one). The results allow differentiation between batches [53]. Similarly, it has been possible to use $^{13}C/^{12}C$ ratios as a basis for classifying randomly selected tablets of confiscated methylenedioxy methamphetamine samples into four groups. $^{15}N/^{14}N$ ratio determinations further differentiated two samples with similar $^{13}C/^{12}C$ ratios.

A more specific study has been reported. Carbon, hydrogen, and oxygen isotope compositions of chemically pure caffeine derived from coffee and tea from different geographic locations are shown in Table 11.4. Although total differentiations of all samples based on isotopic compositions are not likely, the Brazilian D/H ratio can be distinguished from the other two D/H values, as can the Jamaican $^{18}O/^{16}O$ ratio from the other two $^{18}O/^{16}O$ values.

The lower ^{18}O content observed in the Darjeeling sample is expected because of the higher altitude of this mountainous country. The isotopic composition of its water is expected to be depleted in deuterium and ^{18}O. The D/H ratio of this sample is more enriched than those of the other two samples of tea caffeine. The D/H ratio of caffeine derived from tea, –196 to –227%, is distinctly lower than that

Table 11.4. Carbon, hydrogen, and oxygen isotope compositions of caffeine derived from various sources.[a]

Source	Origin	$\delta^{13}C$ [b]		δD [b]		$\delta^{18}O$ [b]	
		Ave	Std dev	Ave	Std dev	Ave	Std dev
Coffee	Jamaica	−28.8	0.6	−132.5	3.8	9.6	1.8
	Kenya	−29.8	0.6	−136.5	3.5	3.6	0.6
	Brazil	−28.2	0.2	−157.3	3.9	4.9	0.7
Tea	Sri Lanka	−31.7	0.8	−223.6	2.8	1.8	0.2
	Darjeeling	−29.6	0.2	−195.9	2.5	−4.3	0.8
	China	−32.4	0.6	−226.8	4.1	1.2	0.3
	BDH lab grade	−35.8	0.2	−237.1	1.7	13.0	0.3

[a] Data are taken from Ref. [51].
[b] These values are reported in $^0/_{00}$ with respect to PDB (calcium carbonate — a fossil of *Belemnitella americana* from the Peedee formation in South Carolina) and Standard Mean Ocean Water (hydrogen and oxygen).

derived from coffee, −132 to −157%. The distinct results obtained from the commercial preparation are considered an indication of synthetic preparation rather than extraction from biological sources.

Stable-Isotope Coding as a Tracing Mechanism: Methamphetamine Example. The isotope-ratio measurement of compounds that have been deliberately tagged with selected stable isotopes can also be used as a mechanism for sample-differentiation purposes. In parallel with the mechanism of labeling medically important research drugs with radioactive and nonradioactive isotopes [76], drugs with a high potential for being channeled into illicit use could be labeled with specific amounts of a nonradioactive isotope during the manufacturing process. A subsequent isotope-ratio measurement could then be used to differentiate samples from different sources, and to further identify the source of the drug and its precursors.

As an example of this concept, a study [77] on the frequently abused drug methamphetamine was conducted in this author's (R. H. Liu's) laboratory. Varying amounts of ^{13}C-labeled methylamine were incorporated during the methamphetamine synthesis process. The ^{13}C to ^{12}C ratios in the resulting products were monitored by an electron impact–quadrupole MS, with the selected ion-monitoring mode measuring the intensity ratio of the $m/z = 59$ and the $m/z = 58$ ions of the $[C_3H_8N]^+$ fragment. Synthesized products were introduced into the MS through the GC inlet without prior cleanup. The measured ratios were in excellent agreement with the calculated ones, as shown in Table 11.5.

Table 11.5. ^{13}C/^{12}C ratios in methamphetamine synthesized with ^{13}C-methylamine.[a]

Sample no.	Methamphetamine yield (%)	^{13}C-Methylamine used (%)	^{13}C to ^{12}C ratio in $[C_3H_8N]^+$		
				Measured	
			Calculated	Ave	Std dev
7	31	0	0.03961	0.03873	0.00068
9	22	0		0.03944	0.00045
14	7.5	0.25	0.04199	0.04140	0.00074
16	6.4	0.25		0.04156	0.00071
17	58	0.25		0.04166	0.00071
13	26	0.50	0.04439	0.04414	0.00066
12	18	1.0	0.04932	0.04844	0.00083
15	40	1.0		0.04871	0.00103
10	31	2.0	0.05932	0.05803	0.00096
11	23	4.0	0.07956	0.07623	0.00143

[a] Data are taken from Ref. [77].

The artificial enrichment of ^{13}C in the compound of interest allows the isotope-ratio measurements to be made with a "direct" approach using a conventional MS. Data shown in Table 11.5 indicate that a variation step of 0.25% in the ^{13}C enrichment of methylamine is sufficient for product differentiation. Calculation indicates that this variation step could be reduced at least 50-fold with the use of an isotope-ratio MS. With the development of on-line GC/combustion/isotope-ratio MS systems [78], the approach presented here may well provide a sensible mechanism for monitoring controlled substances. It should be further noted that the isotope approach exemplified here allows sample differentiations after human consumption. Analysis of appropriate metabolites may provide valuable sample-source information. Although scientific merits and economic feasibility are considered in this study, obvious sociological and legislative concerns remain to be addressed.

Analysis of Diastereoisomers and Enantiomers

The differentiation of diastereoisomers and enantiomers has two significant implications in forensic analysis. These differentiations are essential in cases where one isomer is under regulation while others may not be. For example, *l*-methamphetamine was found as a metabolite [79] of selegiline (an antiparkinson drug) and is used in Vicks inhaler (which contains a sympathomimetic drug for treating sinus disorder) [80, 81], so the differentiation of this compound from its commonly abused *d*-methamphetamine is important. Second, the determination of isomeric compositions can provide valuable information for assisting the investigation process. Certain drugs derived from natural sources occur only in one form of isomer, whereas synthesized counterparts usually result in a mixture of diastereoisomers or enantiomers or both [25]. In this sense, isomeric determination could classify a sample as to being of either synthetic or naturally occurring origin, as illustrated by the example of cocaine in the section "Impurity and Component Identification and Quantification" earlier in this chapter. Furthermore, analysis of the composition of isomers (if present) may differentiate samples prepared under different synthetic conditions. For example, identification of *l*-ephedrine in optically pure *d*-methamphetamine is an indication of the synthesis of *d*-methamphetamine from *l*-ephedrine [82].

Because of differences in their physical properties, diastereoisomers can often be determined by conventional methods. On the other hand, the differentiation of enantiomers requires the development of special methods. Several approaches have been explored in enantiomeric analyses. Earlier procedures used microcrystal tests [83] and optical methods [84, 85]. Recent applications include the use of NMR and chromatographic methods.

NMR Spectrometry

Nuclear magnetic resonance (NMR) spectrometry, which provides definite chemical information, has been applied to the resolution of enantiomers. Four different approaches are being used in the NMR determination of enantiomer composition. Prior to the use of chiral lanthanide shift reagents, enantiomers were derivatized with a chiral reagent [86] or combined with a chiral solvent [87] to induce a chemical shift difference ($\Delta\Delta\delta$) between corresponding enantiomeric groups. The third approach involves the use of a chiral solvent and an achiral lanthanide shift reagent [88]. The fourth approach employs optically active shift reagents to induce nonequivalent NMR spectra of enantiomers [89–91]. The applications of these approaches to the analysis of samples of forensic interest are demonstrated by the use of a chiral solvent [92], (−)α-phenethyl alcohol, and a chiral shift reagent [93], *tris*(3-trifluoroacetyl-α-camphorato)europium(III), in the determination of cocaine enantiomers. A more effective chiral shift reagent, *tris*(*d,d*-dicampholylmethanato)europium(III), has been used in the determination of *d*- and *l*-amphetamine [94] and *d*- and *l*-methamphetamine [95]. Details of these applications are described in Chapter 8.

Recently, enantiomeric compositions have been reported for norpseudoephedrine, norephedrine, and cathinone [96]; and ephedrine, pseudoephedrine, methamphetamine, and methcathinone [97] using NMR with a chiral solvating reagent, (*R*)-(+)-1,1'-bis-2-naphthol. It was advocated [97] that synthesis batches of methcathinone may be differentiated on the basis of the enantiomeric data obtained.

Chromatographic Methods

Both liquid and gas chromatographic techniques can be used to separate enantiomers. Separations can be achieved by the use of chiral or achiral stationary phases [98]. The latter approach requires prior derivatization with a chiral reagent to convert enantiomers to an isolable diastereoisomer. The former approach achieves the desired separation on the basis of in situ formations of transient diastereoisomers between the chiral stationary phase (CSP) and substrates [98]. To increase the stability and volatility of substrates, however, prior derivatization with an achiral derivatizing reagent is generally required, even though enantiomers are separated by CSPs.

GC Separation of Amphetamine and Methamphetamine Enantiomers

Chiral Column. Chiral columns have the advantage of being able to resolve all isomers, and direct calculation of enantiomeric compositions is possible. In cases where an achiral column is used, correction for the coelution of the other enantiomer is needed if the chiral derivatizing reagent is impure. This difficulty may become a serious problem if the exact enantiomeric purity of the chiral derivatizing reagent is not known.

Figure 11.9a shows the chromatograms resulting from the combined use of a chiral derivatizing reagent and a chiral column. The four possible isomers resulting from the reaction of *d*- and *l*-amphetamine with *N*-trifluoroacetyl-*d*-prolyl chloride (*d*-TPC) and *l*-TPC are completely resolved by the Chirasil-Val column [99].

This is important because commercial TPC contains a small amount of *d*-TPC. The elution order of these four isomers in increasing retention time is *N*-TFA-*d*-prolyl-*d*-amphetamine (Da-d), *N*-TFA-*l*-prolyl-*l*-amphetamine (La-l), *N*-TFA-*d*-prolyl-*l*-amphetamine (La-d), and *N*-TFA-*l*-prolyl-*d*-amphetamine (Da-l). The assignments of these four peaks in a chromatogram were based on relative peak sizes. Because the purity of the TPC reagent and the relative concentrations of *d*- and *l*-amphetamine in control samples are known, the relative intensities of Da-d, La-l, La-d, and Da-l are predictable and their corresponding peaks are assigned accordingly. Contrariwise, the four isomers resulting from the reaction of *d*- and *l*-methamphetamine with *d*- and *l*-TPC are resolved into only three peaks (Figure 11.9b). Based on relative intensities and the known quantity injected, these three peaks, in order of increasing retention time, are *N*-TFA-*d*-prolyl-*d*-methamphetamine (Dm-d), *N*-TFA-*l*-prolyl-*l*-methamphetamine/ *N*-TFA-*d*-prolyl-*l*-methamphetamine (Lm-l/Lm-d), and *N*-TFA-*l*-prolyl-*d*-methamphetamine (Dm-l). The inability of the Chirasil-Val column to resolve the four resulting isomers is attributed to the replacement of the active hydrogen atom attached to the nitrogen atom by a methyl group. This replacement reduces the efficiency in forming a transient diastereoisomeric association complex between the substrate and the chiral phase [46].

Achiral Column. Figure 11.9c is a chromatogram of an authentic amphetamine and methamphetamine mixture obtained from the 25-m SP-2100 column. Because Da-l and La-d and Da-l and La-l are enantiomers to each other and are not resolved by the achiral column, only two peaks are observed. By observing the relative intensity of these two peaks, it is concluded that the La-l–Da-d pair elutes first. Similar assignments have been applied to the methamphetamine peaks.

The contribution due to the small amount of *d*-TPC can be corrected using the following equations [99]:

$$A_{a,d} = A(A'_{a,d} - D)/(A - D) \qquad (11.3)$$

$$A_{a,l} = A(A'_{a,l} - D)/(A - D) \qquad (11.4)$$

where $A_{a,d}$ and $A_{a,l}$ are the corrected areas for the *d*- and *l*-enantiomer respectively; $A'_{a,d}$ and, $A'_{a,l}$ are the apparent areas of *d*- and *l*-enantiomer obtained from the chromatograms; $A = (A'_{a,d} + A'_{a,l})/2$; and D is the impurity, Y, of *d*-TPC in units of peak area and is given by $D = Y/100 \times (A'_{a,d} + A'_{a,l})$.

The SP-2100 columns were further used to analyze illicit amphetamine and methamphetamine (White Cross) samples. Chromatograms of the two illicit samples are shown in Figures 11.10a and 11.10c, respectively. Because the impurities present in the illicit amphetamine interfere with quantitative analyses, a single-ion ($m/z = 251$) chromatogram (Figure 11.10b) is added to clearly display the

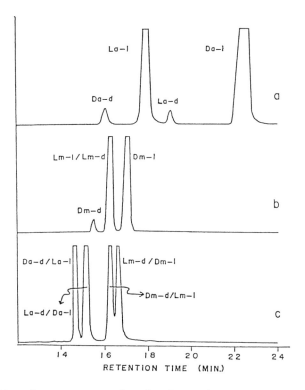

Figure 11.9. Total-ion chromatograms of authentic amphetamine (a), methamphetamine (b), and an amphetamine and methamphetamine mixture (c). (a) and (b) were obtained from a 25-m (0.30-mm i.d.) glass Chirasil-Val column, and (c) was obtained from a 25-m (0.20-mm i.d.) fused-silica SP-2100 column. (Reprinted with permission from Ref. [99].)

amphetamine peaks. For the illicit amphetamine sample, the *d/l* ratio calculated is 51.5:48.5. The illicit methamphetamine sample (White Cross) represents 95.2% *d*-methamphetamine and 4.8% *l*-methamphetamine. The peak preceding the *l*-methamphetamine peak is caffeine, which is commonly found [100] in illicit methamphetamine preparations. Assuming equal extraction efficiencies and response factors, the concentration of caffeine is about 48.3% of *d*-methamphetamine.

Liquid Chromatography

Mode of Separation. Two approaches are used to separate enantiomers using high-pressure liquid chromatography (HPLC). The indirect approach involves pre-column formation of diastereoisomers by reacting the analytes with chiral derivatizing reagents, followed by the subsequent separation of these diastereoisomers on a normal- or reversed-phase HPLC operation. The direct approach requires no prior diastereoisomeric formation; rather, it relies on the formation of transient diastereoisomeric complexes between the drug enantiomers and the CSP or between the drug enantiomers and added chiral molecules in the mobile phase.

The formation of diastereoisomers in the indirect approach uses a homochiral (single-enantiomer) derivatizing reagent (HDA) to react with the analytes. The commonly used reactions for the homochiral derivatization of amines and alcohols are shown in Table 11.6 [101], while the structures of some examples are shown in **1** through **13** (Chart 11.1).

The formation of transient diastereoisomeric complexes between the mobile-phase additive and the analytes is unique to HPLC, but it is not applicable in GC operation. This approach, although reported [102], has not found wide application in drug analysis. On the other hand, separations based on the formation of transient diastereoisomeric complexes between the drug enantiomers and the CSP have been widely used. Complex formations that constitute the differential-retention mechanisms of commercially available HPLC CSPs can be grouped into the following categories [103]:

diastereoisomers. The separation of amphetamine and methamphetamine enantiomers was studied [109] using a chiral derivatizing reagent, *N*-(trifluoroacetyl)-*l*-prolyl chloride, to form diastereoisomer pairs, which were then resolved with CSPs. The Pirkle Type 1A (ionic) column achieved baseline resolution for the derivatized amphetamine enantiomers, whereas the resolution obtained with the Pirkle Covalent (Regis Chemical Co.: Morton Grove, IL) column was not as desirable. Methamphetamine counterparts can only be partially resolved.

A chiral derivatizing agent, 2,3,4,6-tetra-*O*-acetyl-β-glucopyranosyl isothiocyanate (GITC), was successfully used [110, 111] to simultaneously resolve *R*, *R*-(–)-pseudoephedrine, *S*,*S*-(+)-pseudoephedrine, *R*,*S*-(–)-ephedrine, *S*,*R*-(+)-ephedrine, *S*-(+)-methamphetamine, and *R*-(–)-methamphetamine using a standard achiral C-18 stationary phase with tetrahydrofuran (THF)/water (3:7) as the mobile phase. A typical chromatogram is shown in Figure 11.11 [111].

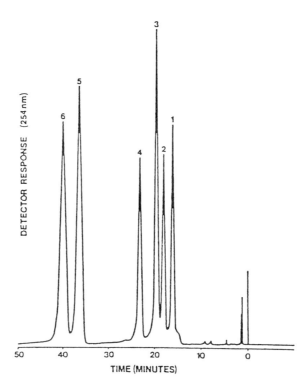

Figure 11.11. HPLC separation of GITC-derivatized amines using THF–water (3:7) mobile phase and a C-18 column: R,R-(–)-pseudoephedrine (1), S,S-(+)-pseudoephedrine (2), R,S-(–)-ephedrine (3), S,R-(+)-ephedrine (4), S-(+)-methamphetamine (5), and R-(–)-methamphetamine (6). (Reprinted with permission from Ref. [111].)

Conclusions

In principle, differences exist in all samples. Differences may be intrinsic, partly because of variations in sample cultivation or synthetic conditions; or extrinsic, attributable to variations in sample history or simply because of inhomogeneity. Each kind of difference requires a unique approach for differentiation. As more approaches are explored and further advanced, more success will be expected in the pursuit of sample differentiation.

Undoubtedly, each method surveyed in this chapter has unique merit and is suitable for a specific kind of sample or a specific analytical goal. However, none of these methods can be considered truly universal. For example, the identification and quantification of minor components or impurities serve the purposes of qualitative identification and possibly source identification. However, this approach is limited by the fact that different sample-storage conditions may influence the rate of decomposition and in this way affect the final composition [20]. Composition may also be changed as the drug is diluted and adulterated by successive dealers. On the other hand, precise determination of isotopic ratio involves extensive sample preparation and instrumentation, which are not normally available. The idea of isomer determinations is excellent because of the need to identify a specific isomer in forensic analysis [55]. Unfortunately, the logical choice of NMR method may not be practical because of the limited resolving window of the NMR spectrometric method.

The best approach should take the advantages of the high resolving power of capillary GC and the high identification capability of MS. The GC component should have the capacity to resolve enantiomers by using either a chiral column or a chiral derivatizing reagent. The MS component should have the selected ion-monitoring capacity to provide adequate isotope-ratio information if needed. In essence, this single analytical procedure will provide information about minor components and impurities, isotopic ratio, and isomer composition for samples of interest.

References

1. Nicol, J. D. *The Bachelor of Science in Criminalistics;* University of Illinois: Chicago, IL, 1972; p 2.
2. Tillson, A. H.; Franzosa, E. S. *Drug Enforcement* **1975**, *2*, 8.
3. Gomm, P. J. *J. Forensic Sci. Soc.* **1975**, *15*, 219.
4. Gomm, P. J.; Humphreys, I. J.; Armstrong, N. A. *J. Forensic Sci. Soc.* **1976**, *26*, 283.
5. Sanger, D. G.; Humphreys, I. J.; Joyce, J. R. *J. Forensic Sci. Soc.* **1979**, *19*, 65.
6. Maehly, A.; Stromberg, L. *Chemical Criminalistics;* Springer-Verlag: New York, 1981; p 30.
7. Liu, J. R. H. *Approachs to Drug Sample Differentiation;* Central Police College: Taipei, Taiwan, 1981.
8. Liu, J. R. H. *J. Forensic Sci.* **1981**, *26*, 651.
9. Baugh, L. D.; Liu, R. J. H. In *Proceedings of the Taipei International Symposium on Forensic Sciences;* Liu, R. J. H.; Chen, H.-S.; Lee, C. I. J.; Meng, H.-H.; Chen, C.-W.; Lin, Y.-S., Eds.; Central Police University: Taipei, Taiwan, 1991; p 121.
10. Neumann, H. *Proceedings of the International Symposium on the Forensic Aspects of Controlled Substances;* FBI Academy: Quantico, VA, 1988; p 121.
11. Casale, J. F.; Waggoner, R. W. *J. Forensic Sci.* **1991**, *36*, 1312.
12. Perillo, B. A.; Klein, R. F. X.; Franzosa, E. S. *Forensic Sci. Int.* **1994**, *69*, 1.
13. Tillson, A. H.; Johnson, D. W. *J. Forensic Sci.* **1974**, *19*, 873.
14. *Drug Enforcement* **1975**, *2*, 8.
15. Barron, R. P.; Kruegel, A. V.; Moore, J. M.; Kram, T. C. *J. Assoc. Off. Anal. Chem.* **1974**, *57*, 1147.
16. Kram, T. C.; Kruegel, A. V. *J. Forensic Sci.* **1977**, *22*, 40.
17. Lomonte, J. N.; Lowry, W. T.; Stone, I. C. *J. Forensic Sci.* **1976**, *21*, 575.
18. Clark, C. C. In *Analytical Methods in Forensic Chemistry;* Ho, M. H., Ed.; Ellis Horwood: New York, 1990; p 173.
19. Ensing, J. G.; de Zeeuw, R. A. *J. Forensic Sci.* **1991**, *36*, 1299.
20. Lukaszewski, T.; Jeffery, W. K. *J. Forensic Sci.* **1980**, *25*, 499.
21. Moore, J. M. *J. Assoc. Off. Anal. Chem.* **1973**, *56*, 1199.
22. Cooper, D. A.; Allen, A. C. *J. Forensic Sci.* **1984**, *29*, 1045.
23. Siegel, J. A.; Cormier, R. A. *J. Forensic Sci.* **1980**, *25*, 357.
24. Allen, A. C.; Cooper, D. A.; Kiser, W. O.; Cottrell, R. C. *J. Forensic Sci.* **1981**, *26*, 12.
25. Hufsey, J.; Cooper, D. *Microgram* **1979**, *12*, 231.
26. Novotny, M.; Lee, M. L.; Low, C.-E.; Raymond, A. *Anal. Chem.* **1976**, *48*, 24.
27. Moore, J. M.; Allen, A. C.; Cooper, D. A. *Anal. Chem.* **1984**, *56*, 642.
28. Moore, J. M.; Allen, A. C.; Cooper, D. A. *Anal. Chem.* **1984**, *56*, 642.
29. Allen, A. C.; Cooper, D. A.; Moore, J. M.; Gloger, M.; Neuman, H. *Anal. Chem.* **1984**, *56*, 2940.
30. van der Slooten, E. P. J.; van der Helm, H. T. *Forensic Sci.* **1975**, *6*, 83.
31. Narayanaswami, G. H. C.; Dua, R. D. *Forensic Sci. Int.* **1979**, *4*, 181.
32. Huizer, H. *J. Forensic Sci.* **1983**, *28*, 40.
33. O'Neil, P. J.; Baker, P. B.; Gough, T. A. *J. Forensic Sci.* **1984**, *29*, 889.
34. Evans, H. K.; Kellett, P. M. *Proceedings of the International Symposium on the Forensic Aspects of Controlled Substances;* FBI Academy: Quantico, VA, 1988; p 151.
35. dal Cason, T. A.; Fox, R.; Frank, R. S. *Anal. Chem.* **1980**, *52*, 804.
36. Verweij, A. M. A. *Forensic Sci. Rev.* **1989**, *1*, 1.

37. Liu, J. R. H. Fitzgerald, M. P.; Smith, G. V. *Anal. Chem.* **1979,** *51,* 1985.
38. Liu, J. R. H.; Fitzgerald, M. P. *J. Forensic Sci.* **1980,** *25,* 815.
39. Neumann, H. *Forensic Sci. Int.* **1994,** *69,* 7.
40. Neumann, H.; Schönberger, T. *Proceedings of the International Symposium of Forensic Science;* Tokyo, 1994; p 117.
41. Haefs, J. *Stable Isotope Geochemistry;* Spring-Verlag: New York, 1973.
42. Smith, B. N.; Epstein, S. *Plant Physiol.* **1971,** *47,* 380.
43. Schoeller, D. A.; Schneider, J. F.; Solomons, N. W.; Watkins, J. B.; Klein, P. D. *J. Lab. Clin. Med.* **1977,** *90,* 412.
44. Nakagawa, A.; Kitagawa, A.; Asami, M.; Nakamura, K.; Schoeller, D. A.; Slater, R.; Minagawa, M.; Kaplan, I. R. *Biomed. Mass Spectrom.* **1985,** *12,* 502.
45. Bricout, J. *Int. J. Mass Spectrom. Ion Phys.* **1982,** *45,* 195.
46. Doner, L. W.; Kushnir, I.; White, J. W., Jr. *Anal. Chem.* **1979,** *51,* 224.
47. Donner, L. W. *Symposium on the Detection of Juice Adulteration;* Association of Food and Drug Adulteration: New Orleans, LA, 1983.
48. Nassenbaum, A. *J. Forensic Sci.* **1975,** *20,* 455.
49. Nakamura, K.; Schoeller, D.; Winkler, F. J.; Schmidt, H.-L. *Biomed. Mass Spectrom.* **1982,** *9,* 390.
50. Liu, J. R. H.; Lin, W. F.; Fitzgerald, M. P.; Sexana, S. C.; Shieh, Y. N. *J. Forensic Sci.* **1979,** *24,* 814.
51. Dunbar, J.; Wilson, A. T. *Anal. Chem.* **1982,** *54,* 590.
52. Desage, M.; Guilluy, R.; Brazier, J. L.; Chaudron, H.; Girard, J.; Cherpin, H.; Jumeau, J. *Anal. Chim. Acta* **1991,** *247,* 249.
53. Bommer, P.; Moser, H.; Stichler, W.; Trimborn, P.; Vetter, W. *Z. Naturforsch., C: Biosci.* **1976,** *31C,* 111.
54. Mas, F.; Beemsterboer, B.; Veltkamp, A. C.; Verweij, A. M. A. *Forensic Sci. Int.* **1995,** *71,* 225.
55. Liu, R. J. H. In *Analytical Methods in Forensic Chemistry;* Ho, M. H., Ed.; Ellis Horwood: New York, 1990.
56. Brazier, J. L. In *Forensic Applications of Mass Spectrometry;* Yinon, J., Ed.; CRC Press: Boca Raton, FL, 1995.
57. Friedman, I.; Hardcastle, K. *Geochim. Cosmochim. Acta* **1972,** *34,* 125.
58. Rittenberg, D.; Ponticorvo, L. *Int. J. Appl. Radiat. Isot.* **1956,** *1,* 208.
59. Preston, T.; Owens, N. J. P. *Biomed. Mass Spectrom.* **1985,** *12,* 510.
60. Preston, T.; Owens, N. J. P. *Analyst (Cambridge, U.K.)* **1983,** *108,* 971.
61. *Analytical News* **1985,** *11,* 1.
62. Becker, J. F.; Sauke, T. B.; Loewenstein, M. *Appl. Opt.* **1992,** *31,* 1921.
63. Schmid, E. R.; Grundmann, H.; Fogy, I.; Papesch, W.; Rank, D. *Biomed. Mass Spectrom.* **1981,** *8,* 496.
64. Monson, K. D.; Hayes, J. M. *J. Biol. Chem.* **1982,** *257,* 5568.
65. Low, I. A.; Liu, R. J. H.; Barker, S. A.; Fish, F.; Settine, R. L.; Piotrowski, E. G.; Damert, W. C.; Liu, J. Y. *Biomed. Mass Spectrom.* **1985,** *12,* 633.
66. Liu, R. J. H.; Smith, F. P.; Low, I. A. *Biomed. Mass Spectrom.* **1985,** *12,* 638.
67. McKinney, C. R.; McCrea, J. M.; Epstein, S.; Allen, H. A.; Urey, H. C. *Rev. Sci. Instrum.* **1950,** *21,* 724.
68. Hayes, J. M. *Spectra* **1982,** *8,* 3.
69. Gonfiantini, R. *Nature (London)* **1978,** *271,* 534.
70. Urey, H. C.; Lowenstam, H. A.; Epstein, S.; McKinney, C. R. *Bull. Geol. Soc. Am.* **1952,** *62,* 399.
71. Craig, H. *Science (Washington, D.C.)* **1961,** *133,* 1833.
72. Junk, G.; Svec, H. *Geochim. Cosmochim. Acta* **1958,** *14,* 234.
73. Krueger, H. W.; Reesman, R. H. *Mass Spectrom. Rev.* **1982,** *1,* 205.
74. Keelin, C. D. *Geochim. Cosmochim. Acta* **1961,** *24,* 299.
75. Liu, R. J. H. In *Proceedings of International Symposium of Forensic Science, Tokyo 1993;* National Institute of Police Science: Tokyo, Japan; 1993; p 127.
76. Wolfe, R. R. *Tracers in Metabolic Research: Radioactive and Stable Isotope/Mass Spectrometry Methods;* Alan R. Liss: New York, 1983.
77. Low, I. A., Liu, R. J. H.; Piotrowski, E. G.; Furner, R. L. *Biomed. Environ. Mass Spectrom.* **1986,** *13,* 531.
78. *Analytical News* **1985,** *11,* 1.
79. Meeker, J. E.; Reynolds, P. C. *J. Anal. Toxicol.* **1990,** *14,* 330.
80. Solomon, M. D.; Wright, J. A. *Clin. Chem.* **1977,** *23,* 1504.
81. Fitzgerald, R. L.; Ramos, J. M., Jr.; Bogema, S. C.; Poklis, A. *J. Anal. Toxicol.* **1988,** *12,* 255.
82. Noggle, F. T., Jr.; DeRuiter, J.; Clark, C. R. *Anal. Chem.* **1986,** *58,* 1643.
83. Fulton, C. C. *Modern Microcrystal Tests for Drugs;* Wiley-Interscience: New York, 1969.
84. Raban, M.; Mislow, K. In *Topics in Stereochemistry;* Allinger, N. L.; Eliel, E. L., Eds.; Interscience: New York, 1967; Vol. 2, p 199.
85. Crabbe, P. *ORD and CD in Chemistry and Biochemistry: An Introduction;* Academic: New York, 1972.
86. Dale, J. A.; Dull, D. L.; Mosher, H. S. *J. Org. Chem.* **1969,** *34,* 2543.

87. Pirkle, W. H.; Beare, S. D. *J. Am. Chem. Soc.* **1969,** *91,* 5151.
88. Jennison, C. P. R.; MacKay, D. *Can. J. Chem.* **1973,** *51,* 3726.
89. MeCreary, M. D.; Lewis, D. W.; Wemick, D. L.; Whitesides, G. M. *J. Am. Chem. Soc.* **1974,** *96,* 1038.
90. Goering, H. C.; Eikenberry, J. N.; Koermer, G. S.; Lattimer, C. J. *J. Am. Chem. Soc.* **1974,** *96,* 1493.
91. Sullivan, G. R. In *Topics in Stereochemistry,* Eliel, E. I.; Allinger, N. L., Eds.; Interscience: New York, 1978; Vol. 10, p 287.
92. Gelsomino, R.; Raney, J. K. *Microgram* **1979,** *12,* 222.
93. Kroll, J. A. *J. Forensic Sci.* **1979,** *24,* 303.
94. Liu, J. R. H.; Tsay, J. T. *Analyst (Cambridge, U.K.)* **1982,** *107,* 544.
95. Liu, J. R. H.; Ramesh, S.; Tsay, J. T.; Ku, W. W.; Fitzgerald, M. P.; Angelos, S. A.; Lins, L. K. *J. Forensic Sci.* **1981,** *26,* 656.
96. Dawson, H. F.; Black, D. B.; Lavoie, A.; LeBelle, M. J. *J. Forensic Sci.* **1994,** *39,* 1026.
97. LeBelle, M. J.; Savard, C.; Dawson, B. A.; Black, D. B.; Katyal, L. K.; Zrcek, F.; By, A. W. *Forensic Sci. Int.* **1995,** *71,* 215.
98. Souther, R. W. *Chromatographic Separation of Stereoisomers;* CRC Press: Boca Raton, FL, 1985.
99. Liu, J. R. H.; Ku, W. W. *Anal. Chem.* **1981,** *53,* 2180.
100. *Microgram* **1974,** *7,* 51.
101. Gal, J. *LC-GC* **1987,** *5,* 106.
102. Mularz, E. A.; Cline-Love, L. J.; Petersheim, M. *Anal. Chem.* **1988,** *60,* 2751.
103. Wainer, I. W. *Chromatogr. Forum* **1986,** 55.
104. Crowther, J. B.; Covey, T. R.; Dewey, E. A.; Denion, J. D. *Anal. Chem.* **1984,** *56,* 2921.
105. Lee, E. D.; Henion, J. D.; Brunner, C. A.; Wainer, I. W.; Doyle, T. D.; Gal, J. *Anal. Chem.* **1988,** *58,* 1349.
106. *Biotext* **1989,** *2,* 6.
107. Hermansson, J. *J. Chromatogr.* **1984,** *298,* 67.
108. Chiarotti, M.; Fucci, N. *Forensic Sci. Int.* **1990,** *44,* 37.
109. Hayes, S. M.; Liu, R. J. H.; Tsang, W.-S.; Legendre, M. G.; Berni, R. J.; Pillion, D. J.; Barnes, S.; Ho, M. H. *J. Chromatogr.* **1987,** *398,* 239.
110. Noggle, F. T., Jr.; DeRuiter, J.; Clark, C. R. *Anal. Chem.* **1986,** *58,* 1643.
111. Noggle, F. T.; Clark, C. R. *J. Forensic Sci.* **1986,** *31,* 732.
112. Nuclide Product Bulletin PUBS 1002-A-0172, State College, PA, 1973.

Chapter 12
Interpretation of Test Results

Sound analytical methodology and good laboratory practice are undoubtedly necessary components for achieving accurate test results. However, end users of a test report may be required to look beyond the scientific aspects associated with test methodology and practice, and to cautiously interpret the test results. Factors such as sample adulteration, the origin of the drug or metabolite found, and quantitative information may require careful consideration in formulating the true meaning of a test result.

The considerations of test-result interpretation in this chapter mainly address issues that are pertinent to workplace urine drug testing—the category of drug-analysis activities with the largest sample volume at this time.

Sample Adulteration

The contents of a test sample may be altered unintentionally or intentionally. A recent review addressed a variety of adulterants and their effects on urine drug-test results by employing various test methodologies [1]. In vivo and in vitro adulteration have both been reported.

In Vivo Adulteration

The intake of certain substances may affect test results in three ways. First, the use of effective diuretics or simply large volumes of liquids has been reported to cause the dilution of drugs being monitored by a factor of up to 10 [2].

Second, the use of certain liquids, such as vinegar and acidic fruit juices, may alter the pH of the urine, thereby indirectly altering the excretion patterns of certain drugs. For example, it has been reported [3–6] that under an alkaline urinary pH, amphetamines are excreted more slowly; thus, these drugs may be present in the urine for a longer period of time but at lower concentrations. It has also been reported [7] that the excretion of phencyclidine (PCP) is enhanced under acidic urine conditions.

Third, the use of certain materials may result in the excretion of components that interfere with testing methods. For example, the literature contains a report [8] that the presence of alkaloids resulting from the use of golden seal root may mask the thin-layer chromatographic characteristics of opiates. It also has been reported [9] that the presence of a large amount of ibuprofen may consume the derivatizing reagent commonly used in the analysis of 11-nor-Δ^9-tetrahydrocannabinol-9-carboxylic acid (9-THC-COOH). Thus, the targeted metabolite may escape detection if handled by an inexperienced analyst utilizing a casual test procedure.

In Vitro Adulteration

Following the universal adoption of urine as the preferred test specimen for various workplace drug-testing programs, numerous adulteration procedures have been publicized. Studies conducted in

Modified with permission from *Forensic Science Review,* 1992, Vol. 4, pages 51–65. Copyright 1992 Central Police University, Taiwan.

drug-testing laboratories have also confirmed that the addition of various adulterants can indeed alter immunoassay data. The effects of adulterants that are commonly available in the sample-collection environment have been examined in several radioimmunoassay (RIA) and enzyme immunoassay (EIA) studies, but these adulterants have not been as thoroughly tested using fluorescence polarization immunoassay (FPIA). Adulterants that were reported to cause significant interference were reviewed in a recent article [1] and are summarized in Table 12.1.

Table 12.1. Effects of adulterants on immunoassays.[a]

Substance	Amount	Method	Reported alteration of test result	Ref.
Ammonia	5–10%	RIA	False negative for benzoylecgonine	[10]
Bleach	10%	RIA	False negative for amphetamine, 9-THC-COOH, morphine	[10]
	One drop/10 mL	EMIT	False negative for 9-THC-COOH, amphetamine, benzodiazepine, barbiturate, benzoylecgonine, opiate	[11]
Blood	One drop/10 mL	EMIT	False negative for 9-THC-COOH	[12]
Drano	All concn	EMIT	False negative for amphetamine, benzodiazepine, barbiturate, benzoylecgonine, opiate, 9-THC-COOH	[14]
	10%	RIA	Indistinguishable results for positive and negative samples for morphine, amphetamine, PCP, barbiturate	[10]
Goldenseal	0.9%	RIA	Reduced 9-THC-COOH	[10]
	30 mg/mL	EMIT	False negative for 9-THC-COOH	[14]
Joy		RIA	False positive for 9-THC-COOH	[13]
		FPIA	False positive for amphetamine, barbiturate, benzodiazepine	[13]
Liq. Detergent	One drop/10 mL	EMIT	False negative for 9-THC-COOH	[12]
Multi-Terge	10%	RIA	False negative for benzoylecgonine	[27]
Salt	50–250 g/L	EMIT	May cause false negative for amphetamine, barbiturate, benzoylecgonine, benzodiazepine, opiate, PCP, 9-THC-COOH, methadone	[12–16]
Soap	10%	RIA	False positive for 9-THC-COOH	[10]
		EMIT	False negative for barbiturate, benzodiazepine, 9-THC-COOH	[14]
Vanish	1–10%	RIA	May cause false positive or false negative	[10]
Vinegar		EMIT	False negative for 9-THC-COOH	[12, 14]
Visine	1–25%	RIA	Reduced 9-THC-COOH	[10]
		EMIT	False negative for benzodiazepine, 9-THC-COOH	[14]

[a] Data are taken from Ref. [1].

Table 12.1 shows disturbing data. Many commonly available adulterants can impact significantly on test results, especially when high concentrations of adulterants are present or when an enzyme multiplied immunoassay technique (EMIT) is used as the screen test procedure. Concern is not so much focused on false positives generated by a preliminary test, because they are identified by a sound confirmatory test procedure when applied properly. Instead, concern centers on false negatives, as such samples are likely to evade detection. Even though a rigorous gas chromatography–mass spectrometry (GC–MS) procedure can correctly identify the drug in the presence of an adulterant, these procedures are not used to test negative samples under most circumstances. Thus, a serious drug-testing program should adopt the most reliable screen test procedure available and take preventive steps to eliminate or minimize false test results caused by sample adulteration.

Positive Results Derived From Nonabuse Exposure

Urinary excretion of morphine and codeine [17] derived from the consumption of poppy seeds, and the excretion of many commonly abused drugs following unconventional means of exposure [18], was recently reviewed. Of particular concern is the detection of drugs and metabolites that are commonly used as the indicators of drug abuse following an unintended exposure to illicit drugs or the consumption of licit food or medicinal items.

Unintended Exposure

Marijuana

The detection of cannabinoids in the urine following passive inhalation is an interesting topic that has been substantially investigated and reported on. These studies have been summarized by Cone and Huestis [18], as shown in Table 12.2. The observation of a urinary 9-THC-COOH concentration sufficient to be called "positive" as a result of passive inhalation may be possible, but it is unlikely.

A 1-h exposure over a period of 6 consecutive days in a closed bathroom-size area with 16 smoking marijuana cigarettes (average weight, 877 mg; THC content, 2.8%) yielded, in 6 passive inhalers, maximum urinary 9-THC-COOH concentrations ranging from 10 to 87 ng/mL [25]. The value of 87 ng/mL was obtained on day 4 of the study. These observations were similar to those found following the active smoking of one or two marijuana cigarettes.

The passive-inhalation condition adopted in the above study is undoubtedly unrealistic. The researchers commented [25] that:

1. One is unlikely to be able to tolerate exposure to the smoke of 16 cigarettes without eye goggles for an extended period of time.

2. The THC level in the air in the room would be reduced >90% if the door to the room were kept open.

3. The results of 10 min of passive-inhalation exposure at the highest THC level would be roughly equivalent to 60 min of passive-inhalation exposure to the smoke of 4 marijuana cigarettes—an exposure (1 h for 6 consecutive days) that resulted in the detection of maximum 9-THC-COOH concentrations ranging from 0 to 12 ng/mL.

It was thus extremely improbable [25] that an individual would unknowingly tolerate a noxious smoke environment for a length of time sufficient to produce urinary metabolites equivalent to 50 or 100 ng/mL of 9-THC-COOH, which is required for a positive immunoassay that utilized 50 or 100 ng/mL of 9-THC-COOH as the cutoff calibration standard. A 1988 report by Mulé et al. [27] concluded that passive inhalation under conditions likely to reflect realistic exposure consistently resulted in <10 ng/mL equivalent of 9-THC-COOH, as determined by the highly 9-THC-COOH specific Cannabis Direct RIA for Urine THC Kit from Amersham Corporation (Arlington Heights, IL), available at the time of the study.

Although these literature data on passive inhalation of marijuana smoke are informative, one should also note the reported differences in interindividual excretion patterns when attempting to interpret or extrapolate the observations from these controlled studies. Their impact on the interpretation of drug-test results needs more thorough study.

Cocaine

Studies concerning unintended exposure to cocaine [28–30] have been reported. Passive inhalation of cocaine vapor generated by heating 200 mg of freebase cocaine and by smokers (12.5–50 mg) in an unventilated room resulted in the detection of peak benzoylecgonine concentrations of 123 and 6 ng/mL, respectively [30]. It was estimated that passive conditions that would result in absorption of cocaine in amounts exceeding 1 mg could result in the production of cocaine-positive urine specimens [30].

Another type of exposure that deserves serious attention is the possibility of urinary detection of drugs and metabolites derived from unintended skin absorption. Although serving mainly as a protective barrier, skin is also accessible to the entry of foreign substances [31]. It has also been reported [32, 33] that low levels of benzoylecgonine were found in the urine samples of laboratory workers who came in contact with cocaine. A recent study [34] further reported the detection of high urinary levels (>1500 ng/mL) of benzoylecgonine from a criminalist with prolonged exposure to airborne cocaine hydrochloride dust. Thus, the possibility of inadvertent exposure by law-enforcement and laboratory workers who handle these substances during seizure and analysis must be seriously considered.

Table 12.2. Summary of studies on the passive inhalation of marijuana.[a]

Approx amt of THC smoked	Room vol (L)	Length of exposure	No. of subjects	Analytical method	Peak urine level (ng/mL)	Time to last positive	Comment	Ref.
"Heavy smoker"	Not given	Close contact intermittently for 15 d; some exposed 6 more d	5 smoking; 1 passive	Chelate/colorimetric analysis	260[b]	>50 ng/mL after 11 d	Earliest report; poorly designed and controlled study	[19]
0.0 mg (placebo)	≈26,000	2 h	1 smoking; 1 passive (16 trials)	EMIT	32[c,d]	—	Morning specimen (after night exposure)	[20]
2.6 mg	≈26,000	2 h	1 smoking; 1 passive (16 trials)	EMIT	21[c]	2.5 h		[20]
5.3 mg	≈26,000	2 h	1 smoking; 1 passive (16 trials)	EMIT	28[c]		Morning specimen (after night exposure)	[20]
47 mg (two 2.5% cigarettes)	15,500	1 h	4 smoking; 2 passive	EMIT	<20[c]	None positive	Urine collected 24 h	[21,22]
52 mg (two 2.8% cigarettes)	15,500	1 h	4 smoking; 2 passive	EMIT	<20[c]	None positive	Urine collected 24 h	[21,22]
52 mg (two 2.8% cigarettes)	3,500 (car)	1 h	4 smoking; 2 passive	EMIT	>20[c]	6 h	1 of 23 urine above 20 ng/mL cutoff in 24 h	[21,22]
105 mg (four 2.8% cigarettes)	15,500	1 h	4 smoking; 2 passive (day 1 and 3) and 3 passive (day 2 only)	EMIT	>20[c]	5 h	1 of 27 urine slightly above 20 ng/mL cutoff in 72 h	[21,22]
103 mg (six 17.1 mg cigarettes)	27,950	3 h	6 smoking; 4 passive	RIA	<10[c]	—	Urine collected only up to 6 h; longer period desirable	[23]
90 mg (twelve 7.5% cigarettes)	1,650 (car)	0.5 h	2 smoking; 3 passive	EMIT and RIA	>20 (38 ng/mL by RIA)[c]	72 h	2 of 3 subjects >20 ng/mL day 1; 1 subject positive for 5 d	[24]
98 mg each day for 6 d (four 2.8% cigarettes)	12,226	1 h	5 passive	EMIT	>20[c,e]	6.5 h	Placebo-controlled, crossover designed study	[25, 26]
398 mg each day for 6 d (sixteen 2.8% cigarettes)	12,226	1 h	5 passive	RIA	>100[c,e]	57.3 h		[25, 26]
398 mg each day for 6 d (sixteen 2.8% cigarettes)	12,226	1 h	2 passive (marijuana naive)	RIA	>100[c,e]	141 h		[25, 26]
108 mg (four 2.8% cigarettes)	21,600	1 h	3 passive	RIA[f]	<10[c] (20–24 h after exposure)	Unknown	Urine not collected prior to 20–24 h	[27]

a With the exceptions of the information in the "Analytical method" column and the footnotes in the "Peak urine level (ng/mL)" column, all data are taken with permission from Ref. [18].

b The specificity of chelation formation toward cannabinoids and their metabolites is not known.

c Unreliable data (false positive results).

d Because immunoassays are used in these studies, the listed concentrations represent total 9-THC-COOH equivalents.

e GC-MS analyses were also used to determine 9-THC-COOH concentrations in selected samples.

f The RIA used was known to be highly specific toward 9-THC-COOH.

A recent study on RIA analysis of hair samples collected from nine undercover narcotics officers failed to generate a concentration exceeding 5 ng/mg [35]. None was detected in proteinase enzyme digest [35] of hair samples that had been washed with 2-propanol and phosphate-buffer washes that were claimed to be effective for removing surface contamination [36].

Others

Unintended exposure to phencyclidine (PCP) [37] has also been reported. Methamphetamine-related cases will likely be reported in the future. An alarming case has been reported [38] in which a Navy pilot wearing PCP-soiled clothes continued to test positively even after the clothes were laundered repeatedly. More-thorough studies on the potential absorption of PCP [39] and tricyclic antidepressants [40] were reported thereafter using the hairless mouse as a model.

Food and Licit Medication

The detection of morphine and codeine following the consumption of commonly used food items, such as poppy seeds, has been reported by several investigators [41–47]. As summarized by ElSohly and Jones [17], the concentrations of morphine and codeine in various sources of poppy seeds available as food items are generally <300 and 5 µg/g, respectively, with a morphine-to-codeine ratio of more than 5. Urinary excretions of morphine and codeine normally peak at 2 to 6 h following the ingestion of poppy seeds. With the consumption of a normal amount of poppy-seed-containing food items, the maximum morphine and codeine concentrations detected in urine are normally below 5000 and 300 ng/mL, respectively. The morphine-to-codeine ratios reported range from 2:1 to 60:1. A later study [48] conducted to verify these "guidelines" reported the observed urinary morphine and codeine concentrations exceeding the respective 5000- and 300-ng/mL limits from individuals consuming Danish pastry and streusel that contained a total of 24 g of poppy seeds from Mexico. However, this study also concluded that no specimen had a morphine-to-codeine ratio of less than 2.

The urinary excretion data as generalized above cannot be used alone as definite indicators in detecting drug abuse. The use of these data in conjunction with other information listed below, however, can assist in differentiating the source of morphine and codeine detected [49]:

1. The presence of 6-monoacetylmorphine can be used as a definite indication of heroin use [50].

2. Because the morphine-to-codeine ratio after the intake of codeine can grow to ≥1 [51, 52] in the late excretion phase or with high codeine dosage, the claim of predominant presence of morphine resulting from the use of codeine-containing medication should be examined carefully.

Thus, while the morphine-to-codeine concentration ratio observed in a urinalysis may be informative, it should not be used as the sole indicator as to what type of opiate the subject was exposed to.

The detection of methamphetamine following the use of Vicks nasal inhaler further exemplifies the necessity for cautious test-result interpretation. Fortunately, the *R*(–)-methamphetamine (*l*-methamphetamine) used in Vicks nasal inhaler is not the sole component of the methamphetamine that is commonly abused. Thus, the predominant presence of the other optical isomer, *S*(+)-methamphetamine (*d*-methamphetamine), can provide the basis for differentiating an inhaler user from a drug abuser. Indeed, detailed studies of the differentiation and determination of amphetamine [53] and methamphetamine [54] enantiomeric compositions have been reported. Data obtained from field specimens [55] have also substantiated the validity of the approach based on optical isomer analysis.

With the understanding that (1) relatively pure *l*-methamphetamine is used in Vicks nasal inhaler, (2) either relatively pure *d*-methamphetamine or a racemic mixture is found in the methamphetamine used for abuse, and (3) the preferential metabolism of either optical isomer and racemization of *l*- or *d*-methamphetamine metabolism is limited, the U.S. Navy Drug Screening Laboratories adopted the policy [56] that a methamphetamine positive result (*d,l*-methamphetamine ≥ 500 ng/mL) should be further tested for its enantiomeric composition if all of the following conditions are met:

1. Methamphetamine < 12,000 ng/mL;

2. Amphetamine < 1000 ng/mL; and

3. Amphetamine < methamphetamine.

This policy was changed later; all methamphetamine positive samples are now tested for their enantiomeric compositions.

A sample will be reported as positive for d-methamphetamine if the enantiomeric composition determination reveals that the d-methamphetamine composition is ≥90% of the total methamphetamine present. (One should note that as a result of coeluting chromatographic peaks [53, 54], the 90% criterion may not be suitable if the purity of the chiral derivatizing reagent is too low.) A sample will also be reported positive for d-methamphetamine if d-methamphetamine is <90% but >10%, and one of the following conditions is met:

1. The d-methamphetamine component alone is ≥500 ng/mL; or
2. The amount of d-amphetamine present is more than 250 ng/mL and is more than the amount of l-amphetamine.

U.S. Air Force and Army drug-testing programs also conduct enantiomer composition determinations for samples testing positive for amphetamines.

In light of the problem associated with the use of Vicks nasal inhaler and the presence of a high concentration of ephedrine or pseudoephedrine, the U.S. Department of Health and Human Services [57] recommended the determination of d- and l-methamphetamine composition for methamphetamine source differentiation. A ≥80% l-methamphetamine finding is considered consistent with the usage of Vicks nasal inhaler; therefore, the finding should be interpreted as negative. To guard against reporting a false methamphetamine positive because of the presence of high levels of ephedrine or pseudoephedrine, a reporting policy that requires that 200 ng/mL of amphetamine be present in order to report a sample as positive for methamphetamine was implemented [58]. Although it is plausible to implement reporting policies that exceed scientific requirements as an additional safeguard, it should be noted that Vicks nasal inhaler contains essentially pure l-methamphetamine, and that this compound will not metabolize to yield a significant amount of d-methamphetamine [4, 55]. One should also note that illicit methamphetamine may be formulated for abuse and result in a urine d-methamphetamine composition that will meet the criteria noted above as an indication of Vicks nasal inhaler use. Furthermore, the 200-ng/mL amphetamine requirement essentially raises the positive cutoff level of methamphetamine beyond the established 500-ng/mL value, because more than 500 ng/mL of methamphetamine will most likely result from the metabolic process that generates 200 ng/mL of amphetamine [4].

Amphetamine and methamphetamine may derive from other medicines besides Vicks nasal inhaler. For example, amphetamine is available as Dexedrine, which consists solely of dextroamphetamine (d-amphetamine), and benzphetamine, which consists of equal amounts of d-amphetamine and racemic (d,l-) amphetamine. The l-isomer is not available in pure form in any pharmaceutical preparation. Desoxyn is pure d-methamphetamine and is prescribed for attention deficit disorder with hyperactivity and for patients in whom obesity is refractory to alternate therapy. The l-form of methamphetamine, which produces primarily peripheral effects, is available over-the-counter in Vicks nasal inhaler, as discussed above [59].

Amphetamine and methamphetamine may also present as metabolites of other drugs (precursors) such as: amphetaminil, benzphetamine, clobenzorex, deprenyl, dimethylamphetamine, ethylamphetamine, famprofazone, fencamine, fenethylline, fenproporex, furfenorex, mefenorex, mesocarb, and prenylamine. Although many of these drugs are not universally available throughout the world, one or more of these drugs are available in essentially every part of the world [60]. Amphetamine and methamphetamine derived from these drugs have certain enantiomeric composition characteristics and may often be differentiated from illicit manufactures.

Quantification

Modern clinical and forensic drug analyses often include quantification of the drug detected. Recent practices in workplace drug testing require determining whether the concentration of the analyte identified is below or above (or at) a threshold level (often called the *cutoff*). Thus, the calibration procedure is a significant part of the overall analytical protocol. One commonly used calibration approach involves use of a multipoint calibration curve. Another popular approach uses the arithmetic mean of two or more calibrators chosen at a single concentration level at or near the cutoff concentration. It is stated that the "assessment of accuracy at a threshold by the arithmetic mean procedure may be quite acceptable at that threshold but not at significantly higher or lower concentrations. Multipoint calibration, on the other hand, can define the accuracy of a method within broader limits, provided the limits of precision are known and are acceptable." [61]

Concentration Near the Cutoff

When used for administrative purposes, drug test results are often reported as either "positive" or "negative". A positive result normally means that the target drug or metabolite was found at a level at or above a preset level (the cutoff concentration). A quantitative result obtained from any scientific measurement does not represent an absolute value; rather, it represents a statistical estimate that is allowed to vary within a predetermined limit, such as ±20%, which is commonly considered acceptable in the drug urinalysis industry. With this in mind, one may argue that a specimen that tests positive, with the drug concentration on the cutoff level, may in fact contain the drug in a range from 20% below or above the cutoff concentration. Specifically, if a specimen is found to contain benzoylecgonine at the level of 150 ng/mL (the cutoff concentration as established by the Mandatory Guidelines [62]), the concentration actually may be as low as 120 ng/mL or as high as 180 ng/mL. Should any action be taken when a positive result is based on a concentration at this level? Further, it is not uncommon to find that no action is taken when a blood alcohol concentration obtained through breath alcohol analysis fails to exceed 0.11%, where the adopted legal limit is 0.10%.

A fundamental difference should be recognized between the analysis of alcohol and abusive drugs, as implied by test results. The ingestion of alcohol by adults is legal, whereas the use of abusive drugs outside a medical context is illegal. In general, the presence of alcohol in a tested person is legally and morally acceptable in today's society. It is when the amount exceeds a level set by the law (for example, while operating a motor vehicle) that the person being tested will be subjected to legal action. Therefore, the cutoff concentration is used as the criterion for establishing the presence of alcohol, and to further certify that the amount present is above a permitted limit. Before any legal action is taken, one must demonstrate with scientific certainty that the reported amount is indeed above the cutoff concentration.

On the other hand, the utilization of abusive drugs outside a medical context is illegal. If it is beyond scientific doubt that the drug is indeed present in the body, the tested person is proven to have violated the law. The significance of an exact concentration is diminished because the amount found depends primarily on the time lapse between drug ingestion and specimen collection. The use of a cutoff concentration is merely for the sake of convenience and in applying a uniform standard for fair practice. If a test result is above the cutoff level, whether this result could be statistically below the cutoff is irrelevant, provided that (1) the statistical variation does not include the possibility of being scientifically indistinguishable from zero, and (2) the same criterion is applied to all persons tested.

In essence, the issue in testing abusive drugs is to determine whether a drug is present, whereas the issue in blood alcohol analysis is to ensure that the alcohol present is above or below a set limit.

Thus, if (1) the testing methodology can establish the presence (with acceptable precision and accuracy) of the drug beyond scientific doubt, and (2) the concentration level is above the cutoff level necessary to comply with the legal technicality requirement, the possibility that the concentration level may be statistically below the cutoff concentration level should not be used as a valid argument for placing the specimen in the "negative" category. It is this author's (R. H. Liu's) opinion that this policy can be adopted to (1) the categorization of a preliminary test result for deciding whether a sample should be submitted for further testing, or (2) the categorization of a confirmatory test result for the final test report.

Quantification Corrected with Creatinine Content and Urinary pH

Quantitative data on detected drugs are normally reported as nanograms of drug per milliliter of urine specimen (ng drug/mL urine). Although this reporting system is generally adequate, circumstances exist under which the variations in urine water content (due to dehydrating exercise or the intake of a large volume of liquid) should be considered, and a reporting system based on nanograms of drug per milligrams of creatinine (ng drug/mg creatinine) will better serve drug-monitoring purposes. For example, the latter reporting system may defeat the attempt to evade a positive test by sample dilution. More significantly, one may avoid unjustly accusing an individual in a drug-intervention program of re-establishing drug use if the observation of higher values (ng drug/mL urine) in a sample subsequently collected can be attributed to the lower water content of the sample in question.

The data shown in Figure 12.1 [63] illustrate the merit of the ng drug/mg creatinine reporting system. When the drug ng/mL data shown in the figure are used, the apparent increase on day 6 may be interpreted as an indication of further marijuana use; however, the creatinine-corrected data show a continuous decrease in the marijuana metabolite level. Similarly, the creatinine-corrected data obtained on day 10 provide an indication of further marijuana use. This creatinine-corrected reporting system was satisfactorily used [64] for monitoring drug-excretion profiles in heavy marijuana users.

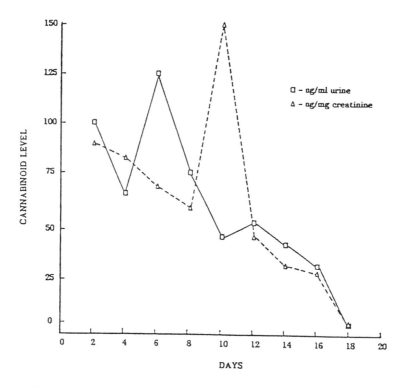

Figure 12.1. Urinary cannabinoid (THC metabolite) levels of specimens taken on alternate days after last use of marijuana, expressed in ng metabolite/mL urine (squares) and in ng metabolite/mg creatinine (triangles). (Reprinted with permission from Ref. [63].)

Although the creatinine content correction approach may be useful for reporting 9-THC-COOH data, the excretion of some drugs, such as amphetamine [65] and methamphetamine [4], is little influenced by urine volume, but far more by urinary pH. The rate of excretion of these drugs increases significantly under acidic urinary pH conditions. Reporting amphetamine test results with urinary pH correction is not a simple matter, however, and it cannot be reasonably done. To the best knowledge of this author (R. H. Liu), reporting test results with urine pH adjustment has not been explicitly proposed.

Time Lapse Between Drug Exposure and Sample Collection

Information on the dosage amount and the time of the alleged drug exposure may be useful for establishing the source of the drug that contributes to the test finding or for relating drug exposure to impaired performance. Often, the testing laboratory is asked to provide estimates of this information. Although numerous pharmacokinetic studies on various drugs may be found, and it may be possible to estimate the dosage amount and time information on the basis of blood or serum drug concentrations, estimating dosage amount and the postexposure time interval is not an easy task on the basis of urinary drug-concentration information alone.

Difficulties in providing reliable dosage amount and time information based on urinary drug concentration are due to the pH dependency of urinary excretion patterns and to the reabsorption that may have taken place during the collection of urine specimens following several hours of accumulation. Nevertheless, the literature reports the time-curve information, which may be of reference value. For example, the plasma concentrations of THC and 9-THC-COOH, and the urinary 9-THC-COOH concentration observed in heavy and light marijuana users, were reported [66] as shown in Figures 5.3 and 5.4 (Chapter 5). Similarly, nomograms (Figure 12.2) for relating urinary benzoylecgonine concentrations to the size of dosage and the time interval since the cocaine use have been reported [67].

Quantitative information concerning the distribution of drug metabolites has been advocated as a basis for estimating the time of the drug exposure. This topic will be further reviewed in the section "Identification and Quantification of Drug Metabolites".

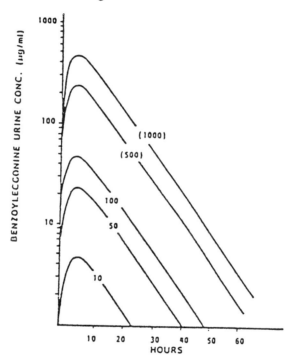

Figure 12.2. A nomogram for relating benzoylecgonine urine concentration to the size of, and time interval since, the cocaine dose. The cocaine dosage may be single or repeated and cumulative if taken within a relatively short period (relative to BZ excretion) of a few hours. Extrapolation between curves is shown for other doses. Doses of 500 mg and above are shown in parentheses to indicate that although the kinetics are probably similar above 200 mg, no data are available at the present time. (Reprinted with permission from Ref. [67].)

Stability

Stability of the analyte during the postcollection period may affect both the quantitative data obtained at the time of analysis and the interpretation of these data. Data from the literature on the stability characteristics of many common drugs in various biological specimens stored under common conditions have recently been reviewed and are summarized in Table 12.3 [68]. This information provides a general guide for handling specimens of concern.

The most important factors that may affect the stability of an analyte include pH variations [106] caused by intrinsic and extrinsic sample characteristics; the physical characteristics of the container [107, 108]; the presence of foreign materials such as bar-code label [109], oxygen [110, 111], and preservatives [106]; and storage conditions such as light [110] and temperature [106].

Table 12.3. Summary of stability studies of drugs and metabolites in biological specimens.[a]

Drug category	Sample	Storage condition and duration[b]	General stability	Ref.
Amphetamines				
Amphetamine	Blood	amb, preservative; up to 4 yr	Slight decrease	[69]
Methamphetamine	Blood	amb, preservative; up to 5 yr	Slight decrease	[69]
Barbiturates	Blood	amb; 2 wk	Increase (<30%)	[70]
	Blood	amb and rfg; 3 mon	>78% detected	[71]
	Tissue	amb and rfg; 2 mon	<25% change	[71]
Pentobarbital	Blood	amb	Progressive decrease	[72]
	Blood	rfg	No significant changes	[72]
Phenobarbital	Serum	rfg; 12 wk	No change	[73]
	Serum	amb; 6 mon	No change	[74]
	Tissue	amb; 90 day	100% Increase	[75]
Benzodiazepines and metabolites				
Chlordiazepoxide	Blood	amb	Decrease	[76]
	Liver		Decrease	[77]
Nitrazepam	Blood; plasma	rfg or frz; 1 wk	No change	[78]
		amb; 3 wk	Decrease	
	Liver	3 day	Decrease	[78]
Clonazepam	Plasma	amb, rfg, frz; up to 35 day	Decrease	[79–81]
Diazepam	Blood; liver	amb or rfg; 5 mon	Stable	[70, 76, 77]
Flurazepam & N-1-desalkylflurazepam	Blood	4–5 mon	<20% decrease	[70, 76]
Cocaine and metabolites				
Cocaine	Blood	amb, preservative; up to 2 yr	Decrease, then not detectable	[69]
	Blood	amb	Decrease	[82]
	Blood	rfg, no preservative	Decrease	[83]
	Blood	amb, preservative	Stable	[84]
	Blood	rfg	Stable	[84]
	Urine	pH 5	No change	[83]
	Urine	pH 8	Decrease	[83]
	Saliva	amb; 7 day	Stable	[85]
	Saliva	rfg	Stable	[85]
Benzoylecgonine	Blood	amb, preservative; up to 2 yr	Decrease, then not detectable	[69]
	Blood	amb or rfg	Stable	[84]
	Urine	Neutral or alkaline, amb	Decrease	[86]
Ethanol	Blood	amb, preservative; 10 day	No significant production	[87]
	Blood	amb, rfg; 15 day	Variable effects	[88]
	Blood	No preservative, increasing temperature and time period	Increased loss	[89]
	Blood	amb, rfg; 14 day	No change	[90]
	Urine	–10 °C; 12 mon	Decrease	[91]
	Urine	rfg; 30 day	No significant change	[92]
	Urine	High glucose concn and presence of Candida albicans	Increase	[93]
Lysergic acid diamide	Urine; serum	amb	80% detected	[93]
	Blood	amb	30% detected	[93]
	Urine	amb; 6 mon	Decrease	[94]
	Urine	amb, rfg; 4 wk	Stable	[95]
Marijuana components and metabolites				
Tetrahydrocannabinol (THC)	Blood; plasma	amb; 2 mon	Stable	[96]
	Blood	rfg and –10 °C; 6 mon	No change	[96]
	Blood	Glass tube	Stable	[97]
	Blood	Plastic tube	Decrease	[97]
9-THC-COOH	Blood	amb; 6 day	No change	[96]
	Urine	amb, preservative; 2 yr	Stable	[98]
	Urine	Storage	Decrease	[99–102]
9-THC-COOH–glucuronide	Urine	frz; 2 yr	Decrease in selected samples	[99, 103]

Table 12.3. (Continued.)

Drug category	Sample	Storage condition and duration[b]	General stability	Ref.
Opiates				
Morphine (total); codeine (total)	Urine	amb, frz; 11 mon	Decrease (10–40%)	[104]
	Urine	amb; 11 mon	Large or 10–40% decrease	[104]
Morphine (free); codeine (free)	Urine	amb, frz; 11 mon	Slight but steady increase	[104]
	Urine	amb; 11 mon	Various patterns	[104]
Morphine (free); codeine (free)	Blood	amb, preservative; up to 5 yr	Various patterns	[69]
Phencyclidine	Blood	amb, preservative; up to 5 yr	Moderate decrease	[69]
	Blood	amb; 18 mon	Quite stable	[105]

[a] This is a revised and expanded version of Table 1 in Ref. [68].

[b] Abbreviations: amb: ambient temperature, rfg: refrigerator, frz: freezer; mon: month; yr: year.

Identification and Quantification of Drug Metabolites

Quantitative information concerning the distributions of drug metabolites has been explored as a basis for estimating the time of marijuana [112] or cocaine [113] exposure. For example, a study of intravenous THC exposure [112] reported that the plasma 9-THC-COOH/O-ester 9-THC-COOH glucuronide ratio was greater than 2 in 2–30 min postinfusion and thereafter less than 2 within the 12-day study period. Furthermore, a total plasma 9-THC-COOH/THC ratio of 1 or less occurred only before the 45-min postinfusion period. As for the differentiation of frequent and infrequent users, urine samples from the former category of users contained small quantities of unconjugated 9-THC-COOH during the first few days, whereas samples from the latter category of users had only conjugated O-ester 9-THC-COOH.

Concerning the time of cocaine exposure, the following generalizations have been reported:

> The relative concentration of cocaine, BZ [benzoylecgonine], and/or EME [ecgonine methyl ester] in a single urine sample may provide an indication of the length of time since the dose of cocaine. Cocaine will generally be detectable only in the first few hours after the last dose. If cocaine (C) is detectable, but the EME/C or the BZ/C ratio is greater than 100, this is also evidence that the dose was more than 10 h ago. Conversely, an EME/C ratio less than 100 would suggest the dose was less than 10 h ago. If EME and BZ concentrations are similar(the EME/BZ ratio is greater than 0.5), the dose was likely taken within the last 20 h. Conversely, an EME/BZ ratio less than 0.5 would indicate more than 20 h since the dose. EME will usually not be detectable more than 48 h after the last dose. If EME is detectable but the EME/BZ ratio is less than 0.1, this is also an indication that the dose was more than 48 h ago. [114]

Informative on the surface, these generalizations must be used with extreme caution because these ratios will not apply if the cocaine dose is repeated over an extended period (several hours) [114]. Furthermore, deviations from these generalizations have been reported [115] by the same group that provided research findings constituting the basis of the generalizations.

Courtroom testimony that emphasizes the detection of EME and the determination of urinary BZ/EME ratios is commonly encountered when issues of alleged time-of-exposure or cocaine contamination in the urine sample are raised. Although the availability of this information may be helpful, the urinary concentration of EME is generally much lower than that of BZ [115]. A recent study [116] on case samples from a medical examiner's office reported that the concentration of BZ was higher than that of EME in 73% of 380 consecutive urine samples analyzed. Ninety-six percent of the cases of cocaine use were identified using BE as the sole screen with EMIT and a 300-ng/mL cutoff, whereas the corresponding identification rate was 93% when EMS was the sole screen with a GC nitrogen–phosphorus detector at a cutoff of 50 ng/mL. Thus, systematic adoption of a policy that requires the detection of EME and the BZ/EME ratio may present a challenge to many laboratories.

The detection of a unique metabolite, cocaethylene, which is supposedly present only in the case of simultaneous administration of alcohol and cocaine, may also provide valuable information [117–119]. Because this metabolite is present in urine at a concentration lower than that of EME, and its presence reportedly may also derive from prolonged exposure of cocaine to alcohol [120], the generation of high-quality data on a routine basis may present a challenge to many laboratories. Furthermore, an additional parameter, namely the simultaneous use of alcohol and cocaine, always complicates the issues that have to be addressed in a court of law.

As a heroin metabolite [50] that cannot derive from morphine (or codeine) in vivo [52], 6-monoacetylmorphine is an unequivocal indicator of heroin usage. Unfortunately, this compound presents only at a low concentration level (0.9 to 1.7% of the administered heroin dose) [50], and can only be detected 2 to 8 h [121] after heroin use. Many samples collected from individuals receiving a single dose of heroin hydrochloride were found to contain a substantial concentration of morphine (>1000 ng/mL), but no detectable 6-monoacetylmorphine. This observation is also true for samples routinely received in drug-testing laboratories. For example, 6-monoacetylmorphine was detected only in 58% of samples ($n = 26$) that gave immunoassay responses equivalent to 5000 to 10,000 ng/mL of morphine [122]. Two other studies on immunoassay opiate-positive samples revealed that 73% ($n = 45$) and 56% ($n = 49$) of these samples were found to contain 6-monoacetylmorphine in concentrations ranging from 5 to 500 ng/mL [52] and 24 to 3235 ng/mL [123].

Recent findings of high 6-monoacetylmorphine levels in the hair of heroin users [124] and of high cocaethylene levels in hair samples collected from persons simultaneously using cocaine and alcohol [125] have greatly enhanced the value of hair as a test specimen as well as the use of 6-monoacetylmorphine and cocaethylene as indicators of drug use.

Although the availability of qualitative and quantitative information concerning drug metabolites that can indisputably establish their excretion from the person tested (EME, for example) and the identity of the parent drug involved (e.g., 6-monoacetylmorphine) are valuable, this type of information should not be considered the sole basis for the establishment of drug exposure. If it is essential to identify these "true" metabolites, then the validity of identifying parent drugs, such as amphetamine and PCP, as indicators of drug exposure will have to be reevaluated. This should serve as a reminder that a successful drug urinalysis program does not rely on technical laboratory testing alone; sound test-result interpretation and chain-of-custody practice, especially at the sample-collection level, are also essential elements.

References

1. Cody, J. T. *Forensic Sci. Rev.* **1990,** *2,* 64.
2. Manno, J. E. In *Urine Testing for Drugs of Abuse;* Hawks, R. L.; Chiang, C. N., Eds.; Department of Health and Human Services, National Institute on Drug Abuse: Rockville, MD, 1986; p 24.
3. Beckett, A. H.; Rowland, M. *J. Pharm. Pharmacol.* **1965,** *17,* 628.
4. Beckett, A. H.; Rowland, M. *J. Pharm. Pharmacol.* **1965,** *7,* 109.
5. Dring, L. G.; Smith, R. L.; Williams, R. T. *Biochem. J.* **1970,** *116,* 425.
6. Sever, P. S.; Caldwell, J.; Dring, L. G.; Williams, R. T. *Eur. J. Clin. Pharmacol.* **1973,** *6,* 177.
7. Done, A. K.; Aronow, R.; Miceli, J. N.; Lin, D. C. K. In *Management of the Poisoned Patient;* Rumack, B. H.; Temple, A. R., Eds.; Science: Princeton, NJ, 1977; p 79.
8. Morgan, J. P. In *Drug Testing Legal Manual;* Zeese, K. B., Ed.; Clark Boardman: New York, 1987; pp B7–1.
9. Brunk, S. D. *J. Anal. Toxicol.* **1988,** *12,* 290.
10. Cody, J. T.; Schwarzhoff, R. *J. Anal. Toxicol.* **1989,** *13,* 277.
11. Schwartz, R. H.; Hayden, G. F.; Bogema, S.; Thorne, M. M.; Kicks, J. *Arch. Pathol. Lab. Med.* **1987,** *111,* 708.
12. Schwartz, R. H.; Hayden, G. F.; Riddile, M. *Am. J. Dis. Child.* **1985,** *139,* 1093.
13. Vu Duc, T. *Clin. Chem.* **1985,** *31,* 658.
14. Mikkelsen, S. L.; Ash, K. O. *Clin. Chem.* **1988,** *34,* 2333.
15. Kim, H. J.; Cerceo, E. *Clin. Chem.* **1976,** *22,* 1935.
16. Warner, M. *Anal. Chem.* **1987,** *59,* 521.
17. ElSohly, H. N.; Jones, A. B. *Forensic Sci. Rev.* **1989,** *1,* 13.
18. Cone, E. J.; Huestis, M. A. *Forensic Sci. Rev.* **1989,** *1,* 122.
19. Zeidenberg, P.; Bourdon, R.; Nahas, G. G. *Am. J. Psychiatry* **1977,** *134,* 76.
20. Ferslew, K. E.; Manno, J. E.; Manno, B. R. *Res. Commun. Subst. Abuse* **1983,** *4,* 289.
21. Perez-Reyes, M.; Di Guiseppi, S.; Davis, K. H. *JAMA J. Am. Med. Assoc.* **1983,** *249,* 475.
22. Perez-Reyes. M.; Di Guiseppi, S.; Mason, A. P.; Davis, K. H. *Clin. Pharmacol. Ther.* **1983,** *34,* 36.

23. Law, B.; Mason, P. A.; Moffat, A. C.; King, L. J.; Marks, V. *J. Pharm. Pharmacol.* **1984,** *36,* 578.

24. Morland, J.; Bugge, A.; Skuterud, B.; Steen, A.; Wethe, G. H.; Kjeldsen, T. *J. Forensic Sci.* **1985,** *30,* 997.

25. Cone, E. J.; Johnson, R. E.; Darwin, W. D.; Yousefnejad, D.; Mell, L. D.; Paul, B. D.; Mitchell, J. *J. Anal. Toxicol.* **1987,** *11,* 89.

26. Cone, E. J.; Johnson, R. E. *Clin. Pharmacol. Ther.* **1986,** *40,* 247.

27. Mulé, S. J.; Lomax, P.; Gross, S. J. *J. Anal. Toxicol.* **1988,** *12,* 113.

28. Baselt, R. C.; Yoshikawa, D. M.; Chang, J. Y. *Clin. Chem.* **1991,** *37,* 2160.

29. Bateman, D. A.; Heagarty, M. *Am. J. Dis. Child.* **1989,** *143,* 25.

30. Cone, E. J.; Yousefnejad, D.; Hillsgrove, M. J.; Holicky, B.; Darwin, W. D. *J. Anal. Toxicol.* **1995,** *19,* 399.

31. *Methods for Skin Absorption;* Kemppainen, B. W.; Reifenrath, W. G. Eds.; CRC Press: Boca Raton, FL, 1990.

32. Baselt, R. C.; Chang, J. Y.; Yoshikawa, D. M. *J. Anal. Toxicol.* **1990,** *14,* 383.

33. ElSohly, M. A. *J. Anal. Toxicol.* **1991,** *15,* 46.

34. Le, S. D.; Taylor, R. W.; Vidal, D.; Lovas, J. J.; Ting, E. *J. Forensic Sci.* **1992,** *37,* 959.

35. Mieczkowski, T. *Microgram* **1995,** *28,* 193.

36. Baumgartner, W.; Cawing, C.; Donahue, T.; Hayes, G.; Hill, V.; Scholtz, H. In *Forensic Application of Mass Spectrometry;* Yinon, J. Ed.; CRC Press: Boca Raton, FL, 1995; p 61.

37. Schwartz, R. H.; Einhorn, A. *Pediatr. Emerg. Care* **1986,** *2,* 238.

38. Welkos, R. *Los Angeles Times* Oct. 1, 1978.

39. Bailey, D. N. *Res. Commun. Subst. Abuse* **1980,** *1,* 443.

40. Bailey, D. N. *J. Anal. Toxicol.* **1990,** *14,* 217.

41. Bjerver, K.; Jonsson, J.; Nilsson, A.; Schuberth, J.; Schuberth, J. *J. Pharm. Pharmacol.* **982,** *34,* 798.

42. ElSohly, H. N.; Standford, D. F.; Jones, A. B.; ElSohly, M. A.; Snyder, H.; Pedresen, C. *J. Forensic Sci.* **1988,** *33,* 347.

43. Fritschi, G.; Prescott, W. R., Jr. *Forensic Sci. Int.* **1985,** *27,* 111.

44. Grove, M. D.; Gayland, F. S.; Wakeman, M. V.; Tookey, H. L. *J. Agric. Food Chem.* **1976,** *27,* 896.

45. Hayes, L. W.; Krasselt, W. G.; Mueggler, P. A. *Clin. Chem.* **1987,** *33,* 806.

46. Pettitt, B. C., Jr.; Dyszel, S. M.; Hood, L. V. S. *Clin. Chem.* **1987,** *33,* 1251.

47. Struempler, R. E. *J. Anal. Toxicol.* **1987,** *11,* 97.

48. Selavka, C. M. *J. Forensic Sci.* **1991,** *36,* 685.

49. National Institute on Drug Abuse *Medical Review Official Manual: A Guide to Evaluating Urine Drug Analysis;* U.S. Government Priting Office: Washington, DC, 1988; pp 21–25.

50. Yeh, S. Y.; Gorodetzky, C. W.; McQuinn, R. L. *J. Pharmacol. Exp. Ther.* **1976,** *196,* 249.

51. Cone, E. J.; Welch, P.; Paul, B. D.; Mitchell, J. M. *J. Anal. Toxicol.* **1991,** *15,* 161.

52. Fehn, J.; Megges, G. *J. Anal. Toxicol.* **1985,** *9,* 134.

53. Liu, J. R. H.; Ku, W. W. *Anal. Chem.* **1981,** *53,* 2180.

54. Liu, J. R. H.; Ku, W. W.; Tsay, J. T.; Fitzgerald, M. P.; Kim, S. *J. Forensic Sci.* **1982,** *27,* 39.

55. Fitzgerald, R. L.; Ramos, J. M., Jr.; Bogema, S. C.; Poklis, A. *J. Anal. Toxicol.* **1988,** *12,* 255.

56. *Navy Drug Screening Laboratory Standard Operating Procedurea;* U.S. Department of Navy: Washington, DC, 1988.

57. U.S. Depatment of Human Services, National Institute on Drug Abuse *Technical Advisory to All HHS/NIDA Certified Laboratories;* 1991.

58. U.S. Department of Human Services, National Institute on Drug Abuse *Notice to All HHS/NIDA Certified Laboratories;* 1991.

59. Green, K. B.; Isenschmid, D. S. *Forensic Sci. Rev.* **1995,** *7,* 41.

60. Cody, J. T. *Forensic Sci. Rev.* **1994,** *6,* 81.

61. Finkle, B. S.; Black, D.; Blanke, R. V.; Butler, T. J.; Jones, G. R.; Sample, R. H. B. *Clin. Chem.* **1991,** *37,* 587.

62. *Fed. Regist.* **1988,** *53,* 11970.

63. Manno, J. E. In *Urine Testing for Drugs of Abuse;* NIDA Research Monograph 73; Hawks, R. L.; Chiang, C. N. Eds.; National Institute on Drug Abuse: Rockville, MD, 1986; pp 54–61.

64. Johansson, E.; Halldin, M. M. *J. Anal. Toxicol.* **1989,** *13,* 218.

65. Davis, J. M.; Kopin, I. J.; Lemberger, L.; Axelrod, J. *Ann. N. Y. Acad. Sci.* **1971,** *179,* 493.

66. Barnett, G.; Willette, R. E. In *Advances in Analytical Toxicology;* Baselt, R. C., Ed.; Year Book Medical: Chicago, IL, 1989; Vol. 2.

67. Ambre, J. *J. Anal. Toxicol.* **1985,** *9,* 241.

68. Levine, B.; Smith, M. L. *Forensic Sci. Rev.* **1990,** *2,* 148.

69. Giorgi, S. N.; Meeker, J. E. *J. Anal. Toxicol.* **1995,** *19,* 392.

70. Garriott, J. C.; Hatchett, D.; Dempsey, J. *Forensic Sci. Gaz.* **1977,** *8,* 1.

71. Levine, B. S.; Blanke, R. V.; Valentour, J. C. *J. Forensic Sci.* **1984,** *29,* 131.

72. Coutselinis, A.; Kiaris, H. *Med. Sci. Law* **1970,** *10,* 47.

73. Schafer, H. R. In *Clinical Pharmacology of Anti-Epileptic Drugs;* Schneider, H., Ed.; Springer-Verlag: New York, 1975; p 307.
74. Wilensky, A. J. *Clin. Chem.* **1978,** *24,* 722.
75. Algeri, E. J. *J. Forensic Sci.* **1957,** *2,* 443.
76. Levine, B. S.; Blanke, R. V.; Valentour, J. C. *J. Forensic Sci.* **1983,** *28,* 102.
77. Stevens, H. M. *J. Forensic Sci. Soc.* **1984,** *24,* 577.
78. Kelly, H.; Huggett, A.; Dawling, S. *Clin. Chem.* **1982,** *28,* 1478.
79. Knop, H. J.; Vander Kleijn, E.; Edmunds, L. C. *Pharm. Weekbl.* **1975,** *110,* 297.
80. Löscher, W.; Al-Tahan, F. *Ther. Drug Monit.* **1983,** *5,* 229.
81. Wad, N. *Ther. Drug Monit.* **1986,** *8,* 358.
82. Jatlow, P. I.; Bailey, D. N. *Clin. Chem.* **1975,** *21,* 1918.
83. Baselt, R. C. *J Chromatogr.* **1983,** *268,* 502.
84. McCurdy, H. H.; Callahan, L. S.; Williams, R. D. *J. Forensic Sci.* **1989,** *34,* 858.
85. Cone, E.; Menchen, S. L. *Clin. Chem.* **1988,** *34,* 1508.
86. Cody, J. T.; Foltz, R. L. *Proceedings of the 24th Meeting of International Association of Forensic Toxicologists;* Banff, Canada, 1987; p 183.
87. Plueckhahn, V. D.; Ballard, B. *Med. J. Aus.* **1968,** *1,* 939.
88. Christopoulos, G.; Kirsh, E. R.; Gearien, J. E. *J. Chromatogr.* **1973,** *87,* 455.
89. Brown, G. A.; Neylan, D.; Reynolds, W. J.; Smalldon, K. W. *Anal. Chim. Acta* **1973,** *66,* 271.
90. Winek, C. L.; Paul, L. *Clin. Chem.* **1983,** *29,* 1959.
91. Neuteboom, W.; Zweipfenning, P. G. M. *J. Anal. Toxicol.* **1989,** *13,* 141.
92. Ball, W.; Lichtenwalner, M. *N. Engl. J. Med.* **1979,** *301,* 614.
93. Ratcliffe, W.; Fletcher, S.; Moffat, A.; Ratcliffe, J.; Harland, W. A.; Levitt, T. *Clin. Chem.* **1977,** *23,* 169.
94. Peel, H.; Boynton, A. *Can. Soc. Forensic Sci. J.* **1980,** *13,* 23.
95. Francom, P.; Andrenyak, D.; Lim, H.-K.; Bridges, R.; Foltz, R.; Jones, R. *J. Anal. Toxicol.* **1988,** *12,* 1.
96. Johnson, J. R.; Jennison, T. A.; Peat, M. A.; Foltz, R. L. *J. Anal. Toxicol.* **1984,** *8,* 202.
97. Christophersen, A. S. *J. Anal. Toxicol.* **1986,** *10,* 129.
98. Fuhrmann, L.; Szasz, V. S. *Abstracts of Papers,* 42nd Annual Meeting of the American Academy of Forensic Sciences, Cincinnati, OH; American Academy of Forensic Sciences: Colorado Springs, CO, 1990.
99. Cody, J. T. In *Forensic Toxicology;* Uges, D. R. A.; deZeeuw, T. A., Eds.; University: London, 1988; 204.
100. Craft, N. E.; Byrd. G. D.; Hilpert, L. R. *Anal. Chem.* **1989,** *61,* 540.
101. Dextraze, T.; Griffiths, W. C.; Camara, P.; Audette, L.; Rosner, M. *Ann. Clin. Lab. Sci.* **1989,** *19,* 133.
102. Joern, W. A. *J. Anal. Toxicol.* **1987,** *11,* 49.
103. Gere, J.; Baylor, M.; Dorsey, G. *Abstracts of Papers,* 40th Annual Meeting of the American Academy of Forensic Sciences, Philadelphia, PA, American Academy of Forensic Sciences: Colorado Springs, CO, 1988.
104. Lin D.-L.; Liu, H.; Chen, C.-Y. *J. Anal. Toxicol.* **1995,** *19,* 275.
105. Clardy, D. O.; Ragle, J. L. *Clin. Toxicol.* **1981,** *18,* 929.
106. Isenschmid, D. S.; Levine, B. S.; Caplan, Y. H. *J. Anal. Toxicol.* **1989,** *13,* 250.
107. Christophersen, A. S. *J. Anal. Toxicol.* **1986,** *10,* 99.
108. Fenimore, D. C.; Davis, C. M.; Whitford, J. H.; Harrington, C. A. *Anal. Chem.* **1976,** *48,* 2289.
109. Bond, G. D., II; Chand, P.; Walia, A. S.; Liu, R. H. *J. Anal. Toxicol.* **1990,** *14,* 389.
110. Fairbairn, J. W.; Liebmann, J. A.; Rowan, M. G. *J. Pharm. Pharmacol.* **1976,** *28,* 1.
111. Turk, R. F.; Manno, J.; Jain, N. C.; Forney, R. B. *J. Pharm. Pharmacol.* **1971,** *23,* 190.
112. Kelly P.; Jones, R. T. *J. Anal. Toxicol.* **1992,** *16,* 228.
113. Ambre, J. J.; Ruo, T.-I.; Smith, G. L.; Backes, D.; Smith, C. M. *J. Anal. Toxicol.* **1982,** *6,* 26.
114. Ambre, J. J. In *Cocaine, Marijuana, Designer Drugs: Chemistry, Pharmacology, and Behavior;* Redda, K. K.; Walker, C. A.; Barnett, G., Eds.; CRC Press: Boca Raton, FL, 1989.
115. Ambre, J.; Ruo, T. I.; Nelson, J.; Belknap, S. *J. Anal. Toxicol.* **1988,** *12,* 301.
116. Ramcharitar, V.; Levine, B.; Smialek, J. E. *J. Forensic Sci.* **1995,** *40,* 99.
117. De la Torre, R.; Farrè, M.; Ortuño, J.; Camì, J.; Segura, J. *J. Anal. Toxicol.* **1991,** *15,* 223.
118. Rafla, F. K.; Epstein, R. L. *J. Anal. Toxicol.* **1979,** *3,* 59.
119. Smith, R. M. *J. Anal. Toxicol.* **1984,** *8,* 38.
120. Janzen, K. E. *J. Forensic Sci.* **1991,** *36,* 1224.
121. Cone, E. J.; Welch, P.; Mitchell, J. M.; Paul, B. D. *J. Anal. Toxicol.* **1991,** *15,* 1.
122. Fortner, N. A.; Moezpoor, E. L.; Warren, M.; Fogerson, R. S.; Wade, N. A. *Abstracts of Papers,* 45th Annual Meeting of the American Academy of Forensic Sciences, Boston, MA, American Academy of Forensic Sciences: Colorado Springs, CO, 1993; pp K–7.
123. Mulé, S. J.; Casella, G. A. *Clin. Chem.* **1988,** *34,* 1427.
124. Goldberger, B. A.; Caplan, Y. H.; Maguire, T.; Cone, E. J. *J. Anal. Toxicol.* **1991,** *15,* 226.
125. Cone, E. J.; Yousefnejad, D.; Darwin, W. D.; Maguire, T. *J. Anal. Toxicol.* **1991,** *15,* 250.

APPENDICES

Appendix I
Drug Classification and Other Pertinent Information

Common name	ACS Registry number	Pharmacological action	Schedule (number)	Basic/ acidic	pK_a	Melting/boiling[a] point (°C)
Acetorphine	25333-77-1	Narcotic analgesic	I (9319)	B		mp 214–217
Acetyldihydrocodeinone	466-90-0	Narcotic analgesic/antitussive	I (9051)			mp 154
Acetylmethadol	509-74-0	Narcotic analgesic/euphoriant	I (9601)	B	8.8	
α-Acetylmethadol	17199-58-5	Narcotic analgesic	I (9603)	B	8.3	
β-Acetylmethadol	17199-59-6	Narcotic analgesic	I (9607)	B		mp 215–217
Acetyl-α-methylfentanyl		Narcotic analgesic	I (9815)			
Alfentanil	71195-58-9	Narcotic analgesic	II (9737)		6.5	
Allylprodine	25384-17-2	Analgesic/CNS depressant	I (9602)	B		
Alprazolam	28981-97-7	Sedative/tranquilizer	IV (2882)		2.4	mp 229–230
Aminorex	2207-50-3	Anorexic	I (1585)	B		mp 136–138
Amitriptyline	50-48-6	Antidepressant		B		
Amobarbital	57-43-2	Hypnotic/sedative	II (2125)	A	7.9	mp 156–158
Amphetamine	300-62-9	CNS stimulant	II (1100)	B	9.94	bp_{760} 200–203
Anileridine	144-14-9	Narcotic analgesic	II (9020)	B	3.7, 7.5	mp 83
Atropine	51-55-8	Anticholinergic		B	4.35	mp 114–116
Barbital	57-44-3	CNS depressant/hypnotic sedative	IV (2145)	A	7.91	mp 188–192
Barbituric acid	67-52-7	No hypnotic properties	III (2100)	A	4.0	248 (dec)
Benzethidine	3691-78-9	Potent narcotic analgesic	I (9606)	B		$bp_{0.5}$ 220
Benzphetamine	156-08-1	Anorexic/CNS stimulant	III (1228)	B	6.6	$bp_{0.02}$ 127
Benzylfentanyl		Narcotic analgesic	I (9818)			
Benzylmorphine	14297-87-1	Narcotic analgesic	I (9052)	B		mp 132
Bezitramide	15301-48-1	Narcotic analgesic	II (9800)	B		mp 145–149
Bromazepam	1812-30-2	Tranquilizer	IV (2748)	B	2.9, 11	237–238.5 (dec)
4-Bromo-2,5-dimethoxyamphetamine	32156-26-6	Hallucinogen	I (7391)	B		
4-Bromo-2,5-dimethoxyphenethylamine			I (7392)	B		
Bufotenine	487-93-4	Hallucinogen	I (7433)	B		mp 146–147
Buprenorphine	52485-79-7	Analgesic	V (9064)	B		mp 209
Butethal	77-28-1	Sedative/hypnotic CNS depressant		A	8.10	mp 124–127
Camazepam	36104-80-0	Tranquilizer	IV (2749)			mp 173–174
Carbromal	77-65-6	Sedative/CNS depressant		N		mp 116–119
Carfentanil	59708-52-0	Narcotic/analgesic	II (9743)			
Cathine	48115-38-4	Anorex agent	IV (1230)			mp 77.5–78
Cathinone			I (1235)	B		
Chloral betaine	2218-68-0	Sedative	IV (2460)			122.5–124.5
Chloral hydrate	302-17-0	Hypnotic/sedative	IV (2465)	A	10.0	mp 57; bp 98
Chlordiazepoxide	58-25-3	Sedative/minor tranquilizer	IV (2744)	B, N	4.76	mp 236–236.5
Chlorhexadol	3563-58-4	Hypnotic	III (2510)			mp 102–104
Chlorphentermine	461-78-9	Anorexic/stimulant	III (1645)	B	9.6	$bp_{2.0}$ 100–102
Chlorpromazine	50-53-3	Tranquilizer		B		mp 57
Clobazam	22316-47-8	Minor tranquilizer	IV (2751)	B		mp 180–182
Clomethiazole	533-45-9	Anticonvulsant/sedative				bp_7 92
Clomipramine	303-49-1	Antidepressant		B		$bp_{0.3}$ 160–170
Clonazepam	1622-61-3	Anticonvulsant	IV (2737)	B, N	1.5	mp 236.5–238.5

Systematic nomenclature	Nominal MW	Empirical formula	Structure[b] designation
[5α,7α(R)]-3-(Acetyloxy)-4,5-epoxy-6-methoxy-α,17-dimethyl-α-propyl-6,14-ethenomorphinan-7-methanol	411	$C_{25}H_{33}NO_4$	4a
6,7-Didehydro-4,5α-epoxy-3-methoxy-17-methylmorphinan-6-ol	341	$C_{20}H_{23}NO_4$	2a
Benzeneethanol, β-[2-(dimethylamino)propyl]-α-ethyl-β-phenyl-, acetate	353	$C_{23}H_{31}NO_2$	71a
Benzeneethanol, -β-[2-(dimethylamino)propyl]-α-ethyl-β-phenyl-, acetate [S-(R*,S*)]-	353	$C_{23}H_{31}NO_2$	71a
Benzeneethanol, β-[2-(dimethylamino)propyl]-α-ethyl-β-phenyl-, acetate [R-(R*,R*)]-	353	$C_{23}H_{31}NO_2$	71a
N-Phenylacetamide, N-[1-(1-methyl-2-phenyl)ethyl-4-piperidyl]-			76c
Propionanilide, N-[1-[2-(4-ethyl-5-oxo-2-tetrazolin-1-yl)ethyl]-4-(methoxymethyl)-4-piperidyl]-	416	$C_{21}H_{32}N_6O_3$	83a
4-Piperidinol, 1-methyl-4-phenyl-3-(2-propenyl), propanoate	287	$C_{18}H_{25}NO_2$	77a
4H-[1,2,4]Triazolo[4,3-a][1,4]benzodiazepine, 8-chloro-1-methyl-6-phenyl-	308	$C_{17}H_{13}ClN_4$	29a
4,5-Dihydro-5-phenyl-2-oxazolamine	162	$C_9H_{10}N_2O$	117
1-Propanamine, 3-(10,11-dihydro-5H-dibenzo[a,d]cyclohepten-5-ylidene)-N,N-dimethyl-	277	$C_{20}H_{23}N$	50a
2,4,6(1H,3H,5H)-Pyrimidinetrione, 5-ethyl-5-(3-methylbutyl)-	226	$C_{11}H_{18}N_2O_3$	12e
(±)-Benzeneethanamine, α-methyl-,	135	$C_9H_{13}N$	62a
4-Piperidinecarboxylic acid, 1-[2-(4-aminophenyl)ethyl]-4-phenyl-, ethyl ester	352	$C_{22}H_{28}N_2O_2$	79c
Benzeneacetic acid, α-(hydroxymethyl)-8-methyl-8-azabicyclo[3.2.1]oct-3-yl ester, endo (±)-	289	$C_{17}H_{23}NO_3$	93a
2,4,6(1H,3H,5H)-Pyrimidinetrione, 5,5-diethyl-	184	$C_8H_{12}N_2O_3$	12j
2,4,6(1H,3H,5H)-Pyrimidinetrione	128	$C_4H_4N_2O_3$	12k
Ethyl 1-(2-benzyloxyethyl)-4phenyl-4-piperidine-4-carboxylate	367	$C_{23}H_{29}NO_3$	79e
(+)- Benzeneethanamine, N,α-dimethyl-N-(phenylmethyl)-,	239	$C_{17}H_{21}N$	62b
N-Phenylpropanamide, N-(1-benzyl-4-piperidyl)-	322	$C_{21}H_{26}N_2O$	83i
Morphinan-6-ol, 7,8-didehydro-4,5-epoxy-17-methyl-3-(phenylmethoxy)-	375	$C_{24}H_{25}NO_3$	3a
2H-Benzimidazol-2-one, 1-[1-(3-cyano-3,3-diphenylpropyl)-4-piperidinyl]-1,3-dihydro-3-(1-oxopropyl)-	492	$C_{31}H_{32}N_4O_2$	60a
2H-1,4-Benzodiazepin-2(1H)-one, 7-bromo-1,3-dihydro-5-(2-pyridinyl)-	315	$C_{14}H_{10}BrN_3O$	34a
Benzeneethanamine, 4-bromo-2,5-dimethoxy-α-methyl-	273	$C_{11}H_{16}BrNO_2$	64c
Benzeneethanamine, 4-bromo-2,5-dimethoxy-	259	$C_{10}H_{14}BrNO_2$	64d
1H-Indol-5-ol, 3-[2-(dimethylamino)ethyl]-	204	$C_{12}H_{16}N_2O$	97a
6,14-Ethenomorphinan-[5α,7α(Σ)]–7-methanol, 17-(cyclopropylmethyl)-α-(1,1-dimethylethyl)-4,5-epoxy-18,19-dihydro-3-hydroxy-6-methoxy-α-methyl	467	$C_{29}H_{41}NO_4$	8
2,4,6(1H,3H,5H)-Pyrimidinetrione, 5-butyl-5-ethyl-	212	$C_{10}H_{16}N_2O_3$	12n
2H-1,4-Benzodiazepin-2-one, 7-chloro-1,3-dihydro-3-hydroxy-1-methyl-5-phenyl-, dimethylcarbamate (ester)	371	$C_{19}H_{18}ClN_3O_3$	23e
Butanamide, N-(aminocarbonyl)-2-bromo-2-ethyl-	236	$C_7H_{13}BrN_2O_2$	106
Methyl 4-[(1-oxopropyl)pheny(amino)-1-(2-ohenylethyl)-4-piperidinecarboxylate	294	$C_{24}H_{30}N_2O_3$	83c
(R*,R*)-α-(1-Aminoethyl)-benzenemethanol	151	$C_9H_{13}NO$	118
(S)-2-Amino-1-phenyl-1-propananone	149	$C_9H_{11}NO$	67a
Methanaminium, 1-carboxy-N,N,N-trimethyl-, hydroxide, inner salt, with chloral hydrate (1:1)	282	$C_7H_{14}Cl_3NO_4$	108a
1,1-Ethanediol, 2,2,2-trichloro-	165	$C_2H_3Cl_3O_2$	108b
3H-1,4-Benzodiazepin-2-amine, 7-chloro-N-methyl-5-phenyl-, 4-oxide	299	$C_{16}H_{14}ClN_3O$	26a
2-Pentanol, 2-methyl-4-(2,2,2-trichloro-1-hydroxyethoxy)-	265	$C_8H_{15}Cl_3O_3$	108c
Benzeneethanamine, 4-chloro-α,α-dimethyl-	183	$C_{10}H_{14}ClN$	62f
10H-Phenothiazine-10-propanamine, 2-chloro-N,N-dimethyl-	318	$C_{17}H_{19}ClN_2S$	46d
1H-1,5-Benzodiazepine-2,4(3H,5H)-dione, 7-chloro-1-methyl-5-phenyl-	300	$C_{16}H_{13}ClN_2O_2$	27
Thiazole, 5-(2-chloroethyl)-4-methyl-	161	C_6H_8ClNS	110
5H-Dibenz[b,f]azepine-5-propanamine, 3-chloro-10,11-dihydro-N,N-dimethyl-	314	$C_{19}H_{23}ClN_2$	49a
2H-1,4-Benzodiazepin-2-one, 5-(2-chlorophenyl)-1,3-dihydro-7-nitro-	315	$C_{15}H_{10}ClN_3O_3$	23j

Common name	ACS Registry number	Pharmacological action	Schedule (number)	Basic/ acidic	pK_a	Melting/boiling[a] point (°C)
Clonitazene	3861-76-5	Narcotic analgesic	I (9612)	B		mp 75–76
Clorazepate	57109-90-7	Minor tranquilizer	IV (2768)	A	3.5, 12.5	
Clortermine	10389-73-8	Anorexic	III (1647)			bp$_{16}$ 116–118
Clotiazepam	33671-46-4	Tranquilizer	IV (2752)			mp 243–246
Cloxazolam	24166-13-0	Minor tranquilizer	IV (2753)			200 (dec)
Cocaine	50-36-2	Narcotic anesthetic	II (9041)	B	8.5	mp 98
Coca leaves			II (9040)			
Codeine	76-57-3	Narcotic analgesic/antitussive	II (9050)	B	8.2	mp 154–156
Codeine methyl bromide	125-27-9	Narcotic analgesic/antitussive	I (9070)	B		mp 260
Codeine-N-oxide	3688-65-1	Antitussive	I (9053)	B		mp 231–232
Cyclizine	82-92-8	Antiemetic/antitussive		B	2.4	mp 105.5–107.5
Cyprenorphine	4406-22-8	Narcotic antagonist	I (9054)	B		mp 234
Delorazepam	2894-67-9	Tranquilizer	IV (2754)			mp 200–201
Desomorphine	427-00-9	Narcotic analgesic	I (9055)	B		mp 189
Dextromoramide	357-56-2	Narcotic analgesic	I (9613)	B	6.60	mp 180–184
Dextropropoxyphene	469-62-5	Analgesic	II (9273)	B	6.3	mp 75–76
Diampromide	552-25-0	Analgesic	I (9615)	B		
Diazepam	439-14-5	Sedative/tranquilizer	IV (2765)	B, N	3.4	mp 125–126
Dichloralphenazone (chloral)	480-30-8	Hypnotic/sedative		B		mp 68
Diethylpropion	90-84-6	Anorexic	IV (1610)	B		
Diethylthiambutene (thiambutene)	86-14-6	Narcotic analgesic	I (9616)	B		bp$_{0.03}$ 122–128
Diethyltryptamine	61-51-8	Hallucinogen	I (7434)	B		
Difenoxin	28782-42-5	Antidiarrheal	I (9168)			
Digoxin	20830-75-5	Cardiotonic		N		230–265 (dec)
Dihydrocodeine	125-28-0	Narcotic analgesic/antitussive	II (9120)	B	8.68	mp 112–113
Dihydrocodeinone bitartrate	143-71-5	Narcotic analgesic/antitussive			6.61	mp 192–193
Dihydromorphine	509-60-4	Analgesic	I (9145)	B	8.6	mp 159
Dimenoxadol	509-78-4	Analgesic	I (9617)	B		
Dimepheptanol	545-90-4	Analgesic	I (9618)			mp 127–128
2,5-Dimethoxyamphetamine		Hallucinogen	I (7396)	B		mp 107–108
2,5-Dimethoxy-4-methylamphetamine		Hallucinogen	I (7395)	B		mp 189–190
N,N-Dimethylamphetamine			I (1480)	B		
Dimethylthiambutene	524-84-5	Narcotic analgesic	I (9619)	B		bp$_3$ 157–158
N,N-Dimethyltryptamine	61-50-7	Hallucinogen	I (7435)	B	8.68	mp 44.6–46.8
Dioxaphetyl butyrate	467-86-7	Analgesic/antispasmodic	I (9621)	B		
Diphenhydramine	58-73-1	Antihistaminic		B	9.1	bp$_{2.0}$ 150–165
Diphenoxylate	915-30-0	Narcotic antiperistaltic	II (9170)	B	7.1	
Dipipanone	467-83-4	Analgesic/sedative hypnotic	I (9622)	B		
Dothiepin	113-53-1	Antidepressant		B		mp 55-57
Doxepin	1668-19-5	Antidepressant		B	9.0	bp$_{0.03}$ 154–157
Dronabinal	1972-08-3	Hallucinogen	II (7369)			
Drotebanol	3176-03-2	Narcotic analgesic/antitussive	I (9335)			mp 165.5–166.5
Ecgonine	481-37-8	Topical anesthetic	II (9180)	B	11.1	mp 198
Ephedrine	50906-05-3	Sympathomimetic		B	9.63	mp 40

Systematic nomenclature	Nominal MW	Empirical formula	Structure[b] designation
2-[(4-Chlorophenyl)methyl]-*N,N*-diethyl-5-nitro-1*H*-benzimidazole-1-ethanamine	386	$C_{20}H_{23}ClN_4O_2$	100a
1*H*-1,4-Benzodiazepine-3-carboxylic acid, 7-chloro-2,3-dihydroxy-5-phenyl-,	332	$C_{16}H_{13}ClN_2O_4$	38
Benzeneethanamine, 2-chloro-α,α-dimethyl-	183	$C_{10}H_{14}ClN$	62g
5-(2-Chlorophenyl)-7-ethyl-1,3-dihydro-1-methyl-2*H*-thieno[2,3,e]-1,4-diazepin-2-one	318	$C_{16}H_{15}ClN_2OS$	36
[3,2-d][1,4]Benzodiazepin-6(5*H*)-one, 10-chloro-11b-(*o*-chlorophenyl)-2,3,7,11b-tetrahydrooxalo-	349	$C_{17}H_{14}Cl_2N_2O_2$	41a
8-Azabicyclo[3.2.1]octane-2-carboxylic acid, 3-(benzoyloxy)-8-methyl-, methyl ester, [1*R*-(*exo,exo*)]-	303	$C_{17}H_{21}NO_4$	93c
Morphinan-6-ol, 7,8-didehydro-4,5-epoxy-3-methoxy-17-methyl-, (5α,6α)-	299	$C_{18}H_{21}NO_3$	3b
Morphinanium, 7,8-didehydro-4,5-epoxy-6-hydroxy-3-methoxy-17,17-dimethyl-, bromide, (5α,6α)-	394	$C_{19}H_{24}BrNO_3$	3d
Morphinan-6-ol, 7,8-didehydro-4,5-epoxy-3-methoxy-17-methyl-, 17-oxide, (5α,6α)-	315	$C_{18}H_{21}NO_4$	3e
Piperazine, 1-(diphenylmethyl)-4-methyl-	266	$C_{18}H_{22}N_2$	59
6,14-Ethenomorphinan-7-methanol, 17-(cyclopropylmethyl)-4,5-epoxy-3-hydroxy-6-methoxy-α,α-dimethyl-, (5α,7α)-	423	$C_{26}H_{33}NO_4$	4b
7-Chloro-5-(2-chlorophenyl)1,3-dihydro-2*H*-1,4-benzodiazepin-2-one	304	$C_{15}H_{10}Cl_2N_2O$	
Morphinan-3-ol, 4,5-epoxy-17-methyl-, (5α)-	271	$C_{17}H_{21}NO_2$	9b
(+)-Pyrrolidine, 1-[3-methyl-4-(4-morpholinyl)-1-oxo-2,3-diphenylbutyl]-, (*S*)-	392	$C_{25}H_{32}N_2O_2$	66e
Benzeneethanol, α-[(2-dimethylamino)-1-methylethyl]-α-phenyl-, propanoate (ester), [*S*-(*R**,*S**)]-	339	$C_{22}H_{29}NO_2$	66h
Propanamide, *N*-[2-[methyl(2-phenylethyl)amino]propyl]-*N*-phenyl-	324	$C_{21}H_{28}N_2O$	80a
2*H*-1,4-Benzodiazepin-2-one, 7-chloro-1,3-dihydro-1-methyl-5-phenyl-	284	$C_{16}H_{13}ClN_2O$	23p
3*H*-Pyrazol-3-one, 1,2-dihydro-1,5-dimethyl-2-phenyl-, compd. with 2,2,2-trichloro-1,1-ethanediol (1:2)	519	$C_{15}H_{18}Cl_6N_2O_5$	108d
1-Propanone, 2-(diethylamino)-1-phenyl-	205	$C_{13}H_{19}NO$	67b
3-Buten-2-amine, *N,N*-diethyl-4,4-di-2-thienyl-	291	$C_{16}H_{21}NS_2$	109a
3-Ethanamine, *N,N*-diethyl-1*H*-indole-	216	$C_{14}H_{20}N_2$	97b
4-Piperidinecarboxylic acid, 1-(3-cyano-3,3-diphenylpropyl)-4-phenyl	424	$C_{28}H_{28}N_2O_2$	60c
Card-20(22)-enolide, 3-[(*O*-2,6-dideoxy-β-D-ribo-hexopyranosyl-(1 -> 4)-*O*-2,6-dideoxy-β-D-ribo-hexopyranosyl-(1 -> 4)-2,6-dideoxy-β-D-ribohexopyranosyl)oxy]-12,14-dihydroxy-, (3β,5β,12β)-	780	$C_{41}H_{64}O_{14}$	115
Morphinan-6-ol, 4,5-epoxy-3-methoxy-17-methyl-, (5α,6α)-, [*R*-(*R**,*R**)]-	301	$C_{18}H_{23}NO_3$	9c
Morphinan-6-one, 4,5-epoxy-3-methoxy-17-methyl-, (5α)-, [*R*-(*R**,*R**)]-2,3-dihydroxybutanedioate (1:1) (salt)	299	$C_{18}H_{21}NO_3$	9c
(5α,6α)-4,5-Epoxy-17-methyl-morphinan-3,6-diol	287	$C_{17}H_{21}NO_3$	9d
Benzeneacetic acid, α-ethoxy-α-phenyl-2-(dimethylamino)ethyl ester	327	$C_{20}H_{25}NO_3$	69a
Benzeneethanol, β-[2-(dimethylamino)propyl]-α-ethyl-β-phenyl-	311	$C_{21}H_{29}NO$	71h
Benzeneethanamine, 2,5-dimethoxy-	195	$C_{11}H_{17}NO_2$	64e
Benzeneethanamine, 2,5-dimethoxy-4-methyl-	209	$C_{12}H_{19}NO_2$	64h
Benzeneethanamine, N,N,-trimethyl-, (S)-	163	$C_{11}H_{17}N$	62i
3-Buten-2-amine, *N,N*-dimethyl-4,4-di-2-thienyl-, (*R*)-	263	$C_{14}H_{17}NS_2$	109b
1*H*-Indole-3-ethanamine, *N,N*-dimethyl-	188	$C_{12}H_{16}N_2$	97c
4-Morpholino-2,2-diphenylbutyrate, ethyl-	353	$C_{22}H_{27}NO_3$	69b
Ethanamine, 2-(diphenylmethoxy)-*N,N*-dimethyl-	255	$C_{17}H_{21}NO$	72c
4-Piperidinecarboxylic acid, 1-(3-cyano-3,3-diphenylpropyl)-4-phenyl-, ethyl ester	452	$C_{30}H_{32}N_2O_2$	60d
3-Heptanone, 4,4-diphenyl-6-(1-piperidinyl)-	349	$C_{24}H_{31}NO$	60e
1-Propanamine, 3-dibenzo[*b,e*]thiepin-11(6*H*)-ylidene-*N,N*-dimethyl-	295	$C_{19}H_{21}NS$	49f
1-Propanamine, 3-dibenz[*b,e*]oxepin-11(6*H*)-ylidene-*N,N*-dimethyl-	279	$C_{19}H_{21}NO$	49g
6a,7,8,10a-Tetrahydro-6,6,9-trimethyl-3-pentyl-6*H*-dibenzo[b,d]pyran-1-ol	314	$C_{21}H_{30}O_2$	55c
Morphinan-6,14-diol, 3,4-dimethoxy-17-methyl-, (6β)-	333	$C_{19}H_{27}NO_4$	10c
8-Azabicyclo[3.2.1]octane-2-carboxylic acid, 3-hydroxy-8-methyl-, [1*R*-(*exo,exo*)]-	185	$C_9H_{15}NO_3$	93d
Propan-1-ol, (−)-2-methylamino-1-phenyl-	165	$C_{10}H_{15}NO$	62k

Common name	ACS Registry number	Pharmacological action	Schedule (number)	Basic/ acidic	pK_a	Melting/boiling[a] point (^{o}C)
Estazolam	29975-16-4	Sedative/hypnotic	IV (2756)			mp 228–229
Ethchlorvynol	113-18-8	Sedative/hypnotic	IV (2540)			bp$_{760}$ 173–174
Ethinamate	126-52-3	Sedative	IV (2545)	N		mp 96–98
Ethoheptazine	77-15-6	Analgesic		B		bp$_1$ 133–134;
N-Ethylamphetamine	33817-13-9	Anorexic/stimulant	I (1475)	B	10.23	
Ethyl loflazepate	29177-84-2	Minor tranquilizer	IV (2758)			mp 193–194
Ethylmethylthiambutene	441-61-2	Narcotic analgesic	I (9623)	B		bp$_{0.01}$ 110–113
Ethylmorphine	76-58-4	Analgesic/antitussive	II (9190)	B	7.88	mp 199–201
N-Ethyl-3-piperidyl benzilate (CH$_3$Br) (pipenzolate bromide)	125-51-9	Anticholinergic	I (7482)			mp 179–180
α-Ethyltryptamine	2235-90-7	Antidepressant/CNS stimulant	I (7249)	B		mp 97–99
Etonitazene	911-65-9	Narcotic analgesic	I (9624)	B		
Etorphine	14521-96-1	Potent narcotic analgesic	I (9056)	B		mp 215–216
Etoxeridine (carbetidine)	469-82-9	Narcotic analgesic	I (9625)	B		bp$_{0.02}$ 170
Fencamfamine	1209-98-9	CNS stimulant	IV (1760)	B		bp$_{0.1}$ 128–130
Fenethylline	3736-08-1	CNS stimulant	I (1503)	B		
Fenfluramine	458-24-2	Anorexic	IV (1670)	B	9.9	bp$_{12}$ 108–112
Fenproporex	15686-61-0	Anorectic/sympathomimetic	IV (1575)	B		bp$_2$ 126–127
Fentanyl	437-38-7	Narcotic analgesic/anesthetic	II (9801)	B	8.4	mp 83–84
Fentanyl citrate	990-73-8					mp 149–151
Fludiazepam	3900-31-0	Minor tranquilizer anxiolytic	IV (2759)	B		mp 88–92
Flunitrazepam	1622-62-4	Sedative/hypnotic	I (2763)	B	1.8	mp 166–167
p-Fluorofentanyl		Narcotic analgesic	I (9812)			
Fluphenazine	69-23-8	Antipsychotic		B		bp$_{0.5}$ 268–274
Flurazepam	17617-23-1	Hypnotic/sedative	IV (2767)	B, N	1.9	mp 77–82
Furethidine	2385-81-1	Narcotic analgesic	I (9626)	B	7.48	mp 28
Glutethimide	77-21-4	Sedative/hypnotic	II (2550)	A, N	9.2	mp 84
Halazepam	23092-17-3	Tranquilizer/sedative	IV (2762)			mp 164–166
Haloperidol	52-86-8	Tranquilizer/sedative		B	8.3	mp 148–149
Haloxazolam	59128-97-1	Sedative/hypnotic	IV (2771)			mp 185
Harmine	442-51-3	Hallucinogen		B	7.6	261 (dec)
Heroin	561-27-3	Narcotic analgesic	I (9200)	B	7.8	mp 173
Hydrocodone	125-29-1	Narcotic analgesic/antitussive	II (9193)	B	6.61	mp 198
Hydromorphinol	2183-56-4	Narcotic analgesic	I (9301)	B		
Hydromorphone	466-99-9	Narcotic analgesic	II (9150)	B	8.2	mp 266–267
β-Hydroxyfentanyl		Narcotic analgesic	I (9830)			
β-Hydroxy-3-methylfentanyl		Narcotic analgesic	I (9831)			
Hydroxypethidine	468-65-4	Narcotic analgesic	I (9627)	B		mp 110
Ibogaine	83-74-9	Antidepressant	I (7260)	B	8.1	mp 152–153
Imipramine	50-49-7	Antidepressant		B		bp$_{0.1}$ 160
Isomethadone	466-40-0	Analgesic/antitussive	II (9226)	B		mp 130–135
Ketazolam	27223-35-4	Minor tranquilizer	IV (2772)	B, N		mp 182–183.5
Ketobemidone	469-79-4	Analgesic	I (9628)	B		mp 156–157

Systematic nomenclature	Nominal MW	Empirical formula	Structure[b] designation
4H-[1,2,4]Triazolo[4,3-a][1,4]benzodiazepine, 8-chloro-6-phenyl-	294	$C_{16}H_{11}ClN_4$	29d
1-Penten-4-yn-3-ol, 1-chloro-3-ethyl-	144	C_7H_9ClO	102a
1-Ethynylcyclohexanol carbamate	167	$C_9H_{13}NO_2$	101a
1H-Azepine-4-carboxylic acid, hexahydro-1-methyl-4-phenyl-, ethyl ester	261	$C_{16}H_{23}NO_2$	92a
Benzeneethanamine, N-ethyl-α-methyl-	163	$C_{11}H_{17}N$	62m
1H-1,4-Benzodiazepine-3-carboxylic acid, 7-chloro-5-(2-fluorophenyl)-2,3-dihydro-2-*oxo*-, ethyl ester	360	$C_{18}H_{14}ClFN_2O_3$	23q
N-Ethyl-N-methyl-4,4-di-2-thienyl-3-buten-2-amine	277	$C_{15}H_{19}NS_2$	109c
Morphinan-6-ol, 7,8-didehydro-4,5-epoxy-3-ethoxy-17-methyl-	313	$C_{19}H_{23}NO_3$	3f
Ethanone, 1-[10-[3-[4-(2-hydroxyethyl)-1-piperidinyl]propyl]-10H-phenothiazin-2-yl]-	433	$C_{22}H_{28}BrNO_3$	70b
3-(2-Aminobuty)indole	314	$C_{12}H_{16}N_2$	97e
1H-Benzimidazole-1-ethanamine, 2-[(4-ethoxyphenyl)methyl]-N,N-diethyl-5-nitro-	396	$C_{22}H_{28}N_4O_3$	100b
6,14-Ethenomorphinan-7-methanol, 4,5-epoxy-3-hydroxy-6-methoxy-α,17-dimethyl-α-propyl-, [5α,7α(R)]-	411	$C_{25}H_{33}NO_4$	4c
4-Piperidinecarboxylic acid, 1-[2-(2-hydroxyethoxy)ethyl]-4-phenyl-, ethyl ester	321	$C_{18}H_{27}NO_4$	79f
N-Ethyl-3-phenyl-bicyclo[2.2.1]-hepran-2-amine	215	$C_{15}H_{21}N$	119
1H-Purine-2,6-dione, 3,7-dihydro-1,3-dimethyl-7-[2-[(1-methyl-2-phenylethyl)amino]ethyl]-, monohydrochloride	341	$C_{18}H_{23}N_5O_2$	17a
Benzeneethanamine, N-ethyl-α-methyl-3-(trifluoromethyl)-	231	$C_{12}H_{16}F_3N$	62o
($\pm$)-3-[(1-Methyl-2-phenylethyl)amino]propanenitrile	188	$C_{12}H_{16}N_2$	64i
Propanamide, N-phenyl-N-[1-(2-phenylethyl)-4-piperidinyl]-	336	$C_{22}H_{28}N_2O$	76a
2H-1,4-Benzodiazepin-2-one, 7-chloro-5-(2-fluorophenyl)-1,3-dihydro-1-methyl-	302	$C_{16}H_{12}ClFN_2O$	23r
2H-1,4-Benzodiazepin-2-one, 5-(2-fluorophenyl)-1,3-dihydro-1-methyl-7-nitro-	313	$C_{16}H_{12}FN_3O_3$	23s
Propanamide, N-[1-(2-phenylethyl)-4-piperidyl]-N-(4-fluorophenyl)-			76b
1-Piperazineethanol, 4-[3-[2-(trifluoromethyl)-10H-phenothiazin-10-yl]propyl]-	437	$C_{22}H_{26}F_3N_3OS$	48b
2H-1,4-Benzodiazepin-2-one, 7-chloro-1-[2-(diethylamino)ethyl]-5-(2-fluorophenyl)-1,3-dihydro-	387	$C_{21}H_{23}ClFN_3O$	23t
4-Piperidinecarboxylic acid, 4-phenyl-1-[2-[(tetrahydro-2-furanyl)methoxy]ethyl]-, ethyl ester	361	$C_{21}H_{31}NO_4$	79g
2,6-Piperidinedione, 3-ethyl-3-phenyl-	217	$C_{13}H_{15}NO_2$	15b
2H-1,4-Benzodiazepin-2-one, 7-chloro-1,3-dihydro-5-phenyl-1-(2,2,2-trifluoroethyl)-	352	$C_{17}H_{12}ClF_3N_2O$	23u
Butyrophenone, 4-[4-(4-chlorophenyl)-4-hydroxypiperidino]-4'-fluoro-	375	$C_{21}H_{23}ClFNO_2$	
[3,2-d][1,4]Benzodiazepin-6(5H)-one, 10-bromo-11b-(2-fluorophenyl)-2,3,7,11b-tetrahydrooxalo-	377	$C_{17}H_{14}BrFN_2O_2$	41b
[3,4-b]Indole, 7-methoxy-1-methyl-9H-pyrido-	212	$C_{13}H_{12}N_2O$	53a
Morphinan-3,6-diol, 7,8-didehydro-4,5-epoxy-17-methyl-(5α,6α)-, diacetate (ester)	369	$C_{21}H_{23}NO_5$	3g
Morphinan-6-one, 4,5-epoxy-3-methoxy-17-methyl-, (5α)-	299	$C_{18}H_{21}NO_3$	6a
Morphinan-3,6,14-triol, 4,5-epoxy-17-methyl-, (5α,6α)-	303	$C_{17}H_{21}NO_4$	9g
Morphinan-6-one, 4,5-epoxy-3-hydroxy-17-methyl-, (5α)-	285	$C_{17}H_{19}NO_3$	6b
N-Phenylpropanamide, N-[1-(2-hydroxy-2-phenyl)ethyl-4-piperidyl]-	352	$C_{22}H_{28}N_2O_2$	
N-Phenylpropanamide, N-[3-methyl-1-(2-hydroxy-2-phenyl)ethyl-4-piperidyl]-	366	$C_{23}H_{30}N_2O_2$	
Piperidine-4-carboxylate, ethyl-4-(3-hydroxyphenyl)-1-methyl-	263	$C_{15}H_{21}NO_3$	79j
Ibogamine, 12-methoxy-	310	$C_{20}H_{26}N_2O$	98
5H-Dibenz[b,f]azepine-5-propanamine, 10,11-dihydro-N,N-dimethyl-	280	$C_{19}H_{24}N_2$	49l
3-Hexanone, 6-(dimethylamino)-5-methyl-4,4-diphenyl-	309	$C_{21}H_{27}NO$	71i
[3,2-d][1,4]Benzodiazepine-4,7(6H)-dione, 11-chloro-8,12b-dihydro-2,8-dimethyl-12b-phenyl-4H-[1,3]-oxazino-	368	$C_{20}H_{17}ClN_2O_3$	40
1-Propanone, 1-[4-(3-hydroxyphenyl)-1-methyl-4-piperidinyl]-	247	$C_{15}H_{21}NO_2$	79k

Common name	ACS Registry number	Pharmacological action	Schedule (number)	Basic/ acidic	pK_a	Melting/boiling[a] point (°C)
Levallorphan	152-02-3	Narcotic antagonist		B		mp 180–182
Levo-α-acethylmethadol	34433-66-4		II (9648)			
Levomethorphan HBr		Antitussive	II (9210)	B		
Levomoramide	5666-11-5	Narcotic analgesic	I (9629)	B	6.60	mp 190
Levophenacylmorphan	10061-32-2		I (9631)	B		mp 173–175
Levorphanol	77-07-6	Narcotic analgesic	II (9220)	B	9.2	mp 198–199
Loprazolam	61197-93-1	Sedative/hypnotic	IV (2773)		6.0	mp 214–215
Lorazepam	846-49-1	Minor tranquilizer	IV (2885)	B, N A	1.3 11.5	mp 166–168
Lormetazepam	848-75-9	Sedative/hypnotic	IV (2774)			mp 205–207
Lysergic acid	82-58-6	Hallucinogen	III (7300)		3.4, 6.3	240 (dec)
Lysergic acid amide (lyseramide)	478-94-4	Hallucinogen	III (7310)	B		242 (dec)
Lysergic acid diethylamide (LSD)	50-37-3	Hallucinogen	I (7315)	B	7.8	mp 80–85
Marijuana (refer to tetrahydrocannabinol(s))			I (7360)			
Mazindol	22232-71-9	Anorexic/CNS stimulant	III (1605)	B	8.6	mp 198–199
Mebutamate	64-55-1	Antihypertensive/CNS depressant	IV (2800)	N		mp 77–79
Mecloqualone	340-57-8	Sedative/hypnotic	I (2572)			mp 126–128
Medazepam	2898-12-6	Minor tranquilizer	IV (2836)	B, N	6.2	mp 95–97
Mefenorex	17243-57-1	Anorexic	IV (1580)			
Meprobamate	57-53-4	Tranquilizer	IV (2820)	N		mp 104–106
α-Meprodine	468-51-9	Narcotic analgesic	I (9604)	B		Sublime in vacuum
β-Meprodine	57401-82-8	Analgesic	I (9608)	B		mp 201–203
Mescaline	54-04-6	Hallucinogen	I (7381)	B		mp 35–36
Metazocine	3734-52-9	Narcotic analgesic	II (9240)	B		mp 232–235
α-Methadol	17199-54-1	Narcotic analgesic	I (9605)	B	7.86	
β-Methadol (see methadol)	545-90-4	Analgesic	I (9609)	B	7.59	mp 127–128
Methadone	76-99-3	Narcotic analgesic	II (9250)	B		mp 78
Methadone synthetic intermediate			II (9254)			
Methamphetamine	537-46-2	CNS stumulant	II (1105)	B	10.1	
Methaqualone	72-44-6	Hypnotic/sedative	I (2565)	B	2.54	mp 120
Methcathinone		Stimulant	I (1237)	B		
Methohexital	151-83-7	Anesthetic	IV (2264)	A	8.3	mp 92–96
4-Methoxyamphetamine	23239-32-9	Hallucinogen	I (7411)	B		
5-Methoxy-3,4-methylenedioxy-amphetamine	13674-05-0	Hallucinogen	I (7401)	B		
Methoxyphenamine	93-30-1	Sympathomimetic		B		bp₂ 97–99
(±)-cis-4-Methylaminorex			I (1590)			
Methyldesorphine	16008-36-9	Narcotic analgesic	I (9302)	B		mp 239–240
Methyldihydromorphine	509-56-8	Narcotic analgesic	I (9304)	B		mp 209–211
3,4-Methylenedioxyamphetamine	4764-17-4	Hallucinogen/sypathomimetic MAO inhibitor	I (7400)	B		
3,4-Methylenedioxy-N-ethylamphetamine			I (7404)	B		
3,4-Methylenedioxymethamphetamine		Hallucinogen	I (7405)	B		mp 147–148
3-Methylfentanyl		Narcotic analgesic	I (9813)	B		mp 164–166

Systematic nomenclature	Nominal MW	Empirical formula	Structure[b] designation
17(2-Propenyl)morphinan-3-ol	283	$C_{19}H_{25}NO$	10m
[S-($R*,R*$)]-β-[2-Dimethylamino)propyl]-α-ethyl-β-phenyl-3-heptanol acetate (ester)	353	$C_{23}H_{31}NO_2$	71f
Morphinan, 3-methoxy-17-methyl-	351	$C_{18}H_{26}BrNO$	10j
Pyrrolidine, 1-[3-methyl-4-(4-morpholinyl)-1-oxo-2,2-diphenylbutyl]-, (*R*)-	392	$C_{25}H_{32}N_2O_2$	66e
Morphinan-3-ol, 17-(2-oxo-2-phenylethyl)-	361	$C_{24}H_{27}NO_2$	10f
Morphinan-3-ol, 17-methyl-	257	$C_{17}H_{23}NO$	10l
[1,2-a][1,4]Benzodiazepin-1-one, 6-(2-chlorophenyl)-2,4-dihydro-2-[(4-methyl-1-piperazinyl)-methylene]-8-nitro-1*H*-imidazo-	464	$C_{23}H_{21}ClN_6O_3$	42
2*H*-1,4-Benzodiazepin-2-one, 7-chloro-5-(2-chlorophenyl)-1,3-dihydro-3-hydroxy-	321	$C_{15}H_{10}Cl_2N_2O_2$	23ba
2*H*-1,4-Benzodiazepin-2-one, 7-chloro-5-(2-chlorophenyl)-1,3-dihydro-3-hydroxy-1-methyl-	335	$C_{16}H_{12}Cl_2N_2O_2$	23bb
Ergoline-8-carboxylic acid, 9,10-didehydro-6-methyl-, (8β)-	268	$C_{16}H_{16}N_2O_2$	56c
Ergoline-8β-carboxamide, 9,10-didehydro-6-methyl-	267	$C_{16}H_{17}N_3O$	56d
Ergoline-8-carboxamide, 9,10-didehydro-*N,N*-diethyl-6-methyl-, (8β)-	323	$C_{20}H_{25}N_3O$	56e
(*refer to* "tetrahydrocannabinol(s)")			
			55a–g
3*H*-Imidazo[2,1-a]isoindol-5-ol, 5-(4-chlorophenyl)-2,5-dihydro-	284	$C_{16}H_{13}ClN_2O$	43a
1,3-Propanediol, 2-methyl-2-(1-methylpropyl)-, dicarbamate	232	$C_{10}H_{20}N_2O_4$	105b
4(3*H*)-Quinazolinone, 3-(2-chlorophenyl)-2-methyl-	270	$C_{15}H_{11}ClN_2O$	114a
1*H*-1,4-Benzodiazepine, 7-chloro-2,3-dihydro-1-methyl-5-phenyl-	270	$C_{16}H_{15}ClN_2$	22b
N-(3-Chloropropyl)-α-methylbenzeneethanamine	211	$C_{12}H_{18}ClN$	64l
1,3-Propanediol, 2-methyl-2-propyl-, dicarbamate	218	$C_9H_{18}N_2O_4$	105c
4-Piperidinol, 3-ethyl-1-methyl-4-phenyl-, propanoate, *cis*-	275	$C_{17}H_{25}NO_2$	77c
(+)-β-3-Ethyl-1-methyl-4-phenyl-4-piperidinol, propanoate (ester)	275	$C_{17}H_{25}NO_2$	77c
Benzeneethanamine, 3,4,5-trimethoxy-	211	$C_{11}H_{17}NO_3$	64m
2,6-Methano-3-benzazocin-8-ol, 1,2,3,4,5,6-hexahydro-3,6,11-trimethyl-	231	$C_{15}H_{21}NO$	11
(+)-Benzeneethanol, β-[2-(dimethylamino)propyl]-α-ethyl-β-phenyl-, [R-($R*,R*$)]-	311	$C_{21}H_{29}NO$	71d
β-6-dimethylamino-4,4-diphenyl-3-heptanol	311	$C_{21}H_{29}NO$	71d
3-Heptanone, 6-(dimethylamino)-4,4-diphenyl-,	309	$C_{21}H_{27}NO$	71j
4-Cyano-2-dimethylamino-4,4-diphenylbutane			
Benzeneethanamine, *N*,α-dimethyl-, (*S*)-	149	$C_{10}H_{15}N$	62v
4(3*H*)-Quinazolinone, 2-methyl-3-(2-methylphenyl)-	250	$C_{16}H_{14}N_2O$	114b
2-(Methylamino)-1-phenylpropanone	163	$C_{10}H_{13}NO$	67c
2,4,6(1*H*,3*H*,5*H*)-Pyrimidinetrione, 1-methyl-5-(1-methyl-2-pentynyl)-5-(2-propenyl)-,	262	$C_{14}H_{17}N_2O_3$	12bb
Benzeneethanamine, 4-methoxy-α-methyl-	165	$C_{10}H_{15}NO$	64n
	209	$C_{11}H_{15}NO_3$	64s
Phenethylamine, 2-methoxy-*N*,α-dimethyl-	179	$C_{11}H_{17}NO$	62w
4,5-Dihydro-4-methyl-5-phenyl-2-oxazolamine	176	$C_{10}H_{12}N_2O$	117a
Morphinan-3-ol, 6,7-didehydro-4,5-epoxy-6,17-dimethyl-, (5α)-	283	$C_{18}H_{21}NO_2$	2b
Morphinan, 4,5-epoxy-3,6-dihydroxy-*N*,6-dimethyl-	301	$C_{18}H_{23}NO_3$	9i
Phenethylamine, α-methyl-3,4-methylenedioxy-	179	$C_{10}H_{13}NO_2$	64p
Phenethylamine, *N*-ethyl-α-methyl-3,4-methylenedioxy-	207	$C_{12}H_{17}NO_2$	64q
Phenethylamine, *N*,α-dimethyl-3,4-methylenedioxy-	193	$C_{11}H_{15}NO_2$	64r
N-Phenylpropanamide, *N*-[3-methyl-1-(2-phenylethyl)-4-piperidyl]-	350	$C_{23}H_{30}N_2O$	76e

Common name	ACS Registry number	Pharmacological action	Schedule (number)	Basic/ acidic	pK_a	Melting/boiling[a] point (°C)
α-Methylfentanyl		Narcotic analgesic	I (9814)			
Methylphenidate	113-45-11	CNS stimulant	II (1724)	B		mp 74–75
Methylphenobarbital (mephobarbital)	115-38-8	Hypnotic/sedative anticonvulsant	IV (2250)	A	7.8	mp 176
1-Methyl-4-phenylpropionoxypiperidine		Meperidine analog/ narcotic analgesic	I (9661)			
N-Methyl-3-piperidyl benzilate	3321-80-0	Hallucinogen	I (7484)			bp$_{0.03}$ 175–176
3-Methylthiofentanyl		Narcotic analgesic	I (9833)			
α-Methylthiofentanyl		Narcotic analgesic	I (9832)			
Methyprylon	125-64-4	Hypnotic/sedative	III (2575)	N, A	12.0	mp 74–77
Metopon	143-52-2	Narcotic analgesic	II (9260)	B	8.1	mp 243–245
Midazolam	59467-70-8	Sedative/anesthetic	IV (2884)		6.2	mp 158–160
Moramide Synthetic intermediate			II (9802)			
Morpheridine	469-81-8	Narcotic analgesic	I (9632)	B		bp$_{0.5}$ 188–192
Morphine	57-27-2	Principal alkaloid/	II (9300)	B		254 (dec)
Morphine methylbromide	125-23-5	Narcotic analgesic	I (9305)	B		mp 265–266
Morphine methylsulfonate		Narcotic analgesic	I (9306)	B		
Morphine N-oxide	639-46-3	Narcotic analgesic	I (9307)	B	4.82	mp 274–275
Myrophine	467-18-5	Narcotic analgesic	I (9308)	B		mp 41
Nabilone	51022-71-0	Tranquilliser	II (7379)			mp 159–160
Nalorphine	62-67-9	Narcotic antagonist	III (9400)	B	7.88	mp 208–209
Nicocodeine	3688-66-2	Narcotic analgesic/antitussive	I (9309)	B		mp 134–135
Nicomorphine	639-48-5	Narcotic analgesic	I (9312)	B		mp 178–178.5
Nimetazepam	2011-67-8	Anticonvulsant sedative	IV (2837)			mp 156.5–157.5
Nitrazepam	146-22-5	Hypnotic/sedative anticonvulsant	IV (2834)	B, N A	3.2 10.8	mp 224–226
Noracymethadol	5633-25-0	Narcotic analgesic	I (9633)	B		
Nordiazepam (Nordazepam)	1088-11-5	Minor tranquilizer	IV (2838)	B	3.5,12.0	mp 216–217
Norlevorphanol	1531-12-0	Narcotic analgesic	I (9634)	B		mp 270–272
Normethadone	467-85-6	Narcotic analgesic/antitussive	I (9635)	B		bp 164–167
Normorphine	466-97-7	Narcotic analgesic	I (9313)	B	9.8	mp 276–277
Norpipanone	561-48-8	Analgesic	I (9636)	B		
Nortriptyline	72-69-5	Antidepressant		B	9.73	
Opium (raw)			II (9600)			
Opium (extracts)			II (9610)			
Opium (fluid)			II (9620)			
Opium (powdered)			II (9639)			
Opium (granulated)			II (9640)			
Opium (tincture of)			II (9630)			
Orphenadrine	83-98-7	Anticholinergic		B		bp$_{12}$ 195
Oxazepam	604-75-1	Tranquilizer	IV (2835)	B, N A	1.8 11.1	mp 205–206
Oxazolam	24143-17-7	Tranquilizer	IV (2839)			mp 186–188
Oxprenolol	6452-71-7	Coronary vasodilator				mp 78–80
Oxycodone	76-42-6	Narcotic analgesic	II (9143)	B	8.5	mp 218–220
Oxymorphone	76-41-5	Narcotic analgesic	II (9652)	B	4.82	248–249 (dec)
Parahexyl (synhexyl)	117-51-1	Psychotomimetic	I (7374)			bp$_{1.0}$ 190–192

Systematic nomenclature	Nominal MW	Empirical formula	Structure[b] designation
Propanamide, *N*-phenyl-*N*-[1-(α-methyl-2-phenylethyl)-4-piperidinyl]-	350	$C_{23}H_{30}N_2O$	76d
2-Piperidineacetic acid, α-phenyl-, (methyl ester)	233	$C_{14}H_{19}NO_2$	85c
2,4,6(1*H*,3*H*,5*H*)-Pyrimidinetrione, 5-ethyl-1-methyl-5-phenyl-	246	$C_{13}H_{14}N_2O_3$	12z
3-Piperidyl benzilate, 1-methyl-	325	$C_{20}H_{23}NO_3$	83e
N-Phenylpropanamide, *N*-[3-methyl-1-(2-(2-thienyl)ethyl-4-piperidyl]-	366	$C_{23}H_{30}N_2S$	
N-Phenylpropanamide, *N*-[1-(1-methyl-2-(2-thienyl)ethyl-4-piperidyl]-	366	$C_{23}H_{30}N_2S$	
Piperidinedione, 3,3-diethyl-5-methyl-2,4-	183	$C_{10}H_{17}NO_2$	13a
Morphinan-6-one, 4,5-epoxy-3-hydroxy-5,17-dimethyl-, (5α)-	299	$C_{18}H_{21}NO_3$	6c
[1,5-a][1,4]Benzodiazepine, 8-chloro-6-(2-fluorophenyl)-1-methyl-4*H*-imidazo-	325	$C_{18}H_{13}ClFN_3$	31d
3-Morpholino-1,1-diphenylpropane carboxylic acid, 2-methyl-			
1-[(2-Morpholinyl)ethyl]-4-phenyl-4-piperidinecarboxylic acid, ethyl ester	346	$C_{20}H_{30}N_2O_3$	79q
Morphinan-3,6-diol, 7,8-didehydro-4,5-epoxy-17-methyl-, (5α,6α)-	285	$C_{17}H_{19}NO_3$	3j
Morphinanium, 7,8-didehydro-4,5-epoxy-3,6-dihydroxy-17,17-dimethyl-, bromide, (5α,6α)-	379	$C_{18}H_{22}BrNO_3$	3l
Morphinan-3,6-diol, 7,8-didehydro-4,5-epoxy-17-methyl-(5α,6α)-, 17-oxide	301	$C_{17}H_{19}NO_4$	3n
Morphinan-6-ol, 7,8-didehydro-4,5-epoxy-17-methyl-3-phenylmethoxy-	585	$C_{38}H_{51}NO_4$	3o
trans-(±)-3-(1,1-Dimethylheptyl)-6-6a-7,8,10,10a-hexahydro-1-hydro-6,6-dimethyl-9*H*-dibenzo[b,d]pyran-9-one	372	$C_{24}H_{36}O_3$	120
Morphinan-3,6-diol, 7,8-didehydro-4,5-epoxy-17-(2-propenyl)-, (5α,6α)-	311	$C_{19}H_{21}O_3$	3p
Morphinan-3,6-diol 3-pyrridinecarboxylate, 4,5-epoxy-3-methoxy-17-methyl- (ester), (5α,6α)-			3q
Morphinan-3,6-diol di-3-pyrridinecarboxylate, 7,8-didehydro-4,5-epoxy-17-methyl-, (5α,6α)-	495	$C_{29}H_{25}N_3O_5$	3r
2*H*-1,4-Benzodiazepin-2-one, 1,3-dihydro-1-methyl-7-nitro-5-phenyl-	295	$C_{16}H_{13}N_3O_3$	23bc
2*H*-1,4-Benzodiazepin-2-one, 1,3-dihydro-7-nitro-5-phenyl-	281	$C_{15}H_{11}N_3O_3$	23bd
Benzeneethanol, α-ethyl-β-[2-(methylamino)propyl]-β-phenyl-, acetate (ester)	339	$C_{22}H_{29}NO_2$	71k
1(2*H*)-1,4-Benzodiazepin-2-one, 7-chloro-1,3-dihydro-5-phenyl-	270	$C_{15}H_{11}ClN_2O$	23be
Morphinan-3-ol	243	$C_{16}H_{21}NO$	10h
3-Hexanone, 6-(dimethylamino)-4,4-diphenyl	295	$C_{20}H_{25}NO$	71l
Morphinan-3,6-diol, 7,8-didehydro-4,5-epoxy-, (5α,6α)-	271	$C_{16}H_{17}NO_3$	3u
3-Hexanone, 4,4-diphenyl-6-(1-piperidinyl)-	335	$C_{23}H_{29}NO$	60g
1-Propanamine, 3-(10,11-dihydro-5*H*-dibenzo[*a*,*d*]cyclohepten-5-ylidene)-*N*-methyl-	263	$C_{19}H_{21}N$	50k
Ethanamine, *N*,*N*-dimethyl-2-[(2-methylphenyl)phenylmethoxy]	269	$C_{18}H_{23}NO$	72f
2*H*-1,4-Benzodiazepin-2-one, 7-chloro-1,3-dihydro-3-hydroxy-5-phenyl-	286	$C_{15}H_{11}ClN_2O_2$	23bf
Oxazolo[3,2-d][1,4]Benzodiazepin-6(5H)-one, 10-chloro-2,3,7,-11b-tetrahydro-2-methyl-11b-phenyl-	328	$C_{18}H_{17}ClN_2O_2$	41c
2-Propanol, 1-[(1-methylethyl)amino]-3-[2-(2-propenyloxy)phenoxy]-	265	$C_{15}H_{23}NO_3$	103
Morphinan-6-one, 4,5-epoxy-14-hydroxy-3-methoxy-17-methyl-, (5α)-	315	$C_{18}H_{21}NO_4$	6f
Morphinan-6-one, 4,5-epoxy-3,14-dihydroxy-17-methyl-, (5α)-	301	$C_{17}H_{19}NO_4$	6g
6*H*-Dibenzo[*b*,*d*]pyran-1-ol, 3-hexyl-7,8,9,10-tetrahydro-6,6,9-trimethyl-	328	$C_{22}H_{32}O_2$	54

Common name	ACS Registry number	Pharmacological action	Schedule (number)	Basic/ acidic	pK_a	Melting/boiling^a point (^{o}C)
Paraldehyde	123-63-7	Hypnotic	IV (2585)			mp 12; bp 124
Pemoline	2152-34-3	CNS stimulant	IV (1530)	B	10.5	256–257 (dec)
Pentazocine	359-83-1	Analgesic	IV (9709)	B	7.6	mp 145.4–147.2
Pentobarbital	76-74-4	Anesthetic/anticonvulsant	II (2270)	A	8.11	mp 129–130
Perphenazine	58-39-9	Tranquilizer/antipsychotic		B	3.7, 7.8	mp 94–100; bp$_1$ 278-281
Pethidine	57-42-1	Narcotic analgesic	II (9230)	B		mp 30
Pethidine Synthetic intermediate A			II (9232)	A, N		
Pethidine Synthetic intermediate B			II (9233)			
Pethidine Synthetic intermediate C			II (9234)	A		
Petrichloral (pentaerythritol chloral)	78-12-6	Hypnotic/sedative	IV (2591)			mp 52–54
Peyote (mescal) (refer to mescaline)		Hallucinogen	I (7415)			
Phenadoxone	467-84-5	Narcotic analgesic	I (9637)	B		mp 75–76
Phenampromide	129-83-9	Analgesic	I (9638)	B		
Phenazocine	127-35-5	Narcotic analgesic/antipyretic	II (9715)	B		mp 181–182
Phencyclidine (PCP)	77-10-1	Anesthetic/analgesic hallucinogen	II (7471)	B	8.5	mp 46–46.5
PCP ethylamine analog		Anesthetic/analgesic hallucinogen	I (7455)			bp$_{0.5}$ 103–105
PCP pyrrolidine analog		Anesthetic/analgesic hallucinogen	I (7458)			
PCP thiophene analog		Anesthetic/analgesic hallucinogen	I (7470)			mp 37–38;
Phendimetrazine	634-03-7	Anorexic	III (1615)	B	7.6	bp$_8$ 122–124
Phenmetrazine	134-49-6	Anorexic	II (1631)	B	8.5	bp$_{12}$ 138–140
Phenobarbital	50-06-6	Hypnotic/anticonvulsant	IV (2285)	A	7.41	mp 174–178
Phenomorphan	468-07-5	Narcotic analgesic	I (9647)	B		mp 243–245
Phenoperidine	562-26-5	Analgesic	I (9641)	B		
Phentermine	122-09-8	Anorexic	IV (1640)	B	10.1	bp$_{750}$ 205
Phenylacetone (phenyl-2-propanone)		Amphetamine synthetic precursor	II (8501)			bp$_{760}$ 214
Phenylbutazone	50-33-9	Analgesic			4.5	mp 105
1-Phenylcyclohexylamine	2201-24-3	Synthetic precursor to PCP	II (7460)	B		mp 80
1-(2-Phenylethyl)-4-phenyl-4-acetyloxy-piperidine		Narcotic analgesic	I (9663)			
Phenytoin	57-41-0	Anticonvulsant/antiepileptic		A	8.33	mp 295–298
Pholcodeine	509-67-1	Antitussive	I (9314)	B	8.0, 9.3	
Piminodine	13495-09-5	Narcotic analgesic	II (9730)	B		
Pinazepam	52463-83-9	Antidepressant	IV (2883)			mp 140–142
1-Piperidinocyclohexanecarbonitrile		Toxic PCP precursor	II (8603)	B		mp 68–70
Pipradrol	467-60-7	CNS stimulant	IV (1750)	B		
Piritramide	302-41-0	Analgesic	I (9642)	B		mp 149–150
Prazepam	2955-38-6	Sedative/tranquilizer	IV (2764)	B, N	2.7	mp 145–146
Primidone	125-33-7	Anticonvulsant/antiepileptic		A		mp 281–282
Prochlorperazine	58-38-8	Antipsychotic		B	3.73, 8.1	
α-Prodine	77-20-3	Narcotic analgesic	II (9010)	B		
β-Prodine	25123-05-1	Narcotic analgesic	I (9611)	B		
Proheptazine	77-14-5	Analgesic	I (9643)	B		bp$_{0.3}$ 126
Promethazine	60-87-7	CNS depressant		B	9.1	mp 60
Properidine	561-76-2	Analgesic/antispasmodic	I (9644)	B		

Systematic nomenclature	Nominal MW	Empirical formula	Structure[b] designation
1,3,5-Trioxane, 2,4,6-trimethyl-	132	$C_6H_{12}O_3$	112
4(5H)-Oxazolone, 2-amino-5-phenyl-	176	$C_9H_8N_2O_2$	113a
2,6-Methano-3-benzazocin-8-ol, 1,2,3,4,5,6-hexahydro-6,11-dimethyl-3-(3-methyl-2-butenyl)-, (2α,6α,11R*)-	285	$C_{19}H_{27}NO$	99a
2,4,6(1H,3H,5H)-Pyrimidinetrione, 5-ethyl-5-(1-methylbutyl)-	226	$C_{11}H_{18}N_2O_3$	12bg
1-Piperazineethanol, 4-[3-(2-chloro-10H-phenothiazin-10-yl)propyl]-	403	$C_{21}H_{26}ClN_3OS$	48c
4-Piperidinecarboxylic acid, 1-methyl-4-phenyl, ethyl ester	247	$C_{15}H_{21}NO_2$	79l
4-Phenylpiperidine, 4-cyano-1-methyl-			
4-Phenylpiperidine-4-carboxylate, ethyl-			
4-Phenylpiperidine-4-carboxylic acid, 1-methyl-			
Ethanol, 1,1'-[2,2-bis[[(2,2,2-trichloro-1-hydroxyethoxy)methyl]-1,3-propanediylbis(oxy)]-bis[2,2,2-trichloro]-	725	$C_{13}H_{16}Cl_{12}O_8$	108e
Flowering head of *Lophophora williamsii* (Lemaire) Coult., *Cactaceae*			
3-Heptanone, 6-(4-morpholinyl)-4,4-diphenyl-	351	$C_{23}H_{29}NO_2$	71m
Propanamide, N-[1-methyl-2-(1-piperidinyl)ethyl]-N-phenyl-	274	$C_{17}H_{26}N_2O$	80b
2,6-Methano-3-benzazocin-8-ol, 1,2,3,4,5,6-hexahydro-6,11-dimethyl-3-(2-phenylethyl)-	321	$C_{22}H_{27}NO$	99b
Piperidine, 1-(1-phenylcyclohexyl)-	243	$C_{17}H_{25}N$	75c
Cyclohexylamine, N-ethyl-1-phenyl- (= cyclohexamine)	203	$C_{14}H_{21}N$	75d
Pyrrolidine, 1-(1-phenylcyclohexyl)-	229		
Piperidine, 1-[1-(2-thienyl)cyclohexyl]-	249	$C_{15}H_{23}NS$	74
Morpholine, 3,4-dimethyl-2-phenyl-, (2S-trans)-	191	$C_{12}H_{17}NO$	94d
Morpholine, 3-methyl-2-phenyl-	177	$C_{11}H_{15}NO$	94e
2,4,6(1H,3H,5H)-Pyrimidinetrione, 5-ethyl-5-phenyl-	232	$C_{12}H_{12}N_2O_3$	12bk
Morphinan-3-ol, 17-(2-phenylethyl)-	347	$C_{24}H_{29}NO$	10i
4-Piperidinecarboxylic acid, 1-(3-hydroxy-3-phenylpropyl)-4-phenyl-, ethyl ester	367	$C_{23}H_{29}NO_3$	79u
Benzeneethanamine, α,α-dimethyl-	149	$C_{10}H_{15}N$	62bc
1-Phenyl-2-peopanone	134	$C9H10O$	65a
3,5-Pyrazolidinedione, 4-butyl-1,2-diphenyl-	308	$C_{19}H_{20}N_2O_2$	21d
Cyclohexaneamine, 1-phenyl-	175	$C_{12}H_{17}N$	75f
Piperidine, 1-(2-phenylethyl)-4-phenyl-4-acetyloxy- (PEPAP)			77e
2,4-Imidazolidinedione, 5,5-diphenyl-	252	$C_{15}H_{12}N_2O_2$	20d
Morphinan-6-ol, 7,8-didehydro-4,5-epoxy-17-methyl-3-[2-(4-morpholinyl)ethoxy]-, (5α,6α)-	398	$C_{23}H_{30}N_2O_4$	3v
4-Piperidinecarboxylic acid, 4-phenyl-1-[3-(phenylamino)propyl]-, ethyl ester	366	$C_{23}H_{30}N_2O_2$	79v
2H-1,4-Benzodiazepin-2-one, 7-chloro-1,3-dihydro-5-phenyl-1-(2-propynyl)-	308	$C_{18}H_{13}ClN_2O$	23bh
1-(1-Cyanocyclohexyl)piperidine	192	$C_{12}H_{20}N_2$	73c
2-Piperidinemethanol, α,α-diphenyl-	267	$C_{18}H_{21}NO$	70c
[1,4'-Biperidine]-4'-carboxamide, 1'-(3-cyano-3,3-diphenylpropyl)-	430	$C_{27}H_{34}N_4O$	60h
2H-1,4-Benzodiazepin-2-one, 7-chloro-1-(cyclopropylmethyl)-1,3-dihydro-5-phenyl-	324	$C_{19}H_{17}ClN_2O$	23bi
4,6(1H,5H)-Pyrimidinedione, 5-ethyldihydro-5-phenyl-	218	$C_{12}H_{14}N_2O_2$	18
10H-Phenothiazine, 2-chloro-10-[3-(4-methyl-1-piperazinyl)-propyl-	373	$C_{20}H_{24}ClN_3S$	48d
(-)-4-Piperidinol, 1,3-dimethyl-4-phenyl-, propanoate (ester) (3S-cis)-	261	$C_{16}H_{23}NO_2$	77b
(–)-4-Piperidinol, 1,3-dimethyl-4-phenyl-, propanoate (3R-trans)-	261	$C_{16}H_{23}NO_2$	77b
1H-Azepin-4-ol, hexahydro-1,3-dimethyl-4-phenyl-, propanoate (ester)	275	$C_{17}H_{25}NO_2$	92b
10H-Phenothiazine-10-ethanamine, N,N,α-trimethyl-	284	$C_{17}H_{20}N_2S$	47d
4-Piperidinecarboxylic acid, 1-methyl-4-phenyl-, 1-methylethyl ester	261	$C_{16}H_{23}NO_2$	79w

Common name	ACS Registry number	Pharmacological action	Schedule (number)	Basic/ acidic	pK_a	Melting/boiling[a] point (^{o}C)
Propiram	15686-91-6	Analgesic	I (9649)			bp 162–163
Propranolol	525-66-6	β-Adrenergic blocker		B		mp 96
Psilocybin	520-52-5	Psychotomimetic	I (7437)	B		mp 185–195
Psilocyn	520-53-6	Psychotomimetic	I (7438)	B		mp 173–176
Pyrovalerone	3563-49-3	CNS stimulant	V (1485)			
Quazepam	36735-22-5	Sedative/hypnotic	IV (2881)			mp 138–139
Quinine	130-95-0	Antimalarial analgesic		B	4.2, 8.8	mp 177
Racemethorphan	510-53-2	Antitussive	II (9732)	B	8.3	
Racemoramide	357-56-2	Narcotic analgesic	I (9645)	B		
Racemorphan	297-90-5	Narcotic analgesic	II (9733)	B		mp 251–253
Secobarbital	76-73-3	Hypnotic/sedative	II (2315)	A	7.9, 12.6	mp 98–100
SPA ((–)-1-dimethylamino-1,2-diphenylethane)		Analgesic	IV (1635)			mp 218–219
Sufentanil/sulfentanyl	56030-54-7	Narcotic analgesic	II (9740)			mp 96.6
Sulfondiethylmethane		Hypnotic	III (2600)			
Sulfonethylmethane	76-20-0	Hypnotic	III (2605)			mp 74–76
Sulfonmethane	115-24-2	Hypnotic	III (2610)			mp 124–126;
Temazepam	846-50-4	Minor tranquilizer	IV (2925)	B, N	1.6	mp 119–121
Tetrahydrocannabinol(s)	1972-08-3	Psychotomimetic	I (7370)		10.6	bp$_{0.02}$ 200
Tetrazepam	10379-14-3	Sedative	IV (2886)			mp 144
Thebacon	466-90-0	Narcotic analgesic/antitussive	I (9315)	B		mp 154
Thebaine	115-37-7	Convulsant/analgesic	II (9333)	B	8.15	mp 191–193
1-[1-(2-Thienyl)cyclohexyl]piperidine	21500-98-1	parasympstholytic	I (7374)	B		mp 37–38
Thienylfentanyl		Narcotic analgesic	I (9834)			
Thiofentanyl		Narcotic analgesic	I (9835)			
Thioridazine	50-52-2	Sedative		B		mp 72–74;
Tiletamine	14176-49-9	Sedative	III (7295)			
Tilidine	20380-58-9	Analgesic	I (9750)			bp$_{0.01}$ 95.5–96.0
Triazolam	28911-01-5	Hypnotic	IV (2887)	B, N		mp 233–235
Trifluoperazine	117-89-5	Sedative		B		bp$_{0.6}$ 202–210
Trimeperidine	64-39-1	Narcotic analgesic	I (9646)	B		
3,4,5-Trimethoxyamphetamine	1082-88-8	Hallucinogen	I (7390)	B		
Trimipramine	739-71-9	Antidepressant		B		mp 45
Zolpidem	82626	Hypnotic	IV (2783)			mp 196

[a] Melting and boiling points may vary because of differences in literature references and exact conditions under which they were determined.

[b] See Appendices III and IV for these structure designations. Structures do not, in general illustrate stereochemistry configurations.

Systematic nomenclature	Nominal MW	Empirical formula	Structure[b] designation
Propanamide, *N*-[1-methyl-2-(1-piperidinyl)ethyl-*N*-2-pyrindinyl-	275	$C_{16}H_{25}N_3O$	68a
Propan-2-ol, (±)-1-isopropylamino-3-(1-naphthyloxy)-	259	$C_{16}H_{21}NO_2$	95e
1*H*-Indol-4-ol, 3-[2-(dimethylamino)ethyl]-, dihydrogen phosphate (ester)	284	$C_{12}H_{17}N_2O_4P$	97o
1*H*-Indol-4-ol, 3-[2-(dimethylamino)ethyl]-	204	$C_{12}H_{16}N_2O$	97n
1-(4-Methylphenyl)-2-(1-pyrrolidinyl)-1-pentanone	245	$C_{16}H_{23}NO$	121
1,4-Benzodiazepin-2-thione, 7-chloro-(2,2,2-trifluoroethyl)-5-*o*-fluorophenyl-1,3-dihydro-2*H*-	386	$C_{17}H_{11}ClF_4N_2S$	23bj
Cinchonan-9-ol, 6'-methoxy-, (8α,9*R*)-	324	$C_{20}H_{24}N_2O_2$	58
(±)-Morphinan, 3-methoxy-17-methyl-,	271	$C_{18}H_{25}NO$	10j
Pyrrolidine, 1-[3-methyl-4-(4-morpholinyl)-1-oxo-2,2-diphenylbutyl]-, (*S*)-	392	$C_{25}H_{32}N_2O_2$	66e
(±)-Morphinan-3-ol, 17-methyl-,	257	$C_{17}H_{23}NO$	10k
2,4,6(1*H*,3*H*,5*H*)-Pyrimidinetrione, 5-(1-methylbutyl)-5-(2-propenyl)-	238	$C_{12}H_{18}N_2O_3$	12bo
(−)-1-Dimethylamino-1,2-diphenylethane	225	$C_{16}H_{19}N$	
Propanamide, *N*-[4-(methoxymethyl)-1-[2-(2-thienyl)ethyl]-4-piperidinyl-*N*-phenyl-	386	$C_{22}H_{30}N_2O_2S$	83h
			107a
Butane, 2,2-bis(ethylsulfonyl)-	242	$C_8H_{18}O_4S_2$	107b
Propane, 2,2-bis(ethylsulfonyl)-	196	$C_7H_{16}O_4S_2$	107c
2*H*-1,4-Benzodiazepin-2-one, 7-chloro-1,3-dihydro-3-hydroxy-1-methyl-5-phenyl-	300	$C_{16}H_{13}ClN_2O_2$	23bk
6*H*-Dibenzo[*b,d*]pyran-1-ol, 6a,7,8,10a-tetrahydro-6,6,9-trimethyl-3-pentyl-, (6a*R*-*trans*)-	314	$C_{21}H_{30}O_2$	55h
2*H*-1,4-Benzodiazepin-2-one, 7-chloro-5-(1-cyclohexen-1-yl)-1,3-dihydro-1-methyl-	288	$C_{16}H_{17}ClN_2O$	25f
Morphinan-6-ol, 6,7-didehydro-4,5-epoxy-3-methoxy-17-methyl-, acetate, (5α)-	341	$C_{20}H_{23}NO_4$	2a
Morphinan, 6,7,8,14-tetradehydro-4,5-epoxy-3,6-dimethoxy-17-methyl, (5α)-	311	$C_{19}H_{21}NO_3$	1b
1-(1-Piperidyl)-1-(2-thienyl)cyclohexane	249	$C_{15}H_{23}NS$	74
N-Phenylpropanamide, *N*-[1-(2-thienyl)methyl-4-piperidyl]-	342	$C_{20}H_{26}N_2OS$	83j
N-Phenylpropanamide, *N*-[1-(2-(2-thienyl)ethyl-4-piperidyl]-			
10*H*-Phenothiazine, 10-[2-(1-methyl-2-piperidinyl)ethyl]-2-(methylthio)-	370	$C_{21}H_{26}N_2S_2$	47g
2-Ethylamino-2-(2-thienyl)cyclohexanone	223	$C_{12}H_{17}NOS$	122
3-Cyclohexene-1-carboxylic acid, 2-(dimethylamino)-1-phenyl-, ethyl ester, *trans*-(±)-	273	$C_{17}H_{23}NO_2$	84c
4*H*-[1,2,4]Triazolo[4,3-a][1,4]benzodiazepine, 8-chloro-6-(2-chlorophenyl)-1-methyl-	343	$C_{17}H_{12}Cl_2N_4$	29j
10*H*-Phenothiazine, 10-[3-(4-methyl-1-piperazinyl)propyl]-2-(trifluoromethyl)-	407	$C_{21}H_{24}F_3N_3S$	48e
4-Piperidinol, 1,2,5-trimethyl-4-phenyl-, propanoate (ester)	275	$C_{17}H_{25}NO_2$	77f
Benzeneethanamine, 3,4,5-trimethoxy-α-methyl	225	$C_{12}H_{19}NO_3$	64u
5*H*-Dibenz[*b,f*]azepine-5-propanamine, 10,11-dihydro-*N,N,*β-trimethyl-	294	$C_{20}H_{26}N_2$	49o
N,N,6-Trimethyl-2-(4-methylphenyl)-imidazo[1,2-a]pyridine-3-acetamide	307	$C_{19}H_{21}N_3O$	123

Appendix II
Metabolites of Common Drugs

Common name	Metabolites[a,b,c]	Structure designation[d]	Ref.[e]
Acetylmethadol	(See α-acetylmethadol)	71a	1
α-Acetylmethadol (LAAM)	Dinoracetylmethadol	71b	2–8
	Dinormethadol (and glucuronide)	71c	
	Methadol (and glucuronide)	71g	
	Noracetylmethadol	71i	
	Normethadol (and glucuronide)	71k	
β-Acetylmethadol	(See α-acetylmethadol)	71a	
Alfentanil	Desmethylalfentanil	83d	9–11
	N-[4-(Hydroxymethyl)-4-piperidinyl]-N-phenylpropanamide	83f	
	N-(4-Hydroxyphenyl)acetamide	82a	
	N-(4-Hydroxyphenyl)propanamide	82b	
	N-[4-(Methoxymethyl)-4-piperidinyl]-N-phenylpropanamide (noralfentanil)	83g	
	[[4-(Phenylamino)-4-piperidinyl]methyl] propanoate	81	
	(Refer also to fentanyl)		
Allobarbital	5-Allyl-5-(2',3'-dihydroxypropyl)barbituric acid (guinea pig and rat)	12b	12
Alphenal	Alphenal (unchanged) (rat)	12h	12
	5-Allyl-5-(4'-hydroxyphenyl)barbituric acid (rat)	12c	
	5-Allyl-5-(3',4'-hydroxyphenyl)barbituric acid (guinea pig)	12d	
	5-(2',3'-Dihydroxypropyl)-5-phenylbarbituric acid (rat)	12r	
Alprazolam	α-Hydroxyalprazolam	29e	13, 14
	4'-Hydroxyalprazolam	29f	
Amitriptyline	10,11-Dihydroxyamitriptyline	50b	15–17
	10,11-Dihydroxydidesmethylamitriptyline	50c	
	1,2-Dihydroxy-1,2-dihydroamitriptyline	51a	
	1,2-Dihydroxy-1,2-dihydronortriptyline	51b	
	2,11-Dihydroxynortriptyline	50d	
	10,11-Dihydroxynortriptyline	50e	
	2-Hydroxyamitriptyline	50f	
	3-Hydroxyamitriptyline	50g	
	10-Hydroxyamitriptyline N-oxide	50h	
	10-Hydroxydidesmethylamitriptyline	50i	
	3-Hydroxynortriptyline	50j	
	Nortriptyline	50k	
	10-Oxoamitriptyline	50l	
	10-Oxonortriptyline	50m	
Amobarbital	3'-Carboxyamobarbital	12o	18–20
	Ethylisopentylacetylurea	104b	
	N-Glucosyl amobarbital	12u	
	3'-Hydroxyamobarbital	12x	
	N-Hydroxyamobarbital	12y	
	Malonuric acid (theoretical)	104a	
Amphetamine	Benzoic acid	63a	18, 21, 22
	Hippuric acid	63d	
	p-Hydroxyamphetamine	62p	
	p-Hydroxynorephedrine	62u	
	Norephedrine	62bc	
	Phenylacetone	65a	
	Benzoylglucuronide		

313

Common name	Metabolites[a,b,c]	Structure designation[d]	Ref.[e]
Anileridine	p-Acetylaminophenylacetic acid		18
	Acetylanileridine	79a	
	Acetylanileridinic acid	79b	
	Anileridinic acid	79d	
	Noranileridine (Normeperidine)	79s	
Aprobarbital	5-Allyl-5-(1'-hydroxypropyl)barbituric acid (rat and guinea pig)	12e	12
	5-Allyl-5-(2'-hydroxypropyl)barbituric acid	12f	
	Aprobarbital	12i	
	5-(2',3'-Dihydroxypropyl)-5-isopropylbarbituric acid (rat and guinea pig)	12q	
Atropine	Tropic acid	63l	18
	Tropine	93a	
Barbital	Barbital	12j	23–25
	5-Ethylbarbituric acid	12t	
	Hydroxybarbital		
Benzphetamine	Amphetamine (rat)	62a	26
	Benzylamphetamine (rat)	62c	
	p-Hydroxyamphetamine (rat)	62p	
	1-(p-Hydroxyphenyl)-2-(N-benzylamino)propane (rat) (p-Hydroxybenzylamphetamine)	62q	
	1-(m-Hydroxyphenyl)-2-(N-methyl-N-benzylamino)propane (rat)	62w	
	1-(p-Hydroxyphenyl)-2-(N-methyl-N-benzylamino)propane (rat) (p-Hydroxybenzphetamine)	62x	
	Methamphetamine (rat)	62y	
	1-(m-Methoxy-p-hydroxyphenyl)-2-(N-benzylamino)propane (rat)	62z	
Benzylamphetamine	Amphetamine (rabbit)	62a	27
	N-Benzylamphetamine (rabbit)	62c	
	N-(1-Benzylethyl)-α-phenylnitrone (rabbit)	62d	
	N-Benzyl-N-hydroxyamphetamine (rabbit)	62e	
	Phenylpropanone (rabbit) (phenylacetone)	65a	
	Phenylpropan-2-ol (rabbit)	65d	
Bezitrimide	3-Cyano-3,3-diphenylpropionic acid	61	28, 29
	1-(4-Piperidinyl)-1,3-dihydro-2H-benzimidazol-2-one	86	
Bromazepam	2-(2-Amino-5-bromobenzoyl)pyridine (ABBP)	45a	30, 31
	2-(2-Amino-5-bromo-3-hydroxybenzoyl)pyridine (OH-ABBP)	45b	
	Bromazepam N_4-oxide	34b	
	3-Hydroxybromazepam	34c	
4-Bromo-2,5-dimethoxyamphetamine			32
Brotizolam	3-Hydroxybrotizolam	35b	33, 34
	1-Methylhydroxybrotizolam	35c	
Bufotenine (5-hydroxy-N-dimethyltrptamine)	Bufotenine (unchanged) (rat)	97a	35, 36
	5-Hydroxyindole acetic acid (rat)	97f	
	5-Hydroxytryptamine (theoretical)	97g	
	O-Methylbufotenine (theoretical) (5-methoxy-N-dimethyltrptamine)	97i	
	5-Methoxytryptamine (theoretical)	97k	
	N-Methylserotonin (theoretical)	97l	
Butallylonal	Ketobutallylonal		23

Common name	Metabolites[a,b,c]	Structure designation[d]	Ref.[e]
Butorphanol	3'-Hydroxybutorphanol	10d	37
	Norbutorphanol	10i	
Chloral betaine	Hydrolyzes to chloral hydrate	108b	23
Chloral hydrate	Trichloroacetic acid	108f	38
	Trichloroethanol	108g	
	Trichloroethanol glucuronide (urochloralic acid)		
Chlordiazepoxide	Demoxepam	24	14, 23,
	Norchlordiazepoxide	26b	39–43
	Nordiazepam	23bf	
	Oxazepam	23bg	
Chlorhexadol	Hemiacetal of chloral hydrate		23
	(*See* chloral hydrate)		
Chlorphentermine	Dimethylnitro-4-chlorophenylethane	62j	20, 44
	Dimethylnitroso-4-chlorophenylethane	62k	
	N-Hydroxychlorphentermine	62r	
Chlorpromazine	Chlorpromazine sulfoxide	46e	45–48
	7,8-Dihydroxychlorpromazine (rat)		
	Dinorchlorpromazine	46f	
	Dinorchlorpromazine sulfoxide	46g	
	7-Hydroxychlorpromazine	46h	
	7-Hydroxychlorpromazine sulfoxide	46i	
	7-Hydroxydinorchlorpromazine	46j	
	7-Hydroxynorchlorpromazine	46k	
	Norchlorpromazine	46l	
	Norchlorpromazine sulfoxide	46m	
Clonazepam	7-Acetamidoclonazepam	23a	49
	7-Aminoclonazepam	23c	
	3-Hydroxyclonazepam		
Clorazepate	Nordiazepam	23bf	50
	Oxazepam	23bg	
Cloxazolam	2-Amino-3-hydroxy-2',5-dichlorobenzophenone	111c	39, 51
	7-Chloro-5-(*o*-chlorophenyl)-1,3-dihydro-3-hydroxy-2*H*-1,4-benzodiazepin-2-one (COX)	23h	
	7-Chloro-2,3-dihydro-5-(*o*-chlorophenyl)-2*H*-1,4-benzodiazepin-2-one	23i	
Cocaine	Benzoylecgonine	93b	52, 53
	Benzoyl norecgonine		
	Norcocaine	93f	
	Ecgonidine		
	Ecgonidine methyl ester		
	Ecgonine	93d	
	Ecgonine methyl ester	93e	
	Norecgonidine methyl ester		
	Norecgonine methyl ester		
	m-Hydroxy benzoylecgonine		
	m-Hydroxycocaine		
	p-Hydroxycocaine		
	Hydroxymethoxybenzoyl ecgonine		

Common name	Metabolites[a,b,c]	Structure designation[d]	Ref.[e]
Codeine	6α-Hydrocodol	9f	54–58
	6β-Hydrocodol	9f	
	Hydrocodone	6a	
	6α-Hydromorphol	9h	
	6β-Hydromorphol	9h	
	Hydromorphone	6b	
	Morphine	3j	
	Norcodeine	3s	
	Normorphine	3u	
Codeine methylbromide	(See codeine)		23
Codeine N-oxide	(See codeine)		23
Cyclobarbital	Cyclobarbital	12p	23
	5-Ethyl-5-(3'-oxo-1'-cyclohexenyl) barbituric acid		
Dextromethorphan	Dextrorphan	10b	20, 23
	3-Hydroxymorphinan	10e	
Dextropropoxyphene	Cyclic dinorpropoxyphene	90	59–61
	Dinorpropoxyphene	66a	
	Dinorpropoxyphene carbinol	66b	
	p-Hydroxynorpropoxyphene	66c	
	p-Hydroxypropoxyphene	66d	
	Norpropoxyphene	66f	
	Norpropoxyphene carbinol	66g	
	Propoxyphene carbinol	66i	
Diazepam	4'-Hydroxydiazepam		13, 14, 22,
	α-Hydroxydiazepam		39–43, 62
	Nordiazepam	23bf	
	Oxazepam	23bg	
	Temazepam (3-Hydroxydiazepam)	23bl	
Diethylpropion	N,N-Diethylnorephedrine	62h	20
	Dinordiethylpropion	67c	
	N-Ethylnorephedrine	62n	
	p-Hydroxylates		
	Nordiethylpropion	67e	
	Norephedrine	62bc	
Diethyltryptamine	(Metabolism similar to N,N-dimethyltryptamine)		23
Difenoxin	(See diphenoxylate)	60d	[63–65]
	Denitrile difenoxine	60b	
	Hydroxydiphenoxylic acid (and conjugates)	60f	
Dihydrocodeinone	(See hydrocodone)	6a	
Dihydromorphine	Dihydromorphine (and 2 conjugated forms) (rats)	9d	66
	Dihydronormorphine (rats)	9e	
N,N-Dimethyltryptamine (DMT)	N,N-DMT N-oxide (rat)	97d	36, 67–71
	5-Hydroxyindoleacetic acid	97f	
	3-Indoleacetic acid	97h	
	2-Methyl-1,2,3,4-tetrahydro-β-carboline (rat)	52a	
	N-Methyltryptamine (rat)	97m	
	1,2,3,4-Tetrahydro-β-carboline (rat)	52b	
	Tryptamine	97p	

Common name	Metabolites[a,b,c]	Structure designation[d]	Ref.[e]
2,5-Dimethoxy-4-methyl-amphetamine (DOM)	1-(2,5-Dimethoxy-4-carboxyphenyl)-2-aminopropane (rats)	64f	22, 72–75
	1-(2,5-Dimethoxy-4-hydroxymethylphenyl)-2-aminopropane (rats)	64g	
	1-(2,5-Dimethoxy-4-methylphenyl)-2-propanone (rats)	63c	
Diphenhydramine	Benzhydrol (monkey)		76
	N-Demethyldiphenhydramine (monkey)	72a	
	N,N-Didemethyldiphenhydramine (monkey)	72b	
	Diphenhydramine	72c	
	Diphenhydramine N-oxide (monkey)	72d	
	(Diphenylmethoxy)acetic acid (monkey) (glutamine conjugate)	72e	
Diphenoxylate	Difenoxin (diphenoxylic acid)	60c	63–65
	Diphenoxylate (unchanged)	60d	
	(See difenoxin)		
Dothiepin (Prothiadene HCl)	Desdimethylprothiadene (rat)	49c	77
	Desmethylprothiadene (rat)	49e	
	Prothiadene sulphoxide (rat)	49n	
Ecgonine	(See cocaine)		
Estazolam	8-Chloro-2,4-dihydro-6-(4-hydroxyphenyl)-1H-s-triazolo[4,3-a][1,4]-benzodiazepin-1-one	30a	78, 79
	5-Chloro-2-(2,3-dihydro-3-oxo-4H-1,2,4-triazol-4-yl)benzophenone	32	
	8-Chloro-2,4-dihydro-6-phenyl-1H-s-triazolo-[4,3-a][1,4]benzodiazepin-1-one	30b	
	8-Chloro-4-hydroxy-6-phenyl-4H-s-triazolo-[4,3-a][1,4]benzodiazepine	29b	
	8-Chloro-6-(4-hydroxyphenyl)-4H-s-triazolo-[4,3-a][1,4]benzodiazepine	29c	
	5-Chloro-2-(4H-1,2,4-triazol-4-yl)benzophenone (Metabolite VII)	33d	
Ethchlorvynol	Hydroxyethchlorvynol	102b	20, 80
Ethinamate	2-Hydroxyethinamate	101b	13
	3-Hydroxyethinamate	101c	
	4-Hydroxyethinamate	101d	
Ethyl Loflazepate	M1 (CM 6913)	23g	82, 83
	M2 (CM 7116)	23l	
N-Ethylamphetamine	Amphetamine (rabbit)	62a	21, 81
	Ethylamphetamine (rabbit)	62m	
	N-Hydroxyethylamphetamine (rabbit)	62s	
	α-Methyl-N-(1'-phenylprop-2'-yl)nitrone (rabbit)	62bb	
	Phenylacetone (rabbit)	65a	
	1-Phenylpropan-2-ol (rabbit)	65d	
Ethylmorphine	Morphine (rat)	3j	84, 85
	Norethylmorphine (rat)	3t	
	(See also morphine)		
Fenethylline	Amphetamine	62a	23, 86
	1,3-Dimethyluric acid	16a	
	3-Methylxanthine	17b	
	1-Methyluric acid	16b	
	Theophylline	17c	
Fenfluramine	Norfenfluramine		20
	m-Trifluoromethylbenzoic acid (glycine conjugate)	63i	
Fentanyl	Despropionylfentanyl	91a	9–11, 87,
	Norfentanyl	83h	88
	N-[1-(2-Phenethyl-4-piperidinyl)]malonanilinic acid (horse)	76f	

Common name	Metabolites[a,b,c]	Structure designation[d]	Ref.[e]
Flunitrazepam	7-Amino flunitrazepam	23d	33, 89
	N-Desmethyl flunitrazepam	23o	
Fluphenazine	N-[α-(2-Trifluoromethylphenothiazin-10-yl)propyl]ethylenediamine	46n	47
	N-[α-(2-Trifluoromethylphenothiazin-10-yl)propyl]-N'-(β-hydroxyethyl)-ethylenediamine (rat)	46o	
	N-[α-(2-Trifluoromethylphenothiazin-10-yl)propyl]piperazine (rat)	48f	
Flurazepam	7-Chloro-5-(2'-fluorophenyl)-2,3-dihydro-2-oxo-1H-1,4-benzodiazepine-1-acetaldehyde	23j	14, 33, 40–42, 62, 90
	N-1-Desalkylflurazepam	23m	
	N-Desethylflurazepam	23n	
	N-Didesethylflurazepam	23p	
	3-Hydroxy-N-Desalkyl-2-oxoquazepam	23w	
	N-1-Hydroxyethylflurazepam	23x	
Glutethimide	3-Dehydroglutethimide	19	91, 92
	Desethylglutethimide	15a	
	(1-Hydroxyethyl)glutethimide (and conjugates)	15c	
	4-Hydroxyglutethimide (and conjugates)	15d	
	p-Hydroxyglutethimide (and conjugates)	15e	
	iso-(1-Hydroxyethyl)glutethimide (and conjugates)	15f	
Halazepam	Desmethyldiazepam	23bf	14
Harmine	Harmol (mouse)	53b	93
	3- or 4-Hydroxy-7-methoxyharman (mouse)	53c	
	6-Hydroxy-7-methoxyharman (mouse)	53d	
Heroin (diacetylmorphine)	6-Monoacetylmorphine	3i	94, 95
	Morphine	3j	
	Normorphine	3u	
Hexobarbital	1,5-Dimethyl-5-(3'-ketocyclohexen-1-yl)barbituric acid	12s	12, 23, 96
	3'-α-Hydroxyhexobarbital	12z	
	3'-β-Hydroxyhexobarbital	12z	
	3'-Ketohexobarbital	12bc	
	5-Methyl-5-cyclohexen-1'-yl-barbituric acid	12bj	
	5-Methyl-5-(3'-ketocyclohexen-1'-yl)barbituric acid	12bk	
Hydrocodone (dihydro-codeinone)	Hydromorphone	6b	37, 58, 97–99
	Norhydrocodone	6d	
	6-Hydrocodol	9f	
	6-Hydromorphol	9h	
Hydromorphone	Hydromorphol (and conjugates)	9h	58, 97, 98, 100
Ibomal	Ketoibomal		23
Imipramine	Desdimethylimipramine	49b	23, 101
	Desmethylimipramine	49d	
	2-Hydroxydesmethylimipramine	49h	
	2-Hydroxyiminodibenzyl	49i	
	2-Hydroxyimipramine	49j	
	Imipramine N-oxide	49m	
Ketobemidone	4'-Hydroxyketobemidone	79h	102, 103
	Ketobemidone	79k	

Common name	Metabolites[a,b,c]	Structure designation[d]	Ref.[e]
Ketobemidone (continued)	1-[4-(4-Hydroxy-3-methoxyphenyl)-1-methyl-4-piperidinyl]-1-propanone (4'-methoxyketobemidone)	79p	
	Norketobemidone	79r	
Lorazepam	6-Chloro-4-(*o*-chlorophenyl)-(1*H*)-quinazolinone (rat)	28a	14, 43, 104
	6-Chloro-4-(*o*-chlorophenyl)-2-quinazoline carboxylic acid (rat)	28b	
	Hydroxylorazepam (rat)	39a	
	Hydroxymethoxylorazepam (rat)	39b	
	Lorazepam dihydrodiol (rat)	39c	
Levorphanol	Levorphanol (rats)	10h	20
	Norlevorphanol (rats)	10j	
Levorphanol (tartrate)	(*See* levorphanol)		
Lysergic acid diethylamide	13-Hydroxy-LSD (monkey)	56a	105–107
	14-Hydroxy-LSD (monkey)	56b	
	Lysergic acid amide (monkey)	56d	
	Lysergic acid monoethylamide (monkey)	56f	
	Nor-LSD (monkey)	56g	
	2-Oxo-LSD (monkey)	57	
Mazindol	2-(*p*-Chlorobenzoyl)-*N*-2-(aminoethyl)benzamide	44a	108, 109
	5-(*p*-Chlorophenyl)-2,5-dihydro-5-hydroxy-3*H*-imidazo[2,1-a]isoindol-3-one (conjugated)	43b	
	5-(*p*-Chlorophenyl)-2,5-dihydro-2,5-dihydro-3*H*-imidazo[2,1-a]isoindol-3-one	43c	
	3-(*p*-Chlorophenyl)-2-glycyl-3-hydroxy-1-isoindolinone	44b	
Medazepam	2-Amino-5-chlorobenzophenone	111a	34, 39–43,
	2-Amino-5-chloro-3-hydroxybenzophenone	111b	100, 111
	N-Desmethyl-1,2-dehydromedazepam	26c	
	N-Desmethyl medazepam	22a	
	N-Desmethyldiazepam	23bf	
	Diazepam	23q	
	3-Hydroxydiazepam (temazepam)	23bl	
	Oxazepam	23bg	
Meperidine (*See* Pethidine)			
Mephobarbital	(*See* methylphenobarbital)		23, 112
Meprobamate	Hydroxymeprobamate	105a	20, 113
	Meprobamate *N*-glucuronide	105d	
Mescaline	*N*-Acetyl-(3,4-dimethoxy-5-hydroxy)phenethylamine	64a	20, 22, 23,
	N-Acetylmescaline	64b	114
	Mescaline	64m	
	3,4,5-Trimethoxybenzoic acid	63j	
	3,4,5-Trimethoxyphenylacetic acid	63k	
Methadol	Metabolite of methadone	71g	2–8
Methadone	2-Ethylidene-1,5-dimethyl-3,3-diphenylpyrrolidine	89a	20, 22,
	2-Ethyl-5-methyl-3,3-diphenylpyrroline	88a	115–117
	Methadol	71g	
	Normethadol	71k	
	p-Hydroxy-2-Ethylidene-1,5-dimethyl-3,3-diphenylpyrrolidine (and glucuronide)	89b	
	p-Hydroxy-2-Ethyl-1,5-dimethyl-3,3-diphenylpyrrolidine (and glucuronide)	88b	

Common name	Metabolites[a,b,c]	Structure designation[d]	Ref.[e]
Methamphetamine	Amphetamine	62a	20–22, 118
	Benzoic acid	63a	
	p-Hydroxyamphetamine	62p	
	p-Hydroxymethamphetamine	62t	
	p-Hydroxynorephedrine	62u	
	Methamphetamine	62y	
	Norephedrine	62bc	
Methaqualone	2-Methyl-3-[2-(hydroxymethyl)phenyl]-4(3H)-quinazolinone	114c	22, 119–
	2-Methyl-3-(2-methyl-3-hydroxyphenyl]-4(3H)-quinazolinone	114d	122
	2-Hydroxymethyl-3-O-tolyl-4(3H)-quinazolinone	114e	
	2-Methyl-3-(2-methyl-4-hydroxyphenyl)-4(3H)-quinazolinone	114f	
	2-Methyl-3-O-tolyl-6-hydroxy-4(3H)-quinazolinone	114g	
Metharbital	Barbital	12j	23
	Hydroxymetharbital		
Methohexital	Hydroxymethohexital		20, 23
	Normethohexital		
4-Methoxyamphetamine (PMA)	p-Hydroxybenzoic acid (monkey)	63	20, 123,
	p-Hydroxyamphetamine	62p	124
	N-Hydroxy-p-methoxyamphetamine	64r	
	p-Hydroxynorephedrine	62u	
5-Methoxy-N,N-dimethyl-tryptamine (5-MeO-DMT)	Bufotenine (free and glucuronide conjugate) (rat)	97a	36, 125
	5-Hydroxyindoleacetic acid (rat)	97f	
	5-Methoxyindoleacetic acid (rat)	97j	
Methoxyphenamine	1-(5-Hydroxy-2-methoxyphenyl)-2-(methylamino)propane	64k	126
	1-(2-Hydroxyphenyl)-2-(methylamino)propane	62v	
	Normethoxyphenamine	62be	
3,4-Methylenedioxy-amphetamine (MDA)	3,4-Dihydroxybenzyl methyl ketone (dog and monkey)	63b	11, 127–
	4-Hydroxy-3-methoxybenzoic acid (dog and monkey)	63f	129
	α-Methyldopamine (dog and monkey)	64o	
	3,4-Methylenedioxybenzoic acid (dog and monkey)	63g	
	3,4-Methylenedioxybenzyl methyl ketone (guinea pig)	63h	
	3,4-Methylenedioxybenzyl methyl ketoxime (guinea pig)		
	3-O-Methyl-α-methyldopamine (dog and monkey)	64t	
α-Methylfentanyl	Despropionylmethylfentanyl (postulated)	91b	11, 130
Methylphenidate	p-Hydroxymethylphenidate	85a	131, 132
	p-Hydroxyritalinic acid	85b	
	6-Oxoritalinic acid	85d	
	Ritalinic acid	85e	
Methylphenobarbital	p-Hydroxymephobarbital	12ba	23, 112
	p-Hydroxyphenobarbital	12bq	
	Mephobarbital	12bd	
	3-O-Methylcatechol of mephobarbital	12bg	
	3-O-Methylcatechol of phenobarbital	12bi	
	Phenobarbital	12bp	
	(See also phenobarbital)		
Methyprylon	5-Carboxypyrithyldione	14a	20, 23
	5-Hydroxymethylpyrithyldione	14b	
	5-Methylpyrithyldione	14c	
	6-Oxomethyprylon	13b	

Common name	Metabolites[a,b,c]	Structure designation[d]	Ref.[e]
Midazolam	1-Hydroxymethyl-4-hydroxymidazolam	31a	33, 133,
	1-Hydroxymethylmidazolam	31b	134
	4-Hydroxymidazolam	31c	
Morphine	Codeine	3b	22, 55, 56,
	Morphine-3,6-diglucuronide		58, 95,
	Morphine-3-ethereal sulfate		135–138
	Morphine-3-glucuronide		
	Morphine-6-glucuronide	3k	
	Morphine N-oxide	3n	
	Norcodeine	3s	
	Normorphine	3u	
Morphine methyl bromide (*See* morphine)			23
Morphine methylfulfonate (*See* morphine)			23
Morphine N-oxide	(*See* morphine)		23
Naltrexone	2-Hydroxy-3-methoxy-6β-naltrexol		
	6β-Naltrexol		139
Nicomorphine	Morphine (rat) (postulated)	3j	140
	6-Nicotinoylmorphine (postulated) (rat)		
Nitrazepam	7-Acetamidonitrazepam	23b	39–42
	7-Aminonitrazepam	23e	
	2-Amino-5-nitrobenzophenone		
	3-Hydroxy-7-aminonitrazepam		
	3-Hydroxy-2-amino-5-nitrobenzophenone		
	Nitrazepam	23be	
Norlevorphanol	(Metabolite of levorphanol)		
Normethadone	(Metabolite of methadone)		
Normorphine	(Metabolite of morphine)		137
Orphenadrine	O-{α-[2-(Dimethylamino)ethoxy]benzyl}benzoic acid	116a	141
	O-{α-[2-(Dimethylamino)ethoxy]benzyl}benzyl alcohol	116b	
	2[(p-Hydroxy-α-O-tolylbenzyl)oxy]-N,N-dimethylethylamine	116c	
	2[(m-Hydroxy-α-O-tolylbenzyl)oxy]-N,N,-dimethylethylamine	116d	
	2-[(5-Hydroxy-2-methyl-α-phenylbenzyl)oxy]-N,N-dimethylethylamine	116e	
	Tofenacine	116f	
Oxazepam	2-Amino-5-chlorobenzophenone (rabbit)	111a	39, 41, 50,
	2-Amino-5-chloro-3-hydroxybenzophenone glucuronide (rabbit)	111b	62, 142
Oxazolam	2-Amino-5-chlorobenzophenone (rabbit)	111a	39, 143
	2-Amino-5-chloro-3-hydroxybenzophenone (rabbit)	111b	
	Desmethyldiazepam (rabbit)	23bf	
Oxycodone	Noroxycodone	6e	98, 144,
	Oxymorphone	6g	145
	Oxycodone	6f	
Oxymorphone	6-α-Oxymorphol		98, 146
	6-β-Oxymorphol		
	Oxymorphone	6g	
Paraldehyde			20
Pemoline	5-Phenyl-2,4-oxazolidinedione	113b	147

Common name	Metabolites[a,b,c]	Structure designation[d]	Ref.[e]
Pentazocine	Carboxypentazocine		20, 148
	Hydroxypentazocine		
Pentobarbital	3'-Carboxypentobarbital	12bo	149
	3'-Hydroxypentobarbital	12bm	
	N-Hydroxypentobarbital	12bn	
Perphenazine	N-[γ-(2-Chlorophenothiazin-10-yl)propyl]ethylenediamine	46a	47
	N-[γ-(2-Chlorophenothiazin-10-yl)propyl]piperazine	48a	
	N-[γ-(2-Chlorophenothiazin-10-yl)propyl]-N'-(β-hydroxyethyl)-ethylenediamine	46b	
Pethidine (meperidine)	p-Hydroxymeperidine (and conjugates)	79i	150–153
	N-Hydroxynormeperidine	78	
	Meperidine	79l	
	Meperidinic acid	79n	
	Meperidine N-oxide	79m	
	Methoxyhydroxymeperidine	79o	
	Normeperidine	79s	
	Normeperidinic acid	79t	
Phencyclidine (PCP)	Acid metabolite of PCP	75a	154–157
	3-Hydroxy PCP	75b	
	4-Phenyl-4-piperidinocyclohexanol	75i	
	1-Phencyclohexylamine	75f	
	1-(1-Phenylcyclohexyl)-4-hydroxypiperidine	75g	
	1-Phenyl-4-hydroxycyclohexyl-4-hydroxypiperidine	75h	
Phendimetrazine	Phenmetrazine	94e	158
	(See phenmetrazine for further metabolites)		
Phenmetrazine	N-Hydroxyphenmetrazine	94a	20, 159
	p-Hydroxyphenmetrazine	94b	
	Nitrone of phenmetrazine	87	
	3-Oxophenmetrazine	94c	
Phenobarbital	Dihydrodiol	12br	20, 23,
	p-Hydroxyphenobarbital	12bq	160, 161
	m-Hydroxyphenobarbital	12bs	
Phenoperidine	Meperidine (Pethidine)	79l	162–164
	Norpethidine (normeperidine)	79s	
	Phenoperidine	79u	
	(See pethidine)		
Phentermine	N-Hydroxyphentermine		20, 165
	p-Hydroxyphentermine		
Phenylacetone	Benzoic acid (rabbit)	63a	166
	Methyl phenyl ketone (rabbit)	65a	
	1-Phenyl-2-butanone (rabbit)	65b	
	1-Phenyl-1,2-propanediol (rabbit)	65c	
	1-Phenyl-2-propanol (rabbit)	65d	
	1-Phenyl-2-propanon-1-ol (rabbit)	65e	
Phenylbutazone	Dihydroxyphenylbutazone (rat)	21a	167
	Hydroxyphenylbutazone (rat)	21b	
	Oxyphenbutazone (rat)	21c	

Common name	Metabolites[a,b,c]	Structure designation[d]	Ref.[e]
Phenylcyclohexylamine	Active metabolite of phencyclidine		22, 154, 155
Phenytoin	3,4-Dihydrodihydroxyphenytoin	20a	20, 145
	m-Hydroxyphenytoin	20b	
	p-Hydroxyphenytoin	20c	
Pinazepam	3-Hydroxypinazepam	23z	168
	Nordiazepam (demethyldiazepam)	23bf	
	Oxazepam	23bg	
Piperidinocyclohexane-caronitrile	1-Cyclohexenyl piperidine (postulated)	73a	169
Prazepam	3-Hydroxyprazepam and glucuronide conjugate	23ba	14, 40–42
	Nordiazepam	23bf	
	Oxazepam	23bg	
Premazepam	N-7-Desmethylpremazepam (rat)	37a	170
	6-Hydroxypremazepam (dog)	37b	
Primidone	Phenobarbital	12bp	15
	Phenylethylmalonamide		
Prochlorperazine	N-[γ-(2-Chlorophenothiazin-10-yl)propyl]ethylenediamine (rat)	46a	47
	N-[γ-(2-Chlorophenothiazin-10-yl)propyl]-N'-methylethylenediamine (rat)	46c	
	N-[γ-(2-Chlorophenothiazin-10-yl)propyl]piperazine (rat)	48a	
α-Prodine (pethidine homolog)	Nor-α-prodine (dogs)	77c	20, 171, 172
	Unidentified conjugates		
β-Prodine	Nor-β-prodine	77c	172
	(*See also* α-prodine and references)		
Propiram	N-2-(2-Pyridylamino)-propyl-N-propionyl-5-aminovalerianic acid	68b	173–175
Propoxyphene	(*See* dextropropoxyphene)	66h	59–61
Propranolol	N-Desisopropylpropranolol (DIP)	95a	176–179
	4-Hydroxypropranolol	95b	
	Methoxyhydroxypropranolol	95c	
	Naphthoxylactic acid	95d	
	α-Naphthoxyacetic acid	96	
	Ring hydroxylated DIP	95f	
	Ring hydroxylated naphthoxylactic acid	95g	
	Ring hydroxylated propranolol glycol	95h	
Quazepam	N-Desalkylflurazepam (and glucuronide)	23m	180, 181
	3-Hydroxy-N-desalkylflurazepam	23w	
	3-Hydroxy-2-oxoquazepam (and glucuronide)	23y	
	2-Oxoquazepam	23bh	
	Quazepam	23bk	
Quinine	Dihydroxyquinine		16, 182
	Quinetin		
	QDP		
Racemethorphan	(*See* dextromethorphan)		
Racemorphan		10m	183
Secobarbital	5-(2,3-Dihydroxypropyl)-5-(1-methylbutyl) barbituric acid	12bv	12, 20, 184
	3'-Hydroxysecobarbital (and glucuronide)	12bu	

Common name	Metabolites[a,b,c]	Structure designation[d]	Ref.[e]
Secobarbital (continued)	5-(1-Methylbutyl)barbituric acid	12bh	
	Secodiol glucuronide	12bw	
Sufentanil/sufentanyl	Desmethyl sufentanil	83e	9, 10, 185
	N-[4-(Hydroxymethyl)-4-piperidinyl]-N-phenylpropanamide	83f	
	N-(4-Hydroxyphenyl)acetamide	82a	
	N-(4-Hydroxyphenyl)propanamide	82b	
	N-[4-(Methoxymethyl)-4-piperidinyl]-N-phenylpropanamide (norsufentanil)	83g	
	[[4-(Phenylamino)-4-piperidinyl]methyl]-propanoate	81	
	(Refer also to fentanyl)		
Temazepam	Oxazepam	23bg	14, 62
Tetrahydrocannabinol	8-α,11-Dihydroxy-Δ^9-tetrahydrocannabinol	55a	22, 186–
	8-β,11-Dihydroxy-Δ^9-tetrahydrocannabinol	55b	190
	8-α-Hydroxy-Δ^9-tetrahydrocannabinol	55d	
	8-β-Hydroxy-Δ^9-tetrahydrocannabinol	55e	
	11-Hydroxy-Δ^9-tetrahydrocannabinol	55f	
	11-Nortetrahydrocannabinol-Δ^9-carboxylic acid	55g	
	(glucuronide conjugates predominate)		
Tetrazepam	3-Hydroxynortetrazepam	25a	14, 191
	3'-Hydroxynortetrazepam	25b	
	3-Hydroxytetrazepam	25c	
	3'-Hydroxytetrazepam	25d	
	Nortetrazepam	25e	
Thebaine	Codeinone (rat)	5a	192–194
	Codeine (and glucuronides) (rat)	3b	
	Codeine-6-glucuronide (rat)	3c	
	14-Hydroxycodeinone (and glucuronides) (rat)	5b	
	14-Hydroxydihydrocodeinone (rat)	7	
	14-Hydroxycodeine (rat)	3h	
	Morphine (and glucuronides) (rat)	3j	
	Morphine-3-glucuronide (rat)		
	Norcodeine (and glucuronides) (rat)	3s	
	Normorphine (and glucuronides) (rat)	3u	
	Oripavine (rat)	1a	
	Thebaine (rat)	1b	
Thioridazine	Disulphone of thioridazine (rat)	47a	15, 195
	Disulfoxide of thioridazine (rat)	47b	
	Northioridazine (rat)	47c	
	Ring sulfoxide of thioridazine (rat)	47e	
	Side chain sulfoxide of thioridazine (rat)	47f	
Tilidine	Bisnortilidine	84a	196–198
	Nortilidine	84b	
	Tilidine	84c	
Triazolam	5-Chloro-2-(3,5-bis(hydroxymethyl)-4H,1,2,4-triazol-4-yl)-2'-chloro-benzophenone	33a	13, 33, 199–203
	5-Chloro-2-(3-hydroxymethyl-5-methyl-4H,1,2,4-triazol-4-yl)-2'-chloro-benzophenone	33b	
	5-Chloro-2-(3-methyl-4H,1,2,4-triazol-4-yl)-2-chlorobenzophenone	33c	
	1-Hydroxymethyl-4-hydroxytriazolam	29g	
	1-Hydroxymethyltriazolam (and glucuronide)	29h	

Common name	Metabolites[a,b,c]	Structure designation[d]	Ref.[e]
Triazolam	4-Hydroxytriazolam (and glucuronide)	29i	
(continued)	Triazolam	29j	
Trifluoperazine	N-[2-Trifluoromethylphenothiazin-10-yl)propyl]ethylenediamine (rat)	46n	47
	N-[2-Trifluoromethylphenothiazin-10-yl)propyl]-N'-methylethylenediamine (rat)		
	N-[2-Trifluoromethylphenothiazin-10-yl)propyl]piperazine (rat)	48f	

[a] Where metabolites in humans have not been investigated, metabolites found in experimental animals have been indicated.

[b] Metabolites are listed according to names used by authors.

[c] Unchanged drugs and conjugates of metabolites are commonly found, but they are not always listed; the cited references should be consulted.

[d] See Appendices III and IV for these structure designations. Structures do not in general illustrate stereochemical configurations. Structures are not illustrated for those designated with asterisks.

[e] References, by no means exhaustive, are cited to provide a basis for looking up further information.

REFERENCES

1. Kochhar, M. M.; Hamrick, M. E.; Bavda, L. T. *J. Pharm. Sci.* **1976,** *65,* 137.
2. Misra, A. L.; Vardy, J.; Bloch, R.; Mule, S. J.; Deneau, G. A. *J. Pharm. Pharmacol.* **1976,** *8,* 316.
3. Billings, R.; Booher, R.; Smith, S.; Pohland, A.; McMahon, R. E. *J. Med. Chem.* **1973,** *16,* 305.
4. Toro-Goyco, E.; Martin, B. R.; Harris, L. S. *Biochem. Pharmacol.* **1980,** *29,* 1897.
5. Kiang, C.-H.; Campos-Flor, S.; Inturrisi, C. E. *J. Chromatogr.* **1981,** *222,* 81.
6. Kaiko, R. F.; Inturrisi, C. E. *J. Chromatogr.* **1973,** *82,* 315.
7. Kaiko, R. F. *J. Chromatogr.* **1975,** *109,* 247.
8. Finkle, B. S.; Jennison, T. A.; Chinn, D. M. *J. Anal. Toxicol.* **1982,** *6,* 100.
9. Gillespie, T. J.; Gandolfi, A. J.; Maiorino, R. M.; Vaughan, R. W. *J. Anal. Toxicol.* **1981,** *5,* 133.
10. Meuldermans, W.; Hendrickx, J.; Lauwers, W.; Hurkmans, R.; Swysen, E.; Thijssen, J.; Timmerman, Ph.; Woestenborghs, R. *Drug Metab. Dispos.* **1987,** *15,* 905.
11. Baum, R. M. *Chem. Eng. News* **1985,** 7–16.
12. Harvey, D. J.; Glazener, L.; Johnson, B.; Butler, C. M.; Horning, M. G. *Drug Metab. Dispos.* **1977,** *5,* 527.
13. Gall, M.; Kamdar, B. V. *J. Med. Chem.* **1978,** *21,* 1290.
14. Sohr, C. J.; Buechel, A. T. *J. Anal. Toxicol.* **1982,** *6,* 286.
15. Cravey, R. H.; Baselt, R. C. *Introduction to Forensic Toxicology*; Biomedical Publications: Davis, CA, 1981.
16. Prox, A.; Breyer-Pfaff, U. *Drug Metab. Dispos.* **1987,** *15,* 890.
17. Breyer-Pfaff, U.; Prox, A.; Wachsmuth, H.; Yao, P. *Drug Metab. Dispos.* **1987,** *15,* 882.
18. Frey, H. H.; Magnussen, M. P. *Arzneim.-Forsch.* **1966,** *16,* 612.
19. Tang, B. K.; Inaba, T.; Kalow, W. *Drug Metab. Dispos.* **1975,** *3,* 479.
20. Baselt, R. C.; Cravey, R. H. *Disposition of Toxic Drugs and Chemicals in Man*, 3rd ed; Year Book Medical: Chicago, IL, 1989.
21. Beckett, A. H.; Van Dyk, J. M.; Chissick, H. M.; Gorrod, J. W. *J. Pharm. Pharmacol.* **1971,** *23,* 560.
22. Foltz, R. L.; Fentiman, A. F., Jr.; Foltz, R. B. *GC/MS Assays for Abused Drugs in Body Fluids*; NIDA Research Monograph 32; National Institute on Drug Abuse: Rockville, MD, 1980.
23. Lowry, W. T.; Garriott, J. C. *Forensic Toxicology: Controlled Substances and Dangerous Drugs*; Plenum: New York, 1979.
24. Maynert, E. W.; van Dyke, H. B. *Drug Metab. Rev.* **1985–6,** *16,* 185.
26. Inoue, T.; Suzuki, S.; Niwaguchi, T. *Xenobiotica* **1983,** *13,* 241.
27. Beckett, A. H.; Gibson, G. G. *Xenobiotica* **1978,** *8,* 73.
28. Van Rooij, H. H.; Soe-Agnie, C. *J. Chromatogr.* **1978,** *156,* 189.
29. Van Rooy, H. H.; Kok, M.; Modderman, E.; Soe Agnie, C. *J. Chromatogr.* **1978,** *148,* 447.
30. Schwartz, M. A.; Postma, E.; Kolis, S. J.; Leon, A. S. *J. Pharm. Sci.* **1973,** *62,* 1973.
31. Schwartz, M. A.; Pool, W. R.; Hane, D. L.; Postma, E. *Drug Metab. Dispos.* **1974,** *2,* 31.

32. Sargent, T.; Kalbhen, D. A.; Shulgin, A. T.; Braun, G.; Stauffer, H.; Kusubov, N. *Neuropharmacology.* **1975,** *14,* 165.
33. Breimer, D. D.; Jochemsen, R. *Brit. J. Clin. Pharmacol.* **1983,** *16,* 277S.
34. Bechtel, W. D. *Brit. J. Clin. Pharmacol.* **1983,** *16,* 279S.
35. Sanders, E.; Bush, M. T. *J. Pharmacol. Exp. Ther.* **1967,** *158,* 340.
36. Ahlborg, U.; Holmstedt, B.; Lindgren, J.-E. *Adv. Pharmacol.* Academic: New York, **1968,** *6B,* 213.
37. Gaver, R. C.; Vasiljev, M.; Wong, H.; Monkovic, I.; Swigor, J. E.; Van Harken, D. R.; Smyth, R. D. *Drug Metab. Dispos.* **1980,** *8,* 230.
38. Breimer, D. D.; Ketelaars, H. C. J.; van Rossum, J. M. *J. Chromatogr.* **1974,** *88,* 55.
39. Sawada, H.; Hara, A.; Asano, S.; Matsumoto, Y. *Clin. Chem.* **1976,** *22,* 1596.
40. Schallek, W.; Schlosser, W.; Randall, L. O. *Adv. Pharmacol. Chemother.* **1972,** *10,* 119.
41. Garattini, S.; Marcucci, F.; Mussini, E. *Drug Metab. Rev.* **1972,** *1,* 291.
42. Schallek, W.; Schlosser, W.; Randall, L. O. In *Advances in Pharmacology and Chemotherapy*; Garattini, S.; Goldin, A.; Hawking, F.; Kopin, I. P., Eds.; Academic Press: New York, 1972.
43. Clifford, J. M.; Smyth, W. F. *Analyst* **1974,** *99,* 241.
44. Beckett, A. H.; Belanger, P. M. *Brit. J. Clin. Pharmacol.* **1977,** *4,* 193.
45. Allender, W. J.; Archer, A. W.; Dawson, A. G. *J. Anal. Toxicol.* **1983,** *7,* 203.
46. Daly, J. W.; Manian, A. A. *Biochem. Pharmacol.* **1967,** *16,* 2131.
47. Gaertner, H. J.; Liomin, G.; Villumsen, D.; Bertele, R.; Breyer, U. *Drug Metab. Dispos.* **1975,** *3,* 437.
48. Sgaragli, G.; Ninci, R.; Della Corte, L.; Valoti, M.; Nardini, M.; Andreoli, V.; Moneti, G. *Drug Metab. Dispos.* **1986,** *14,* 263.
49. Naestoft, J.; Larsen, N.-E. *J. Chromatogr.* **1974,** *93,* 113.
50. Ruelius, H. W.; Tio, C. O.; Knowles, J. A.; McHugh, S. L.; Schillings, R. T.; Sisenwine, S. F. *Drug Metab. Dispos.* **1979,** *7,* 40.
51. Murata, H.; Kougo, K.; Yasumura, A.; Nakajima, E.; Shindo, H. *Chem. Pharm. Bull.* **1973,** *21,* 404.
52. Misra, A. L.; Pontani, R. B. *Drug Metab. Dispos.* **1977,** *5,* 556.
53. Zhang, J. Y.; Foltz, R. L. *J. Anal. Toxicol.* **1990,** *14,* 201.
54. Adler, T. K.; Fujimoto, E.; Leong, W.; Baker, E. M. *J. Pharmacol. Exp. Ther.* **1955,** *114,* 251.
55. Boerner, U.; Abbott, S. *Experientia* **1973,** *29,* 180.
56. Nakamura, G. R.; Griesemer, E. C.; Noguchi, T. T. *J. Forensic Sci.* **1976,** *21,* 518.
57. Solomon, M. D. *Clin. Toxicol.* **1974,** *7,* 255.
58. Cone, E. J.; Darwin, W. D.; Buchwald, W. F. *J. Chromatogr.* **1983,** *275,* 30.
59. Sullivan, H. R.; Emmerson, J. L.; Marshall, F. J.; Wood, P. G.; McMahon, R. E. *Drug Metab. Dispos.* **1974,** *2,* 526.
60. McMahon, R. E.; Sullivan, H. R.; Due, S. L.; Marshall, F. J. *Life Sci.* **1973,** *12,* 463.
61. Due, S. L.; Sullivan, H. R.; McMahon, R. E. *Biomed. Mass Spectrom.* **1976,** *3,* 217.
62. *Progress in Drug Metabolism*; Bridges, J. W.; Chasseaud, L. F., Eds. Taylor and Francis: Philadelphia, PA, 1984, Vol. 8.
63. van Wijngaarden, I. *Arzneim.-Forsch.* **1972,** *22,* 513.
64. Heykants, J.; Brugmans, J.; Verhaegen, H. *Arzneim.-Forsch.* **1972,** *22,* 529.
65. Karim, A.; Garden, G.; Trager, W. *J. Pharmacol. Exp. Ther.* **1971,** *177,* 546.
66. Hug, C. C.; Mellett, L. B. *J. Pharmacol. Exp. Ther.* **1965,** *149,* 446.
67. Barker, S. A.; Monti, J. A.; Christian, S. T. *Biochem. Pharmacol.* **1980,** *29,* 1049.
68. St. Szara, A.; *Experientia* **1956,** *12,* 441.
69. Barker, S. A.; Beaton, J. M.; Christian, S. T.; Monti, J. A.; Morris, P. E. *Biochem. Pharmacol.* **1984,** *33,* 1395.
70. Barker, S. A.; Beaton, J. M.; Christian, S. T.; Monti, J. A.; Morris, P. E. *Biochem. Pharmacol.* **1982,** *31,* 2513.
71. Beaton, J. M.; Barker, S. A. *Pharmacol. Biochem. Behav.* **1982,** *16,* 811.
72. Ho, B. T.; Estevez, V.; Tansey, L. W.; Englert, L. F.; Creaven, P. J.; McIsaac, W. M. *J. Med. Chem.* **1971,** *14,* 158.
73. Ho, B. T.; Tansey, L. W. *J. Med. Chem.* **1971,** *14,* 156.
74. Shulgin, A. T.; Dyer, D. C. *J. Med. Chem.* **1978,** *18,* 1201.
75. Matin, S. B.; Callery, P. S.; Zweig, J. S.; O'Brien, A.; Rapoport, R.; Castagnoli, N., Jr. *J. Med. Chem.* **1974,** *17,* 877.
76. Drach, J. C.; Howell, J. P. *Biochem. Pharmacol.* **1968,** *17,* 2125.
77. Horesovsky, O.; Franc, Z.; Kraus, P. *Biochem. Pharmacol.* **1967,** *16,* 2421.
78. Tanayama, S.; Shirakawa, Y.; Kanai, Y.; Suzuoki, Z. *Xenobiotica* **1974,** *4,* 33.
79. Kanai, Y. *Xenobiotica* **1974,** *4,* 441.
80. Horwitz, J. P.; Brukwinski, W.; Treisman, J.; Andrzejewski, D.; Hills, E. B.; Chung, H. L.; Wang, C. Y. *Drug Metab. Dispos.* **1980,** *8,* 77.

81. Beckett, A. H.; Haya, K. *Xenobiotica* **1978,** *8,* 85.
82. Cautreels, W.; Jeanniot, J. P. *Biomed. Mass Spectrom.* **1980,** *7,* 565.
83. Cano, J. P.; Sumirtapura, Y. C.; Cautreels, W.; Sales, Y. *J. Chromatogr.* **1981,** *226,* 413.
84. Duquette, P. H.; Holtzman, J. L. *J. Pharm. Exp. Ther.* **1979,** *211,* 213.
85. Thompson, J. A.; Holtzman, J. L. *Drug Metab. Dispos.* **1977,** *5,* 9.
86. Ellison, T.; Levy, L.; Bolger, J. W.; Okun, R. *Eur. J. Pharmacol.* **1970,** *13,* 123.
87. Frincke, J. M.; Henderson, G. L. *Drug Metab. Dispos.* **1980,** *8,* 425.
88. de Silva, J. A. F.; Puglisi, C. V.; Munno, N. *J. Pharm. Sci.* **1974,** *63,* 520.
89. Goromaru, T.; Matsuura, H.; Furuta, T.; Baba, S.; Yoshimura, N.; Miyawaki, T.; Sameshima, T. *Drug Metab. Dispos.* **1982,** *10,* 542.
90. Garland, W. A.; Miwa, B. J.; Dairman, W.; Kappell, B.; Chiueh, M. C. C.; Divoll, M.; Greenblatt, D. J. *Drug Metab. Dispos.* **1983,** *11,* 70.
91. Kennedy, A.; Fischer, L. J. *Drug Metab. Dispos.* **1979,** *7,* 319.
92. Ambre, J. J.; Fischer, L. J. *Drug Metab. Dispos.* **1974,** *2,* 151.
93. Tweedie, D. J.; Burke, M. D. *Drug Metab. Dispos.* **1987,** *15,* 74.
94. Elliott, H. W.; Parker, K. D.; Wright, J. A.; Nomof, N. *Clin. Pharm. Ther.* **1971,** *12,* 806.
95. Boerner, U.; Abbott, S.; Roe, R. L. *Drug Metab. Rev.* **1975,** *4,* 39.
96. Van der Graaff, M.; Vermeulen, N. P. E.; Joeres, R. P.; Breimer, D. D. *Drug Metab. Dispos.* **1983,** *11,* 489.
97. Cone, E. J.; Darwin, W. D.; Gorodetzky, C. W.; Tan, T. *Drug Metab. Dispos.* **1978,** *6,* 488.
98. Cone, E. J. *J. Chromatogr.* **1976,** *129,* 355.
99. Findlay, J. W.; Jones, E. C.; Welch, R. M. *Drug Metab. Dispos.* **1979,** *7,* 310.
100. Vallner, J. J.; Stewart, J. T.; Kotzan, J. A.; Kirsten, E. B.; Honigberg, I. L. *J. Clin. Pharmacol.* **1981,** *21,* 152.
101. Bickel, M. H.; Baggiolini, M. *Biochem. Pharmacol.* **1966,** *15,* 1155.
102. Bondesson, U.; Hartvig, P.; Danielsson, B. *Drug Metab. Dispos.* **1981,** *9,* 376.
103. Bondesson, U.; Hartvig, P.; Abrahamsson, L.; Ahnfelt, N.-O. *Biomed. Mass Spectrom.* **1983,** *10,* 283.
104. Schillings, R. T.; Sisenwine, S. F.; Ruelius, H. W. *Drug Metab. Dispos.* **1977,** *5,* 425.
105. Sullivan, A. T.; Twitchett, P. J.; Fletcher, S. M.; Moffat, A. C. *J. Forensic Sci. Soc.* **1978,** *18,* 89.
106. Toda, T.; Oshino, N. *Drug Metab. Dispos.* **1981,** *9,* 108.
107. Faed, E. M.; McLeod, W. R. *J. Chromatogr. Sci.* **1973,** *11,* 4.
108. Dugger, H. A.; Vistacion, O. M.; Talbot, K. C.; Coombs, R. A.; Orwig, B. A. *Drug Metab. Dispos.* **1979,** *7,* 132.
109. Dugger, H. A.; Coombs, R. A.; Schwarz, H. J.; Migdalof, B. H.; Orwig, B. A. *Drug Metab. Dispos.* **1976,** *4,* 262.
110. Arthur, J.; de Silva, F.; Puglisi, C. V. *Anal. Chem.* **1970,** *42,* 1725.
111. Randall, O.; Scheckel, C. L.; Pool, W. *Arch. Int. Pharmacodyn. Ther.* **1970,** *185,* 135.
112. Hooper, W. D.; Kunze, H. E.; Eadie, M. J. *Drug Metab. Dispos.* **1981,** *9,* 381.
113. Shearer, C.; Rulon, P. In *Analytical Profiles of Drug Substances*, Florey, K., Ed.; Academic Press: New York, 1972; Vol 1, p 207.
114. Demisch, L.; Kaczmarczyk, P.; Seiler, N. *Drug Metab. Dispos.* **1978,** *6,* 507.
115. Robinson, A. E.; Williams, F. M. *J. Pharm. Pharmacol.* **1971,** *23,* 353.
116. Beckett, H.; Mitchard, M.; Shihab, A. A. *J. Pharm. Pharmacol.* **1971,** *23,* 347.
117. Gerardy, B. M.; Kapusta, D.; Dumont, P.; Poupaert, J. H. *Drug Metab. Dispos.* **1986,** *14,* 477.
118. Yamamoto, T.; Takano, R.; Egashira, T.; Yamanaka, Y. *Xenobiotica* **1984,** *14,* 867.
119. Kazyak, L.; Anthony, R. M.; Sunshine, I. *J. Anal. Toxicol.* **1979,** *3,* 67.
120. Stillwell, W. G.; Gregory, P. A.; Horning, M. G. *Drug Metab. Dispos.* **1975,** *3,* 287.
121. Ericsson, O.; Danielsson, B. *Drug Metab. Dispos.* **1977,** *5,* 497.
122. Peat, M. A.; Finkle, B. S. *J. Anal. Toxicol.* **1980,** *4,* 114.
123. Hubbard, J. W.; Midha, K. K.; Cooper, J. K. *Drug Metab. Dispos.* **1977,** *5,* 329.
124. Hubbard, J. W.; Bailey, K.; Midha, K. K.; Cooper, J. K. *Drug Metab. Dispos.* **1981,** *9,* 250.
125. Agurell, S.; Holmstedt B.; Lindgren, J. E. *Biochem. Pharm.* **1969,** *8,* 2771.
126. Midha, K. K.; Cooper, J. K.; McGilveray, I. J.; Coutts, R. T.; Dawe, R. *Drug Metab. Dispos.* **1976,** *4,* 568.
127. Midha, K. K.; Cooper, J. K.; By, A.; Ethier, J.-C. *Drug Metab. Dispos.* **1977,** *5,* 143.
128. Midha, K. K.; Hubbard, J. W.; Bailey, K.; Cooper, J. K. *Drug Metab. Dispos.* **1978,** *6,* 623.
129. O'Brien, B. A.; Bonicamp, J. M.; Jones, D. W. *J. Anal. Toxicol.* **1982,** *6,* 143.
130. Gillespie, T. J.; Gandolfi, A. J.; Davis, T. P.; Morano, R. A. *J. Anal. Toxicol.* **1982,** *6,* 139.
131. Egger, H.; Bartlett, F.; Dreyfuss, R.; Karliner, J. *Drug Metab. Dispos.* **1981,** *9,* 415.
132. Redalieu, E.; Bartlett, F.; Waldes, L. M.; Darrow, R.; Egger, H.; Wagner, E. *Drug Metab. Dispos.* **1982,** *10,* 708.
133. Heizmann, P.; Ziegler, W. H. *Arzneim.-Forsch.* **1981,** *31,* 2220.
134. Vree, T. B.; Baars, A. M.; Booij, L. H. D.; Driessen, J. J. *Arzneim.-Forsch.* **1981,** *31,* 2215.
135. Yeh, S. Y. *J. Pharm. Sci.* **1973,** *62,* 1827.
136. Boerner, U.; Roe, R. L.; Becker, C. E. *J. Pharm. Pharmacol.* **1974,** *26,* 393.

137. Yeh, S. Y.; McQuinn, R. L.; Gorodetzky, C. W. *Drug Metab. Dispos.* **1977,** *5,* 335.

138. Svensson, J.-O. *J. Chromatogr.* **1986,** *375,* 174.

139. Verebey, K. *J. Anal. Toxicol.* **1980,** *4,* 33.

140. Lobbezoo, M. W.; Van Rooy, H. H.; Van Wijngaarden, I.; Soudijn, W. *Eur. J. Pharm.* **1982,** *82,* 207.

141. den Besten, W.; Mulder, D.; Funcke, A. B. H.; Nauta, W. Th. *Arzneim.-Forsch.* **1970,** *20,* 538.

142. Sisenwine, S. F.; Tio, C. O. *Drug Metab. Dispos.* **1986,** *14,* 41.

143. Yasumura, A.; Murata, H.; Hattori, K.; Matsuda, K. *Chem. Pharm. Bull.* **1971,** *19,* 1929.

144. Baselt, R. C.; Stewart, C. B. *J. Anal. Toxicol.* **1978,** *2,* 107.

145. *Clarke's Isolation and Identification of Drugs in Pharmaceuticals, Body Fluids, and Post Mortem Material;* Moffat, A. C., Ed.; The Pharmaceutical Press: London, 1986.

146. Cone, E. J.; Darwin, W. D.; Buchwald, W. F.; Gorodetzky, C. W. *Drug Metabol. Dispos.* **1983,** *11,* 446.

147. Vermeulen, N. P. E.; Teunissen, M. W. E.; Breimer, D. D. *J. Chromatogr.* **1978,** *157,* 133.

148. Lynn, R. K.; Smith, R. G.; Leger, R. M.; Deinzer, M. L.; Griffin, D.; Gerber, N. *Drug Metab. Dispos.* **1977,** *5,* 47.

149. Holtzman, J. L.; Thompson, J. A. *Drug Metab. Dispos.* **1975,** *3,* 113.

150. Yeh, S. Y. *J. Pharm. Sci.* **1984,** *73,* 1783.

151. Iorio, M. A.; Casy, A. F. *J. Pharm. Pharmacol.* **1974,** *26,* 553.

152. Guay, D. R. P.; Meatherall, R. C.; Chalmers, J. L.; Grahame, G. R. *Brit. J. Clin. Pharmacol.* **1984,** *18,* 907.

153. Dahlstrom, B. E.; Paalzow, L. K.; Lindberg, C.; Bogentoft, C. *Drug Metab. Dispos.* **1979,** *7,* 108.

154. Holsztynska, E. J.; Domino, E. F. *Drug Metab. Rev.* **1985–6,** *6,* 285.

155. Wong, L. K.; Biemann, K. *Biomed. Mass Spectrom.* **1975,** *2,* 204.

156. Kammerer, R. C.; Schmitz, D. A.; Distefano, E. W.; Cho, A. K. *Drug Metab. Dispos.* **1981,** *9,* 274.

157. Schwartz, M. A.; Pool, W. R.; Hane, D. L.; Postma, E. *Drug Metab. Dispos.* **1974,** *2,* 31.

158. Ward, D. P.; Trevor, A. J.; Kalir, A.; Adams, J. D.; Baillie, T. A.; Castagnoli, N., Jr. *Drug Metab. Dispos.* **1982,** *10,* 690.

159. Franklin, R. B.; Dring, L. G.; Williams, R. T. *Drug Metab. Dispos.* **1977,** *5,* 223.

160. Tang, B. K.; Kalow, W.; Grey, A. A. *Drug Metab. Dispos.* **1979,** *7,* 315.

161. Whyte, M. P.; Dekaban, A. S. *Drug Metab. Dispos.* **1977,** *5,* 63.

162. Milne, L.; Williams, N. E.; Calvey, T. N.; Murray, G. R.; Chan, K. *Brit. J. Anaesth.* **1980,** *52,* 537.

163. Isherwood, C. N.; Murray, G. R.; Chan, K.; Calvey, T. N.; Williams, N. E. *Brit. J. Clin. Pharmacol.* **1982,** *13,* 612P.

164. Calvey, T. N.; Chan, K.; Milne, L. A.; Murray, G. R.; Williams, N. E. *Brit. J. Clin. Pharmacol.* **1981,** *11(suppl 1),* 124P.

165. Cho, A. K.; Lindeke, B.; Sum, C. Y. *Drug Metabol. Dispos.* **1974,** *2,* 1.

166. Kammerer, R. C.; Cho, A. K.; Jonsson, J. *Drug Metabol. Dispos.* **1978,** *6,* 396.

167. Bakke, O. M.; Draffan, G. H.; Davies, D. S. *Xenobiotica* **1974,** *4,* 237.

168. Trebbi, A.; Gervasi, G. B.; Comi, V. *J. Chromatogr.* **1975,** *110,* 09.

169. Bailey, K.; Chow, A. Y. K.; Downie, R. H.; Pike, R. K. *J. Pharm. Pharmacol.* **1976,** *28,* 713.

170. Assandri, A.; Barone, D.; Ferrari, P.; Perazzi, A.; Ripamonti, A.; Tuan, G.; Zerilli, L. F. *Drug Metabol. Dispos.* **1984,** *12,* 257.

171. Abdel-Monem, M. M.; Harris, P. A.; Portoghese, P. S. *J. Med. Chem.* **1972,** *5,* 706.

172. Abdel-Monem, M. M.; Larson, D. L.; Kupferberg, H. J.; Portoghese, P. S. *J. Med. Chem.* **1972,** *15,* 494.

173. von Duhm, B.; Maul, W.; Medenwald, H.; Patzschke, K.; Wegner, L. A. *Arzneim.-Forsch.* **1974,** *24,* 632.

174. von Putter, J.; Kroneberg, G. *Arzneim.-Forsch.* **1974,** *24,* 643.

175. von Horster, F. A.; Duhm, B.; Medenwald, H.; Patzschke, K.; Wegner, L. A. *Arzneim.-Forsch.* **1974,** *24,* 652.

176. Nelson, W. L.; Shetty, H. U. *Drug Metab. Dispos.* **1986,** *14,* 506.

177. Wong, K. K.; Schreiber, E. C. *Drug Metab. Rev.* **1972,** *1,* 101.

178. Walle, T.; Walle, U. K.; Olanoff, L. S. *Drug Metab. Dispos.* **1985,** *13,* 204.

179. Pritchard, J. F.; Schneck, D. W.; Hayes, A. H. *J. Chromatogr.* **1979,** *162,* 47.

180. Zampaglione, N.; Hilbert, J. M.; Ning, J.; Chung, M.; Gural, R.; Symchowicz, S. *Drug Metabol. Dispos.* **1985,** *13,* 25.

181. Hilbert, J.; Pramanik, B.; Symchowicz, S.; Zampaglione, N. *Drug Metab. Dispos.* **1984,** *12,* 452.

182. Mead, J.; Koepfli, J. B. *J. Biol. Chem.* **1944,** *154,* 507.

183. Randall, L. O.; Lehmann, G. *J. Pharmacol. Exp. Ther.* **1950,** *99,* 163.

184. Muhlhauser, R. O.; Watkins, W. D.; Murphy, R. C.; Chidsey, C. A. *Drug Metab. Dispos.* **1974,** *2,* 513.

185. Weldon, S. T.; Perry, D. F.; Cork, R. C.; Gandolfi, A. J. *Anesthesiology* **1985,** *63,* 684.

186. Halldin, M. M.; Carlsson, S.; Kanter, S. L.; Widman, M.; Augrell, S. *Arzneim.-Forsch.* **1982,** *32,* 764.

187. Childs, P. S.; McCurdy, H. H. *J. Anal. Toxicol.* **1984,** *8,* 220.

188. Perez-Reyes, M.; Timmons, M. C.; Lipton, M. A.; Christensen, H. D.; Davis, K. H.; Wall, M. E. *Experientia* **1973,** *29,* 1009.

189. Pallante, S.; Lyle, M. A.; Fenselau, C. *Drug. Metab. Dispos.* **1978,** *6,* 389.
190. Wall, M. E.; Perez-Reyes, M. *J. Clin. Pharmacol.* **1981,** *21,* 178S.
191. Bun, H.; Philip, F.; Berger, Y.; Necciari, J.; Al-Mallah, N. R.; Serradimigni, A.; Cano, J. P. *Arzneim.-Forsch.* **1987,** *37,* 199.
192. Misra, A. L.; Pontani, R. B.; Mule, S. J. *J. Chromatogr.* **1972,** *71,* 554.
193. Misra, A. L.; Pontani, R. B.; Mule, S. J. *Experientia* **1973,** *29,* 1108.
194. Misra, A. L.; Pontani, R. B.; Mule, S. J. *Xenobiotica* **1973,** *4,* 17.
195. Zehnder, K.; Kalberer, F.; Kreis, W.; Rutschmann, J. *Biochem. Pharmacol.* **1962,** *11,* 535.
196. von Vollmer, K. O.; Achenbach, H. *Arzneim.-Forsch.* **1974,** *24,* 1237.
197. von Vollmer, K. O.; Poisson, A. *Arzneim.-Forsch.* **1976,** *26,* 1827.
198. Hengy, H.; Vollmer, K. O.; Gladigau, V. *J. Pharm. Sci.* **1978,** *67,* 1765.
199. Eberts, F. S.; Philopoulos, Y.; Reineke, L. M.; Vliek, R. W. *Clin. Pharmacol. Ther.* St. Lewis, **1981,** *29,* 81.
200. Eberts, F. S. *Drug Metab. Dispos.* **1977,** *5,* 547.
201. Adams, W. J.; Bombardt, P. A.; Code, A. *J. Pharm. Sci.* **1983,** *72,* 1185.
202. Konishi, M.; Mori, Y. *J. Chromatogr.* **1982,** *229,* 355.
203. Coassolo, P.; Aubert, C.; Cano, J. P. *J. Chromatogr.* **1983,** *274,* 161.

Appendix III

Structural Frameworks for Drugs and Metabolite Listed in Appendices I and II

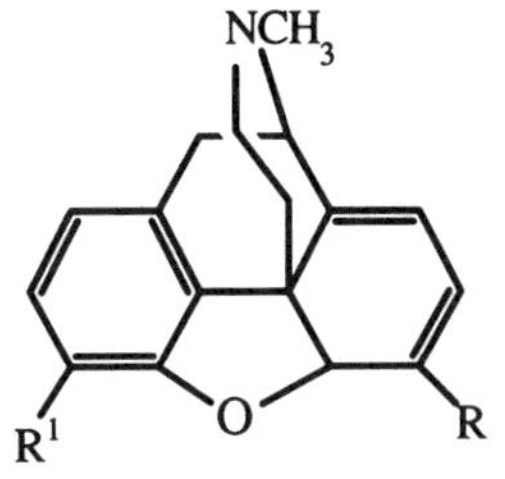

1

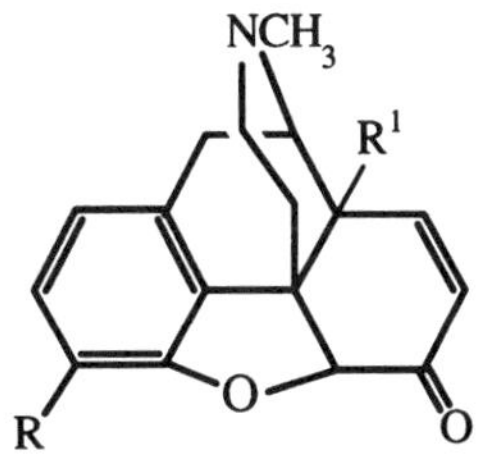

5

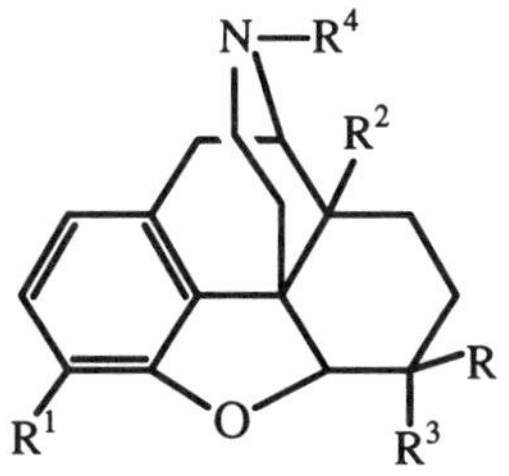

9

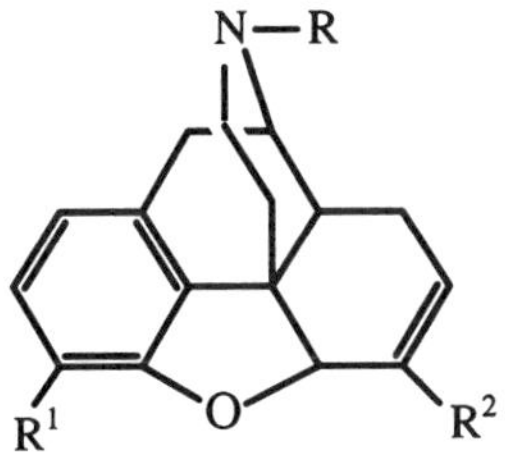

2

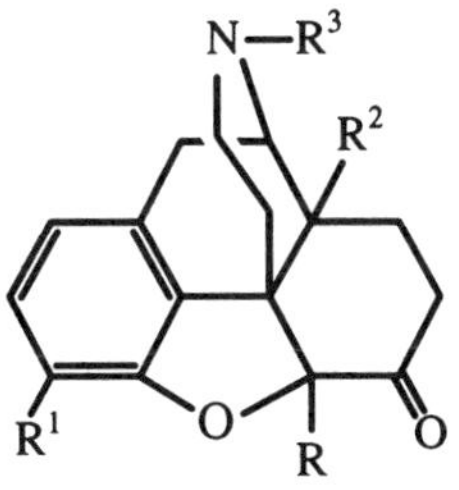

6

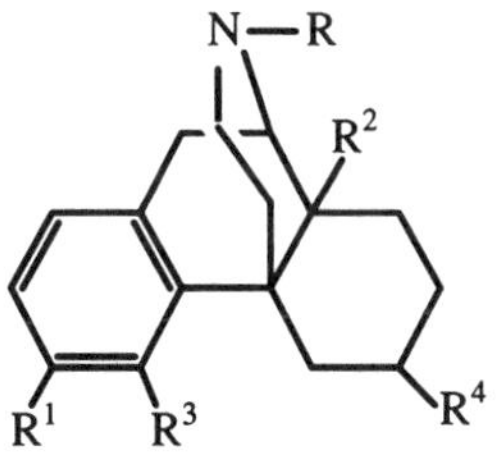

10

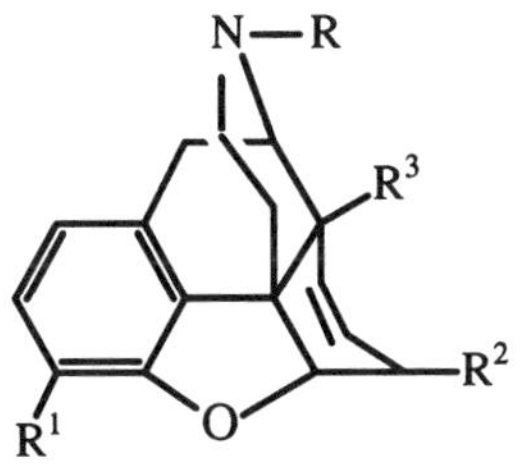

3

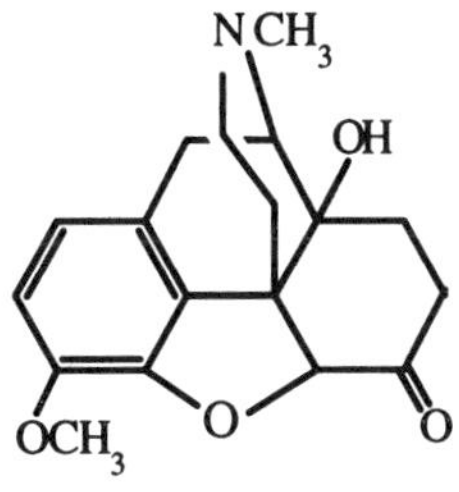

7

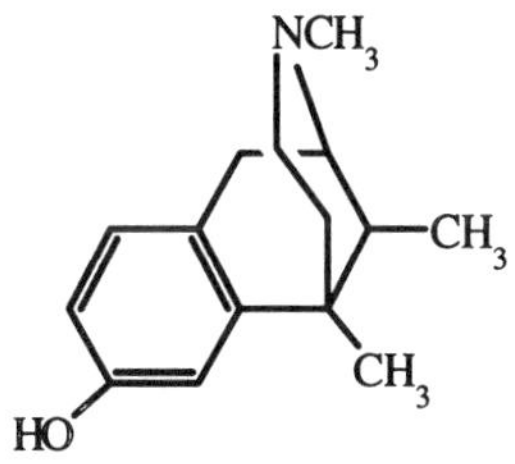

11

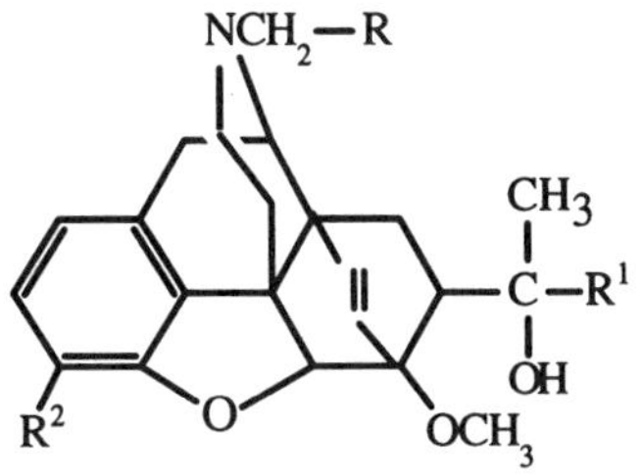

4

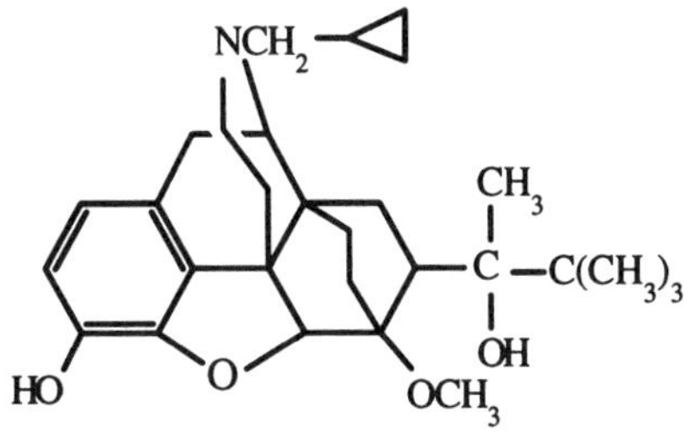

8

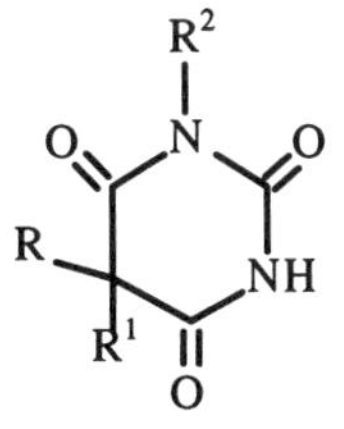

12

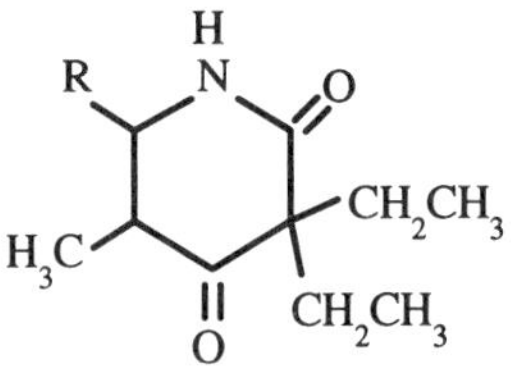

1 3

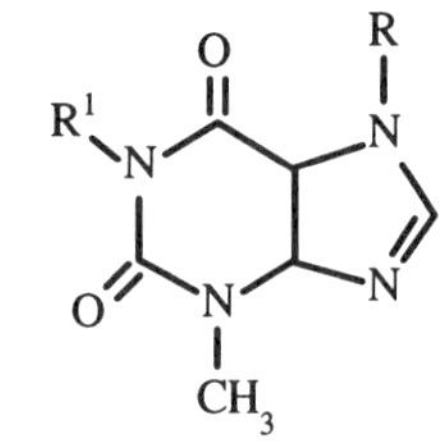

1 4

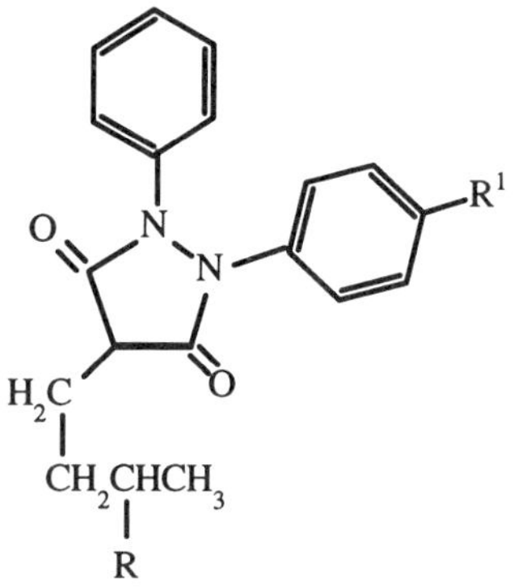

1 5

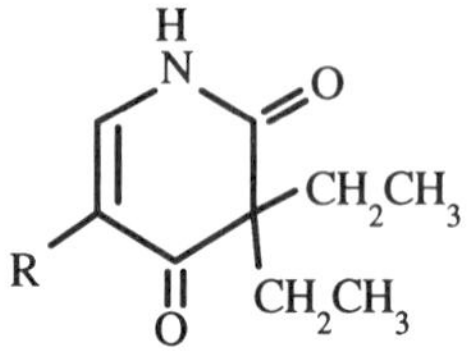

1 6

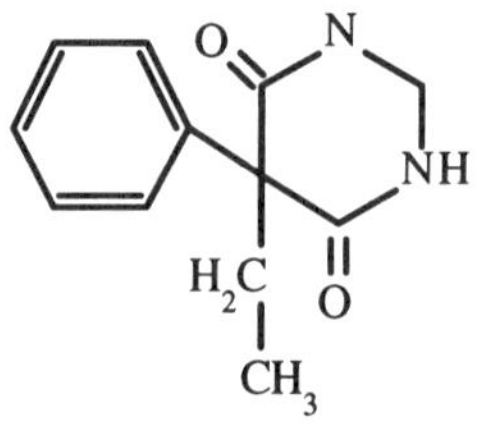

1 7

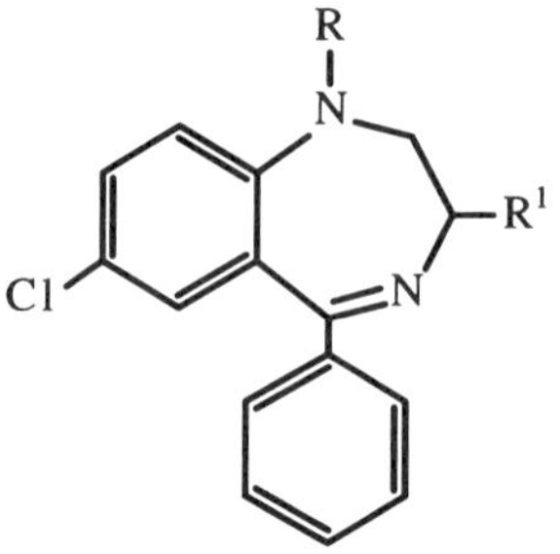

1 8

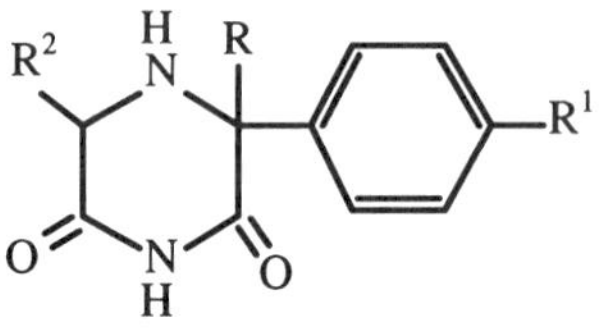

1 9

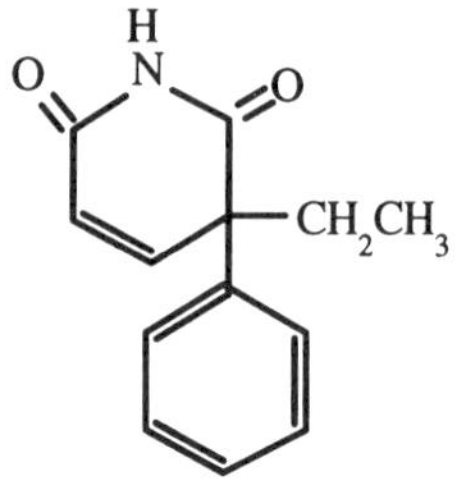

2 0

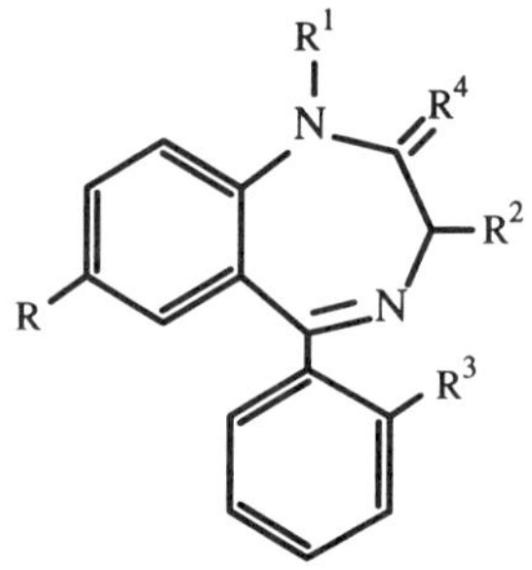

2 1

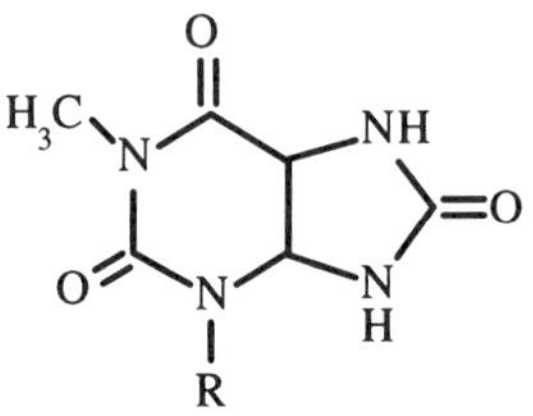

2 2

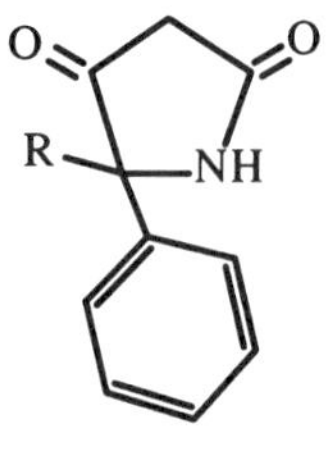

2 3

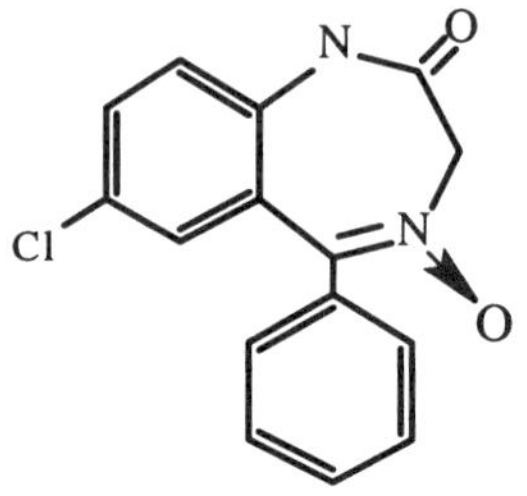

2 4

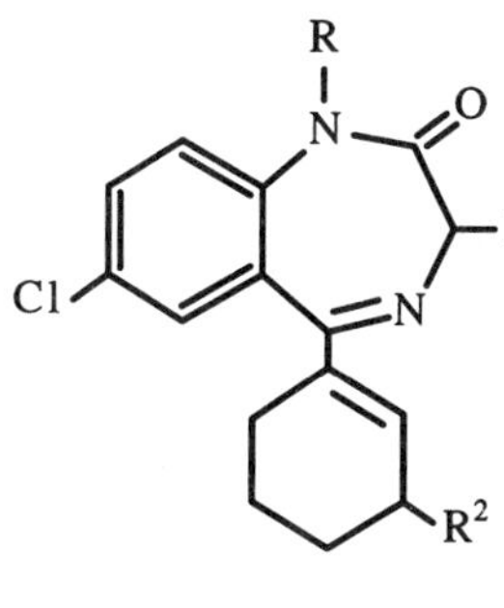

2 5

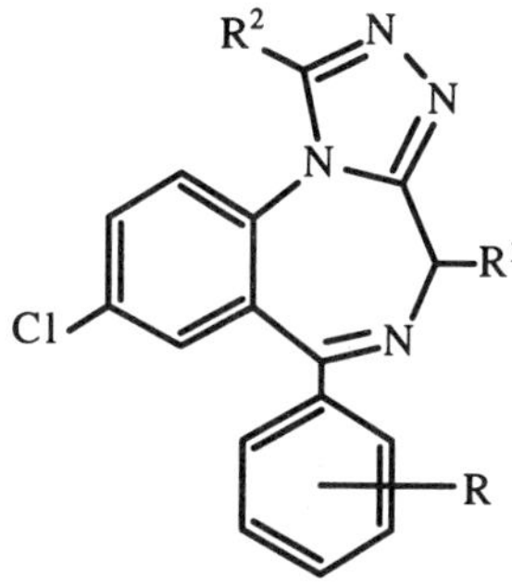

2 9

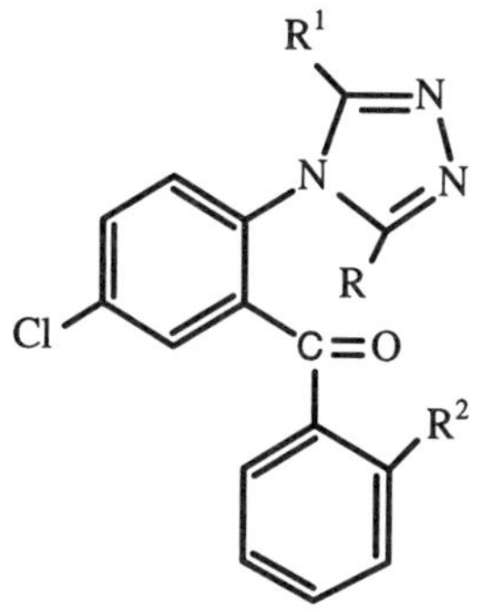

3 3

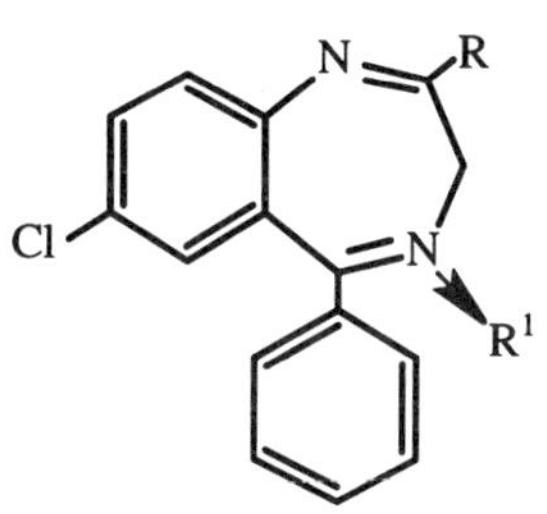

2 6

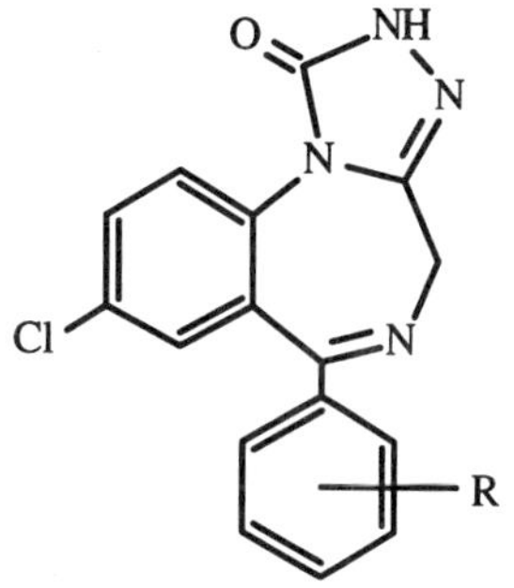

3 0

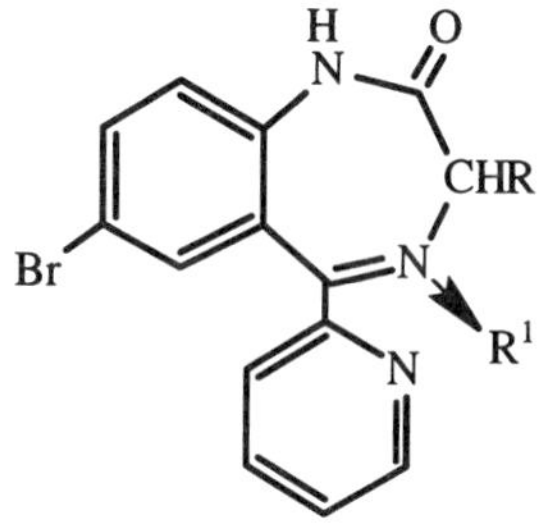

3 4

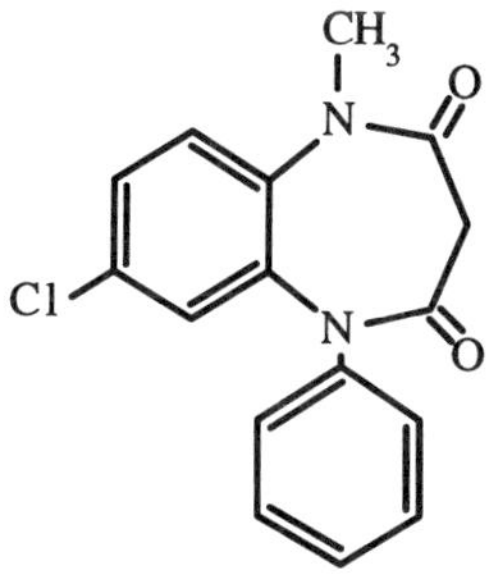

2 7

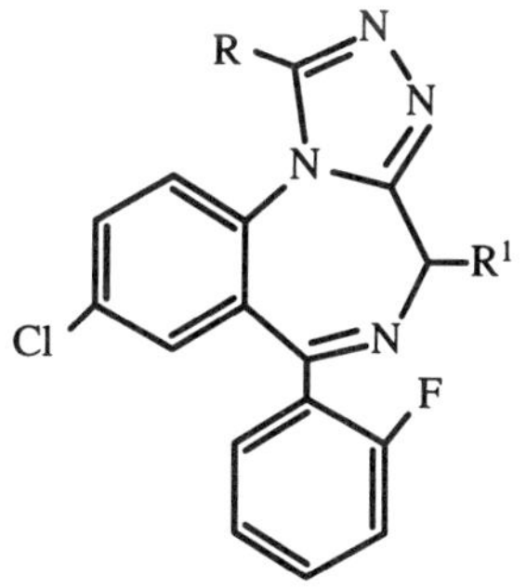

3 1

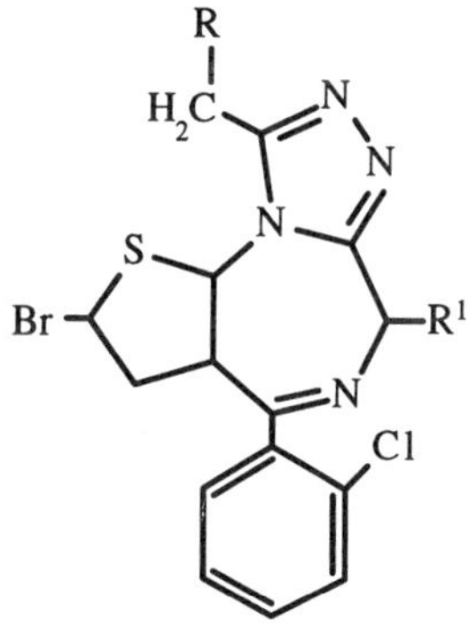

3 5

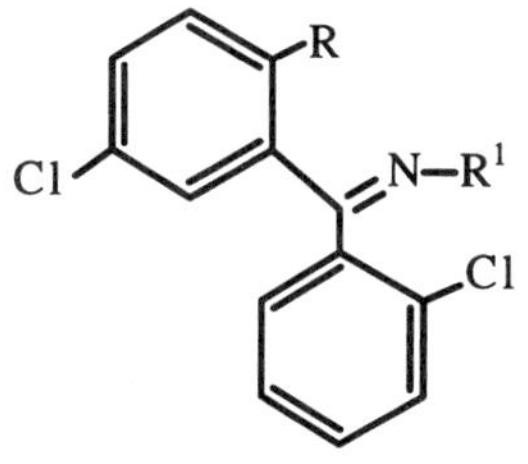

2 8

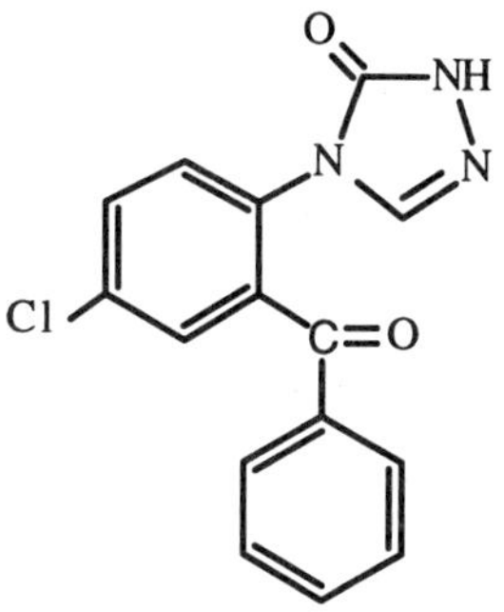

3 2

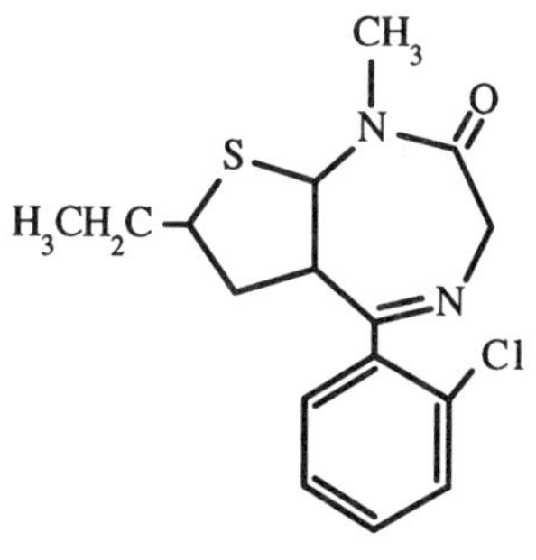

3 6

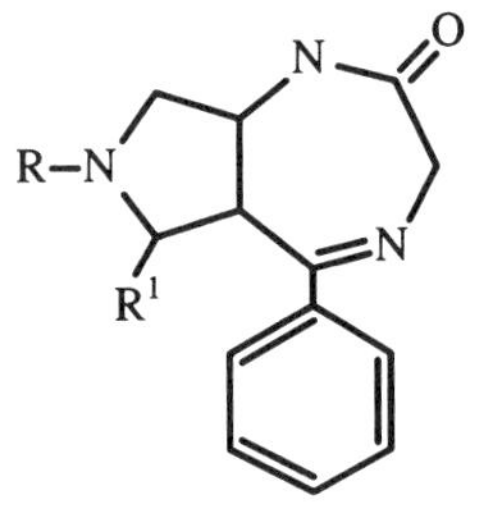

3 7

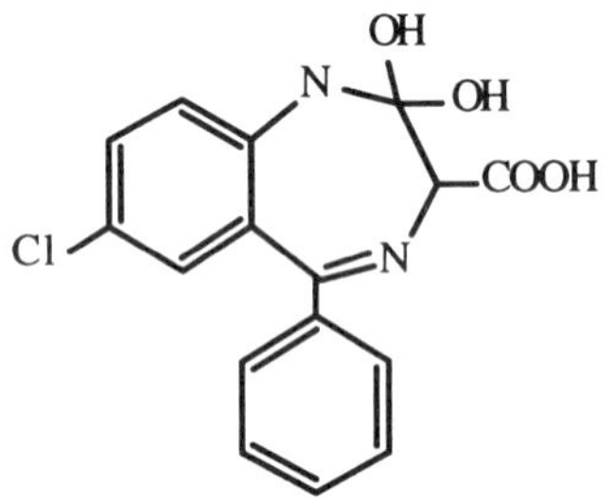

4 1

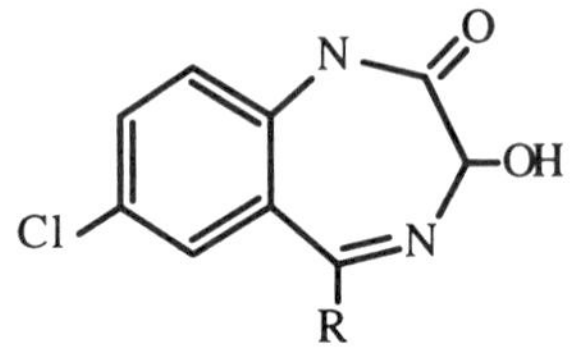

4 5

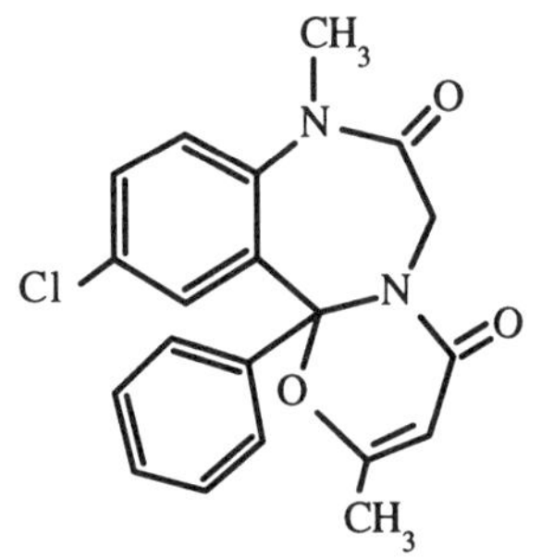

3 8

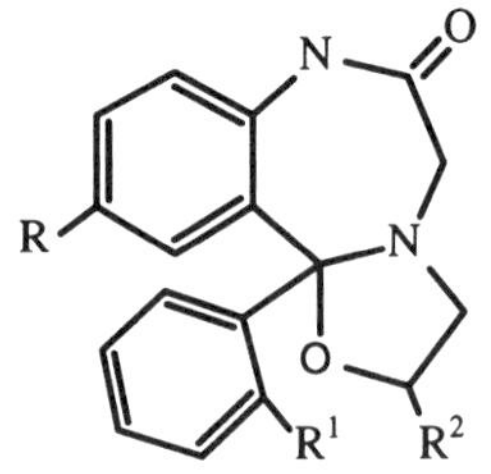

4 2

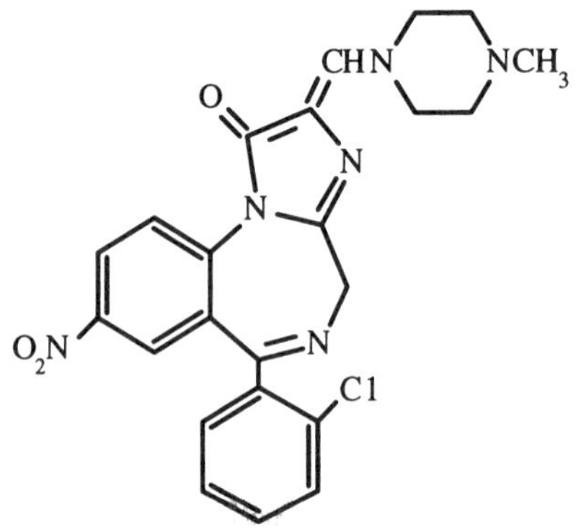

4 6

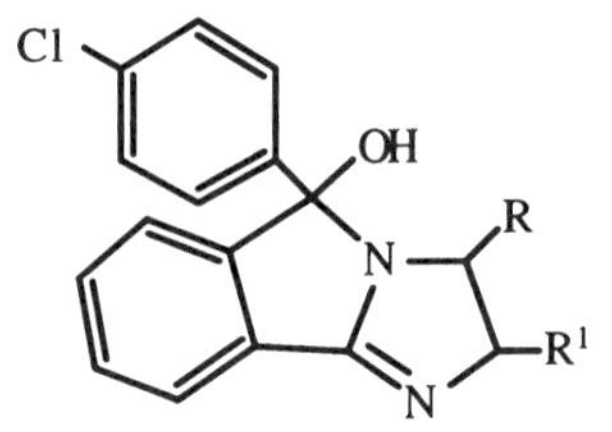

3 9

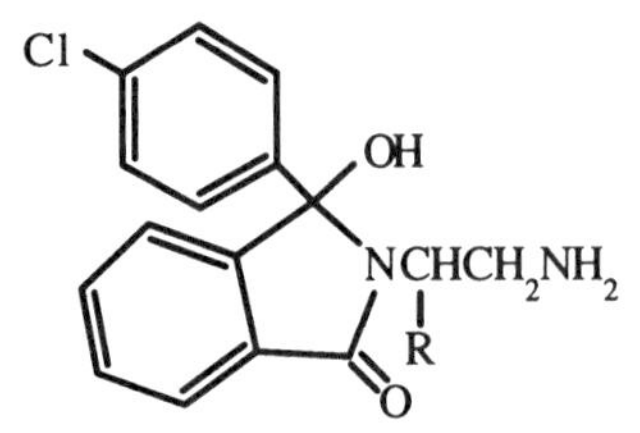

4 3

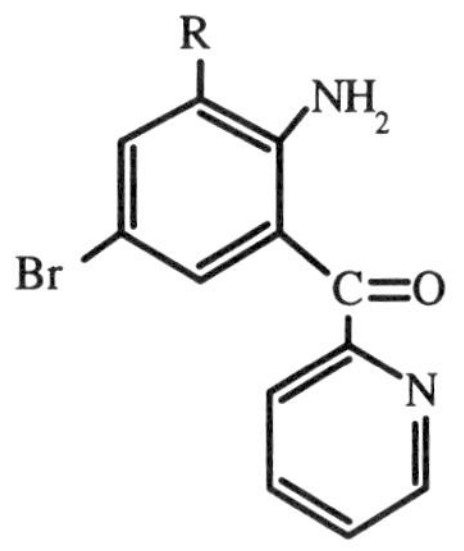

4 7

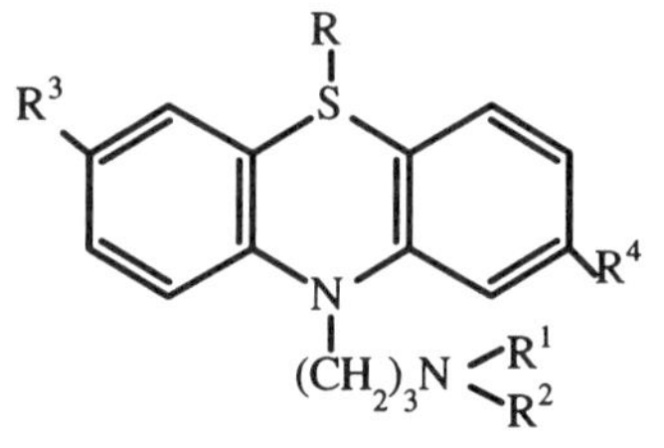

4 0

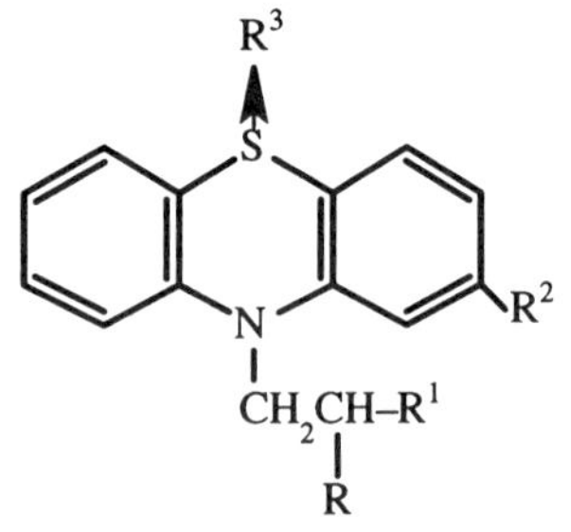

4 4

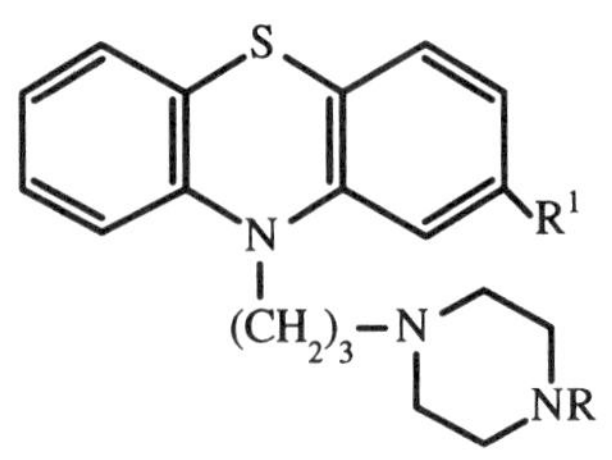

4 8

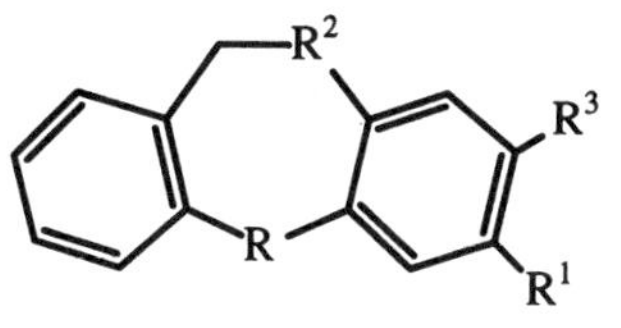

4 9

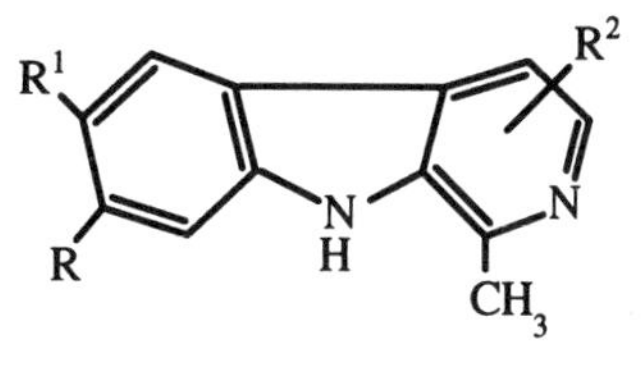

5 0

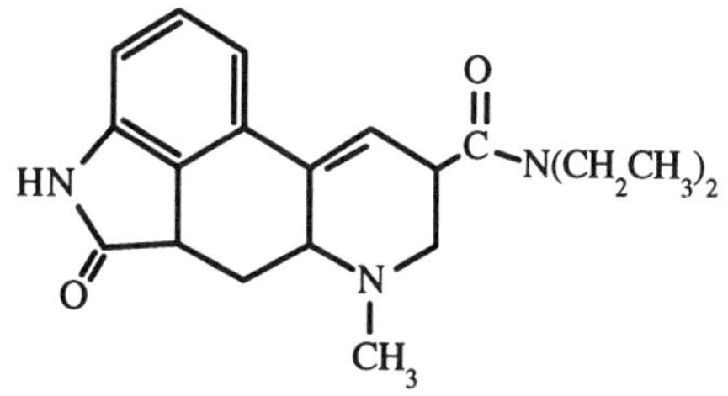

5 1

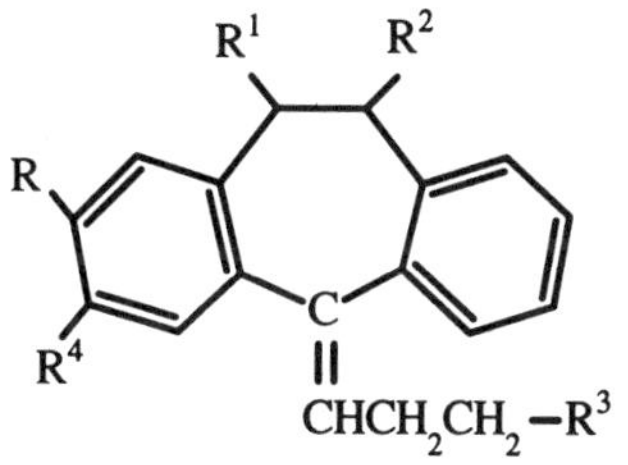

5 2

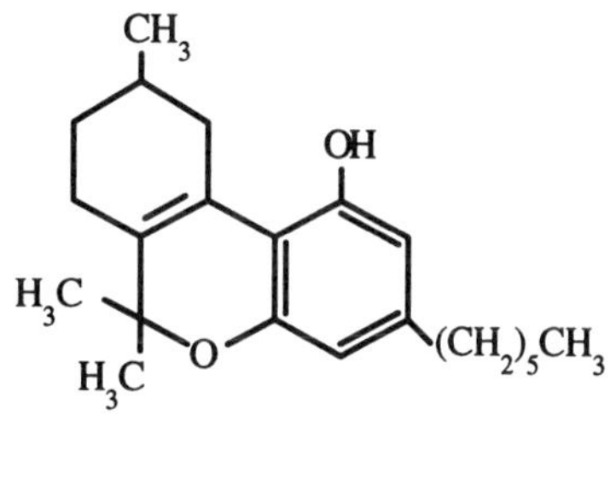

5 3

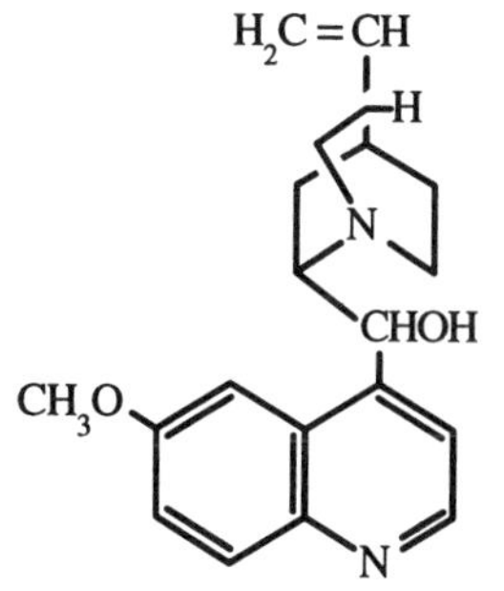

5 4

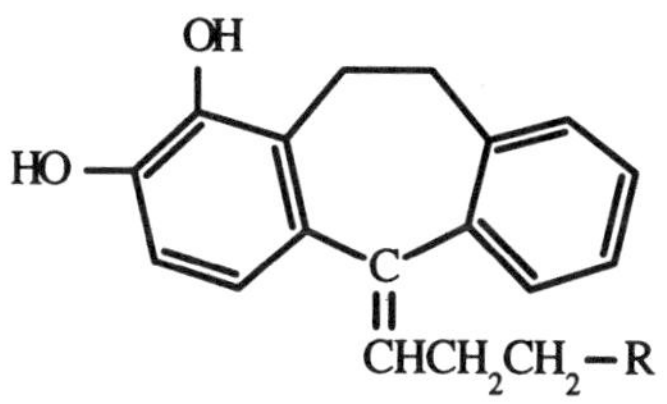

5 5

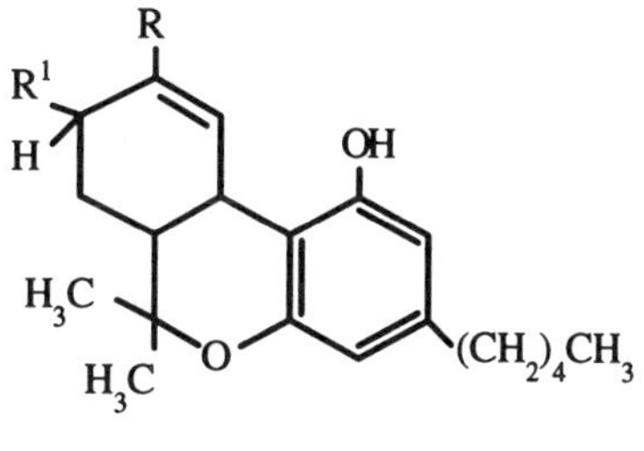

5 6

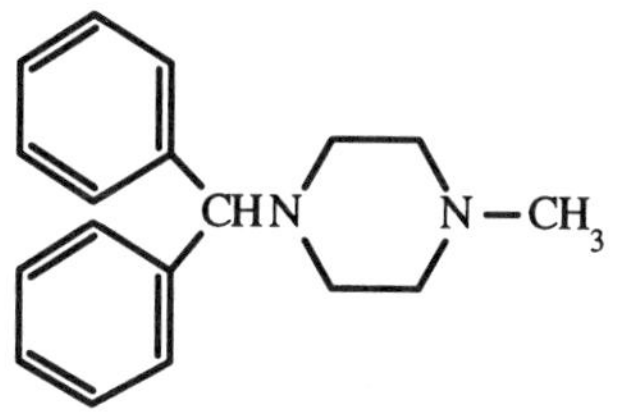

5 7

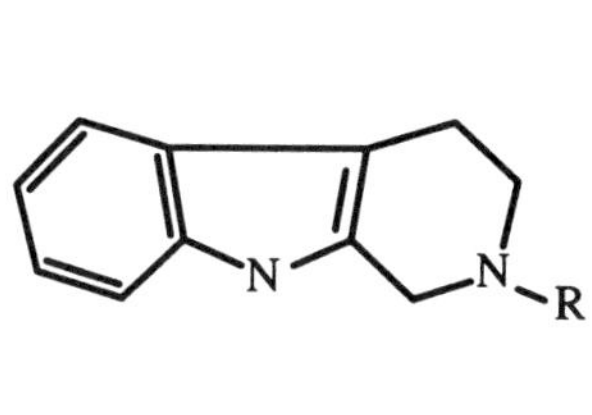

5 8

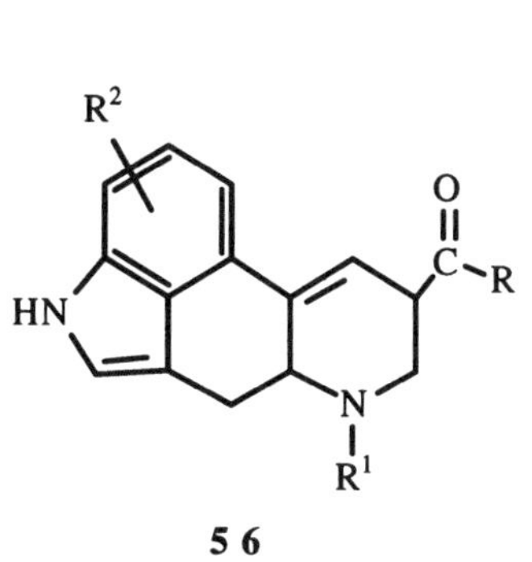

5 9

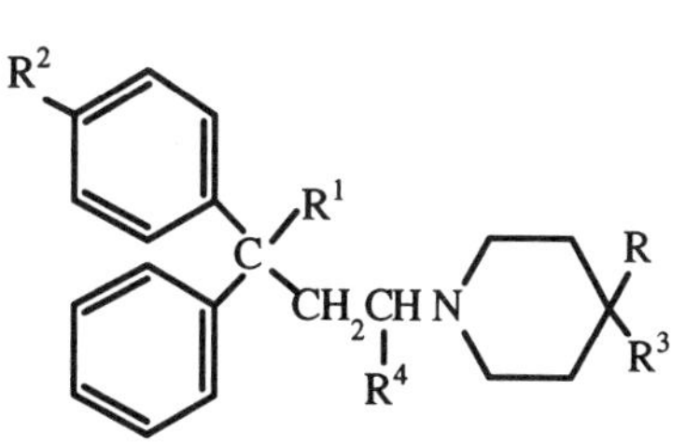

6 0

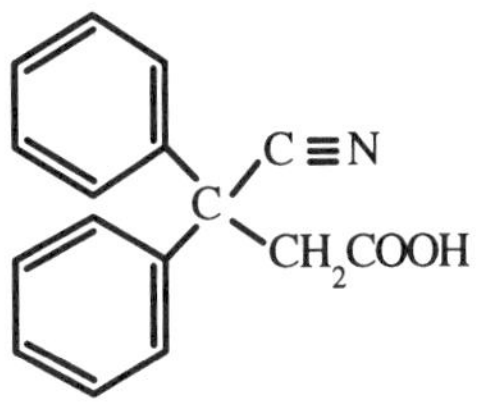

6 1

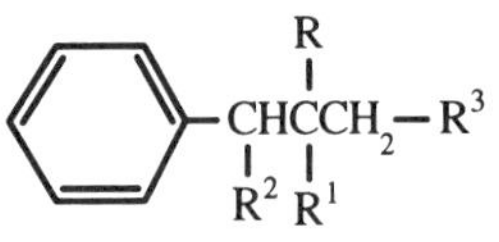

6 5

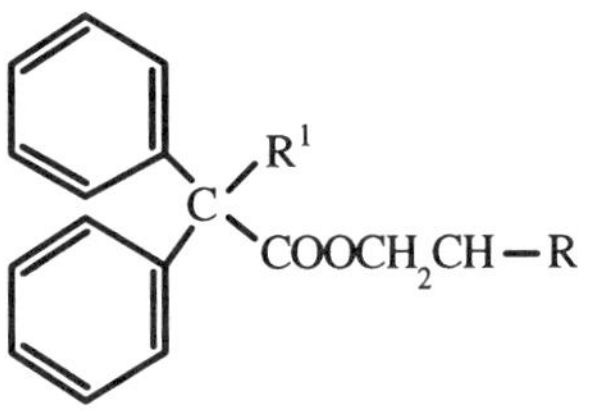

6 9

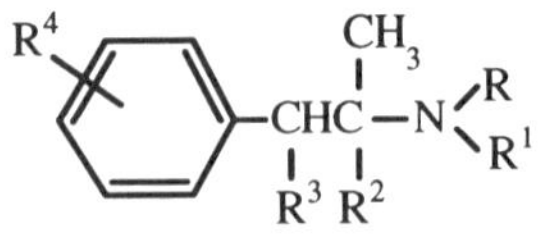

6 2

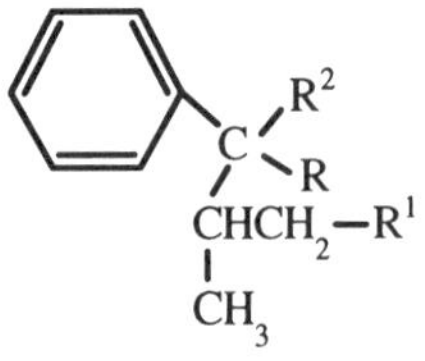

6 6

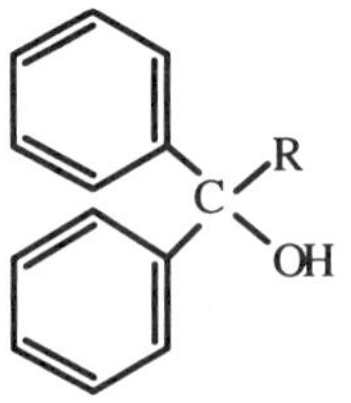

7 0

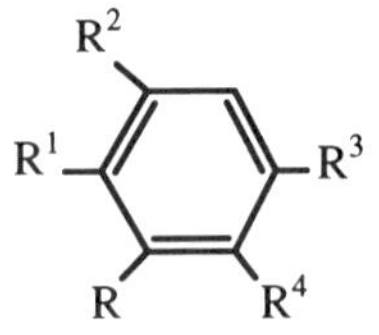

6 3

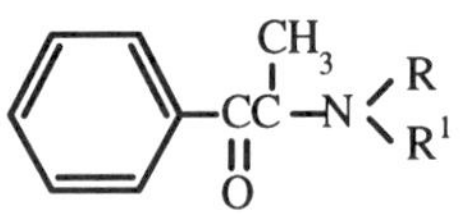

6 7

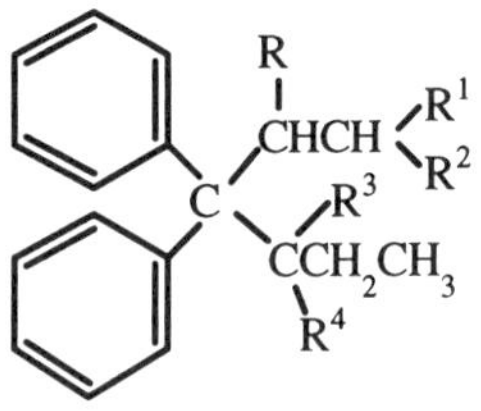

7 1

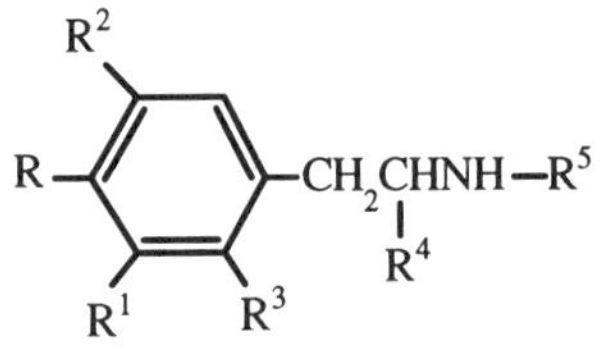

6 4

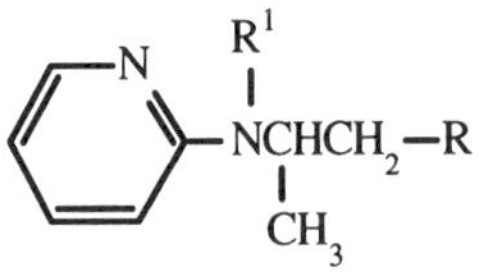

6 8

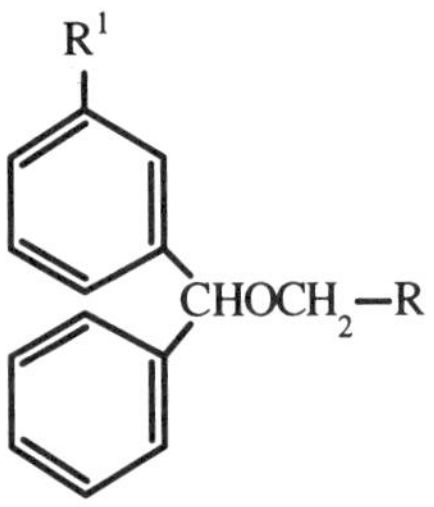

7 2

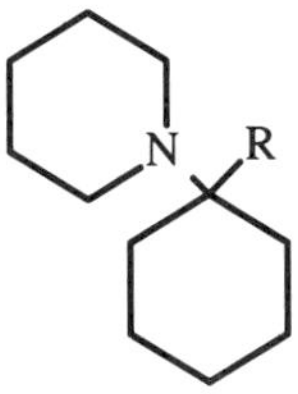

7 3

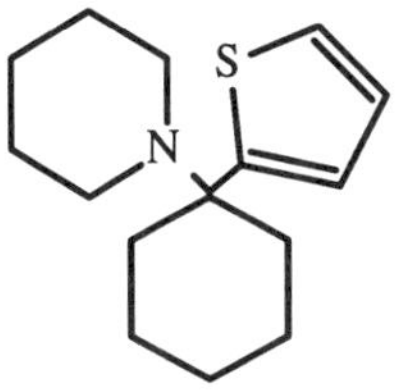

7 7

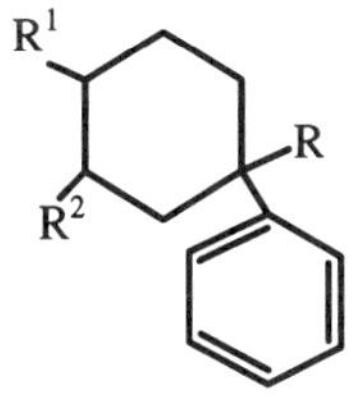

8 1

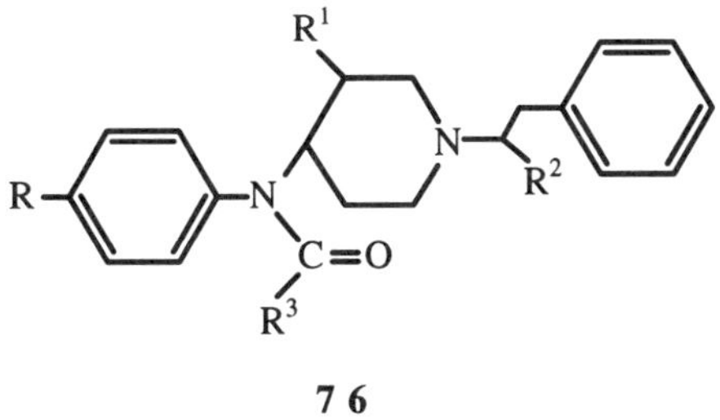

7 4

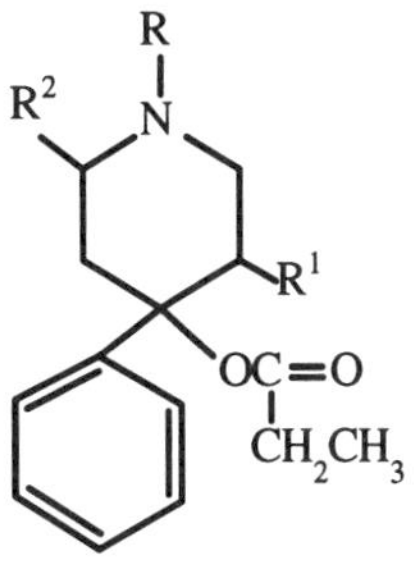

7 8

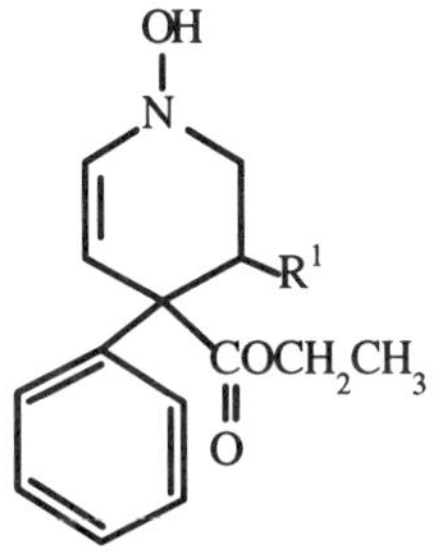

8 2

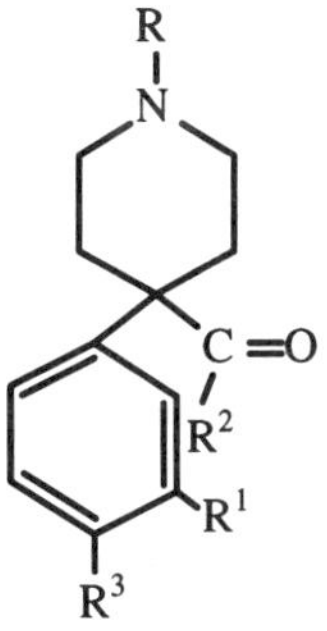

7 5

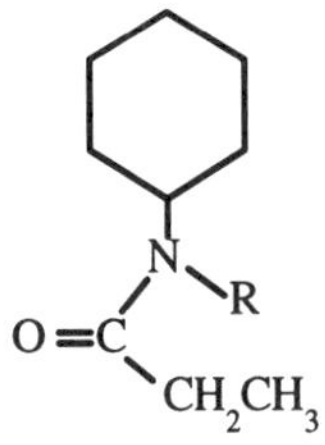

7 9

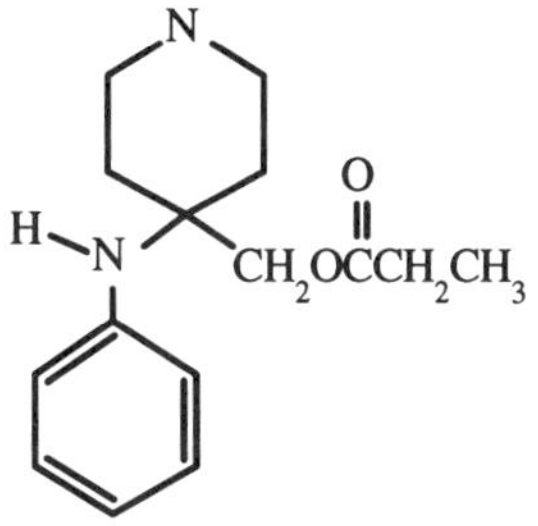

8 3

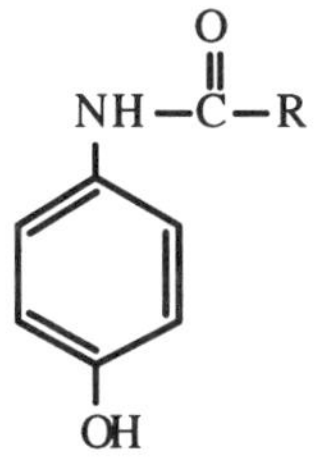

7 6

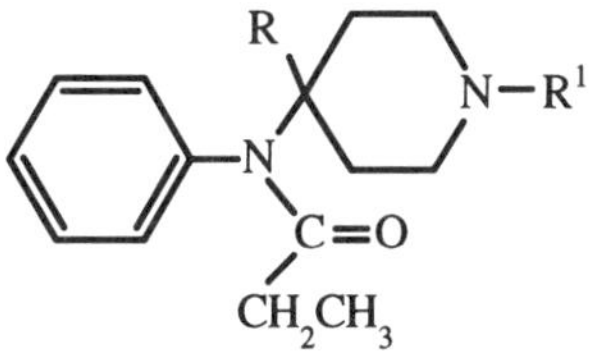

8 0

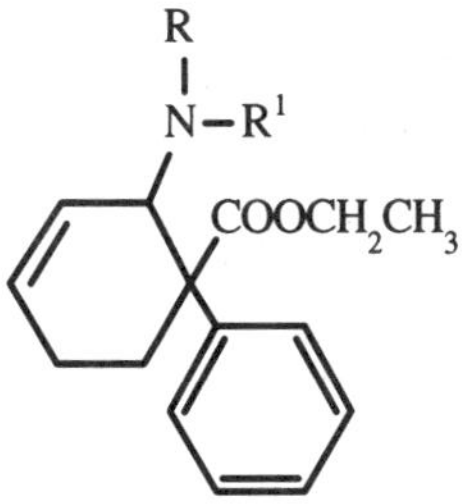

8 4

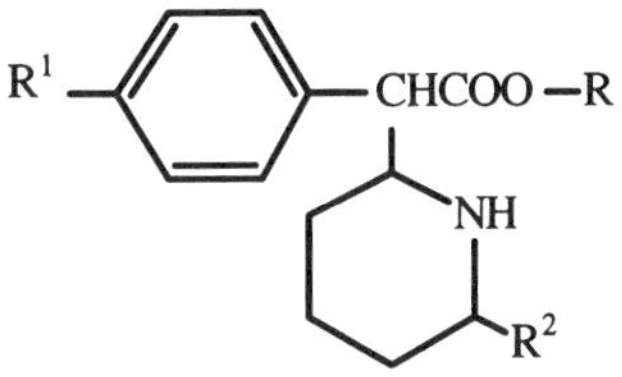

R¹—⟨phenyl⟩—CHCOO—R
NH
R²

85

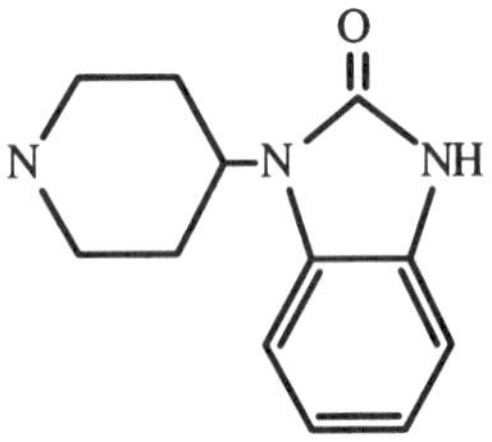

⟨phenyl⟩
=CHCH₃
N—CH₃
R—⟨phenyl⟩
CH₃

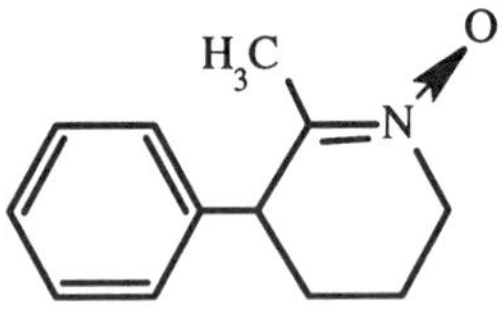

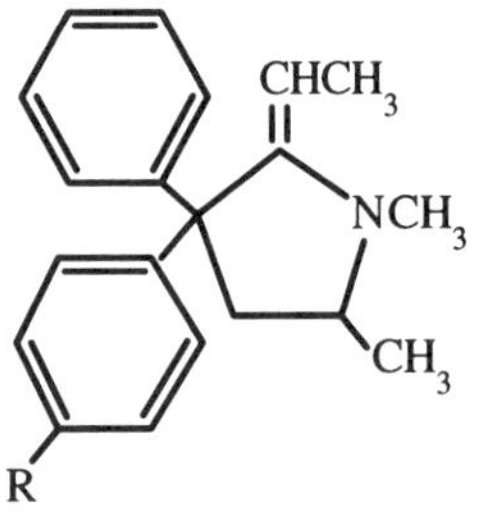

89

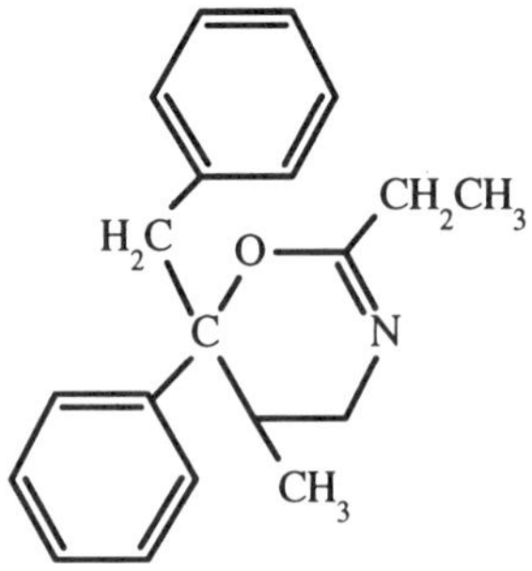

R—N
R¹
OR²

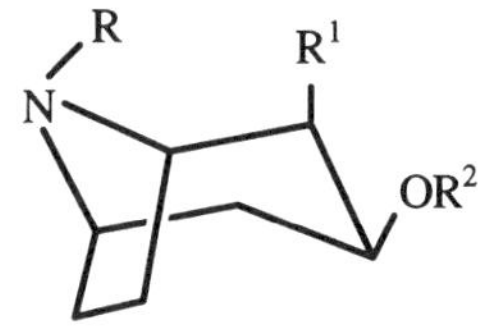

93

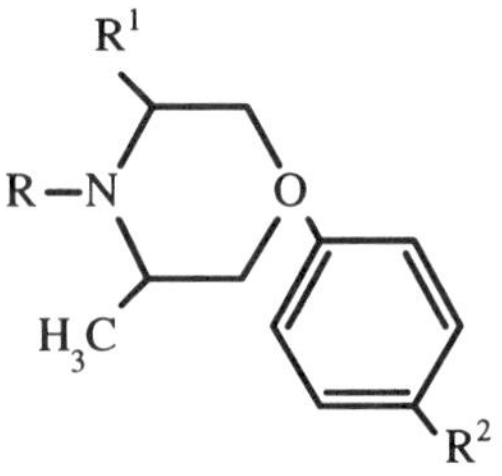

O
⟨piperidine⟩—N   NH
⟨benzimidazole⟩

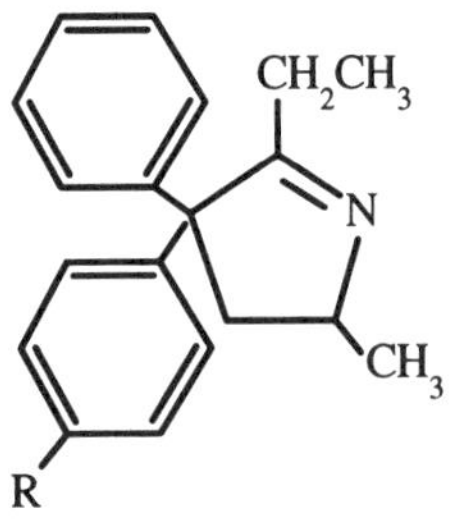

86

⟨phenyl⟩
H₂C
C   O   CH₂CH₃
N
⟨phenyl⟩
CH₃

90

R¹
R—N   O
H₃C
R²

94

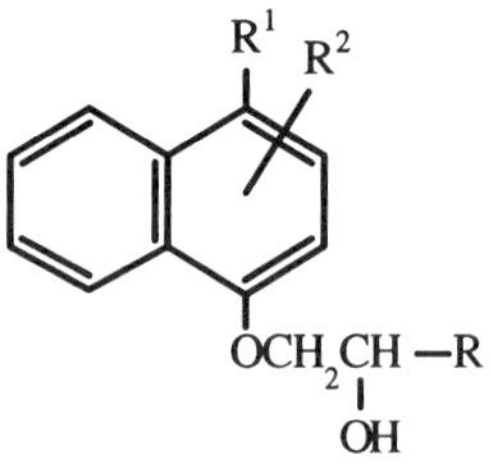

H₃C
⟨phenyl⟩   N→O

87

⟨phenyl⟩—NH—⟨piperidine⟩—N—CH₂—⟨phenyl⟩
R

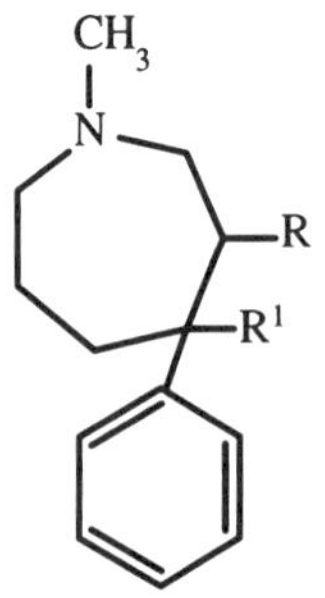

91

R¹ R²
⟨naphthalene⟩
OCH₂CH—R
OH

95

⟨phenyl⟩
CH₂CH₃
N
R—⟨phenyl⟩
CH₃

88

CH₃
N
R
R¹
⟨phenyl⟩

92

⟨isoquinoline⟩
OCH₂COOH

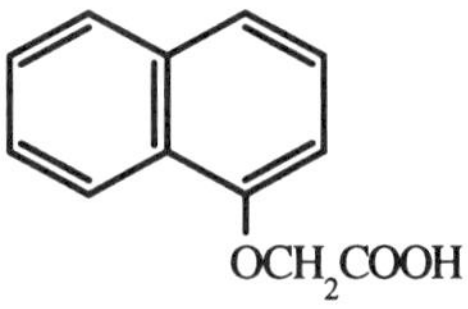

96

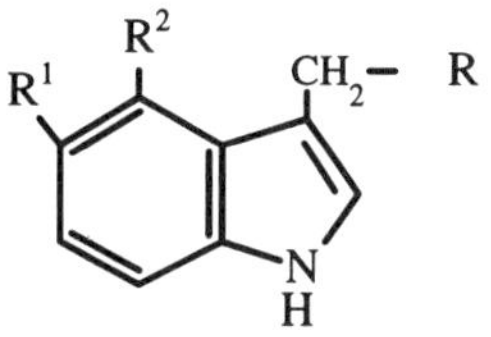

97

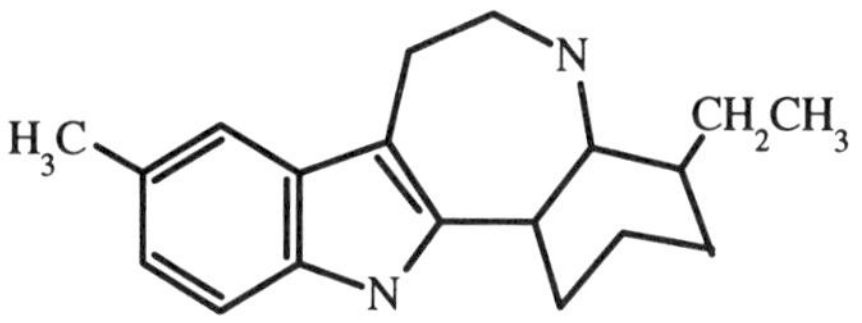

98

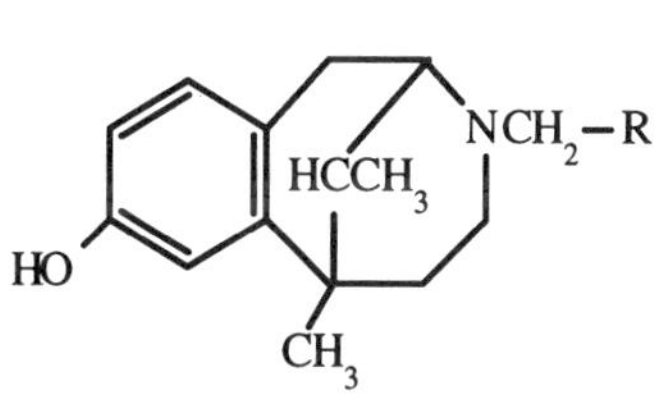

99

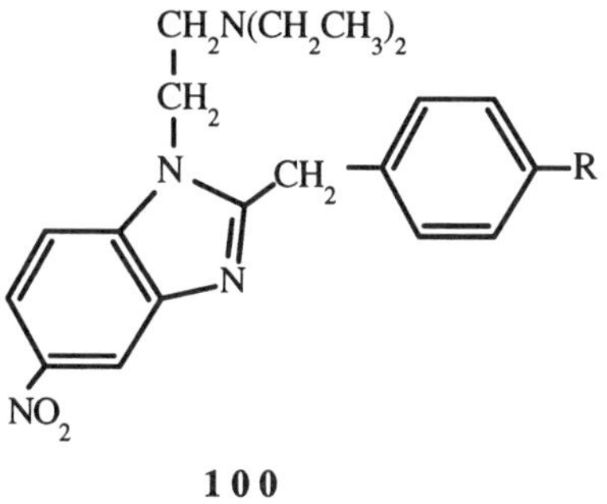

100

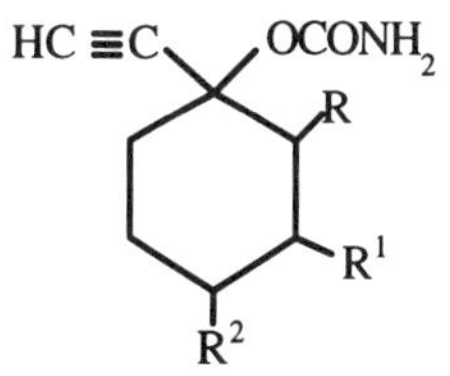

101

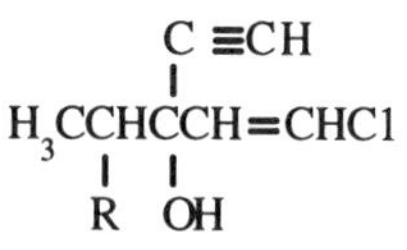

102

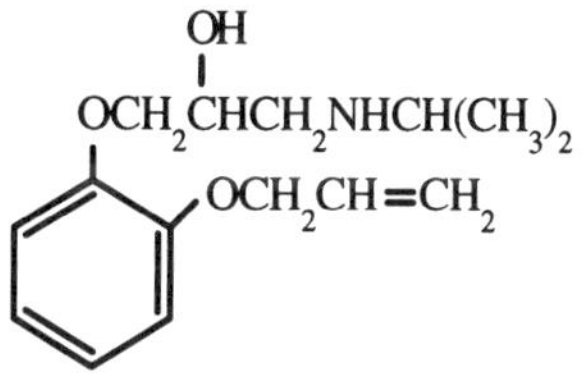

103

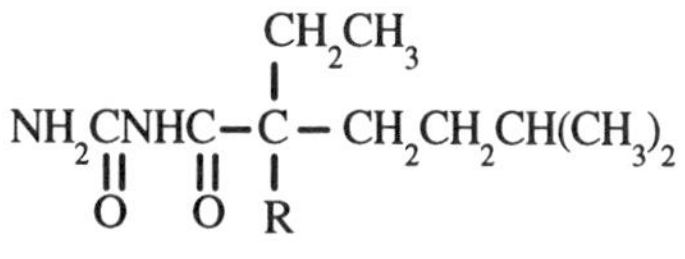

104

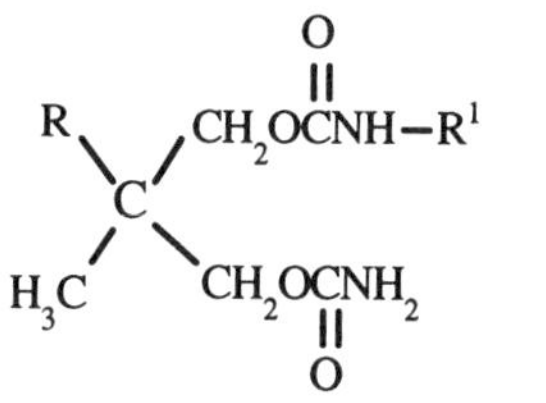

105

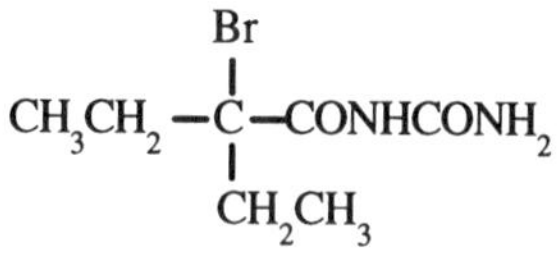

106

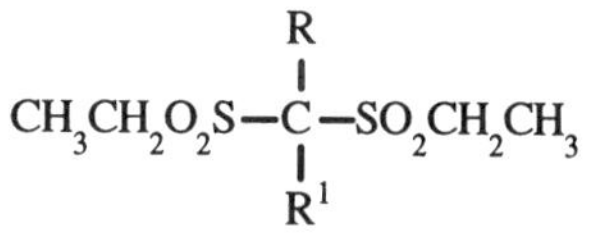

107

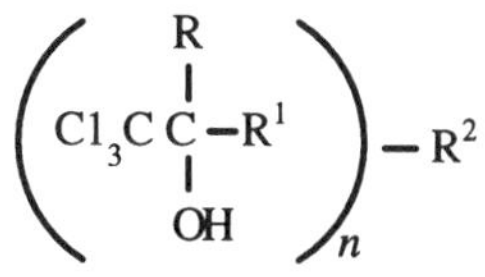

108

Appendix IV

Functional Groups That Complete the Structural Frameworks (Appendix III) for Drugs Listed in Appendices I and II

Figure	Drug or metabolite name (Appendices I and II)	R	R^1	R^2	R^3	R^4	R^5
1a	Oripavine	OCH_3	OH	OH			
1b	Thebaine	OCH_3	OCH_3	OCH_3			
2a	Acetyldihydrocodeinone (thebacon)	CH_3	OCH_3	$OOCCH_3$			
2b	Methyldesorphine	CH_3	OH	CH_3			
3a	Benzylmorphine	CH_3	OCH_2Ph	OH			
3b	Codeine	CH_3	OCH_3	OH			
3c	Codeine-6-glucuronide	CH_3	OCH_3	$OC_6H_9O_6$	H		
3d	Codeine methyl bromide	$(CH_3)CH_3Br$	OCH_3	OH	H		
3e	Codeine N-oxide	$(O)CH_3$	OCH_3	OH	H		
3f	Ethylmorphine	CH_3	OCH_2CH_3	OH	H		
3g	Heroin	CH_3	$OCOCH_3$	$OCOCH_3$	H		
3h	14-Hydroxycodeine	CH_3	OCH_3	OH	OH		
3i	6-Monoacetylmorphine	CH_3	OH	$OOCCH_3$	H		
3j	Morphine	CH_3	OH	OH	H		
3k	Morphine-6-glucuronide	CH_3	OH	$OC_6H_9O_6$	H		
3l	Morphine methyl bromide	$(CH_3)CH_3Br$	OH	OH	H		
3m	Morphine methyl sulfonate	$(CH_3)CH_3(SO_4H)$	OH	OH	H		
3n	Morphine N-oxide	$(O)CH_3$	OH	OH	H		
3o	Myrophine	H	OCH_2Ph	$COO(CH_2)_{12}CH_3$	H		
3p	Nalorphine	$CH_2CH=CH_2$	OH	OH	H		
3q	Nicocodeine	CH_3	OOC–$N\langle morpholine\rangle$	OOC-(pyridyl) structure	H		
3r	Nicomorphine	CH_3	OCH_3	OH	H		
3s	Norcodeine	H	OCH_2CH_3	OH	H		
3t	Norethylmorphine	H	OH	OH	H		
3u	Normorphine	H		OH	H		
3v	Pholcodeine	CH_3	$O(CH_2)_2$–$N\langle morpholine\rangle$	OH	H		
4a	Acetorphine	(cyclopropyl)	$CH_2CH_2CH_3$	$OOCCH_3$			
4b	Cyprenorphine	H	CH_3	OH			
4c	Etorphine	H	$CH_2CH_2CH_3$	OH			
5a	Codeinone	OCH_3	H				
5b	14-Hydroxycodeinone	OCH_3	OH				
6a	Hydrocodone	H	OCH_3	H	CH_3		
6b	Hydromorphone	H	OH	H	CH_3		
6c	Metopon	CH_3	OCH_3	H	CH_3		
6d	Norhydrocodone	H	OCH_3	H	H		
6e	Noroxycodone	H	OCH_3	OH	H		
6f	Oxycodone	H	OCH_3	OH	CH_3		
6g	Oxymorphone	H	OH	OH	CH_3		

Figure	Drug or metabolite name (Appendices I and II)	R	R^1	R^2	R^3	R^4	R^5
9a	Acetyldihydrocodeine	H	OCH_3	H	$OOCCH_3$	CH_3	
9b	Desomorphine	H	OH	H	H	CH_3	
9c	Dihydrocodeine	H	OCH_3	H	OH	CH_3	
9d	Dihydromorphine	H	OH	H	OH	CH_3	
9e	Dihydronormorphine	H	OH	H	OH	H	
9f	Hydrocodol	H	OCH_3	H	OH	CH_3	
9g	Hydromorphinol	H	OH	OH	OH	CH_3	
9h	Hydromorphol	H	OH	OH	OH	CH_3	
9i	Methyldihydromorphine	OH	OH	H	CH_3	CH_3	
10a	Butorphanol	CH_2-◇	OH	OH	H	H	
10b	Dextrorphan	CH_3	OH	OH	H	H	
10c	Drotebanol	CH_3	OCH_3	OH	OCH_3	OH	
10d	Hydroxybutorphanol	CH_2-◇$-OH$	OH	OH	H	H	
10e	3-Hydroxymorphinan	H	OH	OH	H	H	
10f	Levallorphan	$CH_2CH=CH_2$	OH	H	H	H	
10g	Levophenacylmorphan	CH_2COPh	OH	H	H	H	
10h	Levorphanol	CH_3	OH	H	H	H	
10i	Norbutorphanol	H	OH	OH	H	H	
10j	Norlevorphanol	H	OH	OH	H	H	
10k	Phenomorphan	$(CH_2)_2(Ph)$	OH	H	H	H	
10l	Racemethorphan	CH_3	OCH_3	H	H	H	
10m	Racemorphan (same as lavorphanol)						
12a	Allobarbital	$CH_2CH=CH_2$	$CH_2CH=CH_2$				
12b	5-Allyl-5-(2',3'-dihydroxypropyl)barbituric acid	$CH_2CHOHCH_2OH$	$CH_2CH=CH_2$				
12c	5-Allyl-5-(4'-hydroxyphenyl)barbituric acid	$CH_2CH=CH_2$	$Ph-OH$				
12d	5-Allyl-5-(3',4'-hydroxyphenyl)barbituric acid	$CH_2CH=CH_2$	(dihydroxyphenyl, OH, OH)				
12e	5-Allyl-5-(1'-hydroxypropyl)barbituric acid	$CH_2CH=CH_2$	$COH(CH_3)_2$				
12f	5-Allyl-5-(2'-hydroxypropyl)barbituric acid	$CH_2CH=CH_2$	$CH(CH_3)CH_2OH$				
12g	Amobarbital	CH_2CH_3	$CH_2CH_2CH(CH_3)_2$				
12h	Alphenal	$CH_2CH=CH_2$	Ph				
12i	Aprobarbital	$CH_2CH=CH_2$	$CH(CH_3)_2$				
12j	Barbital	CH_2CH_3	CH_2CH_3				
12k	Barbituric acid	H	H				
12l	Butabarbital	CH_2CH_3	$CH(CH_3)CH_2CH_3$				
12m	Butallyonal	$CH_2CBr=CH_2$	$CH(CH_3)CH_2CH_3$				
12n	Butethal	CH_2CH_3	$(CH_2)_3CH_3$				
12o	3-Carboxyamobarbital	CH_2CH_3	$CH_2CH_2CH(CH_3)COOH$				
12p	Cyclobarbital	CH_2CH_3	(cyclohexenyl)				
12q	5-(2',3'-Dihydroxypropyl)-5-isopropylbarbituric acid	$CH_2CHOHCH_2OH$	$CH(CH_3)_2$				
12r	5-(2',3'-Dihydroxypropyl)-5-phenylbarbituric acid	$CH_2CHOHCH_2OH$	Ph				
12s	1,5-Dimethyl-5-(3'-ketocyclohexen-1'-yl)barbituric acid	CH_3	(ketocyclohexenyl)	CH_3			

Figure	Drug or metabolite name (Appendices I and II)	R	R^1	R^2	R^3	R^4	R^5
12t	5-Ethylbarbituric acid	CH_2CH_3	H	H			
12u	N-Glucosylamobarbital	CH_2CH_3	$CH_2CH_2CH(CH_3)_2$	Glucose			
12v	Heptabarbital	CH_2CH_3	(cyclohexenyl)	H			
12w	Hexobarbital	CH_3	(cyclohexenyl)	CH_3			
12x	3-Hydroxyamobarbital	CH_2CH_3	$CH_2CH_2C(OH)(CH_3)_2$	H			
12y	N-Hydroxyamobarbital	CH_2CH_3	$CH_2CH_2CH(CH_3)_2$	OH			
12z	3'-Hydroxyhexobarbital	CH_3	(cyclohexenyl-OH)	CH_3			
12ba	p-Hydroxymephobarbital	CH_2CH_3	Ph–OH	CH_3			
12bb	Ibomal	$CH_2C=CH_2$	$CH(CH_3)_2$	H			
12bc	3'-Ketohexobarbital	CH_3	(keto-cyclohexenyl)	CH_3			
12bd	Mephobarbital	CH_2CH_3	Ph	CH_3			
12be	Metharbital	CH_2CH_3	CH_2CH_3	CH_3			
12bf	Methohexital	$CH_2CH=CH_2$	$CH(CH_3)C≡CCH_2CH_3$	CH_3			
12bg	3-O-Methylcatechol of mephobarbital (OMC-MPB)	CH_2CH_3	(OCH₃, OH aryl)	CH_3			
12bh	5-(1-Methylbutyl)barbituric acid	H	$CH(CH_3)CH_2CH_2CH_3$	H			
12bi	3-O-Methylcatechol of phenobarbital (OMC-PB)	CH_2CH_3	(OCH₃, OH aryl)	H			
12bj	5-Methyl-5-cyclohexen-1'-ylbarbituric acid	CH_3	(cyclohexenyl)	H			
12bk	5-Methyl-5-(3'-ketocyclohexen-1'-yl)-barbituric acid	CH_3	(keto-cyclohexenyl)	H			
12bl	Pentobarbital	CH_2CH_3	$CH(CH_3)CH_2CH_2CH_3$	H			
12bm	3'-Hydroxypentobarbital	CH_2CH_3	$CH(CH_3)CH_2CH(OH)CH_3$	H			
12bn	N-Hydroxypentobarbital	CH_2CH_3	$CH(CH_3)CH_2CH_2CH_3$	OH			
12bo	3'-Carboxypentobarbital	CH_2CH_3	$CHCH_2CH_2COOH$	H			
12bp	Phenobarbital	CH_2CH_3	Ph	H			
12bq	Phenobarbital metabolite a	CH_2CH_3	p-OH–Ph	H			
12br	Phenobarbital metabolite b	CH_2CH_3	(OH, OH aryl)	H			
12bs	Phenobarbital metabolite c	CH_2CH_3	m-OH–Ph	H			
12bt	Secobarbital	$CH_2CH=CH_2$	$CH(CH_3)CH_2CH_2CH_3$	H			
12bu	Secobarbital metabolite a	$CH_2CH=CH_2$	$CH(CH_3)CH_2CH(OH)CH_3$	H			
12bv	Secobarbital metabolite b	$CH_2CH(OH)CH_2OH$	$CH(CH_3)CH_2CH_2CH_3$	H			
12bw	Secobarbital metabolite c	H	$CH(CH_3)CH_2CH_2CH_3$	H			
13a	Methyprylon	2H					
13b	6-Oxomethyprylon	=O					

Figure	Drug or metabolite name (Appendices I and II)	R	R¹	R²	R³	R⁴	R⁵
14a	5-Carboxypyrithyldione	COOH					
14b	5-Hydroxymethylpyrithyldione	CH₂OH					
14c	5-Methylpyrithyldione	CH₃					
15a	Desethylglutethimide	H	H	H			
15b	Glutethimide	CH₂CH₃	H	H			
15c	(1-Hydroxyethyl)glutethimide	CHOHCH₃	H	H			
15d	4-Hydroxyglutethimide	CH₂CH₃	H	OH			
15e	p-Hydroxyglutethimide	CH₂CH₃	OH	H			
15f	Iso(1-hydroxyethyl)glutethimide	CHOHCH₃	H	H			
16a	1,3-Dimethyluric acid (fenethylline metabolite)	CH₃	CH₃				
16b	1-Methyluric acid (fenethylline metabolite)	H	H				
17a	Fenethylline	CH₂CH₂NHCH(CH₃)CH₂-Ph	CH₃				
17b	3-Methylxanthine	H	H				
17c	Theophylline	H	CH₃				
20a	3,4-Dihydrodihydroxyphenytoin	[structure: OH, OH, OH]					
20b	m-Hydroxyphenytoin	[structure: OH]					
20c	p-Hydroxyphenytoin	Ph-OH					
20d	Phenytoin	Ph					
21a	Dihydroxyphenylbutazone	OH	OH				
21b	γ-Hydroxyphenylbutazone	OH	H				
21c	Oxyphenbutazone	H	OH				
21d	Phenylbutazone	H	H				
22a	N-Desmethylmedazepam	H	H				
22b	Medazepam	CH₃	H				
23a	Acetamidoclonazepam	NHCOCH₃	H	H	Cl	O	
23b	Acetamidonitrazepam	NHCOCH₃	H	H	H	O	
23c	Aminoclonazepam	NH₂	H	H	Cl	O	
23d	7-Aminoflunitrazepam	NH₂	CH₃	H	F	O	
23e	Aminonitrazepam	NH₂	H	H	H	O	
23f	Camazepam	Cl	CH₃	OCOON(CH₃)₂	H	O	
23g	Metabolite 1 (CM 6913) (carboxylic acid metabolite of ethyl loflazepate)	Cl	H	COOH	F	O	
23h	7-Chloro-5-(o-chlorophenyl)-1,3-dihydro-3-hydroxy-2H-1,4-benzodiazepin-2-one	Cl	H	OH	Cl	O	
23i	7-Chloro-2,3-dihydro-5-(o-chlorophenyl)-2H-1,4-benzodiazepin-2-one	Cl	H	H	Cl	O	
23j	7-Chloro-5-(2'-fluorophenyl)-2,3-dihydro-2-oxo-1H-1,4-benzodiazepine-1-acetaldehyde	Cl	CH₂CHO	H	F	O	
23k	Clonazepam	NO₂	H	H	Cl	O	

Figure	Drug or metabolite name (Appendices I and II)	R	R¹	R²	R³	R⁴	R⁵
23l	Metabolite 2 (CM 7116) (decarboxylated metabolite of ethyl loflazepate)	Cl	H	H	F	O	
23m	Desalkylflurazepam	Cl	H	H	F	O	
23n	N-Desethylflurazepam	Cl	$CH_2CH_2NHCH_3$	H	F	O	
23o	N-Desmethylflunitrazepam	NO_2	H	H	F	O	
23p	N-Didesethylflurazepam	Cl	$CH_2CH_2NH_2$	H	F	O	
23q	Diazepam	Cl	CH_3	H	H	O	
23r	Ethyl loflazepate	Cl	H	$COOCH_2CH_3$	F	O	
23s	Fludiazepam	Cl	CH_3	H	F	O	
23t	Flunitrazepam	NO_2	CH_3	H	F	O	
23u	Flurazepam	Cl	$CH_2CH_2N(CH_2CH_3)_2$	H	F	O	
23v	Halazepam	Cl	CH_2CF_3	H	H	O	
23w	3-Hydroxy-N-desalkyl-2-oxoquazepam (flurazepam)	Cl	H	OH	F	O	
23x	N-1-Hydroxyethylflurazepam	Cl	CH_2CH_2OH	H	F	O	
23y	3-Hydroxy-2-oxoquazepam	Cl	CH_2CF_3	OH	F	O	
23z	3-Hydroxypinazepam	Cl	$CH_2C{=}CH$	OH	H	O	
23ba	3-Hydroxyprazepam	Cl	$CH_2{-}{\triangleleft}$	OH	H	O	
23bb	Lorazepam	Cl	H	OH	Cl	O	
23bc	Lormetazepam	Cl	CH_3	OH	Cl	O	
23bd	Nimetazepam	NO_2	CH_3	H	H	O	
23be	Nitrazepam	NO_2	H	H	H	O	
23bf	Nordiazepam	Cl	H	H	H	O	
23bg	Oxazepam	Cl	H	OH	H	O	
23bh	2-Oxoquazepam	Cl	CH_2CF_3	H	F	O	
23bi	Pinazepam	Cl	$CH_2C{=}CH$	H	H	O	
23bj	Prazepam	Cl	$CH_2{-}{\triangleleft}$	H	H	O	
23bk	Quazepam	Cl	CH_2CF_3	H	F	S	
23bl	Temazepam	Cl	CH_3	OH	H	O	
25a	3-Hydroxynortetrazepam	H	OH	H			
25b	3-Hydroxynortetrazepam	H	H	OH			
25c	3-Hydroxytetrazepam	CH_3	OH	H			
25d	3-Hydroxytetrazepam	CH_3	H	OH			
25e	Nortetrazepam	H	H	H			
25f	Tetrazepam	CH_3	H	H			
26a	Chlordiazepoxide	$NHCH_3$	O				
26b	Desmethylchlordiazepoxide	NH_2	O				
26c	N-Desmethyl-1,2-dehydromedazepam	H					
28a	6-Chloro-4-(o-chlorophenyl)-(1H)-quinazoline	NH———$C{=}O$ [a]					
28b	6-Chloro-4-(o-chlorophenyl)-2-quinazoline carboxylic acid	N≡≡≡CCOOH [a]					
29a	Alprazolam	H		CH_3			

Figure	Drug or metabolite name (Appendices I and II)	R	R^1	R^2	R^3	R^4	R^5
29b	8-Chloro-4-hydroxy-6-phenyl-4H-s-triazolo-[4,3-a][1,4]benzodiazepine	H	OH	H			
29c	8-Chloro-6-(4-hydroxyphenyl)-4H-s-triazolo-[4,3-a][1,4]benzodiazepine	4'-OH	H	H			
29d	Estazolam	H	H	H			
29e	α-Hydroxyalprazolam	H	H	CH_2OH			
29f	4-Hydroxyalprazolam	H	OH	CH_3			
29g	1-Hydroxymethyl-4-hydroxytriazolam	2-Cl	OH	CH_2OH			
29h	1-Hydroxymethyltriazolam	2-Cl	H	CH_2OH			
29i	4-Hydroxytriazolam	2-Cl	OH	CH_3			
29j	Triazolam	2-Cl	H	CH_3			
30a	8-Chloro-2,4-dihydro-6-(4-hydroxyphenyl)-1H-s-triazolo[4,3-a][1,4]benzodiazepin-1-one	4'-OH	H				
30b	8-Chloro-2,4-dihydro-6-phenyl-1H-s-triazolo-[4,3-a][1,4]-benzodiazepin-1-one	H	H				
31a	1-Hydroxymethyl-4-hydroxymidazolam	CH_2OH	OH				
31b	1-Hydroxymethylmidazolam	CH_2OH	H				
31c	4-Hydroxymidazolam	CH_3	OH				
31d	Midazolam	CH_3	H				
33a	5-Chloro-2-[3,5-bis(hydroxymethyl)-4H,1,2,4-triazol-4-yl]-2'-chlorobenzophenone	CH_2OH	CH_2OH	Cl			
33b	5-Chloro-2-(3-hydroxymethyl-5-methyl-4H,1,2,4-triazol-4-yl)-2'-chlorobenzophenone	CH_2OH	CH_3	Cl			
33c	5-Chloro-2-(3-methyl-4H,1,2,4-triazol-4-yl)-2'-chlorobenzophenone	H	CH_3	Cl			
33d	5-Chloro-2-(4H,1,2,4-triazol-4-yl)-benzophenone	H	H	H			
34a	Bromazepam	H	H				
34b	Bromazepam N_4-oxide	H	O				
34c	3-Hydroxybromazepam	OH	H				
35a	Brotizolam	H	H				
35b	3-Hydroxybrotizolam	H	OH				
35c	1-Methylhydroxybrotizolam	OH	H				
37a	N-7-Desmethylpremazepam	H	CH_3				
37b	6-Hydroxypremazepam	CH_3	CH_2OH				
37c	Premazepam	CH_3	CH_3				
39a	Hydroxylorazepam						
39b	Hydroxymethoxylorazepam						

Figure	Drug or metabolite name (Appendices I and II)	R	R^1	R^2	R^3	R^4	R^5
39c	Lorazepam dihydrodiol	(Cl, OH, HO-substituted ring)					
41a	Cloxazolam	Cl	Cl	H			
41b	Haloxazolam	Br	F	H			
41c	Oxazolam	Cl	H	CH_3			
43a	Mazindol	H	H				
43b	Mazindol metabolite 1	=O	H				
43c	Mazindol metabolite 2	=O	OH				
44a	Mazindol metabolite 3	=O					
44b	Mazindol metabolite 4	H					
45a	2-(2-Amino-5-bromobenzoyl)pyridine	H					
45b	2-(2-Amino-5-bromo-3-hydroxybenzoyl)pyridine	OH					
46a	N-[γ-(2-Chlorophenothiazin-10-yl)propyl]ethylene diamine	None	H	$CH_2CH_2NH_2$	H	Cl	
46b	N-[γ-(2-Chlorophenothiazin-10-yl)propyl]-N'-(β-hydroxy-ethyl)ethylenediamine	None	H	$(CH_2)_2NHCH_2CH_2OH$	H	Cl	
46c	N-[γ-(2-Chlorophenothiazin-10-yl)propyl]-N'-methyl-ethylenediamine)	None	H	$CH_2CH_2NHCH_3$	H	Cl	
46d	Chlorpromazine	None	CH_3	CH_3	H	Cl	
46e	Chlorpromazine sulfoxide	O	CH_3	CH_3	H	Cl	
46f	Dinorchlorpromazine	None	H	H	H	Cl	
46g	Dinorchlorpromazine sulfoxide	O	H	H	H	Cl	
46h	7-Hydroxychlorpromazine	None	CH_3	CH_3	OH	Cl	
46i	7-Hydroxychlorpromazine sulfoxide	O	CH_3	CH_3	OH	Cl	
46j	7-Hydroxydinorchlorpromazine	None	H	H	OH	Cl	
46k	7-Hydroxynorchlorpromazine	None	CH_3	H	OH	Cl	
46l	Norchlorpromazine	None	CH_3	H	H	Cl	
46m	Norchlorpromazine sulfoxide	O	CH_3	H	H	Cl	
46n	N-[α-(2-Trifluoromethylphenothiazin-10-yl)-propyl]-ethylenediamine	None	H	$CH_2CH_2NH_2$	H	CF_3	
46o	N-[α-(2-Trifluoromethylphenothiazin-10-yl)-propyl]-N'-(β-hydroxyethyl)ethylenediamine	None	H	$(CH_2)_2NHCH_2CH_2OH$	H	CF_3	
46p	N-[α-(2-Trifluoromethylphenothiazin-10-yl)-propyl]-N'-(β-methoxyethyl)ethylenediamine	None	H	$(CH_2)_2NHCH_2CH_2OCH_3$	H	CF_3	
47a	Disulfoxide of thioridazine	H	(N–CH₃ piperidinyl)	2(O)			
47b	Disulfoxide of thioridazine	H	(N–CH₃ piperidinyl)	SCH_3 (sulfone), SCH_3–O	O		
47c	Northioridazine	H	(N piperidinyl)	SCH_3–O			
47d	Promethazine	CH_3	$N(CH_3)_2$	H			

Figure	*Drug or metabolite name (Appendices I and II)*	*R*	R^1	R^2	R^3	R^4	R^5
47e	Ring sulfoxide of thioridazine	H	[piperidyl N–CH₃]	SCH_3			O
47f	Side chain sulfoxide of thioridazine	H	[piperidyl N–CH₃]	$\overset{\text{SCH}_3}{\underset{\text{O}}{\vert}}$			
47g	Thioridazine	H	[piperidyl N–CH₃]	SCH_3			
48a	N-[γ-(2-Chlorophenothiazin-10-yl)-propyl]piperazine	H	Cl				
48b	Fluphenazine	CH_2CH_2OH	CF_3				
48c	Perphenazine	CH_2CH_2OH	Cl				
48d	Prochlorperazine	CH_3	Cl				
48e	Trifluoperazine	CH_3	CF_3				
48f	N-[γ-(2-Trifluoromethylphenothiazin-10-yl)-propyl]-piperazine	H	CF_3				
49a	Clomipramine	$N(CH_2)_3N(CH_3)_2$	Cl	C	H		
49b	Desdimethylimipramine	$N(CH_2)_3NH_2$	H	C	H		
49c	Desdimethylprothiadene	$C=CH(CH_2)_2NH_2$	H	S	H		
49d	Desmethylimipramine	$N(CH_2)_3NHCH_3$	H	C	H		
49e	Desmethylprothiadene	$C=CH(CH_2)^2NHCH_3$	H	S	H		
49f	Dothiepin	$C=CHCH_2CH_2N(CH_3)_2$	H	S	H		
49g	Doxepin	$C=CHCH_2CH_2N(CH_3)_2$	H	O	H		
49h	2-Hydroxydesmethylimipramine	$N(CH_2)_3NHCH_3$	H	C	OH		
49i	2-Hydroxyiminodibenzyl	NH	H	C	OH		
49j	2-Hydroxyimipramine	$N(CH_2)_3N(CH_3)_2$	H	C	OH		
49k	Iminobenzyl	NH	H	C	H		
49l	Imipramine	$N(CH_2)_3N(CH_3)_2$	H	C	H		
49m	Imipramine N-oxide	$N(CH_2)_3N(O)(CH_3)_2$	H	C	H		
49n	Prothiadene sulfoxide	$C=CH(CH_2)_2N(CH_3)_2$	H	S=O	H		
49o	Trimipramine	$NCH_2CH(CH_3)CH_2N(CH_3)_2$	H	C	H		
50a	Amitriptyline	H	H	H	$N(CH_3)_2$	H	
50b	10,11-Dihydroxyamitriptyline	H	OH	OH	$N(CH_3)_2$	H	
50c	10,11-Dihydroxydidesmethylamitriptyline	H	OH	OH	NH_2	H	
50d	2,11-Dihydroxynortriptyline	OH	OH	H	$NHCH_3$	H	
50e	10,11-Dihydroxynortriptyline	H	OH	OH	$NHCH_3$	H	
50f	2-Hydroxyamitriptyline	OH	H	H	$N(CH_3)_2$	H	
50g	3-Hydroxyamitriptyline	H	H	H	$N(CH_3)_2$	OH	
50h	10-Hydroxyamitriptyline N-oxide	H	H	OH	$N(O)(CH_3)_2$	H	
50i	10-Hydroxydidesmethylamitriptyline	H	H	OH	NH_2	H	
50j	3-Hydroxynortriptyline	H	H	H	$NHCH_3$	OH	
50k	Nortriptyline	H	H	H	$NHCH_3$	H	
50l	10-Oxoamitriptyline	H	=O	H	$N(CH_3)_2$	H	
50m	10-Oxonortriptyline	H	=O	H	$NHCH_3$	H	
51a	Amitriptyline dihydrodiol	$N(CH_3)_2$					
51b	Nortriptyline dihydrodiol	$NHCH_3$					

Figure	Drug or metabolite name (Appendices I and II)	R	R^1	R^2	R^3	R^4	R^5
52a	2-Methyl-1,2,3,4-tetrahydro-β-carboline	CH_3					
52b	1,2,3,4-Tetrahydro-β-carboline	H					
53a	Harmine	OCH_3	H				
53b	Harmol	OH	H				
53c	3- or 4-Hydroxy-7-methoxyharman	OCH_3	H	3- or 4-OH			
53d	6-Hydroxy-7-methoxyharman	OCH_3	OH	H			
55a	8α,11-Dihydroxy-Δ⁹-tetrahydrocannabinol	CH_2OH	α-OH				
55b	8β,11-Dihydroxy-Δ⁹-tetrahydrocannabinol	CH_2OH	β-OH				
55c	Dronabinal	CH_3	H				
55d	8α-Hydroxy-Δ⁹-tetrahydrocannabinol	CH_3	α-OH				
55e	8β-Hydroxy-Δ⁹-tetrahydrocannabinol	CH_3	β-OH				
55f	11-Hydroxy-Δ⁹-tetrahydrocannabinol	CH_2OH	H				
55g	11-Nortetrahydrocannabinol-Δ⁹-carboxylic acid	COOH	H				
55h	Δ⁹-Tetrahydrocannabinol	CH_3	H				
56a	13-Hydroxy-LSD	$N(CH_2CH_3)_2$	CH_3	13-OH			
56b	14-Hydroxy-LSD	$N(CH_2CH_3)_2$	CH_3	14-OH			
56c	Lysergic acid	OH	CH_3	H			
56d	Lysergic acid amide	NH_2	CH_3	H			
56e	Lysergic acid diethylamide (LSD)	$N(CH_2CH_3)_2$	CH_3	H			
56f	Lysergic acid monoethylamide	$NHCH_2CH_3$	CH_3	H			
56g	Nor-LSD	$N(CH_2CH_3)_2$	H	H			
60a	Bezitramide	H	CN	H	[structure], H_3CH_2CO	H	
60b	Denitrile difenoxin	COOH	H	H	Ph	H	
60c	Difenoxin (diphenoxylic acid)	COOH	CN	H	Ph	H	
60d	Diphenoxylate	$COOCH_2CH_3$	CN	H	Ph	CH_3	
60e	Dipipanone	H	$COCH_2CH_3$	H	H	H	
60f	Hydroxydiphenoxylic acid	COOH	CN	OH	Ph	H	
60g	Norpipanone	H	$COCH_2CH_3$	H	H	H	
60h	Piritramide	$CONH_2$	CN	H	[piperidine N]	H	
62a	Amphetamine	H	H	H	H	H	
62b	Benzphetamine	CH_2–Ph	CH_3	H	H	H	
62c	Benzylamphetamine	CH_2–Ph	H	H	H	H	
62d	N-(1-Benzylethyl)-α-phenylnitrone	O^-	=CH-Ph	H	H	H	
62e	N-Benzyl-N-hydroxyamphetamine	CH_2–Ph	OH	H	H	H	
62f	Chlorphentermine	H	H	CH_3	H	p-Cl	
62g	Clortermine	H	H	CH_3	H	o-Cl	
62h	N,N-Diethylnorephedrine	CH_2CH_3	CH_2CH_3	H	OH	H	
62i	N,N-Dimethylamphetamine	CH_3	CH_3	H	H	H	
62j	Dimethylnitro-4-chlorophenylethane	None	= O	CH_3	H	p-Cl	
62k	Dimethylnitroso-4-chlorophenylethane	O	O	CH_3	H	p-Cl	

Figure	Drug or metabolite name (Appendices I and II)	R	R^1	R^2	R^3	R^4	R^5
62l	Ephedrine	CH_3	H	H	OH	H	
62m	N-Ethylamphetamine	CH_2CH_3	H	H	H	H	
62n	N-Ethylnorephedrine	CH_2CH_3	H	H	OH	H	
62o	Fenfluramine	CH_2CH_3	H	H	H	m-CF_3	
62p	p-Hydroxyamphetamine	H	H	H	H	p-OH	
62q	1-(p-Hydroxyphenyl)-2-(N-benzylamino)propane	CH_2–Ph	H	H	H	p-OH	
62r	N-Hydroxychlorphentermine	H	OH	H	H	p-Cl	
62s	N-Hydroxyethylamphetamine	CH_3	CH_2CH_3	H	H	H	
62t	p-Hydroxymethamphetamine	CH_3	H	H	H	p-OH	
62u	p-Hydroxynorephedrine	H	H	H	OH	p-OH	
62v	1-(2-Hydroxyphenyl)-2-(methylamino)propane (II)	CH_3	H	H	H	2-OH	
62w	1-(m-Hydroxyphenyl)-2-(N-methyl-N-benzylamino)propane	CH_2–Ph	CH_3	H	H	m-OH	
62x	1-(p-Hydroxyphenyl)-2-(N-methyl-N-benzylamino)propane	CH_2–Ph	CH_3	H	H	p-OH	
62y	Methamphetamine	CH_3	H	H	H	H	
62z	1-(m-Methoxy-p-hydroxyphenyl)-2-(N-benzylamino)propane	CH_2–Ph	H	H	H	m-OCH_3	
62ba	Methoxyphenamine (I)	CH_3	H	H	H	2-OCH_3	
62bb	α-Methyl-N-(1'-phenylprop-2-yl)nitrone	O^-	$=CHCH_3$	H	H	H	
62bc	Norephedrine	H	H	H	OH	H	
62bd	Norfenfluramine	H	H	H	H	m-CF_3	
62be	Normethoxyphenamine (VII)	H	H	H	H	2-OCH_3	
62bf	Phentermine	H	H	CH_3	H	H	
63a	Benzoic acid	H	H	H	COOH	H	
63b	3,4-Dihydroxybenzyl methyl ketone	OH	OH	H	CH_2COCH_3	H	
63c	1-(2,5-Dimethoxy-4-methylphenyl)-2-propanone	H	OCH_3	CH_2COCH_3	OCH_3	CH_3	
63d	Hippuric acid	H	H	H	$CONHCH_2COOH$	H	
63e	p-Hydroxybenzoic acid	H	OH	H	COOH	H	
63f	4-Hydroxy-3-methoxybenzoic acid	OCH_3	OH	H	COOH	H	
63g	3,4-Methylenedioxybenzoic acid	—O—CH_2—O—	[a]	H	COOH	H	
63h	3,4-Methylenedioxybenzyl methyl ketone	—O—CH_2—O—	[a]	H	CH_2COCH_3	H	
63i	m-Trifluoromethylbenzoic acid	CF_3	H	H	COOH	H	
63j	3,4,5-Trimethoxybenzoic acid	OCH_3	OCH_3	OCH_3	COOH	H	
63k	3,4,5-Trimethoxyphenylacetic acid	OCH_3	OCH_3	OCH_3	CH_2COOH	H	
63l	Tropic acid	H	H	H	$CH(CH_2OH)COOH$	H	
64a	N-Acetyl-(3,4-dimethoxy-5-hydroxy)-phenethylamine	OCH_3	OCH_3	OH	H	H	$COCH_3$
64b	N-Acetylmescaline	OCH_3	OCH_3	OCH_3	H	H	$COCH_3$
64c	4-Bromo-2,5-dimethoxyamphetamine	Br	H	OCH_3	OCH_3	CH_3	H
64d	4-Bromo-2,5-dimethoxyphenethylamine	Br	H	OCH_3	OCH_3	H	H
64e	2,5-Dimethoxyamphetamine	H	H	OCH_3	OCH_3	CH_3	H
64f	1-(2,5-Dimethoxy-4-carboxyphenyl)-2-aminopropane	COOH	H	OCH_3	OCH_3	CH_3	H
64g	1-(2,5-Dimethoxy-4-hydroxymethylphenyl)-2-aminopropane	CH_2OH	H	OCH_3	OCH_3	CH_3	H
64h	2,5-Dimethoxymethylamphetamine	CH_3	H	OCH_3	OCH_3	CH_3	H
64i	Fenproporex	H	H	H	H	CH_3	CH_2CH_2CN
64j	N-Hydroxy-p-methoxyamphetamine	OCH_3	H	H	H	CH_3	OH

Figure	Drug or metabolite name (Appendices I and II)	R	R^1	R^2	R^3	R^4	R^5
64k	1-(5-Hydroxy-2-methoxyphenyl)-2-methylaminopropane	H	H	OH	OCH_3	CH_3	CH_3
64l	Mefenorex	H	H	H	H	CH_3	C_2H_4Cl
64m	Mescaline	OCH_3	OCH_3	OCH_3	H	H	H
64n	4-Methoxyamphetamine	OCH_3	OCH_3	H	H	CH_3	H
64o	α-Methyldopamine	OH	OH	H	H	CH_3	H
64p	3,4-Methylenedioxyamphetamine	$O\!-\!CH_2\!-\!O$	a	H	H	CH_3	H
64q	3,4-Methylenedioxymethamphetamine	$O\!-\!CH_2\!-\!O$	a	H	H	CH_3	C_2H_5
64r	3,4-Methylenedioxymethamphetamine	$O\!-\!CH_2\!-\!O$	a	H	H	CH_3	CH_3
64s	3,4-Methylenedioxy-2-methoxyamphetamine	$O\!-\!CH_2\!-\!O$	a	OCH_3	H	CH_3	H
64t	3-O-Methyl-α-methyldopamine	OH	OCH_3	H	H	CH_3	H
64u	Trimethoxyamphetamine	OCH_3	OCH_3	H	OCH_3	CH_3	H
65a	Methyl phenyl ketone	=O	=O	H	H		
65b	1-Phenyl-2-butanone	=O	=O	H	CH_3		
65c	1-Phenyl-1,2-propanediol	H	OH	OH	H		
65d	1-Phenyl-2-propanol	H	H	H	H		
65e	1-Phenyl-2-propanon-1-ol	=O	=O	OH	H		
66a	Dinorpropoxyphene	$OCOCH_2CH_3$	NH_2	CH_2–Ph			
66b	Dinorpropoxyphene carbinol	OH	NH_2	CH_2–Ph			
66c	p-Hydroxynorpropoxyphene	$OCOCH_2CH_3$	$NHCH_3$	CH_2–⟨C₆H₄⟩–OH			
66d	p-Hydroxypropoxyphene	$OCOCH_2CH_3$	$N(CH_3)_2$	CH_2–⟨C₆H₄⟩–OH			
66e	Moramide	$O\!=\!C\!-\!N$⟨ ⟩ (pyrrolidine)	$-N$⟨ ⟩O (morpholine)	Ph			
66f	Norpropoxyphene	$OCOCH_2CH_3$	$NHCH_3$	CH_2–Ph			
66g	Norpropoxyphene carbinol	OH	$NHCH_3$	CH_2–Ph			
66h	Propoxyphene	$OCOCH_2CH_3$	$N(CH_3)_2$	CH_2–Ph			
66i	Propoxyphene carbinol	OH	$N(CH_3)_2$	CH_2–Ph			
67a	Cathinone	H	H				
67b	Diethylpropion	CH_2CH_3	CH_2CH_3				
67c	Dinordiethylpropion	H	H				
67d	Methcathinone	CH_3	CH_3				
67e	Nordiethylpropion	CH_2CH_3	H				
68a	Propiram	⟨piperidine ring⟩	$COCH_2CH_3$				
68b	Propiram metabolite (valioneric acid)	$N[CH_2(CH_2)_3COOH]HCOCH_2CH_3$	H				
69a	Dimenoxadol	$N(CH_3)_2$	H				
69b	Dioxaphetyl butyrate	CH_2CH_2	OCH_2CH_3				
70a	Methylpiperidinyl benzolate	COO–⟨piperidine, $N^+(CH_3)_2\ Br^-$⟩					

Figure	Drug or metabolite name (Appendices I and II)	R	R^1	R^2	R^3	R^4	R^5
70b	Pipenzolate bromide	COO— (piperidine) + N—CH$_2$CH$_3$, CH$_3$, Br$^-$					
70c	Pipradrol	piperidine structure					
71a	Acetylmethadol	H	CH$_3$	N(CH$_3$)$_2$	OCOCH$_3$	H	
71b	Dinoracetylmethadol	H	CH$_3$	NH$_2$	OCOCH$_3$	H	
71c	Dinormethadol	H	CH$_3$	NH$_2$	OH	H	
71d	Dimepheptanol	H	CH$_3$	N(CH$_3$)$_2$	H	OH	
71e	Isomethadone	CH$_3$	H	N(CH$_3$)$_2$	=O		
71f	Levo-α-acetylmethadol	H	CH$_3$	N(CH$_3$)$_2$	CH$_3$COO—	H	
71g	Methadol	H	CH$_3$	N(CH$_3$)$_2$	OH	H	
71h	Methadone	H	CH$_3$	N(CH$_3$)$_2$	=O		
71i	Noracetylmethadol	H	CH$_3$	NHCH$_3$	OCOCH$_3$	H	
71j	Noracymethadol	H	CH$_3$	NHCH$_3$	OCOCH$_3$	H	
71k	Normethadol	H	CH$_3$	NHCH$_3$	OH	H	
71l	Normethadone	H	H	N(CH$_3$)$_2$	=O		
71m	Phenadoxone	H	CH$_3$	—N (morpholine, O)	=O		
72a	N-Demethyldiphenhydramine	CH$_2$NHCH$_3$	H	H			
72b	N,N-Didemethyldiphenhydramine	CH$_2$NH$_2$	H	H			
72c	Diphenhydramine	CH$_2$N(CH$_3$)$_2$	H	H			
72d	Diphenhydramine N-oxide	CH$_2$N(O)(CH$_3$)$_2$	H	H			
72e	Diphenylmethoxyacetic acid	COOH	H	H			
72f	Orphenadrine	CH$_2$N(CH$_3$)$_2$	CH$_3$	H			
73	1-Piperidinocyclohexanecarbonitrile	CN					
75a	Acid metabolite of phencyclidine	NH(CH$_2$)$_4$COOH	H	H			
75b	3-Hydroxyphencyclidine	—N (pyrrolidine)	H	OH			
75c	Phencyclidine	—N (pyrrolidine)	H	H			
75d	Cyclohexamine	NHCH$_2$CH$_3$	H	H			
75e	Phenylcyclopentylpiperidine	—N (pyrrolidine)	H	H			
75f	Phenylcyclohexylamine	NH$_2$	H	H			
75g	1-(1-Phenylcyclohexyl)-4-hydroxypiperidine	—N (piperidine, OH)	OH	H			
75h	1-Phenyl-4-hydroxycyclohexyl-4-hydroxypiperidine	—N (piperidine, OH)	OH	H			
75i	4-Phenyl-4-piperidinocyclohexanol	—N (piperidine)	OH	H			
76a	Fentanyl	H	H	H	CH$_2$CH$_3$		
76b	p-Fluorofentanyl	F	H	H	CH$_2$CH$_3$		
76c	α-Methylacetylfentanyl	H	H	CH$_3$	CH$_3$		

Figure	Drug or metabolite name (Appendices I and II)	R	R^1	R^2	R^3	R^4	R^5
76d	α-Methylfentanyl	H	H	CH_3	CH_2CH_3		
76e	3-Methylfentanyl	H	CH_3	H	CH_2CH_3		
76f	N-[1-(2-Phenethyl-4-piperidinyl)]malonanalinic acid	H	H	H	CH_2COOH		
77a	Allylprodine	CH_3	$CH_2CH=CH_2$	H			
77b	Meprodine	CH_3	CH_2CH_3	H			
77c	Nor-α-prodine	H	CH_3	H			
77d	1-(2-phenylethyl)-4-phenyl-4-acetyloxypiperidine	CH_2CH_2-Ph	H	H			
77e	α-Prodine	CH_3	CH_3	H			
77f	Trimeperidine	CH_3	CH_3	CH_3			
79a	Acetylanileridine	CH_2CH_2-Ph–NHCOCH$_3$	H	OCH_2CH_3	H		
79b	Acetylanileridinic acid	CH_2CH_2-Ph–NHCOCH$_3$	H	OH	H		
79c	Anileridine	CH_2CH_2-Ph–NH$_2$	H	OCH_2CH_3	H		
79d	Anileridinic acid	CH_2CH_2-Ph–NH$_2$	H	OH	H		
79e	Benzethidine	CH_2CH_2-O–CH$_2$-Ph	H	OCH_2CH_3	H		
79f	Etoxeridine	$CH_2CH_2OCH_2CH_2OH$	H	OCH_2CH_3	H		
79g	Furethidine	$CH_2CH_2OCH_2–$	H	OCH_2CH_3	H		
79h	4'-Hydroxyketobemidone	CH_3	OH	CH_2CH_3	OH		
79i	p-Hydroxymeperidine	CH_3	H	OCH_2CH_3	OH		
79j	Hydroxypethidine	CH_3	OH	OCH_2CH_3	H		
79k	Ketobemidone	CH_3	OH	CH_2CH_3	H		
79l	Meperidine (Pethidine)	CH_3	H	OCH_2CH_3	H		
79m	Meperidine N-oxide	(O)CH$_3$	H	OCH_2CH_3	H		
79n	Meperidinic acid	CH_3	H	OH	H		
79o	Methoxyhydroxymeperidine	CH_3	OCH_3	OCH_2CH_3	OH		
79p	4'-Methoxyketobemidone	CH_3	OCH_3	CH_2CH_3	OH		
79q	Morpheridine	CH_2CH_2–N (morpholine)	H	OCH_2CH_3	H		
79r	Norketobemidone	H	OH	CH_2CH_3	H		
79s	Normeperidine	H	H	OCH_2CH_3	H		
79t	Normeperidinic acid	H	H	OH	H		
79u	Phenoperidine	$CH_2CH_2CH(Ph)OH$	H	OCH_2CH_3	H		
79v	Piminodine	$CH_2CH_2CH_2NH$-Ph	H	OCH_2CH_3	H		
79w	Properidine	CH_3	H	$OCH(CH_3)_2$	H		
80a	Diampromide	$CH_2CH(CH_3)–N(CH_3)CH_2CH_2$-Ph					
80b	Phenampromide	$CH(CH_3)CH_2$–N (piperidine)					
82a	N-(4-Hydroxyphenyl)acetamide	CH_3					
82b	N-(4-Hydroxyphenyl)propanamide	CH_2CH_3					
83a	Alfentanil	CH_2OCH_3	$-(CH_2)_2$–N (triazolone ring, O=, N–N=N, –CH$_2$CH$_3$)				
83b	Benzylfentanyl	H	CH_2–Ph				

Figure	Drug or metabolite name (Appendices I and II)	R	R¹	R²	R³	R⁴	R⁵
83c	Carfentanil	H	H				
83d	Desmethylalfentanil	CH₂OH					
83e	Desmethylsufentanil	CH₂OH					
83f	N-[4-(Hydroxymethyl)-4-piperidinyl]-N-phenyl-propanamide	CH₂OH	H				
83g	N-[4-(Methoxymethyl)-4-piperidinyl]-N-phenyl-propanamide	CH₂OCH₃	H				
83h	Norfentanyl	H	H				
83i	Sufentanil	CH₂OCH₃					
83j	Thienylfentanyl	H					
84a	Bisnortilidine	H	H				
84b	Nortilidine	CH₃	H				
84c	Tilidine	CH₃	CH₃				
85a	p-Hydroxymethylphenidate	CH₃	OH	H			
85b	p-Hydroxyritalinic acid	H	OH	H			
85c	Methylphenidate	CH₃	H	H			
85d	6-Oxoritalinic acid	H	H	=O			
85e	Ritalinic acid	H	H	H			
88a	2-Ethyl-5-methyl-3,3-diphenylpyrroline	H					
88b	p-Hydroxy-2-ethyl-5-methyl-3,3-diphenylpyrroline	OH					
89a	2-Ethylidene-1,5-methyl-3,3-diphenylpyrroline	H					
89b	p-Hydroxy-2-ethylidene-1,5-methyl-3,3-diphenylpyrroline	OH					
91a	Despropionylfentanyl	H					
91b	Despropionylmethylfentanyl	CH₃					
92a	Ethoheptazine	H	COOCH₂CH₃				
92b	Proheptazine	CH₃	OCOCH₂CH₃				
93a	Atropine	CH₃	H	COCH(CH₂OH)Ph			
93b	Benzoylecgonine	CH₃	COOH	COPh			
93c	Cocaine	CH₃	COOCH₃	COPh			
93d	Ecgonine	CH₃	COOH	H			
93e	Ecgonine methyl ester	CH₃	COOCH₃	H			
93f	Norcocaine	H	COOCH₃	COPh			
93g	Tropine	CH₃	H	H			
94a	N-Hydroxyphenmetrazine	OH	H	H			
94b	p-Hydroxyphenmetrazine	H	H	OH			
94c	3-Oxophenmetrazine	H	=O	H			

Figure	*Drug or metabolite name (Appendices I and II)*	-	R	R^1	R^2	R^3	R^4	R^5
94d	Phendimetrazine		CH_3	H	H			
94e	Phenmetrazine		H	H	H			
95a	N-Desisopropylpropranolol (DIP)		CH_2NH_2	H	H			
95b	4-Hydroxypropranolol		$CH_2NHCH(CH_3)_2$	OH	H			
95c	Methoxyhydroxypropranolol		$CH_2NHCH(CH_3)_2$	OH	OCH_3			
95d	Naphthoxylactic acid		$COOH$	H	H			
95e	Propranolol		$CH_2NHCH(CH_3)_2$	H	H			
95f	Ring hydroxylated DIP		CH_2NH_2	OH	H			
95g	Ring hydroxylated naphthoxylactic acid		$COOH$	OH	H			
95h	Ring hydroxylated propranolol glycol		CH_2OH	OH	H			
97a	Bufotenine		$CH_2N(CH_3)_2$	OH	H			
97b	Diethyltryptamine		$CH_2N(CH_2CH_3)_2$	H	H			
97c	N,N-Dimethyltryptamine		$CH_2N(CH_3)_2$	H	H			
97d	N,N-Dimethyltryptamine N-oxide		$CH_2N(O)(CH_3)_2$	H	H			
97e	α-Ethyltryptamine		$CH(NH_2)C_2H_5$	H	H			
97f	5-Hydroxyindoleacetic acid		$COOH$	OH	H			
97g	5-Hydroxytryptamine		CH_2NH_2	OH	H			
97h	3-Indolacetic acid		$COOH$	H	H			
97i	5-Methoxy-N,N-dimethyltryptamine		$CH_2N(CH_3)_2$	OCH_3	H			
97j	5-Methoxyindoleacetic acid		$COOH$	OCH_3	H			
97k	5-Methoxytryptamine		CH_2NH_2	OCH_3	H			
97l	N-Methylserotonin		CH_2NHCH_3	OH	H			
97m	N-Methyltryptamine		CH_2NHCH_3	H	H			
97n	Psilocin		$CH_2N(CH_3)_2$	H	OH			
97o	Psilocybin		$CH_2N(CH_3)_2$	H	H_2PO_4			
97p	Tryptamine		CH_2NH_2	H	OH			
99a	Pentazocine		$CH=C(CH_3)_2$					
99b	Phenazocine		CH_2-Ph					
100a	Clonitazene		Cl					
100b	Etonitazine		OCH_2CH_3					
101a	Ethinimate		H	H	H			
101b	2-Hydroxyethinimate		OH	H	H			
101c	3-Hydroxyethinimate		H	OH				
101d	4-Hydroxyethinimate		H	H	OH			
102a	Ethchlorvynol		H					
102b	Hydroxyethchlorvynol		OH					
104a	Amobarbital metabolite D		$COOH$					
104b	Amobarbital metabolite E		H					
105a	Hydroxymeprobamate		$CH_2CH(OH)CH_3$	H				
105b	Mebutamate		CH_2CH_3	H				
105c	Meprobamate		$CH_2CH_2CH_3$	H				
105d	Meprobamate N-glucuronide		$CH_2CH_2CH_3$	$C_6H_9O_6$				

Figure	Drug or metabolite name (Appendices I and II)	R	R^1	R^2	R^3	R^4	R^5
107a	Sulfondiethylmethane	CH_2CH_3	CH_2CH_3				
107b	Sulfonethylmethane	CH_3	CH_2CH_3				
107c	Sulfonmethane	CH_3	CH_3				
108a	Chloral betaine ($n=1$)	OH	H	$(CH_3)_3N^+CH_2COO^-$			
108b	Chloral hydrate ($n=1$)	OH	H				
108c	Chlorhexadol ($n=1$)	H	$OCH(CH_3)CH_2C(OH)(CH_3)_2$				
108d	Dichloral phenazone ($n=2$)	OH	H				
108e	Petrichloral ($n=4$)	H	OCH_2				
108f	Trichloroacetic acid ($n=1$)		$=O$				
108g	Trichloroethanol ($n=1$)	H	H				
109a	Diethylthiambutene	CH_2CH_3	CH_2CH_3				
109b	Dimethylthiambutene	CH_3	CH_3				
109c	Ethylmethylthiambutene	CH_2CH_3	CH_3				
111a	2-Amino-5-chlorobenzophenone	H	H				
111b	2-Amino-5-chloro-3-hydroxybenzophenone	OH	H				
111c	2-Amino-3-hydroxy-2',5-dichlorobenzophenone	OH	Cl				
113a	Pemoline	NH					
113b	5-Phenyl-2,4-oxazolidinedione	$=O$					
114a	Mecloqualone	H	Cl	H	H		
114b	Methaqualone	H	CH_3	H	H		
114c	Methaqualone metabolite A	OH	CH_3	H	H		
114d	Methaqualone metabolite B	H	CH_2OH	H	H		
114e	Methaqualone metabolite C	H	CH_3	3'-OH	H		
114f	Methaqualone metabolite D	H	CH_3	4'-OH	H		
114g	Methaqualone metabolite E	H	CH_3	H	OH		
116a	o-{α-[2-(Dimethylamino)ethoxy]benzyl}benzyl alcohol	CH_2OH			CH_3		
116b	o-{α-[2-(Dimethylamino)ethoxy]benzyl}benzoic acid	COOH			CH_3		
116c	2[(p-Hydroxy-α-o-tolylbenzyl)oxy]-N,N-methylethylamine	CH_3	4'-OH	H	CH		
116d	2[(m-Hydroxy-α-o-tolylbenzyl)oxy]-N,N-methylethylamine	CH_3	3'-OH	H	CH_3		
116e	2-[(5-Hydroxy-2-methyl-α-phenylbenzyl)oxy]-dimethyl-ethylamine	CH3	H	5-OH	CH_3		
116f	Tofenacine	CH_3	H	H	H		
117a	Aminorex	H					
117b	(±)cis-4-Methylaminorex	CH_3					

^{a} Long single (or double) lines indicate these two functional groups are connected by single (or double) bonds.

Index

Acquisitions: Barbara Pralle, Laura J. Manicone

Production: Randall Frey, Susan D. Fisher

Copyediting / indexing: Jay C. Cherniak

Cover design: Rhonda L. Rawlings

Typesetting: Constance Grigutis

Printing / binding: Maple Press, York, PA